T0275582

LONDON MATHEMATICAL SOCIETY LECTURE NOTE SERIES

Managing Editor: Professor J.W.S. Cassels, Department of Pure Mathematics and Mathematical Statistics, University of Cambridge, 16 Mill Lane, Cambridge CB2 1SB, England

The titles below are available from booksellers, or, in case of difficulty, from Cambridge University Press.

London Mathematical Society Lecture Note Series. 228

Ergodic Theory of $\mathbb{Z}^d$ Actions

Proceedings of the Warwick Symposium 1993-4

Edited by

Mark Pollicott
University of Manchester

Klaus Schmidt
University of Vienna

CAMBRIDGE
UNIVERSITY PRESS

Published by the Press Syndicate of the University of Cambridge
The Pitt Building, Trumpington Street, Cambridge CB2 1RP
40 West 20th Street, New York, NY 10011-4211, USA
10 Stamford Road, Oakleigh, Melbourne 3166, Australia

First published 1996

Library of Congress cataloging in publication data available

British Library cataloguing in publication data available

ISBN   0 521 57688 1 paperback

Transferred to digital printing 2004

CONTENTS

INTRODUCTION

These notes represent the proceedings of the Warwick Symposium on 'Ergodic Theory of $\mathbb{Z}^d$-actions' in 1993-94.

The classical theory of dynamical systems has tended to concentrate on $\mathbb{Z}$-actions or $\mathbb{R}$-actions (i.e. discrete or continuous 'time evolutions'). However, in recent years there has been considerable progress in the study of higher dimensional actions (i.e. actions of $\mathbb{Z}^d$ or $\mathbb{R}^d$ with $d > 1$). This progress was motivated not only by statistical physics, but also by the remarkable successes of multiple recurrence arguments in certain number theoretic problems, and by the intriguing rigidity properties of some classes of 'geometric' and 'algebraic' $\mathbb{Z}^d$-actions.

Historically, much of the interest in $\mathbb{Z}^d$-actions came from the study of classical lattice gas models (for example, the famous Ising model). In the simplest case where $d = 1$ (i.e. in the case of $\mathbb{Z}$-actions) this led to the development of the thermodynamic approach to the ergodic theory of Anosov and Axiom A systems during the 1960s and 1970s (cf. D. Ruelle's monograph *Thermodynamical Formalism*). By contrast, the corresponding problems of understanding the ergodic theory of even the simplest higher rank actions leads quickly to deep unsolved problems (for example, Furstenberg's conjecture on the invariant measures for ×2 and ×3 on the unit interval, or the undecidability problems associated with higher dimensional shifts of finite type). In the context of statistical mechanics, one of the manifestations of this difference between $d = 1$ and $d > 1$ is the existence of very complicated phase transitions in higher dimensions.

The exploitation of connections between multi-parameter ergodic theory and number theory originated largely in the now famous ergodic theory proof, due to Furstenburg, of Szemeredi's theorem on the existence of arithmetic progressions in sequences of integers of positive density. In 1978 Furstenburg showed how Szemeredi's theorem could be reduced to a problem of recurrence for measurable $\mathbb{Z}^d$-actions (cf. H. Furstenburg, *Recurrence in ergodic theory and combinatorial number theory*). This proof of Szemeredi's theorem initiated a highly successful programme which yielded important results both in the ergodic theory of $\mathbb{Z}^d$ actions and in number theory.

A third source of recent interest in $\mathbb{Z}^d$-actions is the study of certain geometric and algebraic examples of $\mathbb{Z}^d$-actions. The simplest actions of this kind are obtained by considering two or more commuting toral automorphisms. Such actions exhibit some remarkable rigidity properties (such as a very small centralizer of the action, or a scarcity of invariant measures and 'regular' cocycles), which were first encountered in the context of geometric actions of groups like $SL(d, \mathbb{Z})$. Recent work by Katok and others has shown that these rigidity properties are shared by a much wider class of $\mathbb{Z}^d$-actions by commuting diffeomorphisms of manifolds. The class of 'algebraic' $\mathbb{Z}^d$-actions mentioned above originated in a simple example proposed by Ledrappier and consists of all $\mathbb{Z}^d$-actions by commuting automorphisms of compact groups (cf. K. Schmidt, *Dynamical systems of algebraic origin*). This class allows the development of a quite detailed and coherent theory which goes far beyond what can be established for more general systems.

The surveys and original research articles in this volume cover many of these diverse ingredients of the theory of $\mathbb{Z}^d$-actions. The success of the symposium was due to the outstanding survey lecture series and individual research talks by many of the main contributors to these recent developments in multi-parameter ergodic theory, and to the expertise and enthusiasm of the staff of the Warwick Mathematics Research Centre (Peta McAllister and Elaine Coelho Greaves). On behalf of the other co-organizers (W. Parry and P. Walters) we would like to thank them all for their contributions.

During the symposium, Bill Parry celebrated his 60th birthday. As colleagues and friends we would like to dedicate these proceedings to him.

MARK POLLICOTT
DEPARTMENT OF MATHEMATICS
MANCHESTER UNIVERSITY
OXFORD ROAD
MANCHESTER, M13 9PL
ENGLAND
EMAIL: MP@MATHS.WARWICK.AC.UK

KLAUS SCHMIDT
MATHEMATICS INSTITUTE
UNIVERSITY OF VIENNA
A-1090 VIENNA
AUSTRIA
EMAIL: KLAUS.SCHMIDT@UNIVIE.AC.AT

Ergodic Ramsey Theory–an Update

VITALY BERGELSON
The Ohio State University
Columbus, OH 43210 U.S.A.

0. Introduction.

This survey is an expanded version and elaboration of the material presented by the author at the *Workshop on Algebraic and Number Theoretic Aspects of Ergodic Theory* which was held in April 1994 as part of the 1993/1994 *Warwick Symposium on Dynamics of Z^n-actions and their connections with Commutative Algebra, Number Theory and Statistical Mechanics*. The leitmotif of this paper is: Ramsey theory and ergodic theory of multiple recurrence are two beautiful, tightly intertwined and mutually perpetuating disciplines. The scope of the survey is mostly limited to Ramsey-theoretical and ergodic questions about Z^n–partly because of the proclaimed goals of the Warwick Symposium and partly because of the author's hope that Z^n-related combinatorics, number theory and ergodic theory can serve as an ideal lure through which the author's missionary zeal will reach as wide an audience of potential adherents to the subject as possible.

To compensate for the selective neglect of details and for the lack of full generality in some of the proofs, which were imposed by natural time and space limitations, a significant effort was spent on accentuation and motivation of ideas which lead to conjectures and techniques on which the proofs of conjectures hinge.

Here now is a brief description of the content of the five sections constituting the body of this survey. In Section 1 three main principles of Ramsey theory are introduced and their connection with the ergodic theory of multiple recurrence is emphasized. This section contains a lot of discussion and very few proofs. The goal in this section is to help create in the reader a feeling of what Ergodic Ramsey Theory is all about.

Section 2 is devoted to a multifaceted treatment of a special case of the polynomial ergodic Szemerédi theorem recently obtained in [BL1] (Theorem 1.19 of Section 1). Different approaches are discussed and brought to (hopefully) a convincing level of detail.

In Section 3 the somewhat esoteric, but fascinating and very useful object βN, the Stone-Čech compactification of N, is introduced and discussed in some detail. An ultrafilter proof of the celebrated Hindman's theorem is given and some applications of βN and Hindman's theorem to topological dynamics, especially to distal systems, are discussed. This section concludes with a formulation and discussion of an ultrafilter refinement of the

1

Furstenberg-Sárközy theorem on recurrence along polynomials, and a proof of a special case of this refinement.

Section 4 is devoted to ramifications of results brought in previous sections. Most of the discussion is devoted to polynomial ergodic theorems along IP-sets. In addition, the role of a polynomial refinement of the combinatorial Hales-Jewett theorem is emphasized. The flow of this discussion naturally leads to some open problems which are collected and commented on in Section 5.

I was fortunate to be a graduate student at the Hebrew University of Jerusalem at the time of the inception and early development of Ergodic Ramsey Theory. It is both my duty and pleasure to acknowledge the influence of and express my gratitude to Izzy Katznelson, Benji Weiss, and especially my Ph.D. thesis advisor, Hillel Furstenberg.

I wish to express my indebtedness to my friend and co-author Neil Hindman for many useful discussions of ultrafilter lore.

I also owe a large debt to Randall McCutcheon, whose numerous and most pertinent suggestions for improvements of presentation greatly facilitated the preparation of this survey.

In addition, my thanks go to Boris Begun and Paul Larick for their useful remarks on a preliminary version of this paper.

Finally, I would like to thank the organizers and hosts of the Symposium and the editors of these Proceedings, Bill Parry, Mark Pollicott, Klaus Schmidt, Caroline Series, and Peter Walters for creating a great atmosphere and for their efforts to promote and advance the beautiful science of ergodic theory.

1. Three main principles of Ramsey theory and its connection with the ergodic theory of multiple recurrence.

> A mathematician, like a painter
> or a poet, is a maker of patterns.

–G.H. Hardy, [Ha], p. 84.

Van der Waerden's Theorem, one of Khintchine's "Three Pearls of Number Theory" ([K1]), states that whenever the natural numbers are finitely partitioned (or, as it is customary to say, finitely colored), one of the cells of the partition contains arbitrarily long arithmetic progressions. One can reformulate van der Waerden's theorem in the following, "finitistic" form:

Theorem 1.1. For any natural numbers k and r there exists $N = N(k,r)$ such that whenever $m \geq N$ and $\{1, \cdots, m\} = \bigcup_{i=1}^{r} C_i$, one of C_i, $i = 1, \cdots, r$ contains a k-term arithmetic progression.

Exercise 1. Show the equivalence of van der Waerden's theorem and

Theorem 1.1.

Van der Waerden's theorem belongs to the vast variety of results which form the body of *Ramsey theory* and which have the following general form: If V is an infinite, "highly organized" structure (a semi-group, a vector space, a complete graph, etc.) then for any finite coloring of V there exist arbitrarily large (and sometimes even infinite) highly organized monochromatic substructures. In other words, the high level of organization cannot be destroyed by partitioning into finitely many pieces—one of these pieces will still be highly organized. To fit van der Waerden's theorem into this framework, let us call a subset of $\mathbf{Z}$ *a.p.-rich* if it contains arbitrarily long arithmetic progressions. Then van der Waerden's theorem can be reformulated in the following way:

Theorem 1.2. If $S \subset \mathbf{Z}$ is an a.p.-rich set and, for some $r \in \mathbf{N}$, $S = \bigcup_{i=1}^{r} C_i$, then one of C_i, $i = 1, \cdots, r$ is a.p.-rich.

(Since $\mathbf{Z}$ is a.p.-rich, van der Waerden's theorem is obviously a special case of Theorem 1.2. On the other hand, it is not hard to derive Theorem 1.2 from Theorem 1.1.)

We cannot resist the temptation to bring here two more equivalent forms of van der Waerden's theorem, each revealing still another of its facets.

Theorem 1.3. For any finite partition of $\mathbf{Z}$, one of the cells of the partition contains an affine image of any finite set. (An affine image of a set $F \subset \mathbf{Z}$ is any set of the form $a + bF = \{a + bx : x \in F\}$.)

Exercise 2. Show the equivalence of Theorems 1.3 and 1.1.

Theorem 1.4 (A special case of a theorem due to Furstenberg and Weiss, [FW1]). Suppose $k \in \mathbf{N}$ and $\epsilon > 0$. For any continuous self-mapping of a compact metric space (X, ρ), there exists $x \in X$ and $n \in \mathbf{N}$ such that $\rho(T^{in}x, x) < \epsilon$, $i = 1, \cdots, k$.

Theorem 1.3 shows that van der Waerden's theorem is actually a *geometric* rather than number theoretic fact. On the other hand, Theorem 1.4 establishes the seminal connection between partition theorems of van der Waerden type with *topological dynamics*—the link which proved to be extremely useful.

Another example of "unbreakable" structure is given by Hindman's finite sums theorem ([H2]). To formulate Hindman's theorem let us (following notation in [FW1]) call a set $S \subset \mathbf{N}$ an *IP-set* if it consists of an infinite sequence $(x_n)_{n=1}^{\infty} \subset \mathbf{N}$ together with all finite sums of the form $x_\alpha = \sum_{n \in \alpha} x_n$, where α ranges over the finite non-empty subsets of $\mathbf{N}$.

Theorem 1.5 (Hindman). If $E \subset \mathbf{N}$ is an IP-set, then for any finite coloring $E = \bigcup_{i=1}^{r} C_i$, one of C_i, $i = 1, \cdots, r$ contains an IP-set.

We shall return to Hindman's theorem in the discussions of Section 3. We refer the reader to [GRS] for many more examples illustrating the *first principle of Ramsey theory*–the *preservation of structure under finite partitions*.

After being convinced of the validity of this first principle of Ramsey theory, one is led to the next natural question: why is this so? What exactly is responsible for this stubborn tendency of highly organized infinite structures to preserve their (rightly interpreted) replicas in at least one cell of an arbitrary finite partition? The answer is, and this is the *second principle of Ramsey theory*, that there is always an appropriate notion of *largeness* which is behind the scenes and such that *any* large set contains the sought-after highly organized sub-structures. The only other requirement that the notion of largeness should satisfy is that if A is large and $A = \bigcup_{i=1}^{r} C_i$, then one of C_i, $1 = 1, \cdots, r$ is also large. It is the mathematician's task when dealing with this or that result of partition Ramsey theory to guess what the appropriate notion of largeness responsible for the truth of the proposition is. It is the almost intentional vagueness of the approach which allows one to obtain stronger and stronger theorems by modifying and playing with different notions of largeness. To illustrate this second principle of Ramsey theory we shall now give some examples.

Given a set $A \subset \mathbf{N}$, define its *upper density* $\bar{d}(A)$ by

$$\bar{d}(A) = \limsup_{N \to \infty} \frac{|A \cap \{1, \cdots, N\}|}{N}.$$

If the limit (rather than lim sup) exists, we say that A has density, and denote it by $d(A)$. Being of positive upper density is obviously a notion of largeness and it is natural to ask (as P. Erdös and P. Turán did in [ET]) whether this notion of largeness is responsible for the validity of van der Waerden's theorem. Namely, is it true that any set $A \subset \mathbf{N}$ of positive upper density is a.p.-rich?

The question turned out to be very hard. After some partial results were obtained in [Ro] and [Sz1], Szemerédi [Sz2] settled the Erdös-Turán conjecture affirmatively, thus providing a convenient sufficient condition for a set to be a.p.-rich.

Theorem 1.6 (Szemerédi, [Sz2]). Any set $E \subset \mathbf{N}$ having positive upper density is a.p.-rich.

Exercise 3. Derive from Theorem 1.6 the following finitistic version of it:

For any $\epsilon > 0$ and any $k \in \mathbf{N}$ there exists $N = N(\epsilon, k)$ such that if $m > N$ and $A \subset \{1, 2, \cdots, m\}$ satisfies $\frac{|A|}{m} > \epsilon$, then A contains a k-term arithmetic progression.

It follows from Exercise 3 that in the formulation of Theorem 1.6 a somewhat weaker notion of largeness would do, namely, the notion of *upper Banach density*. Given a set $E \subset \mathbf{Z}$ define its upper Banach density $d^*(E)$ by

$$d^*(E) = \limsup_{N-M \to \infty} \frac{|E \cap \{M, M+1, \cdots, N\}|}{N - M + 1}.$$

It is the notion of positive upper Banach density and its natural extensions to $\mathbf{Z}^d$ and, indeed, to any countable *amenable* group which naturally participate in many questions and results of *density Ramsey theory*.

It is easy to check that for any $E \subset \mathbf{Z}$ and any $t \in \mathbf{Z}$ the set $E - t := \{x - t : x \in E\}$ satisfies $d^*(E - t) = d^*(E)$. This shift-invariance of the upper Banach density hints that there is a genuine measure preserving system behind any set $E \subset \mathbf{Z}$ with $d^*(E) > 0$. This is indeed so (see below). On the other hand, the notion of upper Banach density does not provide the right notion of largeness for results like Hindman's theorem. For example, the set $E = 2\mathbf{N}+1$ is large in the sense that $d^*(E) = \frac{1}{2}$ but obviously cannot contain any IP-set, or even any triple of the form $\{x, y, x+y\}$ (see also Exercise 8 in Section 3). We shall see in Section 3 that a notion of largeness relevant for Hindman's theorem is provided by idempotent ultrafilters in $\beta\mathbf{N}$, the Stone-Čech compactification of $\mathbf{N}$. This notion of largeness will also have a mild form of shift invariance which will allow us to prove Hindman's theorem by repeated utilization of a kind of Poincaré recurrence theorem adapted to the situation at hand.

Ergodic Ramsey Theory started with the publication of [F1], in which Furstenberg derived Szemerédi's theorem from a beautiful, far reaching extension of the classical Poincaré recurrence theorem, which corresponds to the case $k = 1$ in the following:

Theorem 1.7 (Furstenberg, [F1]). Let $(X, \mathcal{B}, \mu, T)$ be a probability measure preserving system. For any $k \in \mathbf{N}$ and for any $A \in \mathcal{B}$ with $\mu(A) > 0$ there exists $n \in \mathbf{N}$ such that

$$\mu(A \cap T^{-n}A \cap T^{-2n}A \cap \cdots \cap T^{-kn}A) > 0.$$

In order to derive Szemerédi's Theorem 1.6, Furstenberg introduced a *correspondence principle*, which provides the link between density Ramsey theory and ergodic theory.

Theorem 1.8 *Furstenberg's correspondence principle.* Given a set $E \subset \mathbf{Z}$ with $d^*(E) > 0$ there exists a probability measure preserving system $(X, \mathcal{B}, \mu, T)$ and a set $A \in \mathcal{B}$, $\mu(A) = d^*(E)$, such that for any $k \in \mathbf{N}$ and any $n_1, \cdots n_k \in \mathbf{Z}$ one has:

$$d^*\big(E \cap (E - n_1) \cap \cdots \cap (E - n_k)\big) \geq \mu(A \cap T^{-n_1}A \cap \cdots \cap T^{-n_k}A).$$

Since the set E contains a progression $\{x, x+n, \cdots, x+kn\}$ if and only if $E \cap (E-n) \cap \cdots \cap (E - kn) \neq \emptyset$, it is clear that Furstenberg's multiple recurrence Theorem 1.7 together with the correspondence principle imply Szemerédi's theorem. We remark that as a matter of fact, Theorem 1.7 follows from Szemerédi's theorem using fairly elementary arguments. Alternatively one can utilize the following refinement of the Poincaré recurrence theorem.

Theorem 1.9 ([B1]). For any probability measure preserving system $(X, \mathcal{B}, \mu, T)$ and any $A \in \mathcal{B}$ with $\mu(A) > 0$ there exists a sequence $E = (n_m)_{m=1}^{\infty}$ whose density exists and satisfies $d(E) \geq \mu(A)$ such that for any $m \in \mathbf{N}$

$$\mu(A \cap T^{-n_1} A \cap \cdots \cap T^{-n_m} A) > 0.$$

Van der Waerden's theorem has a natural multidimensional extension which is hinted at by the geometric formulation (Theorem 1.3).

Theorem 1.10 *Multidimensional van der Waerden theorem* (Gallai-Grünwald). For any finite coloring of $\mathbf{Z}^d$, $\mathbf{Z}^d = \bigcup_{i=1}^{r} C_i$, one of C_i, $i = 1, \cdots, r$ contains an affine image of any finite subset $F \subset \mathbf{Z}^d$. In other words, there exists i, $1 \leq i \leq r$, such that for any finite $F \subset \mathbf{Z}^d$, there exists $u \in \mathbf{Z}^d$ and $a \in \mathbf{N}$ such that $u + aF = \{u + ax : x \in F\} \subset C_i$.

Remark. An attribution of Theorem 1.10 to G. Grünwald is made in [Ra], p. 123. As far as we know, Grünwald never published his proof. He later changed his name to Gallai, to whom the result is attributed in [GRS].

In accordance with the second principle of Ramsey theory one should expect that Theorem 1.10 has a density version. This is indeed so and was proved in [FK1]. The multidimensional Szemerédi theorem established by Furstenberg and Katznelson there was the first in a chain of strong combinatorial results ([FK2], [FK4], [BL1]) which were achieved by means of ergodic theory and which so far have no conventional combinatorial proof.

Let us say that a set $S \subset \mathbf{Z}^k$ has positive upper Banach density if for some sequence of parallelepipeds $\Pi_n = [a_n^{(1)}, b_n^{(1)}] \times \cdots \times [a_n^{(k)}, b_n^{(k)}] \subset \mathbf{Z}^k$, $n \in \mathbf{N}$, with $b_n^{(i)} - a_n^{(i)} \to \infty$, $i = 1, \cdots, k$ one has:

$$\frac{|S \cap \Pi_n|}{|\Pi_n|} > \epsilon$$

for some $\epsilon > 0$.

The natural question now is whether it is true that any set of positive upper Banach density in $\mathbf{Z}^k$ contains an affine image of any finite set $F \subset \mathbf{Z}^k$. Furstenberg and Katznelson answered this question affirmatively by deducing the answer from the following generalization of Furstenberg's multiple recurrence theorem.

Theorem 1.11 ([FK1], Theorem A). Let $(X, \mathcal{B}, \mu)$ be a measure space with $\mu(X) < \infty$, let $T_1, \cdots, T_k$ be commuting measure preserving transformations of X and let $A \in \mathcal{B}$ with $\mu(A) > 0$. Then

$$\liminf_{N \to \infty} \frac{1}{N} \sum_{n=0}^{N-1} \mu(T_1^{-n} A \cap T_2^{-n} A \cap \cdots \cap T_k^{-n} A) > 0.$$

Corollary 1.12 ([FK1], Theorem B). Let $S \subset \mathbf{Z}^k$ be a subset with positive upper Banach density and let $F \subset \mathbf{Z}^k$ be a finite configuration. Then there exists a positive integer n and a vector $u \in \mathbf{Z}^k$ such that $u + nF \subset S$.

For the derivation of Corollary 1.12 from Theorem 1.11, the reader is referred to [F2], where the correspondence principle is spelled out for $\mathbf{Z}^k$.

The *third* and last *principle of Ramsey theory* which we want to discuss in this section is the following: *the sought-after configurations always to be found in large sets are abundant.* Abundance in our context means not only that the parameters describing the configurations form large sets in the space of parameters, but also that these parameters are nicely spread in all kinds of families of subsets of integers. Let us consider some examples. Take, for instance, Szemerédi's theorem. Let $E \subset \mathbf{Z}$ with $d^*(E) > 0$. For fixed k the progressions $\{x, x + d, \cdots, x + kd\} \subset E$ are naturally parametrized by pairs (x, d). Let us call a point $x \in E$ a (d, k)-starter if $\{x, x + d, \cdots, x + kd\} \subset E$ and a non-(d, k)-starter otherwise. One can show that for any k, "almost every" point of E is a (d, k)-starter for some d. In other words, for any fixed k, the set of (d, k)-starters in E has upper Banach density equal to $d^*(E)$.

Exercise 4. Show that the set of non-$(d, 2)$-starters of a set $E \subset \mathbf{Z}$ with $d^*(E) > 0$ may be infinite.

Let us turn now to a much more interesting set of those d which appear as differences of arithmetic progressions in E. One of the ways of measuring how well "spread" a subset of integers is, would be to see whether it has a nonempty intersection with different families of subsets of integers (analogy: a set $S \subset [0, 1]$ is dense if for any $0 \le a < b \le 1$, $S \cap (a, b) \ne \emptyset$). We shall need a few definitions. Given a countable abelian group G, a set $S \subset G$ is called *syndetic* if for some finite set $F \subset G$ one has: $S + F = \{x + y : x \in S, y \in F\} = G$. It is easy to see that a set $S \subset \mathbf{Z}$ is syndetic if and only if it has bounded gaps, namely intersects non-trivially any big enough interval. We note that any syndetic set $S \subset \mathbf{Z}$ is a.p.-rich. Indeed, as finitely many shifts of S cover $\mathbf{Z}$ completely, by van der Waerden's theorem one of these shifts is a.p.-rich, and as the property of a.p.-richness is clearly shift invariant, we see that S itself must have this property.

Following the terminology introduced in [F2], given a family $\mathcal{S}$ of subsets of $\mathbf{Z}$ let us call a set $E \subset \mathbf{Z}$ an $\mathcal{S}^*$-set if for any $S \in \mathcal{S}$, $E \cap S \neq \emptyset$. In particular, a set $E \subset \mathbf{Z}$ is an IP*-set if E has non-trivial intersection with any IP-set. It is not hard to see that any IP* set is syndetic. Indeed, if a set E was an IP*-set but not syndetic, its complement would contain a union of intervals $[a_n, b_n]$ with $b_n - a_n \to \infty$. One can easily show that any such union of intervals contains an IP-set which leads to contradiction with the assumed IP*-ness of E.

Exercise 5. Show that for any finite measure preserving system $(X, \mathcal{B}, \mu, T)$ and any $A \in \mathcal{B}$ with $\mu(A) > 0$, the set $\{n : \mu(A \cap T^{-n}A) > 0\}$ is an IP*-set.

On the other hand, it is easy to see that not every syndetic set is an IP*-set: take, for example, the odd integers.

Now, IP*-sets are large in a few different senses. Besides having positive lower density (the lower density of a set S is defined as $\liminf\limits_{N \to \infty} \frac{|S \cap [1, \cdots, N]|}{N}$), they, for example, have a finite intersection property.

Lemma 1.13. If $S_1, S_2, \cdots, S_k$ are IP*-sets then $\bigcap_{i=1}^k S_i$ is also an IP*-set.

Proof. It is enough to prove the result for $k = 2$. Let E be an IP-set. Consider the following partition of E: $E = (E \cap S_1) \cup (E \cap S_1^c)$. By Hindman's theorem at least one of $E \cap S_1$, $E \cap S_1^c$ contains an IP-set E_1. Since S_1 is an IP*-set, $E_1 \cap S_1 \neq \emptyset$, hence $E_1 \subset E \cap S_1$. Also S_2 is an IP*-set, hence $E_1 \cap S_2 \neq \emptyset$, which implies that $E \cap (S_1 \cap S_2) \neq \emptyset$. As E was an arbitrary IP-set, the lemma is proved.

$\square$

Given $E \subset \mathbf{Z}$ with $d^*(E) > 0$ let

$$R_k(E) = \{d \in \mathbf{Z} : \{x, x + d, \cdots, x + kd\} \subset E \text{ for some } x \in \mathbf{Z}\}.$$

The question of how well spread the sets $R_k(E)$ are in $\mathbf{Z}$ is interesting already for $k = 1$. The illustrative results about sets $R_1(E)$ which we collect here are special cases of sometimes very far reaching generalizations. Notice that $R_1(E) = E - E = \{x - y : x, y \in E\}$. It follows immediately from Exercise 5 via Furstenberg's correspondence principle that $R_1(E)$ is an IP*-set. We remark that this result has also a simple completely elementary proof: given an IP-set, generated, say, by $n_1, n_2, \cdots$, one considers the sets

$E_i = E - (n_1 + \cdots + n_i)$, $i = 1, 2, \cdots$ and observes that since $d^*(E) > 0$ we have, for some $1 \le i < j \le \frac{1}{d^*(E)} + 1$,

$$d^*(E_i \cap E_j) = d^*\left(E \cap \left(E - (n_{i+1} + \cdots + n_j)\right)\right) > 0.$$

This implies that the set of differences $E - E$ contains the element $n_{i+1} + \cdots + n_j$ from our IP-set.

We sketch now a curious application of this circle of ideas to the theory of almost periodic functions. For the sake of simplicity we shall deal only with functions on **Z**, remarking that easy modifications of these arguments would apply to almost periodic functions on an arbitrary topological group. Recall that, according to H. Bohr ([Bo1]), a function $f : \mathbf{Z} \to \mathbf{C}$ is called almost periodic if for any $\epsilon > 0$ the set of "ϵ-periods",

$$E(\epsilon, f) = \{h \in \mathbf{Z} : |f(x + h) - f(x)| < \epsilon \text{ for all } x\}$$

is syndetic. Later Bogolioùboff, [Bo2], and Følner, [Fø] showed that the condition of syndeticity in the definition may be replaced by the weaker condition of positive upper Banach density. This result is contained in the following proposition:

Theorem 1.14. For a function $f : \mathbf{Z} \to \mathbf{C}$ the following conditions are equivalent:

(i) For any $\epsilon > 0$ the set $E(\epsilon, f)$ has positive upper Banach density.

(ii) For any $\epsilon > 0$ the set $E(\epsilon, f)$ is syndetic.

(iii) For any $\epsilon > 0$ the set $E(\epsilon, f)$ is an IP*-set.

Proof. It is enough to show that (i)→(iii). But this follows immediately from two facts:

(1) $E(\frac{\epsilon}{2}, f) - E(\frac{\epsilon}{2}, f) \subset E(\epsilon, f)$.

(2) If $d^*(E) > 0$ then $E - E$ is an IP*-set.

□

As a byproduct one obtains the following fact, which is not obvious from Bohr's definition (but is obvious from some other equivalent definitions of almost periodicity).

Corollary 1.15. If f, g are almost periodic functions then $f + g$ is also an almost periodic function.

Proof. Observe that $E(\frac{\epsilon}{2}, f) \cap E(\frac{\epsilon}{2}, g) \subset E(\epsilon, f + g)$. The result follows from Lemma 1.13.

□

Following Furstenberg ([F3]) let us call a set $R \subset \mathbf{Z}$ a *set of recurrence* if for any invertible finite measure preserving system $(X, \mathcal{B}, \mu, T)$ and any $A \in \mathcal{B}$ with $\mu(A) > 0$ there exists $n \in R$, $n \neq 0$, with $\mu(A \cap T^n A) > 0$.

Exercise 6. Show that the following are sets of recurrence:

(i) Any $E \subset \mathbf{Z}$ with $d^*(E) = 1$.

(ii) $a\mathbf{N} = \{an : n \in \mathbf{N}\}$, for any $0 \neq a \in \mathbf{Z}$.

(iii) $E - E$, for any infinite set $E \subset \mathbf{Z}$.

(iv) Any IP-set.

(v) Any set of the form $\bigcup_{n=1}^{\infty} \{a_n, 2a_n, \cdots, na_n\}$, $a_n \in \mathbf{N}$.

Denote by $\mathcal{R}$ the family of all sets of recurrence in $\mathbf{Z}$. According to our adopted conventions, a set E is $\mathcal{R}^*$ if it intersects nontrivially any set of recurrence. Similarly to IP-sets, sets of recurrence possess the Ramsey property: if a set of recurrence R is finitely partitioned, $R = \bigcup_{i=1}^{r} C_i$, then one of C_i, $i = 1, \cdots, r$ is itself a set of recurrence. To see this, assume that this is not true. So for a set of recurrence R and some partition $R = \bigcup_{i=1}^{r} C_i$ one can find measure preserving systems $(X_i, \mathcal{B}_i, \mu_i, T_i)$ and sets $A_i \in \mathcal{B}_i$, $i = 1, \cdots, r$, with $\mu_i(A_i) > 0$, such that $\mu_i(A_i \cap T_i^n A_i) = 0$ for all $n \in C_i$. Let $(X, \mathcal{B}, \mu, T)$ be the product system of $(X_i, \mathcal{B}_i, \mu_i, T_i)$, $i = 1, \cdots, r$ and take $A = A_1 \times \cdots \times A_r \in \mathcal{B}_1 \otimes \cdots \otimes \mathcal{B}_r$. Since R is a set of recurrence, there exists $n \in R$ such that $\mu(A \cap T^n A) > 0$, where $\mu = \mu_1 \times \cdots \times \mu_r$, $T = T_1 \times \cdots \times T_r$. This implies that for $i = 1, \cdots, r$, $\mu_i(A_i \cap T_i^n A_i) > 0$ which is a contradiction. The discussion above together with the fact that for any $E \subset \mathbf{Z}$ with $d^*(E) > 0$ the set $E - E$ is an $\mathcal{R}^*$-set imply the following combinatorial fact (cf. [F2], p. 75).

Theorem 1.16. Given sets $E_i \subset \mathbf{Z}$ with $d^*(E_i) > 0$, $i = 1, \cdots, k$, the set $D = (E_1 - E_1) \cap (E_2 - E_2) \cap \cdots \cap (E_k - E_k)$ is $\mathcal{R}^*$. In particular, D is IP* and hence syndetic.

The following fact, due independently to Furstenberg ([F2]) and Sárközy ([S]), provides an example of a set of recurrence of a quite different nature than those of Exercise 6.

Theorem 1.17. Assume that $p(t) \in \mathbf{Q}[t]$ with $p(\mathbf{Z}) \subset \mathbf{Z}$, $\deg p(t) > 0$, and $p(0) = 0$. The the set $\{p(n) : n \in \mathbf{N}\}$ is a set of recurrence.

For more examples and further discussion of sets of recurrence the reader is referred to [F3], [B1], [B2], [BHå], [Fo], and [M]. We comment now on some extensions of these results to multiple recurrence.

Given a finite invertible measure preserving system $(X, \mathcal{B}, \mu, T)$ and a set $A \in \mathcal{B}$ with $\mu(A) > 0$ consider the set

$$R_k(A) = \{n \in \mathbf{Z} : \mu(A \cap T^n A \cap \cdots \cap T^{kn} A) > 0\}.$$

According to the third principle of Ramsey theory, the sets $R_k(A)$ should be "well spread" in $\mathbf{Z}$. They are. The fact that $R_k(A)$ is syndetic was contained already in the pioneering paper of Furstenberg, [F1]: he established Theorem 1.7 by actually showing that

$$\liminf_{N-M\to\infty} \frac{1}{N-M} \sum_{n=M}^{N-1} \mu(A\cap T^n A\cap\cdots\cap T^{kn}A) > 0.$$

The next question to ask about the sets $R_k(A)$ is whether they are always IP*-sets. This turned out to be true and is a special case of a deep and highly nontrivial "IP-Szemerédi theorem" due to Furstenberg and Katznelson, [FK2]. We give a formulation of a more general fact than IP*-ness of $R_k(A)$ which still is quite a special case of the main theorem in [FK2].

Theorem 1.18 ([FK2]). For any finite measure space $(X,\mathcal{B},\mu)$, any $k\in\mathbf{N}$, any commuting invertible measure preserving transformations $T_1,\cdots,T_k$ of $(X,\mathcal{B},\mu)$ and any $A\in\mathcal{B}$ with $\mu(A)>0$ the set

$$\{n : \mu(A\cap T_1^n A\cap\cdots\cap T_k^n A) > 0\}$$

is IP*.

Another desirable refinement of Furstenberg's Szemerédi theorem is hinted upon by Theorem 1.17, an equivalent form of which is that for any invertible measure preserving system $(X,\mathcal{B},\mu,T)$, any $A\in\mathcal{B}$ with $\mu(A)>0$, and any polynomial $p(t)\in\mathbf{Q}[t]$ with $p(\mathbf{Z})\subset\mathbf{Z}$ and $p(0)=0$, the set $R_1(A)=\{n\in\mathbf{Z}:\mu(A\cap T^n A)>0\}$ intersects non-trivially the set $p(\mathbf{N})=\{p(n):n\in\mathbf{N}\}$. Is it true, for example, that the sets

$$R_k(A) = \{n\in\mathbf{Z} : \mu(A\cap T^n A\cap\cdots\cap T^{kn}A) > 0\}$$

also have such an intersection property? Or, even more ambitiously, does a joint extension of Theorems 1.11 and 1.17 hold? Namely, given any k polynomials $p_i(t)\in\mathbf{Q}[t]$ with $p_i(\mathbf{Z})\subset\mathbf{Z}$ and $p_i(0)=0$, $i=1,\cdots,k$ and any k commuting invertible measure preserving transformations $T_1,\cdots,T_k$ of a probability measure space $(X,\mathcal{B},\mu)$, is it true that for any $A\in\mathcal{B}$ with $\mu(A)>0$ one has

$$\liminf_{N\to\infty} \frac{1}{N} \sum_{n=0}^{N-1} \mu(A\cap T_1^{p_1(n)}A\cap\cdots\cap T_k^{p_k(n)}A) > 0?$$

It turns out that the answers to these questions are positive and that a stronger, more general result holds.

Theorem 1.19 (Polynomial Szemerédi theorem, [BL1]). Let $(X, \mathcal{B}, \mu)$ be a probability space, let $T_1, \cdots, T_t$ be commuting invertible measure preserving transformations of X, let $p_{i,j}(n)$ be polynomials with rational coefficients, $1 \leq i \leq k$, $1 \leq j \leq t$, satisfying $p_{i,j}(0) = 0$ and $p_{i,j}(\mathbf{Z}) \subset \mathbf{Z}$. Let $A \in \mathcal{B}$ with $\mu(A) > 0$. Then

$$\liminf_{N \to \infty} \frac{1}{N} \sum_{n=0}^{N-1} \mu\Big(A \cap \Big(\prod_{j=1}^{t} T_j^{p_{1,j}(n)} \Big) A \cap \cdots \cap \Big(\prod_{j=1}^{t} T_j^{p_{k,j}(n)} \Big) A \Big) > 0.$$

As a corollary we have:

Theorem 1.20. Let $P : \mathbf{Z}^r \to \mathbf{Z}^l$, $r, l \in \mathbf{N}$ be a polynomial mapping satisfying $P(0) = 0$, let $F \subset \mathbf{Z}^r$ be a finite set and let $S \subset \mathbf{Z}^l$ be a set of positive upper Banach density. Then for some $n \in \mathbf{N}$ and $u \in \mathbf{Z}^l$ one has

$$u + P(nF) = \big\{ u + P(nx_1, nx_2, \cdots, nx_r) : (x_1, \cdots, x_r) \in F \big\} \subset S.$$

The proof of Theorem 1.19 in [BL1] is, in a sense, a *polynomialization* of the proof of Furstenberg's and Katznelson's multidimensional Szemerédi theorem (Theorem 1.11). We shall try to convey some of the flavor of the proof of Theorem 1.19 in the next section, where we shall treat a special case of it. See also [BM1] where a concise proof of the following refinement of a special case of Theorem 1.19 is given.

Theorem 1.21. Suppose that $(X, \mathcal{B}, \mu, T)$ is an invertible measure preserving system, $k \in \mathbf{N}$, $A \in \mathbf{B}$ with $\mu(A) > 0$ and $p_i(x) \in \mathbf{Q}[x]$ with $p_i(\mathbf{Z}) \subset \mathbf{Z}$ and $p_i(0) = 0$, $1 \leq i \leq k$. Then

$$\liminf_{N-M \to \infty} \frac{1}{N-M} \sum_{n=M}^{N-1} \mu(A \cap T^{p_1(n)} A \cap \cdots \cap T^{p_k(n)} A) > 0.$$

Motivated by the third principle of Ramsey theory, (and by Theorem 1.21), one should expect that Theorem 1.19 has an IP-refinement, similar to the way in which Theorem 1.18 refines Theorem 1.11. This again turns out to be true ([BM2]) and will be discussed in more detail in Section 4. Notice that even in the case of single recurrence along polynomials, it is not obvious at all whether, say, the set $\{n : \mu(A \cap T^{n^2} A) > 0\}$ is an IP*-set. (It is. See Theorem 3.11.)

2. Special case of polynomial Szemerédi theorem: single recurrence.

In this section we shall discuss in detail the following special case of Theorem 1.19, which corresponds to the case $k = 1$.

Theorem 2.1. Suppose $t \in \mathbf{N}$. For any t invertible commuting measure preserving transformations $T_1, T_2, \cdots T_t$ of a probability space $(X, \mathcal{B}, \mu)$, for any $A \in \mathcal{B}$ with $\mu(A) > 0$, and for any polynomials $p_i(x) \in \mathbf{Q}[x]$, $i = 1, \cdots, t$ satisfying $p_i(\mathbf{Z}) \subset \mathbf{Z}$ and $p_i(0) = 0$ one has:

$$\liminf_{N \to \infty} \frac{1}{N} \sum_{n=0}^{N-1} \mu(A \cap T_1^{p_1(n)} T_2^{p_2(n)} \cdots T_t^{p_t(n)} A) > 0.$$

Before embarking on the proof we wish to make some remarks and formulate the facts that will be instrumental to the proof.

The first remark that we want to make is that Theorem 2.1, being a result about *single recurrence*, is *Hilbertian* in nature in the sense that it follows from (more general) facts shared by all unitary $\mathbf{Z}^t$-actions, rather than specifically those induced by measure preserving transformations. One, and from some point of view natural way of proving Theorem 2.1 would be to employ the spectral theorem for unitary $\mathbf{Z}^t$-actions. This is done for $t = 1$ in [F2] and [F3] and the extension for general $t \in \mathbf{N}$ goes through with no problems. Unfortunately, the spectral theorem is of no use when one has to deal with multiple recurrence. That is why we prefer to use a "softer", spectral theorem-free approach, which, on the one hand, looks more involved, but on the other hand is more easily susceptible to generalization and refinement.

The main idea which will govern our approach is that in order to prove this or that sort of ergodic theorem, one looks for a suitable splitting of the underlying Hilbert space into orthogonal invariant subspaces on which the behavior of the studied unitary action along, say, polynomials can be well understood and controlled. Consider, for example, the classical mean ergodic theorem of von Neumann for a unitary operator U on a Hilbert space $\mathcal{H}$. The (almost trivial) splitting in this case is $\mathcal{H} = \mathcal{H}_{inv} \oplus \mathcal{H}_{erg}$, where

$$\mathcal{H}_{inv} = \{f \in \mathcal{H} : Uf = f\}, \text{ and}$$

$$\mathcal{H}_{erg} = \overline{\{f \in \mathcal{H} : \text{ there exists } g \in \mathcal{H} \text{ with } f = g - Ug\}}$$

$$= \{f \in \mathcal{H} : \left\| \frac{1}{N} \sum_{n=0}^{N-1} U^n f \right\| \to 0\}.$$

The $\mathbf{Z}$-action generated by U is trivial on the subspace $\mathcal{H}_{inv}$ of invariant elements, whereas it is easily manageable on the *ergodic* subspace $\mathcal{H}_{erg}$.

This splitting is too trivial to help with ergodic theorems along polynomials, but it hints that if one enlarges $\mathcal{H}_{inv}$ just a little bit (and at the same time appropriately shrinks $\mathcal{H}_{erg}$) the situation will be suitable for a polynomial ergodic theorem. Indeed, consider the splitting $\mathcal{H}_{rat} \oplus \mathcal{H}_{tot.erg}$, where

$$\mathcal{H}_{rat} = \overline{\{f \in \mathcal{H} : \text{ there exists } i \in \mathbf{N} \text{ with } U^i f = f\}}, \text{ and}$$

$$\mathcal{H}_{tot.erg} = \{f \in \mathcal{H} : \text{ for all } i \in \mathbf{N}, \|\frac{1}{N}\sum_{n=0}^{N-1} U^{in} f\| \to 0\}.$$

Exercise 7. Show that $\mathcal{H} = \mathcal{H}_{rat} \oplus \mathcal{H}_{tot.erg}$. Attempt not to use the spectral theorem. Show that

$$\mathcal{H}_{tot.erg} = \left\{f \in \mathcal{H} : \text{ for all } i \in \mathbf{N}, \left\|\frac{1}{N}\sum_{n=0}^{N-1} U^{-in} f\right\| \to 0\right\}.$$

(This fact is related to N. Wiener's speculations about "observable past" and "unattainable future". See [W], Section 1.4.)

Let $p(x) \in \mathbf{Q}[x]$ with $p(\mathbf{Z}) \subset \mathbf{Z}$ and $\deg p(x) > 0$. Let us show how the splitting $\mathcal{H} = \mathcal{H}_{rat} \oplus \mathcal{H}_{tot.erg}$ allows one to establish the existence of the limit

$$\lim_{N\to\infty} \frac{1}{N}\sum_{n=0}^{N-1} U^{p(n)} f.$$

If f belongs to the *rational spectrum* subspace $\mathcal{H}_{rat}$, there is almost nothing to prove: indeed, it is enough to check the case when for some i, $U^i f = f$. If, say, $p(n) = n^2$ and $i = 6$, then

$$\lim_{N\to\infty} \frac{1}{N}\sum_{n=0}^{N-1} U^{n^2} f = \left(\frac{I + 2U + U^3 + 2U^4}{6}\right) f.$$

On the other hand, on the *totally ergodic* subspace $\mathcal{H}_{tot.erg}$, the following theorem applies and does the job.

Theorem 2.2 (van der Corput trick). If $(u_n)_{n \in \mathbf{N}}$ is a bounded sequence in a Hilbert space $\mathcal{H}$ and if for any $h \in \mathbf{N}$

$$\lim_{N\to\infty} \frac{1}{N}\sum_{n=1}^{N} \langle u_{n+h}, u_n \rangle = 0,$$

then $\|\frac{1}{N}\sum_{n=1}^{N} u_n\| \to 0$.

Proof. Notice that for any $\epsilon > 0$ and any $H \in \mathbf{N}$, if N is large enough one has

$$\left\| \frac{1}{N} \sum_{n=1}^{N} u_n - \frac{1}{N} \frac{1}{H} \sum_{n=1}^{N} \sum_{h=0}^{H-1} u_{n+h} \right\| < \epsilon.$$

But

$$\limsup_{N \to \infty} \left\| \frac{1}{N} \frac{1}{H} \sum_{n=1}^{N} \sum_{h=0}^{H-1} u_{n+h} \right\|^2 \leq \limsup_{N \to \infty} \frac{1}{N} \sum_{n=1}^{N} \left\| \frac{1}{H} \sum_{h=0}^{H-1} u_{n+h} \right\|^2$$

$$= \limsup_{N \to \infty} \frac{1}{N} \sum_{n=1}^{N} \frac{1}{H^2} \sum_{h_1,h_2=0}^{H-1} \langle u_{n+h_1}, u_{n+h_2} \rangle \leq \frac{B}{H},$$

where $B = \sup_n \|u_n\|^2$. Since H was arbitrary, we are done.

$\square$

Remark. The classical van der Corput difference theorem in the theory of uniform distribution says that if $(x_n)_{n \in \mathbf{N}}$ is a sequence of real numbers such that for any $h \in \mathbf{N}$ the sequence $(x_{n+h} - x_n)_{n \in \mathbf{N}}$ is uniformly distributed mod 1 then the sequence $(x_n)_{n \in \mathbf{N}}$ is also uniformly distributed mod 1. It is easy to see that this result follows from the Hilbertian van der Corput trick, applied to the 1-dimensional Hilbert space $\mathbf{C}$. Indeed, by Weyl's criterion, a sequence $(x_n)_{n \in \mathbf{N}}$ is uniformly distributed mod 1 if and only if for any $m \in \mathbf{Z} \setminus \{0\}$,

$$\frac{1}{N} \sum_{n=1}^{N} e^{2\pi i m x_n} \to 0.$$

Writing $u_n^{(m)} = e^{2\pi i m x_n}$, we see that the assumption of van der Corput's difference theorem is that for any $m \in \mathbf{Z} \setminus \{0\}$,

$$\frac{1}{N} \sum_{n=1}^{N} \langle u_{n+h}^{(m)}, u_n^{(m)} \rangle \to 0.$$

Hence, $\frac{1}{N} \sum_{n=1}^{N} u_n^{(m)} \to 0$, which gives the uniform distribution of $(x_n)_{n \in \mathbf{N}}$.

Now, if $f \in \mathcal{H}_{tot.erg}$, then an induction on the degree of the polynomial $p(x)$ reduces the situation to the classical von Neumann theorem. Indeed, if $d_p = \deg p(x) > 1$, then writing $u_n = U^{p(n)} f$, we have:

$$\langle u_{n+h}, u_n \rangle = \langle U^{p(n+h)} f, U^{p(n)} f \rangle = \langle U^{p(n+h)-p(n)} f, f \rangle.$$

Notice that for any $h \in \mathbf{N}$ the degree of $p(n+h) - p(n)$ equals $d_p - 1$. Since weak convergence follows from strong convergence, we have by the induction hypothesis:

$$\lim_{N \to \infty} \frac{1}{N} \sum_{n=1}^{N} \langle u_{n+h}, u_n \rangle = \lim_{N \to \infty} \frac{1}{N} \sum_{n=1}^{N} \langle U^{p(n+h)-p(n)} f, f \rangle = 0,$$

and Theorem 2.2 implies that $\left\| \frac{1}{N} \sum_{n=1}^{N} U^{p(n)} f \right\| \to 0$.

Let us explain now how the splitting $\mathcal{H} = \mathcal{H}_{rat} \oplus \mathcal{H}_{tot.erg}$ can be used to establish the existence of the limit

$$\lim_{N \to \infty} \frac{1}{N} \sum_{n=0}^{N-1} U_1^{p_1(n)} U_2^{p_2(n)} \cdots U_t^{p_t(n)} f$$

for $t > 1$. Take for simplicity $t = 2$ (it will be clear from the discussion that the proof for general t is completely analogous). Let U_1, U_2 be commuting unitary operators generating a unitary $\mathbf{Z}^2$-action on a Hilbert space $\mathcal{H}$. Consider the splittings $\mathcal{H} = \mathcal{H}_{rat}^{(i)} \oplus \mathcal{H}_{tot.erg}^{(i)}$, $i = 1, 2$ which correspond to U_1 and U_2. We have the decomposition into invariant subspaces $\mathcal{H} = \mathcal{H}_{rr} \oplus \mathcal{H}_{rt} \oplus \mathcal{H}_{tr} \oplus \mathcal{H}_{tt}$, where $\mathcal{H}_{rr} = \mathcal{H}_{rat}^{(1)} \cap \mathcal{H}_{rat}^{(2)}$, $\mathcal{H}_{rt} = \mathcal{H}_{rat}^{(1)} \cap \mathcal{H}_{tot.erg}^{(2)}$, $\mathcal{H}_{tr} = \mathcal{H}_{tot.erg}^{(1)} \cap \mathcal{H}_{rat}^{(2)}$, and $\mathcal{H}_{tt} = \mathcal{H}_{tot.erg}^{(1)} \cap \mathcal{H}_{tot.erg}^{(2)}$. For $f \in \mathcal{H}$ we have the corresponding decomposition $f = f_{rr} + f_{rt} + f_{tr} + f_{tt}$, where $f_{\alpha\beta} \in \mathcal{H}_{\alpha\beta}$, $\alpha, \beta \in \{r, t\}$. Now, in each of the subspaces $\mathcal{H}_{rr}, \mathcal{H}_{rt}$, and $\mathcal{H}_{tr}$, the problem of establishing the existence of $\lim_{N \to \infty} \frac{1}{N} \sum_{n=0}^{N-1} U_1^{p_1(n)} U_2^{p(n)} f$ is reducible to the case of $\mathbf{Z}$-actions, already discussed above. Consider, for example, $f \in \mathcal{H}_{tr}$. Since $f \in \mathcal{H}_{rat}^{(2)}$ we can assume without loss of generality that for some $a \in \mathbf{N}$, $U_2^a f = f$. Notice that

$$\lim_{N \to \infty} \frac{1}{N} \sum_{n=0}^{N-1} U_1^{p_1(n)} U_2^{p_2(n)} f$$

exists if each of

$$\lim_{N \to \infty} \frac{1}{N} \sum_{n=0}^{N-1} U_1^{p_1(an+r)} U_2^{p_2(an+r)} f; \quad r = 0, 1, \cdots, a-1$$

exists. As $U_2^{p(an+r)} f$ does not depend on n (since we assumed that $U_2^a f = f$), we see that in this case (i.e. on the invariant subspace $\mathcal{H}_{tr}$) the problem is reduced to that of the $\mathbf{Z}$-action generated by U_1.

It remains to show the existence of the limit in question on $\mathcal{H}_{tt}$. Let $f \in \mathcal{H}_{tt}$. First of all, assume without loss of generality that there do not exist non-zero $a, b \in \mathbf{Z}$ and $g \in \mathcal{H}_{tt}$, $g \neq 0$, with $U_1^a U_2^b g = g$ (the set of such $g \in \mathcal{H}_{tt}$ comprises a U_1 and U_2-invariant subspace on which the situation again reduces to $\mathbf{Z}$-actions). Under this assumption, one shows that the limit is zero. The result follows by induction on $d = \max\{\deg p_1(x), \deg p_2(x)\}$. For $d = 1$ one has $p_1(n) = c_1 n$, $p_2(n) = c_2 n$ where at least one of c_1, c_2 is non-zero. In this case we are done by von Neumann's ergodic theorem. If $d > 1$ then, as before, van der Corput's trick reduces the case to $d - 1$.

The case of general $t > 1$ is treated in a similar fashion. We have 2^t invariant subspaces; on all but one of them at least one of the generators $U_1, \cdots, U_t$ has rational spectrum. On these $2^t - 1$ spaces the situation is naturally reduced to that of a $\mathbf{Z}^n$-action, $n < t$. On the remaining subspace, call it $\mathcal{H}_t$, on which all of $U_1, \cdots, U_t$ are totally ergodic, one disposes first with the potential degeneration caused by "linear dependence" between $U_1, \cdots, U_t$; again, on any subspace of $\mathcal{H}_t$ where such a dependence exists, the situation reduces to that of a lower-dimensional unitary action. Finally, if no degeneration occurs, one shows that the limit is zero by induction on $\max_{1 \le i \le t} \deg p_i(x)$ with the help of the van der Corput trick.

To establish Theorem 2.1, it suffices to show that for the unitary $\mathbf{Z}^t$-action induced by measure preserving transformations T_i, $i = 1, \cdots, t$ on $\mathcal{H} = L^2(X, \mathcal{B}, \mu)$ and for the characteristic function $f = 1_A$, where $\mu(A) > 0$, the limit

$$\lim_{N \to \infty} \frac{1}{N} \sum_{n=0}^{N-1} \mu(A \cap T_1^{p_1(n)} T_2^{p_2(n)} \cdots T_t^{p_t(n)} A)$$

$$= \lim_{N \to \infty} \frac{1}{N} \sum_{n=0}^{N-1} \langle U_1^{p_1(n)} U_2^{p_2(n)} \cdots U_t^{p_t(n)} f, f \rangle$$

is positive.

Here is one possible way to see this (we will offer below another proof of the positivity of the limit). Let $L^2(X, \mathcal{B}, \mu) = \mathcal{H} = \mathcal{H}_{rat}^{(i)} \oplus \mathcal{H}_{tot.erg}^{(i)}$, $i = 1, 2, \cdots t$, be the splittings corresponding to the unitary operators U_i, $i = 1, 2, \cdots, t$, defined by $(U_i f)(x) := f(T_i x)$, $f \in L^2(X, \mathcal{B}, \mu)$. We have then a natural splitting of $\mathcal{H}$ into 2^t invariant subspaces each having the form

$$\mathcal{H}_C = \left(\bigcap_{i \in C} \mathcal{H}_{rat}^{(i)} \right) \cap \left(\bigcap_{i \in \{1,2,\cdots,t\} \setminus C} \mathcal{H}_{tot.erg}^{(i)} \right),$$

where C is any of the 2^t subsets of $\{1, 2, \cdots, t\}$. Let $f = \sum_{C \subset \{1,2,\cdots,t\}} f_C$ be the corresponding orthonormal decomposition of $f = 1_A$.

We claim that without loss of generality one can assume that the polynomials $p_1(n), \cdots, p_t(n)$ all have distinct (and positive) degrees. Indeed one can always arrive at such a situation by regrouping and, possibly, collapsing some of the U_i. (Example: $U_1^{2n^2+3n} U_2^{5n^2-n} U_3^{-n^2} = \tilde{U}_1^{n^2} \tilde{U}_2^{n}$, where $\tilde{U}_1 = U_1^2 U_2^5 U_3^{-1}$, $\tilde{U}_2 = U_1^3 U_2^{-1}$.)

Now, if the polynomials $p_1(n), \cdots, p_t(n)$ have pairwise distinct positive degrees, then one can check, by carefully examining the effect of the van der Corput trick, that on each of the subspaces $\mathcal{H}_C$, C a proper subset of $\{1, 2, \cdots, t\}$, the limit in question is zero, so that

$$\lim_{N \to \infty} \frac{1}{N} \sum_{n=0}^{N-1} \mu\left(A \cap T_1^{p_1(n)} T_2^{p_2(n)} \cdots T_t^{p_t(n)} A\right)$$

$$= \lim_{N \to \infty} \frac{1}{N} \sum_{n=0}^{N-1} \langle U_1^{p_1(n)} U_2^{p_2(n)} \cdots U_t^{p_t(n)} f, f \rangle$$

$$= \lim_{N \to \infty} \frac{1}{N} \sum_{n=0}^{N-1} \langle U_1^{p_1(n)} U_2^{p_2(n)} \cdots U_t^{p_t(n)} \overline{f}, \overline{f} \rangle,$$

where $\overline{f} = f_{\{1,2,\cdots,t\}} = Pf$ and P is the projection onto the subspace $\bigcap_{i=1}^t \mathcal{H}_{rat}^{(i)} = \overline{\bigcup_{a \in \mathbf{Z}^t} \mathcal{H}_a}$, where the (potentially containing constants only) subspaces $\mathcal{H}_a$, $a = (a_1, \cdots, a_t) \in \mathbf{Z}^t$ are defined by

$$\mathcal{H}_a = \{f \in \mathcal{H} : U_i^{a_i} f = f, \ i = 1, \cdots, t\}.$$

Now, since $f = 1_A \geq 0$, and since $f \not\equiv 0$, one has $\overline{f} \geq 0$, $\overline{f} \not\equiv 0$. Indeed, $\overline{f}$ minimizes the distance from $\bigcap_{i=1}^t \mathcal{H}_{rat}^{(i)}$ to f and $\max\{\overline{f}, 0\}$ would do at least as well in minimizing this distance (cf. [F2], Lemma 4.23). By the same token, for any $a \in \mathbf{Z}^t$, the projection f_a of $f = 1_A$ onto $\mathcal{H}_a$ satisfies $f_a \geq 0$, $f_a \not\equiv 0$. Since the limit in question is strictly positive for any such f_a (and since the subspaces $\mathcal{H}_a$ span $\bigcap_{i=1}^t \mathcal{H}_{rat}^{(i)}$) we are done.

We shall consider now still another, and from the author's point of view, most important splitting theorem. An appropriate generalization of this splitting plays a significant role in the proofs of Theorems 1.11 and 1.19. Again, the splitting which we are going to introduce follows easily from the spectral theorem, but may be proved without resorting to it. A form of it appears for the first time in [KN].

Theorem 2.3. For any unitary $\mathbf{Z}$-action $(U^n)_{n \in \mathbf{Z}}$ on a Hilbert space $\mathcal{H}$ one has a decomposition $\mathcal{H} = \mathcal{H}_c \oplus \mathcal{H}_{wm}$, where

$$\mathcal{H}_c = \{f \in \mathcal{H} : \text{the orbit } (U^n f)_{n \in \mathbf{Z}} \text{ is precompact in norm topology}\},$$

$$\mathcal{H}_{wm} = \{f \in \mathcal{H} : \text{ for all } g \in \mathcal{H}, \frac{1}{N} \sum_{n=0}^{N-1} |\langle U^n f, g \rangle| \to 0\}.$$

Remark. One can show that the space $\mathcal{H}_c$ of *compact elements* coincides with the span of the eigenvectors of U:

$$\mathcal{H}_c = \overline{\text{span}\{f \in \mathcal{H} : \text{ there exists } \lambda \in \mathbf{C} \text{ with } Uf = \lambda f\}}.$$

The orthocomplement of $\mathcal{H}_c$, the space $\mathcal{H}_{wm}$ on which U acts in a *weakly mixing* manner can be characterized as follows:

$$\mathcal{H}_{wm} = \{f \in \mathcal{H} : \text{ there exists } S \subset \mathbf{N}, d(S) = 1$$

$$\text{such that for all } g \in \mathcal{H}, \lim_{n \to \infty, n \in S} \langle U^n f, g \rangle = 0\}.$$

(This terminology comes from measurable ergodic theory where one normally deals with operators induced by measure preserving transformations.)

Let us show now how the splitting $\mathcal{H} = \mathcal{H}_c \oplus \mathcal{H}_{wm}$ allows one to prove the existence of the limit in Theorem 2.1. Consider first of all the case $t = 1$. Since $\mathcal{H}_{wm} \subset \mathcal{H}_{tot.erg}$ and since on $\mathcal{H}_{tot.erg}$, as we saw above, the van der Corput trick, together with von Neumann's ergodic theorem, does the job, we have to care only about the space $\mathcal{H}_c$. One possibility is to use the characterization of $\mathcal{H}_c$ given in the remark above. Upon momentary reflection the reader will agree that without loss of generality one has to consider only

$$\lim_{N \to \infty} \frac{1}{N} \sum_{n=0}^{N-1} U^{p(n)} f,$$

where f satisfies $Uf = \lambda f$, $\lambda \in \mathbf{C}$, $|\lambda| = 1$. Since the situation on $\mathcal{H}_{rat}$ was already discussed above, assume additionally that $f \in \mathcal{H}_c \setminus \mathcal{H}_{rat}$, i.e., assume that for no integer $k \neq 0$ is $\lambda^k = 1$. Then, by Weyl's theorem on equidistribution of polynomials mod 1 one gets $\lim_{N \to \infty} \frac{1}{N} \sum_{n=0}^{N-1} \lambda^{p(n)} = 0$ and we are done. The extension of the proof to the case of general $t > 0$ is done in complete similarity to the proof above in which the splitting $\mathcal{H} = \mathcal{H}_{rat} \oplus \mathcal{H}_{tot.erg}$ was utilized.

Having in mind generalizations in the direction of multiple recurrence we want to indicate now still another approach which, while utilizing the same splitting $\mathcal{H} = \mathcal{H}_c \oplus \mathcal{H}_{wm}$, is "soft" enough to be susceptible to generalization. This is the gain; the loss is that, unlike the approaches discussed above, this approach leads only to results like

$$\liminf_{N \to \infty} \frac{1}{N} \sum_{n=1}^{N} \mu(A \cap T_1^{p_1(n)} T_2^{p_2(n)} \cdots T_t^{p_t(n)} A) > 0$$

instead of

$$\lim_{N \to \infty} \frac{1}{N} \sum_{n=1}^{N} \mu(A \cap T_1^{p_1(n)} T_2^{p_2(n)} \cdots T_t^{p_t(n)} A) > 0.$$

Theorem 2.4. Assume that $U_1, \cdots, U_t$ are commuting unitary operators on a Hilbert space $\mathcal{H}$. Let $p_i(x) \in \mathbf{Q}[x]$ with $p_i(\mathbf{Z}) \subset \mathbf{Z}$ and $p_i(0) = 0$, $i = 1, \cdots, t$. If for any $f \in \mathcal{H}$ and for any i, $i = 1, \cdots, t$ the orbit $(U_i^n f)_{n \in \mathbf{Z}}$ is precompact in the strong topology, then for any $\epsilon > 0$ the set

$$\{n \in \mathbf{Z} : \|U_1^{p_1(n)} U_2^{p_2(n)} \cdots U_t^{p_t(n)} f - f\| \le \epsilon\}$$

is IP* and hence syndetic.

In the proof of Theorem 2.4 we shall utilize the Gallai-Grünwald theorem (Theorem 1.10) in the following refined form, which can be derived from Theorem 3.2 in [FW1] as well as from the Hales-Jewett theorem (see Exercise 16 in Section 4). See also theorem 2.18 in [F2].

Theorem 2.5. If $\mathbf{Z}^t = \bigcup_{i=1}^r C_i$ is an r-coloring of $\mathbf{Z}^t$, then one of C_i, $i = 1, \cdots, r$ contains a "t-cube" of the form

$$K(n_1, n_2, \cdots, n_t; h)$$
$$= \{(n_1 + \epsilon_1 h, n_2 + \epsilon_2 h, \cdots, n_t + \epsilon_t h) : \epsilon_i \in \{0, 1\}, i = 1, 2, \cdots, t\}.$$

The set of $h \in \mathbf{Z}$ such that for some $(n_1, n_2, \cdots, n_t)$ the t-cube $K(n_1, n_2, \cdots, n_t; h)$ is contained in one of the C_i is an IP*-set.

Proof of Theorem 2.4. For $f \in \mathcal{H}$, let X be the closure in the strong topology of $\mathcal{H}$ of the orbit $(U_1^{n_1} U_2^{n_2} \cdots U_t^{n_t} f)_{(n_1, \cdots, n_t) \in \mathbf{Z}^t}$. It is easy to see that one can assume without loss of generality that $p_i(x) \in \mathbf{Z}[x]$, $i = 1, 2, \cdots, t$. By the increasing if needed the number of commuting operators involved, we may and will assume that the polynomials $p_i(n)$, $i = 1, \cdots, t$, have the form $p_i(n) = n^{b_i}$, $b_i \ge 1$. (For example, if $p_1(n) = 5n^2 - 17n^3$, we would rewrite $U_1^{p_1(n)}$ as $T^{n^2} S^{n^3}$ where $T = U_1^5$, $S = U_1^{-17}$.) Given $\epsilon > 0$, let

$$\epsilon_0 = \frac{\epsilon}{\sum_{i=1}^t 2^{b_i+1}}$$

and let $(g_j)_{j=1}^r$ be an ϵ_0-net in the (compact metric) space X. For each i, $i = 1, 2, \cdots, t$ consider the r-coloring of $\mathbf{Z}^{b_i}$ defined by

$$\chi_i(n_1, n_2, \cdots, n_{b_i}) = \min\{j : \|U_i^{n_1 n_2 \cdots n_{b_i}} f - g_j\| \le \epsilon_0\}.$$

By Theorem 2.5, for each i there exist χ_i-monochrome b_i-cubes $K(n_1^{(i)}, n_2^{(i)},$ $\cdots, n_{b_i}^{(i)}; h_i)$. Let S_i, $i = 1, 2, \cdots, t$ be the sets of "sizes" h of such monochrome cubes:

$$S_i = \{h : \text{there exists } (n_1, \cdots, n_{b_i}) \in \mathbf{Z}^{b_i}$$
$$\text{with } K(n_1, \cdots, n_{b_i}; h) \; \chi_i - \text{monochrome}\}.$$

Again by Theorem 2.5 the sets S_i are IP* and by Lemma 1.13 the set $S = \bigcap_{i=1}^{t} S_i$ is also an IP*-set.

We shall utilize the following identity:

$$h^b = \sum_{a=0}^{b} \sum_{A \subset \{1,2,\cdots,b\}, |A|=a} (-1)^a \prod_{i \in A} n_i \prod_{i \notin A} (n_i + h).$$

(For example, for $b = 3$ one has:

$$h^3 = (n_1 + h)(n_2 + h)(n_3 + h) - n_1(n_2 + h)(n_3 + h)$$
$$- (n_1 + h)n_2(n_3 + h) - (n_1 + h)(n_2 + h)n_3 + n_1 n_2(n_3 + h)$$
$$+ n_1(n_2 + h)n_3 + (n_1 + h)n_2 n_3 - n_1 n_2 n_3.)$$

Notice that for any b the sum of coefficients

$$\sum_{a=0}^{b} \sum_{A \subset \{1,2,\cdots,b\}} (-1)^a$$

equals zero and the number of them equals 2^b.

Since for every two vertices $v' = (v_1', \cdots, v_{b_i}')$, $v'' = (v_1'', \cdots, v_{b_i}'')$ of a monochrome b_i-cube $K(n_1^{(i)}, \cdots, n_{b_i}^{(i)}; h)$ one has

$$\|U_i^{v_1' v_2' \cdots v_{b_i}'} f - U_i^{v_1'' v_2'' \cdots v_{b_i}''} f\| \leq 2\epsilon_0,$$

it follows from the identity above that for any $h \in S_i$, $i = 1, 2, \cdots, t$

$$\|U_i^{h^{b_i}} f - f\| \leq 2^{b_i+1} \epsilon_0.$$

It is clear then that for any $h \in \bigcap_{i=1}^{t} S_i$ one has:

$$\|U_1^{h^{b_1}} U_2^{h^{b_2}} \cdots U_t^{h^{b_t}} f - f\| \leq \sum_{i=1}^{t} 2^{b_i+1} \epsilon_0 = \epsilon.$$

This finishes the proof of Theorem 2.4.

□

Let us show now how Theorem 2.4 can be used for a proof of Theorem 2.1. As before, let U_i, $i = 1, 2, \cdots, t$ be unitary operators on $\mathcal{H} = L^2(X, \mathcal{B}, \mu)$ defined by $(U_i f)(x) = f(T_i x)$.

Let $\mathcal{H} = \mathcal{H}_c^{(i)} \oplus \mathcal{H}_{wm}^{(i)}$ be the corresponding splittings. For each $B \subset \{1, 2, \cdots, t\}$ we have the $U_1, \cdots, U_t$-invariant subspace

$$\mathcal{H}_B = \left(\bigcap_{i \notin B} \mathcal{H}_c^{(i)} \right) \cap \left(\bigcap_{i \in B} \mathcal{H}_{wm}^{(i)} \right).$$

Then $\mathcal{H} = \bigoplus_{B \subset \{1, 2, \cdots, t\}} \mathcal{H}_B$. For the same reasons as before we shall assume without loss of generality that on each $\mathcal{H}_B$ no degeneration caused by "linear dependence" between $U_1, U_2, \cdots, U_t$ occurs.

Let $f = 1_A$ and consider the corresponding decomposition

$$f = \sum_{B \subset \{1, 2, \cdots, t\}} f_B,$$

where $f_B \in \mathcal{H}_B$.

Taking into account that for distinct B the spaces $\mathcal{H}_B$ are mutually orthogonal, we have:

$$\mu(A \cap T_1^{p_1(n)} T_2^{p_2(n)} \cdots T_t^{p_t(n)} A) = \langle f, U_1^{p_1(n)} U_2^{p_2(n)} \cdots U_t^{p_t(n)} f \rangle$$

$$= \sum_{B \subset \{1, 2, \cdots, t\}} \langle f_B, U_1^{p_1(n)} U_2^{p_2(n)} \cdots U_t^{p_t(n)} f_B \rangle.$$

One can show with the help of the van der Corput trick that under our assumptions, for any $B \neq \emptyset$

$$\lim_{N \to \infty} \frac{1}{N} \sum_{n=0}^{N-1} \langle f_B, U_1^{p_1(n)} U_2^{p_2(n)} \cdots U_t^{p_t(n)} f_B \rangle = 0,$$

so that

$$\liminf_{N \to \infty} \frac{1}{N} \sum_{n=0}^{N-1} \mu(A \cap T_1^{p_1(n)} T_2^{p_2(n)} \cdots T_t^{p_t(n)} A)$$

$$= \liminf_{N \to \infty} \frac{1}{N} \sum_{n=0}^{N-1} \langle \tilde{f}, U_1^{p_1(n)} U_2^{p_2(n)} \cdots U_t^{p_t(n)} \tilde{f} \rangle,$$

where $\tilde{f}$ is the component of f in the subspace $\bigcap_{i=1}^{t} \mathcal{H}_c^{(i)}$.

As with the component $\overline{f} \in \bigcap_{i=1}^{t} \mathcal{H}_{rat}^{(i)}$ above, and for the same reasons, the function $\tilde{f}$ satisfies $\tilde{f} \geq 0$, $\tilde{f} \not\equiv 0$. Also, by Theorem 2.4, for any $\epsilon > 0$ the set

$$S_\epsilon = \{n \in \mathbf{Z} : \|U_1^{p_1(n)} U_2^{p_2(n)} \cdots U_t^{p_t(n)} \tilde{f} - \tilde{f}\| < \epsilon\}$$

is syndetic. Therefore, if ϵ is small enough, we shall have

$$\liminf_{N \to \infty} \frac{1}{N} \sum_{n=0}^{N-1} \mu(A \cap T_1^{p_1(n)} T_2^{p_2(n)} \cdots T_t^{p_t(n)} A)$$

$$= \liminf_{N \to \infty} \frac{1}{N} \sum_{n=0}^{N-1} \langle \tilde{f}, U_1^{p_1(n)} U_2^{p_2(n)} \cdots U_t^{p_t(n)} \tilde{f} \rangle$$

$$\geq \liminf_{N \to \infty} \frac{1}{N} \sum_{n \in S_\epsilon \cap [0, N-1]} \langle \tilde{f}, U_1^{p_1(n)} U_2^{p_2(n)} \cdots U_t^{p_t(n)} \tilde{f} \rangle > 0.$$

3. Discourse on $\beta \mathbf{N}$ and some of its applications.

In this section we shall address among other things the question of what the density version of Hindman's theorem (Theorem 1.5) is. It will turn out, somewhat surprisingly, that, appropriately formulated, Hindman's theorem is its own density version! First of all, recall from Section 1 that positive upper Banach density does not provide us with the right notion of largeness for Hindman's theorem. One of the reasons that upper Banach density seems to have nothing to do with Hindman's theorem is that it, unlike Szemerédi's theorem, which deals with *shift-invariant* configurations, deals with configurations which are not shift-invariant. On the other hand, any set of positive upper Banach density contains plenty of configurations of the form

$$t + FS(x_j)_{j=1}^n = t + \left\{ \sum_{j \in \alpha} x_j : \ \emptyset \neq \alpha \subset \{1, 2, \cdots, n\} \right\}.$$

Here "plenty" means, in particular, that if $E \subset \mathbf{Z}$ with $d^*(E) > 0$, then for any n, there are $x_1, \cdots, x_n$ such that

$$d^*\{t : t + FS(x_j)_{j=1}^n \subset E\} > 0.$$

This result has a very simple proof which we shall describe now. Let $E \subset \mathbf{Z}$, $d^*(E) > 0$. As observed above (see the speculations following the proof of Lemma 1.13), there exists x_1 with $1 \leq x_1 \leq \frac{1}{d^*(E)} + 1$ satisfying $d^*(E \cap (E - x_1)) > 0$. Repeating this argument with the set $E \cap (E - x_1)$ leads to finding $x_2 > x_1$ with

$$d^*\left((E \cap (E - x_1)) \cap \big((E \cap (E - x_1)) - x_2 \big) \right)$$

$$= d^*\left(E \cap (E - x_1) \cap (E - x_2) \cap (E - (x_1 + x_2)) \right) > 0.$$

It is clear that after $n-2$ additional steps we will arrive at

$$d^*\left(E \cap \bigcap_{x \in FS(x_j)_{j=1}^n} (E-x)\right) > 0.$$

Any t belonging to this intersection will satisfy $t + FS(x_j)_{j=1}^n \subset E$ and we are done.

Being as simple as it is, the result proved just now is a strengthening of a key lemma needed by D. Hilbert in his famous paper [H1] where he showed among other things that if $f(x,y) \in \mathbf{Z}[x,y]$ is an irreducible polynomial then for some $x_0 \in \mathbf{Z}$ the polynomial of one variable $f(x_0, y)$ is also irreducible. The interesting thing is that although Hilbert's lemma is weaker than the density version of it just proved, namely he shows that for any finite coloring $\mathbf{N} = \bigcup_{i=1}^r C_i$ and for any n, one of C_i, $i = 1, \cdots, r$ has the property that for some x_j, $j = 1, 2, \cdots, n$ and for infinitely many t one has $t + FS(x_j)_{j=1}^n \subset C_i$, his proof is two pages long. Another interesting thing is that the proof we have indicated contains in embryo the main idea behind the proof of the much stronger Hindman's theorem that we are going to pass to now. Before starting, we want to formulate an Exercise which shows that it is hopeless to look for infinite configurations of the form $t + FS(x_j)_{j=1}^\infty$ in arbitrary sets of positive upper density.

Exercise 8. Show that for any $\epsilon > 0$ there exists a set $E \subset \mathbf{N}$ with $d(E) > 1 - \epsilon$ and such that E does not contain a subset of the form $t + FS(x_j)_{j=1}^\infty$. (The reader may try first to produce E with $d^*(E) > 1 - \epsilon$. This is much simpler.)

Though the original proof of Hindman's theorem in [H2] (as well as that in [Ba]) was combinatorial in nature, the real key to understanding Hindman's theorem lies in $\beta\mathbf{N}$–the Stone-Čech compactification of $\mathbf{N}$. More precisely, it is the algebraic structure of $\beta\mathbf{N}$, naturally inherited from addition in $\mathbf{N}$, that is behind a proof of Hindman's theorem which we want to present in this section. We shall also indicate in this section some other applications of $\beta\mathbf{N}$.

Since according to the author's experience most mathematicians (unless they are logicians or set-theoretical topologists) are not overly knowledgeable about or thrilled by $\beta\mathbf{N}$ in general and its algebraic structure in particular, we shall start with some generalities. For more details the reader is encouraged to consult, say [GJ], [C], or Sections 6-9 in [H3]. It was the paper [H3] which convinced the author of the effectiveness of ultrafilters in solving Ramsey-theoretical questions and indeed in the ergodic theory of multiple recurrence.

We take $\beta\mathbf{N}$, the Stone-Čech compactification of $\mathbf{N}$ to be the set of *ultrafilters* on $\mathbf{N}$. Recall that a filter p on $\mathbf{N}$ is a set of subsets of $\mathbf{N}$ satisfying

(i) $\emptyset \notin p$,

(ii) $A \in p$ and $A \subset B$ implies $B \in p$, and

(iii) $A \in p$ and $B \in p$ implies $A \cap B \in p$.

A filter is an ultrafilter if, in addition, it satisfies

(iv) if $r \in \mathbf{N}$ and $\mathbf{N} = A_1 \cup A_2 \cup \cdots \cup A_r$ then $A_i \in p$ for some i, $1 \leq i \leq r$.

In other words, an ultrafilter is a maximal filter. An alternative way of looking at ultrafilters (and actually the one that we shall adopt) is to identify each ultrafilter $p \in \beta\mathbf{N}$ with a finitely additive, $\{0,1\}$-valued probability measure μ_p on the power set $\mathcal{P}(\mathbf{N})$. The measure μ_p is naturally defined by the condition $\mu_p(A) = 1$ if and only if $A \in p$. Without saying so explicitly, we will always think of ultrafilters as such measures, but will prefer to write $A \in p$ instead of $\mu_p(A) = 1$. It is their $\{0,1\}$-valuedness and finite additivity (as well as abundance of ultrafilters with diversified properties) which makes ultrafilters so suitable for Ramsey theory: the property of being a member of an ultrafilter fits nicely into the notion of largeness which the second principle of Ramsey theory encourages us to look for.

Any element $n \in \mathbf{N}$ is naturally identified with an ultrafilter $\{A \subset \mathbf{N} : n \in A\}$. Such (and only such) ultrafilters are called *principal*. A natural question which the shrewd reader may ask at this point is: are there any less dull examples of ultrafilters? The answer is yes ... modulo Zorn's Lemma, which the reader is kindly encouraged to accept. The reader is also asked not to attempt to produce a non-principal ultrafilter without the use of Zorn's Lemma (it will not work). See the discussion on pp. 161-162 of [CN].

Suppose that $\mathcal{C}$ is a family of subsets of $\mathbf{N}$ which has the finite intersection property. Then there is some $p \in \beta\mathbf{N}$ such that $C \in p$ for each $C \in \mathcal{C}$. Indeed, let

$$\tilde{\mathcal{C}} = \{\mathcal{B} \subset \mathcal{P}(\mathbf{N}) : \mathcal{B} \text{ has the finite intersection property and } \mathcal{C} \subset \mathcal{B}\}.$$

Clearly, $\tilde{\mathcal{C}} \neq \emptyset$ (since $\mathcal{C} \in \tilde{\mathcal{C}}$). Also, the union of any chain in $\tilde{\mathcal{C}}$ is a member of $\tilde{\mathcal{C}}$. By Zorn's Lemma there is a maximal member p of $\tilde{\mathcal{C}}$, which is actually maximal with respect to the finite intersection property and hence a member of $\beta\mathbf{N}$. To see that non-principal ultrafilters exist take for example

$$\mathcal{C} = \{A \subset \mathbf{N} : A^c = \mathbf{N} \setminus A \text{ is finite }\}.$$

Clearly $\mathcal{C}$ has the finite intersection property, so there is an ultrafilter $p \in \beta\mathbf{N}$ such that $C \in p$ for all $C \in \mathcal{C}$. It is easy to see that such p cannot be principal.

For another example, take $\mathcal{D} = \{A \subset \mathbf{N} : d(A) = 1\}$. Again, $\mathcal{D}$ clearly satisfies the finite intersection property. If p is any ultrafilter for which

$\mathcal{D} \subset p$, then any member of p has positive upper density. (If $d(A) = 0$, then $A^c = (\mathbf{N} \setminus A) \in \mathcal{D}$.) These examples hint that the space $\beta\mathbf{N}$ is quite large. It is indeed: the cardinality of $\beta\mathbf{N}$ equals that of $\mathcal{P}(\mathcal{P}(\mathbf{N}))$ ([GJ], 6.10 (a)).

Now a few words about topology in $\beta\mathbf{N}$. Given $A \subset \mathbf{N}$, let $\overline{A} = \{p \in \beta\mathbf{N} : A \in p\}$. The set $\mathcal{G} = \{\overline{A} : A \subset \mathbf{N}\}$ forms a basis for the open sets (and a basis for the closed sets). To see that $\mathcal{G}$ is indeed a basis for a topology on $\beta\mathbf{N}$ observe that if $A, B \subset \mathbf{N}$, then $\overline{A} \cap \overline{B} = \overline{A \cap B}$. Also, $\overline{\mathbf{N}} = \beta\mathbf{N}$ and hence $\bigcup_{\overline{A} \in \mathcal{G}} \overline{A} = \beta\mathbf{N}$. (Notice also that $\overline{A} \cup \overline{B} = \overline{A \cup B}$.) The crucial fact for us is that, with this topology, $\beta\mathbf{N}$ satisfies the following.

Theorem 3.1. $\beta\mathbf{N}$ is a compact Hausdorff space.

Proof. Let $\mathcal{K}$ be a cover of $\beta\mathbf{N}$ by sets belonging to the base $\mathcal{G} = \{\overline{A} : A \subset \mathbf{N}\}$. Let $\mathcal{C} \subset \mathcal{P}(\mathbf{N})$ be such that $\mathcal{K} = \{\overline{A} : A \in \mathcal{C}\}$. Assume that $\mathcal{K}$ has no finite subcover. Consider the family $\mathcal{D} = \{A^c : A \in \mathcal{C}\}$. There are two possibilites (each leading to a contradiction):

(i) $\mathcal{D}$ has the finite intersection property. Then, as shown above, there exists an ultrafilter p such that $A^c \in p$ for each $A^c \in \mathcal{D}$. Since p is an ultrafilter, $A^c \in p$ if and only if $A \notin p$. On the other hand, since $\mathcal{K}$ covers $\beta\mathbf{N}$, for some element $\overline{A}$ of the cover $p \in \overline{A}$, or equivalently $A \in p$, a contradiction. (ii) $\mathcal{D}$ does not have the finite intersection property. Then for some $A_1, \cdots, A_r \in \mathcal{C}$ one has $\bigcap_{i=1}^{r} A_i^c = \emptyset$, or $\bigcup_{i=1}^{r} A_i = \mathbf{N}$, which implies that $\bigcup_{i=1}^{r} \overline{A_i} = \beta\mathbf{N}$. Again, this is a contradiction, as we assumed that $\mathcal{K}$ has no finite subcover.

As for the Hausdorff property, notice that if $p, q \in \beta\mathbf{N}$ are distinct ultrafilters then since each of them is maximal with respect to the finite intersection property, neither of them is contained in the other. If $A \in p \setminus q$, then $A^c \in q \setminus p$, which means that $\overline{A}$ and $\overline{A^c}$ are disjoint neighborhoods of p and q.

□

Remark. Being a nice compact Hausdorff space, $\beta\mathbf{N}$ is in many respects quite a strange object. We mentioned already that its cardinality is that of $\mathcal{P}(\mathcal{P}(\mathbf{N}))$. It follows that $\beta\mathbf{N}$ is not metrizable, as otherwise, being a compact and hence separable metric space, it would have cardinality not exceeding that of $\mathcal{P}(\mathbf{N})$. Another curious feature of $\beta\mathbf{N}$ is that any infinite closed subset of $\beta\mathbf{N}$ contains a copy of all of $\beta\mathbf{N}$.

Since $\overline{\mathbf{N}} = \beta\mathbf{N}$, it is natural to attempt to extend the operation of addition from (the densely embedded) $\mathbf{N}$ to $\beta\mathbf{N}$. Since ultrafilters are measures (principal ultrafilters being just the *point measures* corresponding to the elements of $\mathbf{N}$), it comes as no surprise that the extension we look for takes the form of a *convolution*. What is surprising, however, is that the algebraic structure of $\beta\mathbf{N}$ was explicitly introduced only about 35 years ago (in [CY]).

In the following definition, $A - n$ (where $A \subset \mathbf{N}$, $n \in \mathbf{N}$) is the set of all m for which $m + n \in A$. For $p, q \in \mathbf{N}$, define

$$p + q = \{A \subset \mathbf{N} : \{n \in \mathbf{N} : (A - n) \in p\} \in q\}.$$

Exercise 9. Check that for principal ultrafilters the operation + corresponds to addition in $\mathbf{N}$.

Remarks. 1. Despite the somewhat repelling phrasing of the operation just introduced in set-theoretical terms, the perspicacious reader will notice the direct analogy between this definition and the usual formulas for convolution of measures μ, ν on a locally compact group G (cf. [HR], 19.11):

$$\mu * \nu(A) = \int_G \nu(x^{-1}A) \, d\mu(x) = \int_G \mu(Ay^{-1}) \, d\nu(y).$$

2. Before checking the correctness of the definition, a word of warning: the introduced operation + (which will turn out to be well defined and associative) is badly noncommutative. This seems to contradict our intuition since $(\mathbf{N}, +)$ is commutative and in the case of σ-additive measures on abelian semi-groups convolution *is* commutative. The explanation: our ultrafilters, being only *finitely* additive measures, do not obey the Fubini theorem, which is behind the commutativity of the usual convolution.

Let us show that $p+q$ is an ultrafilter. Clearly $\emptyset \notin p+q$. Let $A, B \in p+q$. This means that $\{n \in \mathbf{N} : (A - n) \in p\} \in q$ and $\{n \in \mathbf{N} : (B - n) \in p\} \in q$. Since p and q are ultrafilters, we have:

$$\{n \in \mathbf{N} : (A \cap B) - n \in p\}$$
$$= \{n \in \mathbf{N} : (A - n) \in p\} \cap \{n \in \mathbf{N} : (B - n) \in p\} \in q.$$

Assume now that $A \subset \mathbf{N}$, $A \notin p + q$. We want to show that $A^c \in p + q$. Since $A \notin p + q$, we know that $\{n \in \mathbf{N} : (A - n) \in p\} \notin q$, or, equivalently, $\{n \in \mathbf{N} : (A - n) \in p\}^c \in q$. But this is true precisely when $\{n \in \mathbf{N} : (A^c - n) \in p\} \in q$, which is the same as $A^c \in p + q$. It follows that $p + q \in \beta\mathbf{N}$.

Let us now check associativity of the operation +. Let $A \subset \mathbf{N}$ and $p, q, r \in \beta\mathbf{N}$. One has:

$$A \in p + (q + r) \Leftrightarrow \{n \in \mathbf{N} : (A - n) \in p\} \in q + r$$
$$\Leftrightarrow \{m \in \mathbf{N} : (\{n \in \mathbf{N} : (A - n) \in p\} - m) \in q\} \in r$$
$$\Leftrightarrow \{m \in \mathbf{N} : \{n \in \mathbf{N} : (A - m - n) \in p\} \in q\} \in r$$
$$\Leftrightarrow \{m \in \mathbf{N} : (A - m) \in p + q\} \in r \Leftrightarrow A \in (p + q) + r.$$

Theorem 3.2. For any fixed $p \in \beta\mathbf{N}$ the function $\lambda_p(q) = p + q$ is a continuous self map of $\beta\mathbf{N}$.

Proof. Let $q \in \beta\mathbf{N}$ and let $\mathcal{U}$ be a neighborhood of $\lambda_p(q)$. We will show that there exists a neighborhood $\overline{B}$ of q such that for any $r \in \overline{B}$, $\lambda_p(r) \in \mathcal{U}$. Let $A \subset \mathbf{N}$ be such that $\lambda_p(q) = p + q \in \overline{A} \subset \mathcal{U}$. Then $A \in p + q$. Let us show that the set

$$B = \{n \in \mathbf{N} : (A - n) \in p\}$$

will do for our purposes. Indeed, by the definition of $p + q$, $B \in q$, or, in other words, $q \in \overline{B}$. If $r \in \overline{B}$ then $B = \{n \in \mathbf{N} : (A - n) \in p\} \in r$. This means that $A \in p + r = \lambda_p(r)$, or $\lambda_p(r) \in \overline{A} \in \mathcal{U}$.

$\square$

With the operation $+$, $\beta\mathbf{N}$ becomes, in view of Theorem 3.2, a compact *left topological* semigroup. Such semigroups are known to have idempotents, which is the last preliminary result that we need for the proof of Hindman's theorem. For compact *topological semigroups* (i.e. with an operation which is continuous in both variables), this result is due to Numakura, [N]; for left topological semigroups the result is due to Ellis, [E1].

Theorem 3.3. If $(G, *)$ is a compact left topological semigroup (i.e. for any $x \in G$ the function $\lambda_x(y) = x * y$ is continuous) then G has an idempotent.

Proof. Let

$$\mathcal{G} = \{A \subset G : A \neq \emptyset, A \text{ is compact}, A * A = \{x * y : x, y \in A\} \subset A\}.$$

Since $G \in \mathcal{G}$, $\mathcal{G} \neq \emptyset$. By Zorn's Lemma, there exists a minimal element $A \in \mathcal{G}$. If $x \in A$, then $x * A$ is compact and satisfies

$$(x * A) * (x * A) \subset (x * A) * (A * A) \subset (x * A) * A \subset x * (A * A) \subset x * A.$$

Hence $x * A \in \mathcal{G}$. But $x * A \subset A * A \subset A$, which implies that $x * A = A$. Thus $x \in x * A$, which implies that $x = x * y$ for some $y \in A$. Now consider $B = \{z \in A : x * z = x\}$. The set B is closed (since $B = \lambda_x^{-1}(\{x\})$), and we have just shown that B is non-empty. If $z_1, z_2 \in B$ then $z_1 * z_2 \in A * A \subset A$ and $x * (z_1 * z_2) = (x * z_1) * z_2 = x * z_2 = x$. So $B \in \mathcal{G}$. But $B \subset A$ and hence $B = A$. So $x \in B$ which gives $x * x = x$.

$\square$

For a fixed $p \in \beta\mathbf{N}$ we shall call a set $C \subset \mathbf{N}$ *p-big* if $C \in p$. Clearly, being a member of an ultrafilter is a notion of largeness in the sense discussed in Section 1. The notion of largeness induced by idempotent ultrafilters is special (and promising) in that it inherently has a shift-invariance property. Indeed, if $p \in \beta\mathbf{N}$ with $p + p = p$ then

$$A \in p \Leftrightarrow A \in p + p \Leftrightarrow \{n \in \mathbf{N} : (A - n) \in p\} \in p.$$

A way of interpreting this is that if p is an idempotent ultrafilter, then A is p-big if and only if for p-many $n \in \mathbf{N}$ the shifted set $(A - n)$ is p-big. Or, still somewhat differently: $A \subset \mathbf{N}$ is p-big if for p-almost all $n \in \mathbf{N}$ the set $(A - n)$ is p-big. This is the reason why specialists in ultrafilters called such idempotent ultrafilters "almost shift invariant" in the early seventies (even before the existence of such ultrafilters was established).

Exercise 10. Let $p \in \beta\mathbf{N}$ with $p + p = p$.
(i) Show that p cannot be a principal ultrafilter.
(ii) Show that for any $a \in \mathbf{N}$, $a\mathbf{N} \in p$.

Each idempotent ultrafilter $p \in \beta\mathbf{N}$ induces a "measure preserving dynamical system" with the phase space $\mathbf{N}$, σ-algebra $\mathcal{P}(\mathbf{N})$, measure p, and "time" being the "p-preserving" $\mathbf{N}$-action induced by the shift. The two peculiarites about such a measure-preserving system are that the phase space is countable and that the "invariant measure" is only finitely additive and is preserved by our action not for all but for almost all instances of "time". Notice that the "Poincaré recurrence theorem" trivially holds: If $A \in p$ then, since there are p-many n for which $(A - n) \in p$, one has, for any such n, $A \cap (A - n) \in p$. We are now in position to prove Hindman's theorem which we rephrase slightly as follows.

Theorem 3.4. Let $p \in \beta\mathbf{N}$ be an idempotent and let $\mathbf{N} = \bigcup_{i=1}^{r} C_i$. If $C = C_i$ is p-big, then there exists an infinite sequence $(x_j)_{j=1}^{\infty} \subset C$ such that

$$FS\big((x_j)_{j=1}^{\infty}\big) = \Big\{ \sum_{j \in \alpha} x_j : \emptyset \neq \alpha \subset \mathbf{N}, |\alpha| < \infty \Big\} \subset C.$$

Proof. Let $p \in \beta\mathbf{N}$ be an idempotent ultrafilter. Assume that $\mathbf{N} = \bigcup_{i=1}^{r} C_i$ and choose $i \in \{1, \cdots, r\}$ such that $C = C_i \in p$. Since $C \in p$ and $p + p = p$, one has $R_1 = \{n \in \mathbf{N} : (C - n) \in p\} \in p$. Choose any $x_1 \in R_1 \cap C \in p$. Then $x_1 \in C$ and $A_1 = C \cap (C - x_1) \in p$. Consider

$$R_2 = \{n \in \mathbf{N} : A_1 - n = \big((C \cap (C - x_1)) - n\big) \in p\} \in p.$$

Choose any $x_2 \in R_2 \cap A_1 \in p$. Since p is an idempotent, it is a non-principal ultrafilter and its members are infinite sets. This allows us always to assume in the course of this proof that an element chosen from a member of p lies outside of any chosen finite set. In particular, we may assume that $x_2 > x_1$. Notice that

$$x_2 \in A_1 = C \cap (C - x_1) \Rightarrow \{x_1, x_2, x_1 + x_2\} = FS(x_j)_{j=1}^2 \subset C.$$

Now, since $x_2 \in R_2$,

$$A_2 = A_1 \cap (A_1 - x_2) = C \cap (C - x_1) \cap \big(C \cap (C - x_1) - x_2\big)$$

$$= C \cap (C - x_1) \cap (C - x_2) \cap (C - x_1 - x_2) \in p.$$

And so on! At the nth step we will have

$$A_k = A_{k-1} \cap (A_{k-1} - x_k) = C \cap \left(\bigcap_{t \in FS(x_j)_{j=1}^k} (C - t) \right) \in p,$$

and defining

$$R_k = \{n \in \mathbf{N} : (A_k - n) \in p\} \in p$$

and choosing $x_{k+1} \in R_{k+1} \cap A_k$, $x_{k+1} > x_1 + \cdots + x_k$, we will have:

$$x_{k+1} \in \bigcap_{t \in FS(x_j)_{j=1}^k} (C - t),$$

$$A_{k+1} = A_k \cap (A_k - x_{k+1}) = C \cap \left(\bigcap_{t \in FS(x_j)_{j=1}^{k+1}} (C - t) \right) \in p.$$

The sequence $(x_j)_{j=1}^\infty$ thus created has the property that for any $n \in \mathbf{N}$, $FS(x_j)_{j=1}^n \subset C$. Hence $FS(x_j)_{j=1}^\infty \subset C$ and we are done.

$\square$

Remarks. 1. The proof of Theorem 3.4 which we have presented here is essentially due to S. Glaser, who never published it. For an account of the story behind the discovery of Glaser's proof, see [H5].

2. Theorem 3.4 tells us that a notion of largeness which is behind Hindman's theorem is that of being a member of an idempotent ultrafilter. In a sense, it is *the* notion. Indeed, though the IP-set $FS(x_j)_{j=1}^\infty$ which in the course of the proof was found inside a p-big set C is not necessarily itself p-big, one can show that for any IP-set S there exists an idempotent p such that S is p-big. One can say that Hindman's theorem is the density version of itself!

One can show that $\beta\mathbf{N}$ has 2^c idempotents, where c is the power of continuum (this follows from the fact that $\beta\mathbf{N}$ has 2^c disjoint closed subsemigroups. See for example [D]). One can expect that some of these idempotents are good not only for Hindman's theorem but, say, for van der Waerden's theorem too. As we remarked above, any IP-set is a member of an idempotent. Since there are "thin" IP-sets that do not contain length 3 arithmetic progressions (take for example $FS(5^n)_{n=1}^\infty$), it is clear that not every idempotent may reveal something about van der Waerden's theorem. But, there is an important and natural class of idempotents, namely *minimal* ones, which do (this became clear after Furstenberg and Katznelson put to use in [FK3] the minimal idempotents in Ellis' enveloping semigroup–an object similar in many ways to $\beta\mathbf{N}$). An idempotent $p \in \beta\mathbf{N}$ is called minimal if it belongs to a minimal right ideal in $\beta\mathbf{N}$. One can show that any minimal right ideal in $\beta\mathbf{N}$ has the form $q + \beta\mathbf{N}$ (warning: not every right ideal has this form).

This terminology fits well with the usual definition of minimality in topological dynamics. Recall that a topological dynamical system (X, T), where X is a compact space and T is a continuous self-map of X, is called minimal if for any $x \in X$ one has $\overline{(T^n x)_{n \in \mathbf{N}}} = X$. Notice that for each $p \in \beta\mathbf{N}$ the right ideal $p + \beta\mathbf{N}$ is compact (since $p + \beta\mathbf{N}$ is the continuous image of $\beta\mathbf{N}$ under the continuous function λ_p–see Theorem 3.2). Let $X_p = p + \beta\mathbf{N}$ and define $T : X_p \to X_p$ by $Tx = x + 1$, $x \in X_p$ (here 1 is the principal ultrafilter of sets containing the integer 1). It is not hard to show that the ideal $p + \beta\mathbf{N}$ is minimal if and only if the dynamical system (X_p, T) is minimal.

One can also show that any minimal idempotent $p \in \beta\mathbf{N}$ has the property that if $C \in p$ then C is *piecewise syndetic*, namely the intersection of a syndetic set with a union of intervals $[a_n, b_n]$, where $b_n - a_n \to \infty$.

Exercise 11. (i) Show that a set $S \subset \mathbf{N}$ is piecewise syndetic if and only if there exist $t_1, \cdots, t_k \in \mathbf{N}$ such that $d^*\left(\bigcup_{i=1}^k (S - t_i)\right) = 1$.

(ii) Derive from (i) that any piecewise syndetic set is a.p.-rich.

Theorem 3.5. If $\mathbf{N} = \bigcup_{1=1}^r C_i$ then one of C_i, $i = 1, \cdots, r$ is a.p.-rich and contains an IP-set.

Proof. Take any minimal idempotent $p \in \beta\mathbf{N}$. If $C_i \in p$ then by Theorem 3.4, C_i contains an IP-set. Also, since p is minimal, C_i is piecewise syndetic and hence a.p.-rich.

$\square$

The combinatorial results mentioned in this section so far can be obtained also by methods of topological dynamics introduced in [FW1] and further developed in [F2] and [FK3]. As a matter of fact, one can show

([BH1]) that so-called *central sets*, which are defined dynamically and which are shown in [F2] to be combinatorially rich, are exactly the members of minimal idempotents in $\beta\mathbf{N}$. The following is an example of a result for whose proof the space of ultrafilters seems better suited.

Theorem 3.6 ([BH1]). For any finite coloring of $\mathbf{N}$, one of the cells contains arbitrarily long arithmetic progressions, arbitrarily long geometric progressions, an additive IP-set, and a multiplicative IP-set.

A *multiplicative* IP-set is, in analogy with the additive case, any infinite sequence together with all finite products of its distinct elements. The reason that $\beta\mathbf{N}$ is an appropriate tool for proving the foregoing result lies with the fact that it has also another semigroup structure–the one inherited from the usual multiplication in $\mathbf{N}$, and is a left topological compact semigroup with respect to this structure, too. In particular, there are (many) *multiplicative* idempotents, and in complete analogy to the proof of Theorem 3.4 one can show that any member of a multiplicative idempotent contains a multiplicative IP-set. It follows that for any finite coloring $\mathbf{N} = \bigcup_{i=1}^{r} C_i$, there exists a monochromatic additive IP-set and a monochromatic multiplicative IP-set, but not necessarily of the same color. It was shown by Hindman in [H4] that a single C_i, $i = 1, \cdots, r$ always contains both an additive and a multiplicative IP-set. This as well as the other assertions of Theorem 3.6 are consequences of the following result combining the two structures in $\mathbf{N}$.

Theorem 3.7 ([BH1], Corollary 5.5). For any finite partition $\mathbf{N} = \bigcup_{i=1}^{r} C_i$, one of C_i, $i = 1, \cdots, r$, is a member of a minimal additive idempotent and also a member of a minimal multiplicative idempotent.

It would be interesting to find a dynamical proof of Theorem 3.6.

The discussion above shows that ultrafilters are a convenient tool in partition Ramsey theory. We shall try to explain now why idempotent ultrafilters (and IP-sets which are intrinsically related to them) are helpful in topological dynamics and ergodic theory, and through this, in density Ramsey theory. This discussion will continue into the next section where we shall touch upon issues of multiple recurrence and further refinements of the polynomial Szemerédi theorem. Throughout this section we shall be concerned with single recurrence (both topological and measure theoretical). Since we are mainly concerned with $\mathbf{Z}$ and $\mathbf{Z}^t$ actions we shall confine ourselves to dealing exclusively with the additive structure of $\beta\mathbf{N}$.

Let X be a topological space and let $p \in \beta\mathbf{N}$. Given a sequence $(x_n)_{n \in \mathbf{N}}$ we shall write $p\text{-}\lim_{n \in \mathbf{N}} x_n = y$ if for every neighborhood U of y one has $\{n : x_n \in U\} \in p$. It is easy to see that $p\text{-}\lim_{n \in \mathbf{N}} x_n$ exists and is unique in any compact Hausdorff space.

The following is a special case of 6.10 in [BH1].

Theorem 3.8. Let X be a compact Hausdorff space, let $p, q \in \beta\mathbf{N}$ and let $(x_n)_{n\in\mathbf{N}}$ be a sequence in X. Then

$$(q+p)\text{-}\lim_{r\in\mathbf{N}} x_r = p\text{-}\lim_{t\in\mathbf{N}} q\text{-}\lim_{s\in\mathbf{N}} x_{s+t}.$$

In particular if p is an idempotent and $p = q$, one has

$$p\text{-}\lim_{r\in\mathbf{N}} x_r = p\text{-}\lim_{t\in\mathbf{N}} p\text{-}\lim_{s\in\mathbf{N}} x_{s+t}.$$

Proof. Recall that

$$q + p = \big\{A \subset \mathbf{N} : \{n \in \mathbf{N} : (A - n) \in q\} \in p\big\}.$$

Let $x = (q+p)\text{-}\lim_{r\in\mathbf{N}} x_r$. It will suffice for us to show that for any neighborhood U of x, we have that for p-many t, $q\text{-}\lim_{s\in\mathbf{N}} x_{s+t} \in U$. Fix such a U. We have $\{r : x_r \in U\} \in q + p$, so that

$$\{t : \{s : x_s \in U\} - t \in q\} = \{t : \{s : x_{s+t} \in U\} \in q\} \in p.$$

This implies, in particular, that for p-many t, $q\text{-}\lim_{s\in\mathbf{N}} x_{s+t} \in U$.

$\square$

Here is a simple application of Theorem 3.8. Recall that a continuous self-map of a compact metric space X is called *distal* if

$$\inf_{n\geq 1} d(T^n x, T^n y) = 0 \Rightarrow x = y.$$

The innocent looking fact that distal transformations are invertible is not transparent from the definition. Using Theorem 3.8 we can prove this as follows.

Let $T : X \to X$ be a distal transformation of a compact metric space X. Since a distal map is clearly injective we need only show that it is onto. Fix an idempotent $p \in \beta\mathbf{N}$. We shall show something stonger, namely that for any $x \in X$ $p\text{-}\lim_{n\in\mathbf{N}} T^n x = x$ and hence $x = T(p\text{-}\lim_{n\in\mathbf{N}} T^{n-1}x)$. Let $y = p\text{-}\lim_{n\in\mathbf{N}} T^n x$. By Theorem 3.8 one has

$$p\text{-}\lim_{n\in\mathbf{N}} T^n y = p\text{-}\lim_{n\in\mathbf{N}} T^n (p\text{-}\lim_{k\in\mathbf{N}} T^k x)$$

$$= p\text{-}\lim_{n\in\mathbf{N}} p\text{-}\lim_{k\in\mathbf{N}} T^{n+k} x = p\text{-}\lim_{n\in\mathbf{N}} T^n x = y.$$

Since T is distal, the relations $p\text{-}\lim_{n\in\mathbf{N}} T^n x = y = p\text{-}\lim_{n\in\mathbf{N}} T^n y$ imply $x = y$. We are done.

The conventional proof of the fact that distal transformations are invertible goes by way of the Ellis enveloping semigroup (see for example [Br], p. 60, or [E2]), which is typically non-metrizable. Our proof uses $\beta\mathbf{N}$, which is also non-metrizable, but a modification of the proof using the notion of IP-convergence (which will be explained below) in place of p-limits can serve to place this fact entirely within the scope of metric spaces.

Recall that a point $x \in X$ is uniformly recurrent with respect to T : $X \to X$ if for any neighborhood U of x the set $\{n : T^n x \in U\}$ is syndetic. It is well known that if x is a uniformly recurrent point then the orbit closure $\overline{(T^n x)}_{n \in \mathbf{N}}$ is a minimal T-invariant subset of X. It is not hard to show (see 6.9 in [BH1]) that if $p \in \beta\mathbf{N}$ is a minimal idempotent, and $p\text{-lim}_{n \in \mathbf{N}} T^n y = y$ then y is a uniformly recurrent point. It follows that in a distal system every point is uniformly recurrent. This in its turn implies by a routine argument (see for instance Thm. 1.17 in [F2]) that any distal system is *semisimple*, namely, the disjoint union of minimal subsets.

Let us indicate now how one can modify proofs involving limits along idempotent ultrafilters so that there will be no reference to non-metrizable spaces. The notation that we will introduce will also be used in the next section.

Let $\mathcal{F}$ denote the family of non-empty finite subsets of $\mathbf{N}$. Following the notation in [F2] and [FK2], let us call any sequence indexed by $\mathcal{F}$ an $\mathcal{F}$-*sequence*. Notice that an IP-set in $\mathbf{Z}$ is nothing but an $\mathcal{F}$-sequence $(n_\alpha)_{\alpha \in \mathcal{F}}$ with the property that $n_\alpha + n_\beta = n_{\alpha \cup \beta}$ whenever $\alpha \cap \beta = \emptyset$. For $\alpha, \beta \in \mathcal{F}$ we shall write $\alpha < \beta$ if $\max \alpha < \min \beta$.

If a collection of sets $\alpha_i \in \mathcal{F}$, $i \in \mathbf{N}$ has the property $\alpha_i < \alpha_{i+1}$, $i = 1, 2, \cdots$, then the set

$$\mathcal{F}^{(1)} = \{\bigcup_{i \in \beta} \alpha_i : \beta \in \mathcal{F}\}$$

is called an *IP-ring*. Note that the mapping $\varphi : \mathcal{F} \to \mathcal{F}^{(1)}$, $\varphi(\beta) = \bigcup_{i \in \beta} \alpha_i$ is bijective and structure preserving.

Since elements of $\mathcal{F}^{(1)}$ are naturally indexed by elements of $\mathcal{F}$, any sequence indexed by $\mathcal{F}^{(1)}$ may itself be viewed as an $\mathcal{F}$-sequence. The following exercise is an equivalent form of Hindman's theorem.

Exercise 12. If $\mathcal{F} = \bigcup_{i=1}^{r} C_i$, then one of C_i, $i = 1, \cdots, r$ contains an IP-ring.

Let $(x_\alpha)_{\alpha \in \mathcal{F}}$ be an $\mathcal{F}$-sequence in a topological space X, let $x \in X$, and let $\mathcal{F}^{(1)}$ be an IP-ring. One writes

$$\text{IP-}\lim_{\alpha \in \mathcal{F}^{(1)}} x_\alpha = x$$

if for any neighborhood U of x there exists $\alpha_U \in \mathcal{F}^{(1)}$ such that for any $\alpha \in \mathcal{F}^{(1)}$ with $\alpha > \alpha_U$ one has $x_\alpha \in U$.

The following theorem is Theorem 8.14 in [F2] and is proved by a diagonal procedure with the help of Hindman's theorem as formulated in Exercise 12.

Theorem 3.9. If $(x_\alpha)_{\alpha \in \mathcal{F}}$ is an $\mathcal{F}$-sequence with values in a compact space X, then there exists $x \in X$ and an IP-ring $\mathcal{F}^{(1)}$ such that

$$\operatorname*{IP\text{-}lim}_{\alpha \in \mathcal{F}^{(1)}} \ x_\alpha = x.$$

The following result is a special case of Lemma 8.15 in [F2].

Theorem 3.10. Let T be a continuous self map of a compact metric space X and let $(n_\alpha)_{\alpha \in \mathcal{F}} \subset \mathbf{N}$ be an IP-set. Then for any $x \in X$ there exists an IP-ring $\mathcal{F}^{(1)}$ and a point $y \in X$ such that

$$\operatorname*{IP\text{-}lim}_{\alpha \in \mathcal{F}^{(1)}} \ T^{n_\alpha} x = \operatorname*{IP\text{-}lim}_{\alpha \in \mathcal{F}^{(1)}} \ T^{n_\alpha} y = y.$$

We leave it now safely to the reader to verify that the proof of the invertibility of distal transformations given above can be rewritten in the language of IP-limits with no significant changes.

As we shall see in the next section, besides being more constructive, IP-limits allow refined formulations and non-trivial strengthenings of polynomial ergodic theorems. On the other hand, limits along idempotent ultrafilters, when applicable, have the advantage of there being no need to constantly be passing to convergent subsequences, thereby making the statements and proofs cleaner and more albebraic.

We shall conclude this section by presenting an ultrafilter proof of a special case of the following refinement of the Furstenberg-Sárközy theorem, Theorem 1.17. (At the same time this refinement is a special case of a more general theorem proved in [BFM] which will be discussed in the next section.)

Theorem 3.11. Assume that $p(t) \in \mathbf{Q}[t]$ with $p(\mathbf{Z}) \subset \mathbf{Z}$ and $p(0) = 0$. Then for any invertible probability measure preserving system $(X, \mathcal{B}, \mu, T)$, any $A \in \mathcal{B}$, $\mu(A) > 0$, and any IP-set $(n_\alpha)_{\alpha \in \mathcal{F}} \subset \mathbf{N}$, there exists an IP-ring $\mathcal{F}^{(1)} \subset \mathcal{F}$ such that

$$\operatorname*{IP\text{-}lim}_{\alpha \in \mathcal{F}^{(1)}} \ \mu(A \cap T^{p(n_\alpha)} A) \geq \mu(A)^2.$$

In particular, if $\deg p(n) > 0$ then $\{p(n) : n \in (n_\alpha)_{\alpha \in \mathcal{F}}\}$ is a set of recurrence for any IP-set $(n_\alpha)_{\alpha \in \mathcal{F}}$.

In a situation similar to that of Section 2, Theorem 3.11 follows from a Hilbertian result which we now formulate in the language of ultrafilters:

Theorem 3.12. Let $q(t) \in \mathbf{Q}[t]$ with $q(\mathbf{Z}) \subset \mathbf{Z}$ and $q(0) = 0$. Let U be a unitary operator on a Hilbert space $\mathcal{H}$. Let $p \in \beta\mathbf{N}$ be an idempotent. Then, letting $p\text{-}\lim_{n\in\mathbf{N}} U^{q(n)}f = P_p f$, where the limit is in the weak topology, P_p is an orthogonal projection onto a subspace of $\mathcal{H}$.

To see that Theorem 3.11 follows from Theorem 3.12, take $\mathcal{H} = L^2(X, \mathcal{B}, \mu)$, and take U to be the unitary operator induced by T, that is, $(Ug)(x) := g(Tx)$, and let $f = 1_A$. One then has

$$p\text{-}\lim_{n\in\mathbf{N}} \mu(A \cap T^{q(n)}A) = p\text{-}\lim_{n\in\mathbf{N}} \langle U^{q(n)}f, f\rangle = \langle P_p f, f\rangle$$

$$= \langle P_p f, P_p f\rangle\langle 1, 1\rangle \geq \langle P_p f, 1\rangle^2 = \langle f, 1\rangle^2 = \langle 1_A, 1\rangle^2 = (\mu(A))^2.$$

We leave to the reader to justify the replacement of $p\text{-}\lim_{n\in\mathbf{N}} \mu(A \cap T^{q(n)}A)$ by

$$\text{IP-}\lim_{\alpha\in\mathcal{F}^{(1)}} \mu(A \cap T^{q(n)}A).$$

(Hint: any member of any idempotent contains an IP-set, and any IP-set is a member of some idempotent.)

Proof of Theorem 3.12 (for $q(n) = n^2$). As the reader (after reading Section 2) may guess, the proof boils down to finding an appropriate splitting of $\mathcal{H}$. This guess is true but the situation is more complicated when compared with the one which was encountered in Section 2. As we saw in Section 2, when one is interested in studying limits of the form

$$\lim_{N\to\infty} \frac{1}{N} \sum_{n=0}^{N-1} U^{q(n)}f,$$

the splitting $\mathcal{H} = \mathcal{H}_c \oplus \mathcal{H}_{wm}$ works *for all* $q(t) \in \mathbf{Z}[t]$. Our case here is more delicate. It may occur, for example, that for some $f \in \mathcal{H}$, $p\text{-}\lim_{n\in\mathbf{N}} U^{an}f = f$ weakly for all $a \in \mathbf{N}$ but $p\text{-}\lim_{n\in\mathbf{N}} U^{n^2}f = 0$ weakly.

It follows that the splittings which enable one to distinguish between different kinds of asymptotic behaviour of $U^{q(n)}$ along $p \in \beta\mathbf{N}$ may depend on the polynomial $q(n)$. To convey the gist of the proof we shall utilize a splitting which allows one to prove the theorem for $q(n) = n^2$ (and indeed for any other quadratic polynomial). Recall that any ball of fixed radius is compact in the weak topology, which will assure the existence of the weak p-limits that we shall deal with in the course of the proof.

Let $P_p f = p\text{-}\lim_{n\in\mathbf{N}} U^{n^2}f$, $f \in \mathcal{H}$. Put

$$\mathcal{H}_r = \overline{\{f \in \mathcal{H} : \text{ there exists } a \in \mathbf{N} \text{ with } p\text{-}\lim_{n\in\mathbf{N}} U^{an}f = f\}}.$$

Notice that since $||U^{an}f|| = ||f||$, the relation $p\text{-}\lim_{n\in\mathbb{N}} U^{an}f = f$ may be thought of as valid in the sense of both strong and weak convergence. It is not hard to see that the orthocomplement of $\mathcal{H}_r$, which we shall denote by $\mathcal{H}_m$, may be described as follows:

$$\mathcal{H}_m = \{f \in \mathcal{H} : \text{ for all } a \in \mathbb{N}, \; p\text{-}\lim_{n\in\mathbb{N}} U^{an}f = 0 \text{ weakly }\}.$$

Let us start by showing that $P_p f = 0$ for $f \in \mathcal{H}_m$. To this end we shall need the following IP-version of the van der Corput trick:

If $(x_n)_{n\in\mathbb{N}} \subset \mathcal{H}$ is a bounded sequence and if for any $h \in \mathbb{N}$ we have

$$p\text{-}\lim_{n\in\mathbb{N}} \langle x_{n+h}, x_n \rangle = 0,$$

then $p\text{-}\lim_{n\in\mathbb{N}} \langle x_n, g \rangle = 0$ for any $g \in \mathcal{H}$.

The proof of this is almost trivial: without loss of generality consider $g \in \text{span}(x_n)_{n\in\mathbb{N}}$ and use the hypotheses. The fact that for $f \in \mathcal{H}_m$, $p\text{-}\lim_{n\in\mathbb{N}} U^{n^2}f = 0$ weakly follows routinely now with the help of the IP van der Corput trick.

Let us consider now the space $\mathcal{H}_r$. As $||P_p f|| \leq ||f||$ for all $f \in \mathcal{H}$, in order to show that P_p is a projection, all we have left to establish is that $P_p^2 f = P_p f$ for all $f \in \mathcal{H}_r$. Since $\mathcal{H}_r = \overline{\bigcup_{a\in\mathbb{N}} \mathcal{H}_a}$, where the potentially trivial spaces $\mathcal{H}_a$, $a \in \mathbb{N}$, are defined by

$$\mathcal{H}_a = \{f \in \mathcal{H} : p\text{-}\lim_{n\in\mathbb{N}} U^{an}f = f\},$$

we may assume without loss of generality that for some $a \in \mathbb{N}$,

$$p\text{-}\lim_{n\in\mathbb{N}} U^{an}f = f.$$

Notice that this implies that for any b dividing a one also has

$$p\text{-}\lim_{n\in\mathbb{N}} U^{bn}f = f$$

(since the convergence in this case is strong as well as weak). We shall need to make use of Exercise 10 (ii), which enables us to take p-limits along those $n \in \mathbb{N}$ which are divisible by any prescribed integer. This will be expressed by the notation $p\text{-}\lim_{n\in\mathbb{N},a|n}$. We also remark that since for any m with $a|m$, $p\text{-}\lim_{n\in\mathbb{N}} U^{mn}f = f$, for any such m we have $p\text{-}\lim_{n\in\mathbb{N}} U^{n^2}f = p\text{-}\lim_{n\in\mathbb{N}} U^{n^2+mn}f$.

We now may write

$$P_p^2 f = p\text{-}\lim_{\substack{m \in \mathbf{N}}} U^{m^2} \left(p\text{-}\lim_{\substack{n \in \mathbf{N}}} U^{n^2} f \right) = p\text{-}\lim_{\substack{m \in \mathbf{N}, a \mid m}} U^{m^2} \left(p\text{-}\lim_{\substack{n \in \mathbf{N}}} U^{n^2} f \right)$$

$$= p\text{-}\lim_{\substack{m \in \mathbf{N}, a \mid m}} U^{m^2} \left(p\text{-}\lim_{\substack{n \in \mathbf{N}}} U^{n^2 + 2mn} f \right) p\text{-}\lim_{\substack{m \in \mathbf{N}}} p\text{-}\lim_{\substack{n \in \mathbf{N}}} U^{(n+m)^2} f$$

$$= p\text{-}\lim_{\substack{n \in \mathbf{N}}} U^{n^2} f = P_p f.$$

(We have used at one stage weak continuity of the operator U^{m^2}, and at another stage Theorem 3.8.) This finishes the proof in the case $q(n) = n^2$.

4. IP-polynomials, recurrence, and polynomial Hales-Jewett theorem.

The results discussed in the previous sections, especially Theorems 1.18, 1.19, 1.21, and 3.11 indicate that there are two types of subsets of $\mathbf{Z}$ which stand out as being related to particularly nice refinements of ergodic theorems pertaining to single and multiple recurrence: IP-sets and *polynomial sets*, namely sets of the form $p(\mathbf{Z}) = \{p(n) : n \in \mathbf{Z}\}$, where $p(n) \in \mathbf{Q}[n]$, $p(\mathbf{Z}) \subset \mathbf{Z}$, and $p(0) = 0$. As regards IP-sets, these may be viewed as kinds of generalized additive subsemigroups of $\mathbf{Z}$. It is so since given an IP-set $(x_\alpha)_{\alpha \in \mathcal{F}}$ generated by the sequence $(x_j)_{j=1}^\infty \subset \mathbf{Z}$ (where $\mathcal{F}$ is the set of nonempty finite subsets of $\mathbf{N}$ and $x_\alpha := \sum_{j \in \alpha} x_j$, $\alpha \in \mathcal{F}$), the commutative and associative partial operation defined by the formula $x_\alpha + x_\beta = x_{\alpha \cup \beta}$ has only one flaw: it is only valid when $\alpha \cap \beta = \emptyset$. However, since one generally deals in treatments such as ours with limits of expressions in which the parameter (in this case $\alpha \in \mathcal{F}$) goes to infinity, this limitation turns out not to hinder us, particularly in light of the fact that in *IP ergodic theory* one deals with IP-convergence rather than with the Cesaro convergence typical of classical ergodic theory.

On the other hand, a fundamental property of polynomials is that after finitely many applications of the difference operator they become linear. This often allows one to apply an inductive procedure deducing results for polynomials of a certain degree from similar results for polynomials of lesser degree.

The following weakly mixing PET (Polynomial Ergodic Theorem) is obtained by an application of this procedure. (It is at the same time an important special case of Theorem 1.19. See also [BM1].) Recall that a measure preserving system $(X, \mathcal{B}, \mu, T)$ is weakly mixing if the only eigenfunctions are the constants, that is, if $f(Tx) = \lambda f(x)$ for $f \in L^2(X, \mathcal{B}, \mu)$, $\lambda \in \mathbf{C}$ implies that $f =$const a.e.

Theorem 4.1 ([B4]). Suppose that $(X, \mathcal{B}, \mu, T)$ is a weakly mixing system and let $p_i(t) \in \mathbf{Q}[t]$, $i = 1, 2, \cdots, k$ be polynomials satisfying $p_i(\mathbf{Z}) \subset$

$\mathbf{Z}$, $\deg p_i(t) > 0$, and $\deg\big(p_i(t) - p_j(t)\big) > 0$, $1 \le i \ne j \le k$. Then for any $f_1, \cdots, f_k \in L^\infty(X, \mathcal{B}, \mu)$ one has

$$\lim_{N \to \infty} \left\| \frac{1}{N} \sum_{n=1}^N f_1(T^{p_1(n)}x) f_2(T^{p_2(n)}x) \cdots f_k(T^{p_k(n)}x) \right.$$

$$\left. - \prod_{i=1}^k \int f_i \, d\mu \right\|_{L^2(X)} = 0.$$

Returning to IP-sets, notice that whereas $\mathbf{Z}$ has only countably many additive subsemigroups (and only countably many polynomial subsets of the form $p(\mathbf{Z})$ with $p(t) \in \mathbf{Q}[t]$), an IP-set has uncountably many subsets which are themselves IP-sets. This fact lends additional peculiarity to and hints at greater diversity of the ergodic phenomena inherent in IP-ergodic theory.

We want now to address the question as to whether or not there exists a viable joint framework for ergodic theory along polynomial sets and along IP-sets. Theorem 3.11, for example, would be in this vein. It turns out, in fact, that Theorem 3.11 admits of significant strengthening even in the case of single recurrence. The goal of our present discussion is to provide some motivation for such a generalization. As we shall see, polynomial images of IP-sets (namely sets of the form $\{p(x_\alpha) : \alpha \in \mathcal{F}\}$), which are dealt with in Theorem 3.11, are only a special case of a much wider family of subsets of $\mathbf{Z}$ which deserve to be called *polynomial IP-sets*.

Let us return for a moment to usual polynomials belonging to $\mathbf{Z}[t]$ (we leave to the reader the verification of the fact that the whole discussion is extendable to polynomials from $\mathbf{Q}[t]$ taking on integer values on integers). More specifically, let us examine some of the (numerous) ways of arriving at quadratic polynomials with zero constant term. One possibility is exemplified by formulas like $n^2 = 1 + 3 + \cdots + (2n - 1)$, $n \in \mathbf{N}$. Another approach is to seek solutions to functional equations such as

$$p(a+b+c) - p(a+b) - p(b+c) - p(a+c) + p(a) + p(b) + p(c) = 0, \ a, b, c \in \mathbf{Z}.$$

Still another possibility is to take the "diagonal" of a bilinear form. For example, if $g(n, m) = anm + bn + cm$, then putting $n = m$ one gets $p(n) = an^2 + (b + c)n$.

The latter two approaches make perfect sense for $\mathbf{Z}$-valued functions of $\mathcal{F}$-variables. Call a function $f : \mathcal{F} \times \mathcal{F} \to \mathbf{Z}$ bilinear if the $\mathcal{F}$-sequence obtained by fixing one of the arguments of g satisfies the IP-equation $x_\alpha + x_\beta = x_{\alpha \cup \beta}$, $\alpha \cap \beta = \emptyset$. Then the "diagonal" $p(\alpha) = g(\alpha, \alpha)$ is a natural analog of a quadratic polynomial. One will arrive at the same family of functions $g : \mathcal{F} \to \mathbf{Z}$ by solving functional equations of the form

$$g(\alpha \cup \beta \cup \gamma) + g(\alpha) + g(\beta) + g(\gamma) = g(\alpha \cup \beta) + g(\beta \cup \gamma) + g(\alpha \cup \gamma),$$
$$\alpha, \beta, \gamma \in \mathcal{F}, \alpha \cap \beta = \emptyset, \alpha \cap \gamma = \emptyset, \beta \cap \gamma = \emptyset.$$

The *IP-quadratic* functions which one obtains this way include (but are not limited to) expressions like $g(\alpha) = n_\alpha k_\alpha + l_\alpha m_\alpha$, $\alpha \in \mathcal{F}$, where $(n_\alpha)_{\alpha \in \mathcal{F}}$, $(k_\alpha)_{\alpha \in \mathcal{F}}$, $(l_\alpha)_{\alpha \in \mathcal{F}}$ and $(m_\alpha)_{\alpha \in \mathcal{F}}$ are IP-sets. A natural subclass of IP-polynomials may be obtained in the following way. Let $q(t_1, t_2, \cdots, t_k) \in \mathbf{Z}[t_1, \cdots, t_k]$ and let $(n_\alpha^{(i)})_{\alpha \in \mathcal{F}}$, $i = 1, 2, \cdots k$ be IP-sets. Then $g(\alpha) = q(n_\alpha^{(1)}, n_\alpha^{(2)}, \cdots, n_\alpha^{(k)})$ is an example of an IP-polynomial. If, say, $\deg q(t_1, \cdots, t_k) = 2$, then $g(\alpha)$ will typically look like

$$g(\alpha) = \sum_{i=1}^{s} n_\alpha^{(i)} m_\alpha^{(i)} + \sum_{i=1}^{r} k_\alpha^{(i)}.$$

For IP-polynomials of the type just described we have the following joint refinement of Theorems 1.17 and 3.11.

Theorem 4.2 ([BFM]). For any polynomial $p(x_1, \cdots, x_k) \in \mathbf{Z}[x_1, \cdots, x_k]$ satisfying $p(0, \cdots, 0) = 0$, and for any k IP-sets $(n_\alpha^{(1)})_{\alpha \in \mathcal{F}}, \cdots,$ $(n_\alpha^{(k)})_{\alpha \in \mathcal{F}}$, the set $\{p(n_\alpha^{(1)}, \cdots, n_\alpha^{(k)}) : \alpha \in \mathcal{F}\}$ is a set of recurrence.

While generalizing Theorem 1.17, Theorem 4.2 does not contain as a special case the more general Theorem 2.1. To formulate a result which would contain Theorem 2.1 as well, we need the following definition. Suppose that $\mathbf{T} = \{T_w : w \in W\}$ is an indexed family of measure preserving transformations of a probability space $(X, \mathcal{B}, \mu)$. One says that $\mathbf{T}$ has the R-property if for any $A \in \mathcal{B}$ with $\mu(A) > 0$ there exists $w \in W$ such that $\mu(A \cap T_w^{-1} A) > 0$. Theorem 4.2 tells us that the family $\{T^{p(n_\alpha^{(1)}, \cdots, n_\alpha^{(k)})} : \alpha \in \mathcal{F}\}$ has the R-property. This is a special case of the following.

Theorem 4.3 ([BFM]). Suppose that $p_i(x_1, \cdots, x_k) \in \mathbf{Z}[x_1, \cdots, x_k]$ satisfy $p_i(0, \cdots, 0) = 0$, $1 \leq i \leq m$ and that $(n_\alpha^{(i)})_{\alpha \in \mathcal{F}}$ are IP-sets, $1 \leq i \leq k$. Let $p_\alpha^{(j)} := p_j(n_\alpha^{(1)}, \cdots, n_\alpha^{(k)})$, $1 \leq j \leq m$. Then for any commuting invertible measure preserving transformations $T_1, \cdots, T_m$, the family $\{\prod_{i=1}^{m} T_i^{p_\alpha^{(i)}} : \alpha \in \mathcal{F}\}$ has the R-property.

Exercise 13. Derive Theorem 2.1 from Theorem 4.3.

Theorem 4.3 has the following combinatorial corollary, which can be obtained by applying Furstenberg's correspondence principle for $\mathbf{Z}^t$-actions.

Theorem 4.4. Suppose that a set $E \subset \mathbf{Z}^t$ has positive upper Banach density and let $p_i(x_1, \cdots, x_k) \in \mathbf{Z}[x_1, \cdots, x_k]$ satisfy $p_i(0, \cdots, 0) = 0$, $1 \leq i \leq t$. Let $(n_\alpha^{(i)})_{\alpha \in \mathcal{F}}$ be IP-sets in $\mathbf{N}$, $1 \leq i \leq k$. Then for some $(x_1, \cdots, x_t)$, $(y_1, \cdots, y_t) \in E$ and $\alpha \in \mathcal{F}$ one has

$$x_1 - y_1 = p_1(n_\alpha^{(1)}, \cdots, n_\alpha^{(k)})$$
$$x_2 - y_2 = p_2(n_\alpha^{(1)}, \cdots, n_\alpha^{(k)})$$

$$\vdots$$

$$x_t - y_t = p_t(n_\alpha^{(1)}, \cdots, n_\alpha^{(k)}).$$

The next natural question is whether Theorem 4.4 is a special case of a general *IP-polynomial* Szemerédi theorem, which would bear to Theorem 4.4 the same relation as Theorem 1.19 bears to Theorem 2.1. The answer is yes and is provided by the following theorem.

To formulate it we need the notion of an IP*-set in $\mathbf{Z}^k$. This notion (which actually can be defined in any semigroup) is introduced in a way completely analogous to that for $\mathbf{Z}$ in Section 1. Given any infinite sequence $G = \{g_i : i \in \mathbf{N}\} \subset \mathbf{Z}^k$, the IP-set generated by G is the set $\Gamma = \{g_\alpha\}_{\alpha \in \mathcal{F}}$, where $g_\alpha := \sum_{i \in \alpha} g_i$. A subset $S \subset \mathbf{Z}^k$ is called IP* if for any IP-set $\Gamma \subset \mathbf{Z}^k$ one has $S \cap \Gamma \neq \emptyset$.

Exercise 14. Check that Lemma 1.13 holds for IP*-sets in $\mathbf{Z}^t$.

Theorem 4.5 ([BM2]). Suppose that $T_1, \cdots, T_r$ are commuting invertible measure preserving transformations of a probability space $(X, \mathcal{B}, \mu)$. Suppose that $k, t \in \mathbf{N}$ and that we have polynomials $p_{ij}(n_1, \cdots, n_k) \in \mathbf{Z}[n_1, \cdots, n_k]$, $1 \leq i \leq r$, $1 \leq j \leq t$ having zero constant term. Then for every $A \in \mathcal{B}$ with $\mu(A) > 0$ the set

$$R_A = \left\{ (n_1, \cdots, n_k) \in \mathbf{Z}^k : \mu\left(\bigcap_{i=1}^{t} \left(\prod_{j=1}^{r} T_j^{p_{ij}(n_1, \cdots, n_k)} \right) A \right) > 0 \right\}$$

is an IP*-set in $\mathbf{Z}^k$.

Some of the corollaries of Theorem 4.5 are collected in the following list.

(i) Already the case $k = 1$ of Theorem 4.5 gives a refinement of the polynomial Szemerédi theorem (Theorem 1.19 above) as well as a strengthened form of its special case, Theorem 1.21. Indeed, Theorem 4.5 says that, when $k = 1$, the set

$$R_A = \{n \in \mathbf{Z} : \mu(A \cap T_1^{p_{11}(n)} T_2^{p_{12}(n)} \cdots T_r^{p_{1r}(n)} A \cap \cdots$$
$$\cap T_1^{p_{t1}(n)} T_2^{p_{t2}(n)} \cdots T_r^{p_{tr}(n)} A) > 0\}$$

is IP*, hence syndetic, hence of positive lower density.

(ii) In addition, Theorem 4.5 enlarges the family of configurations which can always be found in sets of positive upper Banach density in $\mathbf{Z}^n$. For example, since for any IP-sets $(n_\alpha^{(i)})_{\alpha \in \mathcal{F}}$, $i = 1, 2, \cdots, k$, any measure preserving system $(X, \mathcal{B}, \mu, T)$ and any $A \in \mathcal{B}$ with $\mu(A) > 0$ there exists $\alpha \in \mathcal{F}$ such that

$$\mu(A \cap T^{n_\alpha^{(1)}} A \cap T^{n_\alpha^{(1)} n_\alpha^{(2)}} A \cap \cdots \cap T^{n_\alpha^{(1)} n_\alpha^{(2)} \cdots n_\alpha^{(k)}} A) > 0,$$

one obtains (via Furstenberg's correspondence principle) the fact that for any $E \subset \mathbf{Z}$ with $d^*(E) > 0$ there exists $x \in \mathbf{Z}$ and $\alpha \in \mathcal{F}$ such that

$$\left\{ x, x + n_\alpha^{(1)}, x + n_\alpha^{(1)} n_\alpha^{(2)}, \cdots, x + n_\alpha^{(1)} n_\alpha^{(2)} \cdots n_\alpha^{(k)} \right\} \subset E.$$

The following theorem gives a more general corollary of Theorem 4.5 (which includes Theorem 1.20 as a special case).

Theorem 4.6. Let $P : \mathbf{Z}^r \to \mathbf{Z}^l$, $r, l \in \mathbf{N}$ be a polynomial mapping satisfying $P(0) = 0$, let $F \subset \mathbf{Z}^r$ be a finite set, let $S \subset \mathbf{Z}^l$ be a set of positive upper Banach density and let $(n_\alpha^{(i)})_{\alpha \in \mathcal{F}}$, $i = 1, 2, \cdots, r$ be arbitrary IP-sets. Then for some $u \in \mathbf{Z}^l$ and $\alpha \in \mathcal{F}$ one has:

$$\left\{ u + P(n_\alpha^{(1)} x_1, n_\alpha^{(2)} x_2, \cdots, n_\alpha^{(r)} x_r) : (x_1, x_2, \cdots, x_r) \in F \right\} \subset S.$$

(iii) The following exercise serves as a good reaffirmation of the third principle of Ramsey theory.

Exercise 15. Given k commuting invertible measure preserving transformations $T_1, \cdots, T_k$ of a probability space $(X, \mathcal{B}, \mu)$, and polynomials $p_i(n, m) \in \mathbf{Z}[n, m]$ with $p_i(0, 0) = 0$, $i = 1, 2, \cdots, k$, and a set $A \in \mathcal{B}$ with $\mu(A) > 0$, let

$$R_A = \left\{ (n, m) : \mu(A \cap T_1^{p_1(n,m)} A \cap \cdots \cap T_k^{p_k(n,m)} A) > 0 \right\}.$$

Show that for any polynomials $q_1, q_2 \in \mathbf{Z}[n]$ satisfying $q_1(0) = q_2(0) = 0$, the set $\{n : (q_1(n), q_2(n)) \in R_A\}$ is an IP*-set in $\mathbf{Z}$.

We want to conclude this section with a brief discussion of the combinatorial tool which is instrumental in the proof of Theorem 4.5, namely the polynomial extension of Hales-Jewett theorem ([HJ]), which was recently obtained in [BL2]. The by now classical Hales-Jewett theorem which deals with finite sequences formed from a finite alphabet rather than with integers, may be regarded as an abstract extension of van der Waerden's theorem. To formulate the Hales-Jewett theorem we introduce some definitions.

Let A be a finite alphabet, $A = \{a_1, \cdots, a_k\}$. The set $W_n(A)$ of words of length n over A can be viewed as an abstract n-dimensional vector space (over A). Having agreed on such a geometric terminology, we look now for an appropriate notion of a (combinatorial) line. Since A is not supposed to have any algebraic structure, the only candidates for combinatorial lines in $W_n(A)$ are sets of n-tuples of elements from A which obey the equations $x_i = x_j$ or $x_j = a$, $a \in A$. For example, if $A = \{1, 2, 3\}$ and $n = 2$, there are 4 lines: $\{11, 22, 33\}$, $\{11, 12, 13\}$, $\{21, 22, 23\}$ and $\{31, 32, 33\}$. Another way of introducing lines is the following. Let $W_n(A, t)$ be the set of words of length n from the alphabet $A \cup \{t\}$, where t is a letter not belonging to A which will serve as a variable. If $w(t) \in W_n(A, t)$ is a word in which the variable t actually occurs, then the set $\{w(t)\}_{t \in A} = \{w(a_1), \cdots, w(a_k)\}$ is a combinatorial line.

Theorem 4.7 ([HJ]). Given any alphabet $A = \{a_1, \cdots, a_k\}$ and $r \in \mathbf{N}$ there exists $N = N(k, r) \in \mathbf{N}$ such that if $n > N$ and $W_n(A)$ is partitioned into r classes then at least one of these classes contains a combinatorial line.

Remarks. (i) Taking $A = \{0, 1, 2, \cdots, s-1\}$ and interpreting $W_n(A)$ as integers to base s having n or less digits in their s-expansion one sees that in this situation the elements of a combinatorial line form an arithmetic progression (with difference of the form $d = \sum_{i=0}^{n-1} a_i s^i$ where $a_i = 0$ or 1). Thus van der Waerden's theorem is a corollary of the Hales-Jewett theorem.

(ii) If one takes A to be a finite field F, then $W_n(F) = F^n$ has a natural structure of an n-dimensional vector space over F. In this case a combinatorial line is an affine linear one-dimensional subspace of F^n.

An interesting feature of the Hales-Jewett theorem (and the one showing that it is, in a sense, the "right" result) is that one can easily derive from it its multidimensional version. Let $t_1, \cdots, t_m$ be m variables and let $W_n(A; t_1, \cdots, t_m)$ be the set of words of length n over the alphabet $A \cup \{t_1, \cdots, t_m\}$. If for some n $w(t_1, \cdots, t_m) \in W_n(A; t_1, \cdots t_m)$ is a word in which all the variables appear, the result of the substitution

$$\left\{w(t_1, \cdots, t_m)\right\}_{(t_1, \cdots, t_m) \in A^m} = \left\{w(a_{i_1}, \cdots, a_{i_m}) : a_{i_j} \in A, j = 1, 2, \cdots m\right\}$$

is called a combinatorial m-space. It easily follows from Theorem 4.7 that for any $r, m \in \mathbf{N}$ there exists $N = N(k, r, m)$ such that if $n > N$ and $A = \{a_1, \cdots, a_k\}$ is partitioned into r classes then at least one of these classes contains a combinatorial m-space.

Exercise 16. Derive from the multidimensional (or, rather, multiparameter) version of the Hales-Jewett theorem just described the Gallai-Grünwald theorem (Theorem 1.10 above). (With a little extra effort one should also be able to get Theorem 2.5.)

We are going to formulate now one more version of the Hales-Jewett theorem. Given an infinite set M and $k \in \mathbf{N}$, denote by $P_f^{(k)}(M)$ the set of k-tuples of finite (potentially empty) subsets of M and let us call any $(k+1)$-element subset of $P_f^{(k)}(M)$ of the form

$$\{(\alpha_1, \alpha_2, \cdots, \alpha_k), (\alpha_1 \cup \gamma, \alpha_2, \cdots, \alpha_k), (\alpha_1, \alpha_2 \cup \gamma, \cdots, \alpha_k),$$
$$\cdots, (\alpha_1, \alpha_2, \cdots, \alpha_k \cup \gamma)\},$$

where γ is non-empty and disjoint from $\alpha_1, \cdots, \alpha_k$, a *simplex* and denote it by
$S(\alpha_1, \alpha_2, \cdots, \alpha_k; \gamma)$ (a familiar Euclidian simplex with vertices

$$(s_1, \cdots, s_k), (s_1 + h, s_2, \cdots, s_k), (s_1, s_2 + h, \cdots, s_k), \cdots, (s_1, s_2, \cdots, s_k + h)$$

should come to the reader's mind).

Theorem 4.8. For any $k \in \mathbf{N}$ and any finite coloring of $P_f^{(k)}(\mathbf{N})$, there exists a monochome simplex.

Let us show that Theorem 4.8 follows from Theorem 4.7. Let

$$\chi : P_f^{(k)}(\mathbf{N}) \to \{1, 2, \cdots, r\}$$

be a finite coloring. Let $W = \bigcup_{i=1}^{\infty} W_i(A)$ denote the set of all finite words over the alphabet $A = \{0, 1, \cdots, k\}$. Let $\varphi : W \to P_f^{(k)}(\mathbf{N})$ be the mapping which corresponds to any word $w = w_1 w_2 \cdots w_m \in W$ a k-tuple $(\alpha_1, \cdots, \alpha_k) \in P_f^{(k)}(\mathbf{N})$ by the rule:

$$\alpha_i = \{j : w_j = i\}, \quad i = 1, 2, \cdots, k.$$

Notice that this induces a coloring $\tilde{\chi}$ of W defined by $\tilde{\chi} = \chi \circ \varphi$. By Theorem 4.7 there exists a $\tilde{\chi}$-monochrome combinatorial line $\{w(t)\}_{t \in A}$. One easily checks that the χ-monochrome image of this line under φ forms a simplex. The following example should make it completely clear. Let $A = \{0, 1, 2, 3, 4\}$ and assume that

$$l = \{2t1t213\}_{t \in A}$$
$$= \{(2010213), (2111213), (2212213), (2313213), (2414213)\}$$

is a $\tilde{\chi}$-monochrome combinatorial line. Letting $\gamma = \{2, 4\}$, one observes that

$$\varphi(2010213) = (\{3, 6\}, \{1, 5\}, \{7\}, \emptyset)$$
$$\varphi(2111213) = (\{3, 6\} \cup \gamma, \{1, 5\}, \{7\}, \emptyset)$$
$$\varphi(2212213) = (\{3, 6\}, \{1, 5\} \cup \gamma, \{7\}, \emptyset).$$
$$\varphi(2313213) = (\{3, 6\}, \{1, 5\}, \{7\} \cup \gamma, \emptyset)$$
$$\varphi(2414213) = (\{3, 6\}, \{1, 5\}, \{7\}, \gamma)$$

This gives us a monochrome simplex $S(\{3,6\},\{1,5\},\{7\},\emptyset;\gamma)$.

Exercise 17. Show that Theorem 4.8 implies Theorem 4.7.

To give the reader a feeling of what the polynomial Hales-Jewett theorem is about we shall bring now two equivalent formulations of one of its simplest cases, the "quadratic" nature of which is self-evident. The first formulation is a natural refinement of Theorem 4.8. The second one shows the connection between the polynomial Hales-Jewett theorem and topological dynamics, by means of which the polynomial Hales-Jewett theorem is proved in [BL2].

Theorem 4.9. For any $k \in \mathbf{N}$ and any finite coloring of $P_f^{(k)}(\mathbf{N} \times \mathbf{N})$ there exists a monochrome simplex of the form

$$\{(\alpha_1, \alpha_2, \cdots, \alpha_k), (\alpha_1 \cup (\gamma \times \gamma), \alpha_2, \cdots, \alpha_k), (\alpha_1, \alpha_2 \cup (\gamma \times \gamma), \cdots, \alpha_k),$$

$$\cdots, (\alpha_1, \alpha_2, \cdots, \alpha_k \cup (\gamma \times \gamma))\},$$

where γ is a finite non-empty subset of $\mathbf{N}$ and the Cartesian square $\gamma \times \gamma$ is disjoint from $\alpha_1, \cdots, \alpha_n$.

Theorem 4.10. Let (X, ρ) be a compact metric space. For some $k \in \mathbf{N}$, let

$$T(\alpha_1, \alpha_2, \cdots, \alpha_k) = T^{(\alpha_1, \alpha_2, \cdots, \alpha_k)}, \quad (\alpha_1, \alpha_2, \cdots, \alpha_k) \in P_f^{(k)}(\mathbf{N} \times \mathbf{N})$$

be a family of self-mappings of X satisfying the condition

$(*)$ for any finite sets $\alpha_i, \beta_i \subset \mathbf{N} \times \mathbf{N}$ satisfying $(\alpha_i \cap \beta_i) = \emptyset$, $i = 1, \cdots, k$, $T^{(\alpha_1 \cup \beta_1, \cdots, \alpha_k \cup \beta_k)} = T^{(\alpha_1, \cdots, \alpha_k)} T^{(\beta_1, \cdots, \beta_k)}$.

Then for any $\epsilon > 0$ and for any $x \in X$ there exist a non-empty set $\gamma \subset \mathbf{N}$ and finite sets $\alpha_1, \cdots, \alpha_k \subset \mathbf{N} \times \mathbf{N}$ such that $\alpha_i \cap (\gamma \times \gamma) = \emptyset$, $i = 1, 2, \cdots, k$ and

$$diam\{T^{(\alpha_1, \alpha_2, \cdots, \alpha_k)}x, T^{(\alpha_1 \cup (\gamma \times \gamma), \alpha_2, \cdots, \alpha_k)}x, T^{(\alpha_1, \alpha_2 \cup (\gamma \times \gamma), \cdots, \alpha_k)}x,$$

$$\cdots, T^{(\alpha_1, \alpha_2, \cdots, \alpha_k \cup (\gamma \times \gamma))}x\} < \epsilon.$$

Before discussing some applications of Theorems 4.9 and 4.10, let us show their equivalence.

$(4.9) \to (4.10)$. Given the compact space (X, ρ) together with the family of its self-mappings

$$T^{(\alpha_1, \alpha_2, \cdots, \alpha_k)}, \quad (\alpha_1, \alpha_2, \cdots, \alpha_k) \in P_f^{(k)}(\mathbf{N} \times \mathbf{N}),$$

a point $x \in X$ and $\epsilon > 0$, let $\{x_1, \cdots, x_r\}$ be an $\frac{\epsilon}{2}$-net in X. This net naturally defines a coloring $\chi : P_f^{(k)}(\mathbf{N} \times \mathbf{N}) \to \{1, 2, \cdots, r\}$ by the rule:

$$\chi\big((\alpha_1, \alpha_2, \cdots, \alpha_k)\big) = \min\Big\{i : \rho(T^{(\alpha_1, \alpha_2, \cdots, \alpha_k)}x, x_i) < \frac{\epsilon}{2}\Big\}.$$

Let $S(\alpha_1, \alpha_2, \cdots, \alpha_k; \gamma \times \gamma)$ be a monochrome simplex as guaranteed by Theorem 4.9. Then clearly,

$$diam\big\{T^{(\alpha_1, \alpha_2, \cdots, \alpha_k)}x, T^{(\alpha_1 \cup (\gamma \times \gamma), \alpha_2, \cdots, \alpha_k)}x, T^{(\alpha_1, \alpha_2 \cup (\gamma \times \gamma), \cdots, \alpha_k)}x,$$

$$\cdots T^{(\alpha_1, \alpha_2, \cdots, \alpha_k \cup (\gamma \times \gamma))}x\big\} < \epsilon.$$

$(4.10) \to (4.9)$ For fixed $r, k \in \mathbf{N}$ let $\Omega_r^{(k)}$ be the space of all r-colorings of $P_f^{(k)}(\mathbf{N} \times \mathbf{N})$ (namely, the set of mappings $\chi : P_f^{(k)}(\mathbf{N} \times \mathbf{N}) \to \{1, 2, \cdots, r\}$ equipped with the metric

$$\rho(\chi_1, \chi_2) = \inf\Big\{\frac{1}{N+1} : \chi\big((\alpha_1, \cdots, \alpha_k)\big) = \chi_2\big((\alpha_1, \cdots, \alpha_k)\big) \text{ for any}$$

$$(\alpha_1, \cdots, \alpha_k) \in \{1, 2, \cdots, N\} \times \{1, 2, \cdots, N\}\Big\}.$$

Clearly, $(\Omega_r^{(k)}, \rho)$ is a compact metric space. Note that $\rho(\chi_1, \chi_2) < 1$ if and only if $\chi_1(\emptyset) = \chi_2(\emptyset)$. Define mappings

$$T^{(\alpha_1, \alpha_2, \cdots, \alpha_k)} : \Omega_r^{(k)} \to \Omega_r^{(k)}, \quad (\alpha_1, \alpha_2, \cdots, \alpha_k) \in P_f^{(k)}(\mathbf{N} \times \mathbf{N})$$

by

$$\big(T^{(\alpha_1, \alpha_2, \cdots, \alpha_k)}\chi\big)(\beta_1, \beta_2, \cdots, \beta_k) = \chi\big((\alpha_1 \cup \beta_1, \alpha_2 \cup \beta_2, \cdots, \alpha_k \cup \beta_k)\big).$$

Applying Theorem 4.10 for $\epsilon = 1$ and given r-coloring χ, we get $(\alpha_1, \cdots, \alpha_k)$ $\in P_f^{(k)}(\mathbf{N} \times \mathbf{N})$ and a finite non-empty set $\gamma \subset \mathbf{N}$, such that $\alpha_i \cap (\gamma \times \gamma) = \emptyset$, $i = 1, 2, \cdots, k$ and such that

$$diam\big\{T^{(\alpha_1, \alpha_2, \cdots, \alpha_k)}\chi, T^{(\alpha_1 \cup (\gamma \times \gamma), \alpha_2, \cdots, \alpha_k)}\chi, T^{(\alpha_1, \alpha_2 \cup (\gamma \times \gamma), \cdots, \alpha_k)}\chi,$$

$$\cdots T^{(\alpha_1, \alpha_2, \cdots, \alpha_k \cup (\gamma \times \gamma))}\chi\big\} < 1.$$

By the remark above we get

$$T^{(\alpha_1,\alpha_2,\cdots,\alpha_k)}\chi(\emptyset) = T^{(\alpha_1\cup(\gamma\times\gamma),\alpha_2,\cdots,\alpha_k)}\chi(\emptyset)$$
$$= T^{(\alpha_1,\alpha_2\cup(\gamma\times\gamma),\cdots,\alpha_k)}\chi(\emptyset)$$
$$\vdots$$
$$= T^{(\alpha_1,\alpha_2,\cdots,\alpha_k\cup(\gamma\times\gamma))}\chi(\emptyset),$$

or, equivalently,

$$\chi((\alpha_1,\alpha_2,\cdots,\alpha_k)) = \chi((\alpha_1\cup(\gamma\times\gamma),\alpha_2,\cdots,\alpha_k))$$
$$= \chi((\alpha_1,\alpha_2\cup(\gamma\times\gamma),\cdots,\alpha_k))$$
$$\vdots$$
$$= \chi((\alpha_1,\alpha_2,\cdots,\alpha_k\cup(\gamma\times\gamma))).$$

Let us derive now some combinatorial consequences from the "quadratic" Hales-Jewett theorem. Let $k,r \in \mathbf{N}$ be given and let $\chi_{\mathbf{N}} : \mathbf{N} \to \{1,2,\cdots,r\}$ be a coloring of $\mathbf{N}$. Induce a coloring $\chi : P_f^{(k)}(\mathbf{N}) \to \{1,2,\cdots,r\}$ in the following way. First of all, for any finite non-empty set $\alpha \subset \mathbf{N}\times\mathbf{N}$ define

$$\chi(\alpha) = \chi_{\mathbf{N}}\Big(\sum_{(t,s)\in\alpha} ts\Big).$$

Notice that if $\alpha = \gamma\times\gamma$, then $\sum_{(t,s)\in\alpha} ts$ is a perfect square. If $\alpha = \emptyset$, let $\chi(\alpha) = 1$. For any $(\alpha_1,\alpha_2,\cdots,\alpha_k) \in P_f^{(k)}(\mathbf{N}\times\mathbf{N})$ let

$$\chi((\alpha_1,\alpha_2,\cdots,\alpha_k)) := \chi_{\mathbf{N}}\Big(\sum_{(t,s)\in\alpha_1} ts + 2\sum_{(t,s)\in\alpha_2} ts + \cdots + k\sum_{(t,s)\in\alpha_k} ts\Big).$$

Let $S(\alpha_1,\cdots,\alpha_k;\gamma\times\gamma)$ be a χ-monochrome simplex whose existence is guaranteed by Theorem 4.10. Then, letting $v = \sum_{(t,s)\in\alpha_1} ts$ and $c^2 = \sum_{(t,s)\in\gamma\times\gamma} ts$, we see that

$$\chi_{\mathbf{N}}(v) = \chi_{\mathbf{N}}(v+c^2) = \chi_{\mathbf{N}}(v+2c^2) = \cdots = \chi_{\mathbf{N}}(v+kc^2).$$

We have obtained the following "quadratic" van der Waerden theorem.

Theorem 4.11. For any $k,r \in \mathbf{N}$, if $\mathbf{N} = \bigcup_{i=1}^r C_i$ then one of C_i, $i = 1,2,\cdots,r$ contains a configuration of the form $\{v, v+c^2, \cdots, v+kc^2\}$.

Exercise 18. Derive from Theorem 4.9 the following combinatorial result. For any finite coloring of $\mathbf{Z}^n$ there exist $(a_1, \cdots, a_n) \in \mathbf{Z}^n$ and $c \in \mathbf{N}$ so that the configuration

$$\{(a_1, a_2, \cdots, a_n), (a_1 + c^2, a_2, \cdots, a_n),$$
$$(a_1, a_2 + c^2, \cdots, a_n), \cdots, (a_1, a_2, \cdots, a_n + c^2)\}$$

is monochromatic.

The general theorem proved in [BL2] allows one to derive many more combinatorial results as well as results belonging to the realm of topological dynamics, all of which have intrinsic polynomial features. To get a feeling how the polynomial Hales-Jewett theorem may be utilized in the course of a proof of a result pertaining to measurable multiple recurrence, the reader is referred to [BM1].

We conclude this section by formulating a topological recurrence theorem which is, in a sense, the most general "commutative topological quadratic" recurrence result.

Theorem 4.12. Let (X, ρ) be a compact metric space. For a fixed $k \in \mathbf{N}$, let $(T_{ij}^{(l)})_{(i,j) \in \mathbf{N} \times \mathbf{N}}$, $l = 1, 2, \cdots, k$ be commuting continuous self-mappings of X. For any finite nonempty $\alpha \in \mathbf{N} \times \mathbf{N}$ define

$$T_\alpha^{(l)} = \prod_{(i,j) \in \alpha} T_{ij}^{(l)}, \quad l = 1, 2, \cdots, k.$$

Then for any $\epsilon > 0$ there exists $x \in X$ and non-empty finite $\gamma \subset \mathbf{N}$, such that for $l = 1, 2, \cdots, k$ one has

$$\rho(T_{\gamma \times \gamma}^{(l)} x, x) < \epsilon.$$

5. Some open problems and conjectures.

Our achievements on the theoretical front will be very poor indeed if...we close our eyes to problems and can only memorize isolated conclusions or principles...

–Mao Tsetung, "Rectify the Party's style of work", [Mao], p. 212.

A mathematical discipline is alive and well if it has many exciting open problems of different levels of difficulty. This section's goal is to show that this is the case with Ergodic Ramsey Theory.

To warm up we shall start with some results and problems related to single recurrence. The following result ([K2]) is usually called Khintchine's recurrence theorem (cf. [Pa], p. 22; [Pe], p. 37).

Theorem 5.1. For any invertible probability measure preserving system $(X, \mathcal{B}, \mu, T)$, $\epsilon > 0$, and any $A \in \mathcal{B}$ the set $\{n \in \mathbf{Z} : \mu(A \cap T^n A) \geq \mu(A)^2 - \epsilon\}$ is syndetic.

One possible way of proving Theorem 5.1 is to use the uniform version of von Neumann's ergodic theorem: if U is a unitary operator acting on a Hilbert space $\mathcal{H}$, then for any $f \in \mathcal{H}$

$$\lim_{N-M \to \infty} \frac{1}{N-M} \sum_{n=M}^{N-1} U^n f = P_{inv} f = f^*,$$

where the convergence is in norm and P_{inv} is the orthogonal projection onto the subspace of U-invariant elements.

Noting that $\langle f^*, f \rangle = \langle f^*, f^* \rangle$ and taking $\mathcal{H} = L^2(X, \mathcal{B}, \mu)$, $(Ug)(x) := g(Tx)$, $g \in L^2(X, \mathcal{B}, \mu)$, and $f = 1_A$ one has

$$\lim_{N-M \to \infty} \frac{1}{N-M} \sum_{n=M}^{N-1} \mu(A \cap T^n A)$$

$$= \lim_{N-M \to \infty} \frac{1}{N-M} \sum_{n=M}^{N-1} \langle U^n f, f \rangle = \langle f^*, f \rangle$$

$$= \langle f^*, f^* \rangle = \langle f^*, f^* \rangle \langle 1, 1 \rangle \geq \left(\langle f^*, 1 \rangle \right)^2 = \left(\langle f, 1 \rangle \right)^2 = \mu(A)^2$$

(cf. [Ho]).

The following alternative way of proving Theorem 5.1 is more elementary and has two additional advantages: it enables one to prove a stronger fact, namely the IP*-ness of the set $\{n \in \mathbf{Z} : \mu(A \cap T^n A) \geq \mu(A)^2 - \epsilon\}$ and is easily adjustable to measure preserving actions of arbitrary (semi)groups.

Note first that if A_k, $k = 1, 2, \cdots$ are sets in a probability measure space such that $\mu(A_k) \geq a > 0$ for all $k \in \mathbf{N}$ then for any $\epsilon > 0$ there exist $i < j$ such that $\mu(A_i \cap A_j) \geq a^2 - \epsilon$. Indeed, if this would not be the case, the following inequality would be contradictive for sufficiently large n:

$$n^2 a^2 \leq \left(\int \sum_{i=1}^n 1_{A_i} \right)^2 \leq \int \left(\sum_{i=1}^n 1_{A_i} \right)^2 = \sum_{i=1}^n \mu(A_i) + 2 \sum_{1 \leq i < j \leq n} \mu(A_i \cap A_j)$$

(cf. [G]).

To show that $\{n \in \mathbf{Z} : \mu(A \cap T^n A) \geq \mu(A)^2 - \epsilon\}$ is an IP*-set, let $(n_i)_{i=1}^{\infty}$ be an arbitrary sequence of integers and let $A_k = T^{n_1 + \cdots + n_k} A$, $k \in \mathbf{N}$. By the above remark, there exist $i < j$ such that

$$a^2 - \epsilon \leq \mu(A_i \cap A_j) = \mu(T^{n_1 + \cdots + n_i} A \cap T^{n_1 + \cdots + n_j} A) = \mu(A \cap T^{n_{i+1} + \cdots + n_j} A).$$

This shows that

$$\{n \in \mathbf{Z} : \mu(A \cap T^n A) \geq \mu(A)^2 - \epsilon\} \cap FS(n_i)_{i=1}^{\infty} \neq \emptyset$$

and we are done. We remark also that the IP*-ness of the set $\{n \in \mathbf{Z} : \mu(A \cap T^n A) \geq \mu(A)^2 - \epsilon\}$ is equivalent to the "linear" case of Theorem 3.11.

Since in mixing measure preserving systems for any $A \in \mathcal{B}$ one has $\lim_{n \to \infty} \mu(A \cap T^n A) = \mu(A)^2$, we see that in a sense, Khintchine's recurrence theorem is the best possible. We have however the following.

Question 1. Is it true that for any invertible mixing measure preserving system $(X, \mathcal{B}, \mu, T)$ there exists $A \in \mathcal{B}$ with $\mu(A) > 0$ such that for all $n \neq 0$, $\mu(A \cap T^n A) < \mu(A)^2$? How about the reverse inequality $\mu(A \cap T^n A) > \mu(A)^2$?

Definition 5.1. A set $R \subset \mathbf{Z}$ is called a set of *nice recurrence* if for any invertible probability measure preserving system $(X, \mathcal{B}, \mu, T)$ and any $A \in \mathcal{B}$ one has $\limsup_{n \to \infty, n \in R} \mu(A \cap T^n A) \geq \mu(A)^2$.

Exercise 19. Check that all the sets of recurrence mentioned in Sections 1 through 4 are sets of nice recurrence.

A natural question arises whether any set of recurrence at all is actually a set of nice recurrence. Forrest showed in [Fo] that this is not always so. See also [M] for a shorter proof.

We saw in Section 1 that sets of recurrence have the Ramsey property: if R is a set of recurrence and $R = \bigcup_{i=1}^{r} C_i$ then at least one of C_i, $i = 1, \cdots, r$ is itself a set of recurrence.

Question 2. Do sets of nice recurrence possess the Ramsey property?

A natural necessary condition for a set $R \subset \mathbf{Z} \setminus \{0\}$ to be a set of recurrence is that for any $a \in \mathbf{Z}$, $a \neq 0$, $R \cap a\mathbf{Z} \neq \emptyset$. In particular, the set $\{2^n 3^k : n, k \in \mathbf{N}\}$ is not a set of recurrence. But what if one restricts oneself to some special classes of systems?

Question 3. Is it true that for any invertible weakly mixing system $(X, \mathcal{B}, \mu, T)$ and any $A \in \mathcal{B}$ with $\mu(A) > 0$ there exist $n, k \in \mathbf{N}$ such that $\mu(A \cap T^{2^n 3^k} A) > 0$?

Some sets of recurrence have an additional property that the ergodic averages along these sets exhibit regular behavior. For example, we saw in Section 2 that for any $q(t) \in \mathbf{Z}[t]$ and for any unitary operator $U :$ $\mathcal{H} \to \mathcal{H}$ the norm limit $\lim_{N\to\infty} \frac{1}{N} \sum_{n=1}^{N} U^{q(n)} f$ exists for every $f \in \mathcal{H}$. The following theorem, due to Bourgain, shows that much more delicate pointwise convergence also holds along the polynomial sets.

Theorem 5.2 ([Bo3]). For any measure preserving system $(X, \mathcal{B}, \mu, T)$, for any polynomial $q(t) \in \mathbf{Z}[t]$ and for any $f \in L^p(X, \mathcal{B}, \mu)$, where $p > 1$, $\lim_{N\to\infty} \frac{1}{N} \sum_{n=1}^{N} f(T^{q(n)} x)$ exists almost everywhere.

Question 4. Does Theorem 5.2 hold true for any $f \in L^1(X, \mathcal{B}, \mu)$?

Another interesting question related to ergodic averages along polynomials is concerned with uniquely ergodic systems. A topological dynamical system (X, T), where X is a compact metric space and T is a continuous self mapping of X is called uniquely ergodic if there is a unique T-invariant probability measure on the σ-algebra of Borel sets in X. The following well known result appeared for the first time in [KB]:

Theorem 5.3. A topological system (X, T) is uniquely ergodic if and only if for any $f \in C(X)$ and any $x \in X$ one has

$$\lim_{N\to\infty} \frac{1}{N} \sum_{n=0}^{N-1} f(T^n x) = \int f \, d\mu,$$

where μ is the unique T-invariant Borel measure.

Question 5. Assume that a topological dynamical system (X, T) is uniquely ergodic and let $p(t) \in \mathbf{Z}[t]$ and $f \in C(X)$. Is it true that for all but a first category set of points $x \in X$ $\lim_{N\to\infty} \frac{1}{N} \sum_{n=0}^{N-1} f(T^{p(n)} x)$ exists?

The next question that we would like to pose is concerned with the possibility of extending results like Theorem 4.2 and 4.3 to polynomial expressions involving infinitely many commuting operators. We shall formulate it for a special "quadratic" case which is a measure theoretic analogue of Theorem 4.12 for $k = 1$. Recall that an indexed family $\{T_w : w \in \mathcal{W}\}$ of measure preserving transformations of a probability measure space $(X, \mathcal{B}, \mu)$ is said to have the R-property if for any $A \in \mathcal{B}$ with $\mu(A) > 0$ there exists $w \in \mathcal{W}$ such that $\mu(A \cap T_w^{-1} A) > 0$.

Question 6. Let $(T_{ij})_{(i,j)\in\mathbf{N}\times\mathbf{N}}$ be commuting measure preserving transformations of a probability measure space $(X, \mathcal{B}, \mu)$. For any finite non-empty set $\alpha \subset \mathbf{N} \times \mathbf{N}$ let $T_\alpha = \prod_{(i,j)\in\alpha} T_{ij}$. Is it true that the family of measure preserving transformations

$$\{T_{\gamma\times\gamma} : \emptyset \neq \gamma \subset \mathbf{N}, \gamma \text{ finite}\}$$

has the R-property?

We move on now to questions related to multiple recurrence.

Definition 5.2. Let $k \in \mathbf{N}$. A set $R \subset \mathbf{Z}$ is a set of *k-recurrence* if for every invertible probability measure preserving system $(X, \mathcal{B}, \mu, T)$ and any $A \in \mathcal{B}$ with $\mu(A) > 0$ there exists $n \in R$, $n \neq 0$, such that $\mu(A \cap T^n A \cap T^{2n} A \cap \cdots \cap T^{kn} A) > 0$.

One can show that items (i), (ii), (iv) and (v) of Exercise 6 are examples of sets of k-recurrence for any k. On the other hand, an example due to Furstenberg ([F], p. 178) shows that not every infinite set of differences (item (iii) of Exercise 6) is a set of 2-recurrence (although every such is a set of 1-recurrence).

Question 7. Given $k \in \mathbf{N}$, $k \geq 2$, what is an example of a set of k-recurrence which is not a set of $(k+1)$-recurrence?

Question 8. Given a set of 2-recurrence S, is it true that for any pair T_1, T_2 of invertible commuting measure preserving transformations of a probability measure space $(X, \mathcal{B}, \mu)$ and a set $A \in \mathcal{B}$ with $\mu(A) > 0$ there exists $n \in S$ such that $\mu(A \cap T_1^n A \cap T_2^n A) > 0$? (The answer is very likely *no*.) Same question for S a set of k-recurrence for any k.

Question 9. Let $k \in \mathbf{N}$, let $T_1, T_2, \cdots, T_k$ be commuting invertible measure preserving transformations of a probability measure space $(X, \mathcal{B}, \mu)$ and let $p_1(t), p_2(t), \cdots, p_k(t) \in \mathbf{Z}[t]$. Is it true that for any $f_1, \cdots, f_k \in L^\infty(X, \mathcal{B}, \mu)$

$$\lim_{N \to \infty} \frac{1}{N} \sum_{n=1}^N f(T_1^{p_1(n)} x) f_2(T_2^{p_2(n)} x) \cdots f_k(T_k^{p_k(n)} x)$$

exists in L^2-norm? Almost everywhere?

Remark. The following results describe the status of current knowledge: The answer to the question about L^2-convergence is *yes* in the following cases:

(i) $k = 2$, $p_1(t) = p_2(t) = t$ ([CL1]).
(ii) $k = 2$, $T_1 = T_2$, $p_1(t) = t$, $p_2(t) = t^2$ ([FW2]).
(iii) $k = 3$, $T_1 = T_2 = T_3$, $p_1(t) = at$, $p_2(t) = bt$, $p_3(t) = ct$, $a, b, c \in \mathbf{Z}$ ([CL2], [FW2]).

The answer to the question about almost everywhere convergence is *yes* for $k = 2$, $T_1 = T_2$, $p_1(t) = at$, $p_2(t) = bt$, $a, b \in \mathbf{Z}$ ([Bo4]).

Question 10. Let $k \in \mathbf{N}$. Assume that $(X, \mathcal{B}, \mu, T)$ is a totally ergodic system (i.e. $(X, \mathcal{B}, \mu, T^k)$ is ergodic for any $k \neq 0$). Is it true that for any set of polynomials $p_i(t) \in \mathbf{Z}[t]$, $i = 1, 2, \cdots, k$ having pairwise distinct (non-zero) degrees, and any $f_1, \cdots, f_k \in L^\infty(X, \mathcal{B}, \mu)$ one has:

$$\lim_{N \to \infty} \Big\| \frac{1}{N} \sum_{n=1}^{N} f_1(T^{p_1(n)}x) f_2(T^{p_2(n)}x) \cdots f_k(T^{p_k(n)}x)$$

$$- \prod_{i=1}^{k} \int f_i d\mu \Big\|_{L^2(X, \mathcal{B}, \mu)} = 0?$$

Remark. It is shown in [FW2] that the answer is *yes* when $k = 2$, $p_1(t) = t$, $p_2(t) = t^2$. See also Theorem 4.1.

We now formulate a few problems related to partition Ramsey theory. A unifying property that many configurations of interest (such as arithmetic progressions or sets of the form $FS(x_j)_{j=1}^n$) have is that they constitute sets of solutions of (not necessarily linear) diophantine equations or systems thereof. A system of diophantine equations is called *partition regular* if for any finite coloring of $\mathbf{Z} \setminus \{0\}$ (or of $\mathbf{N}$) there is a monochromatic solution. For example, the following systems of equations are partition regular:

$$\begin{array}{ll}
x_1 + x_3 = 2x_2 & x + y = t \\
x_2 + x_4 = 2x_3 & x + z = u \\
x_3 + x_5 = 2x_4 & z + y = v \\
x_4 + x_6 = 2x_5 & x + y + z = w
\end{array}$$

A general theorem due to Rado gives necessary and sufficient conditions for a system of *linear* equations to be partition regular (cf. [Ra], [GRS] or [F2]). The results involving polynomials brought forth in Sections 1-4 hint that there are some nonlinear equations that are partition regular too. For example, the equation $x - y = p(z)$ is partition regular for any $p(t) \in \mathbf{Z}[t]$, $p(0) = 0$. To see this, fix $p(t)$ and let $\mathbf{N} = \bigcup_{i=1}^{r} C_i$ be an arbitrary partition. Arguing as in [B2] one can show that one of the cells C_i, call it C, has the property that it contains an IP-set and has positive upper density. Let $\{n_\alpha\}_{\alpha \in \mathcal{F}}$ be an IP-set in C. According to Theorem 3.11, $\{p(n_\alpha)\}_{\alpha \in \mathcal{F}}$ is a set of recurrence. This together with Furstenberg's correspondence principle gives that for some $\alpha \in \mathcal{F}$,

$$\bar{d}\Big(C_i \cap \big(C_i - p(n_\alpha) \big) \Big) > 0.$$

If $y \in \big(C_i \cap (C_i - p(n_\alpha)) \big)$ then $x = y + p(n_\alpha) \in C_i$. This establishes the partition regularity of $x - y = p(z)$. In accordance with the third principle of Ramsey theory one should expect that there are actually many x, y, z having the same color and satisfying $x - y = p(z)$. This is indeed so: using the fact that $\{p(n_\alpha)\}_{\alpha \in \mathcal{F}}$ is a set of nice recurrence one can show, for example, that for any $\epsilon > 0$ and any partition $\mathbf{N} = \bigcup_{i=1}^r C_i$ one of C_i, $i = 1, 2, \cdots, r$ satisfies

$$\bar{d}\left(\left\{ z \in C_i : \bar{d}(C_i \cap (C_i - p(z))) \geq \left(\bar{d}(C_i)\right)^2 - \epsilon\right\}\right) > 0$$

(cf. [B2], see also Theorem 0.4 in [BM1]).

Question 11. Are the following systems of equations partition regular?
(i) $x^2 + y^2 = z^2$.
(ii) $xy = u$, $x + y = w$.
(iii) $x - 2y = p(z)$, $p(t) \in \mathbf{Z}[t]$, $p(0) = 0$.

The discussion in this survey so far has concentrated mainly on topological and measure preserving $\mathbf{Z}^n$-actions. Ergodic Ramsey theory of actions of more general, especially non-abelian groups is much less developed and offers many interesting problems.

In complete analogy with the case of the group $\mathbf{Z}$, given a semigroup G call a set $R \subset G$ a set of recurrence if for any measure preserving action $(T_g)_{g \in G}$ of G on a finite measure space $(X, \mathcal{B}, \mu)$ and for any $A \in \mathcal{B}$ with $\mu(A) > 0$ there exists $g \in R$, $g \neq e$, such that $\mu(A \cap T_g^{-1} A) > 0$. Different semigroups have all kinds of peculiar sets of recurrence. For example, one can show that the set $\{1 + \frac{1}{k} : k \in \mathbf{N}\}$ is a set of recurrence for the multiplicative group of positive rationals. Sets of the form $\{n^\alpha : n \in \mathbf{N}\}$, where $\alpha > 0$, are sets of recurrence for $(\mathbf{R}, +)$. As a matter of fact, one can show (see [BBB]) that for any measure preserving $\mathbf{R}$-action $(S^t)_{t \in \mathbf{R}}$ on a probability space $(X, \mathcal{B}, \mu)$ one has for every $A \in \mathcal{B}$ that

$$\lim_{N \to \infty} \frac{1}{N} \sum_{n=1}^{N} \mu(A \cap S^{n^\alpha} A) \geq \mu(A)^2.$$

On the other hand one has the following negative result.

Theorem 5.4 ([BBB]). Let $(S^t)_{t \in \mathbf{R}}$ be an ergodic measure preserving flow acting on a probability Lebesgue space $(X, \mathcal{B}, \mu)$. For all but countably many $\alpha > 0$ (in particular for all positive $\alpha \in (\mathbf{Q} \setminus \mathbf{Z})$) one can find an L^∞-function f for which the averages $\frac{1}{N} \sum_{n=1}^{N} f(S^{n^\alpha} x)$ fail to converge for a set of x of positive measure.

It is possible that the countable set of "good" α coincides with $\mathbf{N}$. Such a result would follow from a positive answer to the following number-theoretical question which we believe is of independent interest.

Question 12. Let us call an increasing sequence $\{a_n : n \in \mathbf{N}\} \subset \mathbf{R}$ *weakly independent* over $\mathbf{Q}$ if there exists an increasing sequence $(n_i)_{i=1}^\infty \subset \mathbf{N}$

having positive upper density such that the sequence $\{a_{n_i} : i \in \mathbf{N}\}$ is linearly independent over $\mathbf{Q}$. Is it true that for every $\alpha > 0$, $\alpha \notin \mathbf{N}$, the sequence $\{n^\alpha : n \in \mathbf{N}\}$ is weakly independent over $\mathbf{Q}$? (It is known that the answer is yes for all but countably many α.)

Definition 5.3. Given a (semi)group G, a set $R \subset G$ is called a set of topological recurrence if for any minimal action $(T_g)_{g \in G}$ of G on a compact metric space X and for any open, non-empty set $U \subset X$ there exists $g \in R$, $g \neq e$, such that $(U \cap T_g^{-1} U) \neq \emptyset$.

Exercise 20. Prove that in an amenable group any set of (measurable) recurrence is a set of topological recurrence.

An interesting result due to Kriz ([Kr], see also [Fo], [M]) says that in $\mathbf{Z}$ there are sets of topological recurrence which are not sets of measurable recurrence. While the same kind of result ought to hold in any abelian group, and while for any amenable group sets of measurable recurrence are, according to Exercise 20, sets of topological recurrence, the situation for more general groups is far from clear. We make the following

Conjecture. A group G is amenable if and only if any set of measurable recurrence $R \subset G$ is a set of topological recurrence.

An intriguing question is, what is the right formulation of the Szemerédi (or van der Waerden) theorem for general group actions. In this connection we want to mention a very nice noncommutative extension of Theorem 1.19 which was recently obtained by Leibman in [L2]: he was able to show that the conclusion of Theorem 1.19 holds if one replaces the assumption about the commutativity of the measure preserving transformations T_i by the demand that they generate a nilpotent group. He also proved earlier in [L1] a topological van der Waerden-type theorem of a similar kind. This should be contrasted with an example due to Furstenberg of a pair of homeomorphisms T_1, T_2 of a compact metric space X generating a metabelian group such that no point of X is simultaneously recurrent for T_1, T_2 (this implies that for metabelian groups one should look for another formulation of a Szemerédi-type theorem).

A possible way of extending multiple recurrence theorems to a situation involving non-commutative groups is to consider a finite family of pairwise commuting actions of a given group. Results obtained within such framework ought to be called *semicommutative*. We have the following

Conjecture. Assume that G is an amenable group with a Følner sequence $(F_n)_{n=1}^\infty$. Let $(T_g^{(1)})_{g \in G}, \cdots, (T_g^{(k)})_{g \in G}$ be k pairwise commuting measure preserving actions of G on a measure space $(X, \mathcal{B}, \mu)$ ("pairwise commuting" means here that for any $1 \leq i \neq j \leq k$ and any $g, h \in G$ one has $T_g^{(i)}T_h^{(j)} = T_h^{(j)}T_g^{(i)}$). Then for any $A \in \mathcal{B}$ with $\mu(A) > 0$ one has:

$$\liminf_{n \to \infty} \frac{1}{|F_n|} \sum_{g \in F_n} \mu(A \cap T_g^{(1)} A \cap T_g^{(1)} T_g^{(2)} A \cap \cdots \cap T_g^{(1)} T_g^{(2)} \cdots T_g^{(k)} A) > 0.$$

Remarks. (i) We have formulated the conjecture for amenable groups for two major reasons. First of all, the conjecture is known to hold true for $k = 2$ ([BMZ], see also [BR]). Second, in case the group G is countable, a natural analogue of Furstenberg's correspondence principle, which was formulated in Section 1, holds and allows one to obtain combinatorial corollaries which, should the conjecture turn out to be true for any k, contain Szemerédi's theorem as quite a special case.

(ii) The "triangular" expressions

$$A \cap T_g^{(1)} A \cap T_g^{(1)} T_g^{(2)} A \cap \cdots \cap T_g^{(1)} T_g^{(2)} \cdots T_g^{(k)} A$$

appearing in the formulation of the conjecture seem to be the "right" configurations to consider. See the discussion and counterexamples in [BH2] where a topological analogue of the conjecture is treated (but not fully resolved). We suspect that the answer to the following question is, in general, negative.

Question 13. Given an amenable group G and a Følner sequence $(F_n)_{n=1}^{\infty}$ for G, let $(T_g)_{g \in G}$ and $(S_g)_{g \in G}$ be two commuting measure preserving actions on a probability space $(X, \mathcal{B}, \mu)$. Is it true that for any $A \in \mathcal{B}$ the following limit exists:

$$\lim_{n \to \infty} \frac{1}{|F_n|} \sum_{g \in F_n} \mu(T_g A \cap S_g A)?$$

We want to conclude by formulating a conjecture about a density version of the polynomial Hales-Jewett theorem which would extend both the partition results from [BL2] and the density version of the ("linear") Hales-Jewett theorem proved in [FK4]. For $q, d, N \in \mathbf{N}$ let $\mathcal{M}_{q,d,N}$ be the set of q-tuples of subsets of $\{1, 2, \cdots, N\}^d$:

$$\mathcal{M}_{q,d,N} = \{(\alpha_1, \cdots, \alpha_q) : \alpha_i \subset \{1, 2, \cdots, N\}^d, \, i = 1, 2, \cdots, q\}.$$

Conjecture. For any $q, d \in \mathbf{N}$ and $\epsilon > 0$ there exists $C = C(q, d, \epsilon)$ such that if $N > C$ and a set $S \subset \mathcal{M}_{q,d,N}$ satisfies $\frac{|S|}{|\mathcal{M}_{q,d,N}|} > \epsilon$ then S contains a "simplex" of the form:

$$\{(\alpha_1, \alpha_2, \cdots, \alpha_q), (\alpha_1 \cup \gamma^d, \alpha_2, \cdots, \alpha_q), (\alpha_1, \alpha_2 \cup \gamma^d, \cdots, \alpha_q),$$
$$\cdots, (\alpha_1, \alpha_2, \cdots, \alpha_q \cup \gamma^d)\},$$

where $\gamma \subset \mathbf{N}$ is a non-empty set and $\alpha_i \cap \gamma^d = \emptyset$ for all $i = 1, 2, \cdots, q$.

Remark. For $d = 1$ the conjecture follows from [FK4]. This paper contains a wealth of related material and is strongly recommended for rewarding reading.

REFERENCES

[Ba] J.E. Baumgartner, A short proof of Hindman's theorem, *J. Combinatorial Theory (A)* **17** (1974), 384-386.

[B1] V. Bergelson, Sets of recurrence of $\mathbf{Z}^m$-actions and properties of sets of differences in $\mathbf{Z}^m$, *J. London Math. Soc.* (2) **31** (1985), 295-304.

[B2] V. Bergelson, A density statement generalizing Schur's theorem, *J. Combinatorial Theory (A)* **43** (1986), 338-343.

[B3] V. Bergelson, Ergodic Ramsey Theory, *Logic and Combinatorics* (edited by S. Simpson), *Contemporary Mathematics* **65** (1987), 63-87.

[B4] V. Bergelson, Weakly mixing PET, *Ergodic Theory and Dynamical Systems*, **7** (1987), 337-349.

[BBB] V. Bergelson, M. Boshernitzan, and J. Bourgain, Some results on non-linear recurrence, *Journal d'Analyse Mathématique*, **62** (1994), 29-46.

[BeR] V. Bergelson and J. Rosenblatt, Joint ergodicity for group actions, *Ergodic Theory and Dynamical Systems* (1988) **8**, 351-364.

[BFM] V. Bergelson, H. Furstenberg and R. McCutcheon, IP-sets and polynomial recurrence, *Ergodic Theory and Dynamical Systems* (to appear).

[BHå] V. Bergelson and I.J. Håland, Sets of recurrence and generalized polynomials, to appear in *The Proceedings of the Conference on Ergodic Theory and Probability, June 1993*, Ohio State University Math. Research Institute Publications, de Gruyter, 1995.

[BH1] V. Bergelson and N. Hindman, Nonmetrizable topological dynamics and Ramsey theory, *Trans. AMS* **320** (1990), 293-320.

[BH2] V. Bergelson and N. Hindman, Some topological semicommutative van der Waerden type theorems and their combinatorial consequences, *J. London Math. Soc.* (2) **45** (1992) 385-403.

[BL1] V. Bergelson and A. Leibman, Polynomial extensions of van der Waerden's and Szemerédi's theorems, *Journal of AMS* (to appear).

[BL2] V. Bergelson and A. Leibman, Set polynomials and a polynomial extension of Hales-Jewett theorem. Preprint.

[BM1] V. Bergelson and R. McCutcheon, Uniformity in polynomial Szemerédi theorem, these Proceedings.

[BM2] V. Bergelson and R. McCutcheon, An IP polynomial Szemerédi theorem for finite families of commuting transformations, in preparation.

[BMZ] V. Bergelson, R. McCutcheon and Q. Zhang, A Roth theorem for amenable groups (to be submitted).

[Bo1] H. Bohr, Zur theorie der fastperiodischen functionen, I, *Acta Math.* **45** (1924), 29-127.

[Bo2] N. Bogolioùboff, Sur quelques propriétés arithmetiques des presque-périodes, *Ann. Phy. Math. Kiew* **4** (1939).

[Bo3] J. Bourgain, Pointwise ergodic theorems for arithmetic sets, *Publ. Math. IHES* **69** (1989), 5-45.

[Bo4] J. Bourgain, Double recurrence and almost sure convergence, *J. Reine Angew. Math.* **404** (1990), 140-161.

[Br] J.R. Brown, *Ergodic Theory and Topological Dynamics*, Academic Press, 1976.

[C] W. Comfort, Ultrafilters–some old and some new results, *Bull. AMS* **83** (1977), 417-455.

[CL1] J. Conze and E. Lesigne, Théorèmes ergodiques pour des mesures diagonales, *Bull. Soc. Math. Fr.* **112** (1984), 143-175.

[CL2] J. Conze and E. Lesigne, Sur un théorème ergodique pour des mesures diagonales, *C. R. Acad. Sci. Paris, t.* **306**, I (1988), 491-493.

[CN] W. Comfort and S. Negrepontis, *The Theory of Ultrafilters*, Springer-Verlag, Berlin and New York, 1974.

[CY] P. Civin and B. Yood, The second conjugate space of a Banach algebra as an algebra, *Pac. J. Math.* **11** (1961), 847-870.

[D] E. van Douwen, The Čech-Stone compactification of a discrete groupoid, *Topology and its applications* **39** (1991), 43-60.

[E1] R. Ellis, Distal transformation groups, *Pac. J. Math.* **8** (1958), 401-405.

[E2] B. Ellis, Solution of Problem 6612, *Amer. Math. Monthly* **98** (1991), 957-959.

[ET] P. Erdös and P. Turán, On some sequences of integers, *J. London Math. Soc.* **11** (1936), 261-264.

[F1] H. Furstenberg, Ergodic behavior of diagonal measures and a theorem of Szemerédi on arithmetic progressions, *J. d'Analyse Math.* **31** (1977), 204-256.

[F2] H. Furstenberg, *Recurrence in Ergodic Theory and Combinatorial Number Theory*, Princeton University Press, 1981.

[F3] H. Furstenberg, Poincaré recurrence and number theory, *Bull. AMS (New Series)* **5** (1981), 211-234.

[FK1] H. Furstenberg and Y. Katznelson, An ergodic Szemerédi theorem for commuting transformations, *J. d'Analyse Math.* **34** (1978), 275-291.

[FK2] H. Furstenberg and Y. Katznelson, An ergodic Szemerédi theorem for IP-systems and combinatorial theory, *J. d'Analyse Math.* **45** (1985), 117-168.

[FK3] H. Furstenberg and Y. Katznelson, Idempotents in compact semigroups and Ramsey theory, *Israel J. Math.* **68** (1990), 257-270.

[FK4] H. Furstenberg and Y. Katznelson, A density version of the Hales-Jewett theorem, *J. d'Analyse Math.* **57** (1991), 64-119.

[Fo] A.H. Forrest, The construction of a set of recurrence which is not a set of strong recurrence, *Israel J. Math.* **76** (1991), 215-228.

[Fø] E. Følner, Bemaerkning om naestenperiodieitetus definition, *Mat. Tidsskrift B*, 1944.

[FW1] H. Furstenberg and B. Weiss, Topological dynamics and combinatorial number theory, *J. d'Analyse Math.* **34** (1978), 61-85.

[FW2] H. Furstenberg and B. Weiss, A mean ergodic theorem for $\frac{1}{N}\sum_{n=1}^{N} f(T^n x)g(T^{n^2} x)$, to appear in *The Proceedings of the Conference on Ergodic Theory and Probability, June 1993*, Ohio State University Math. Research Institute Publications, de Gruyter, 1995.

[G] J. Gillis, Note on a property of measurable sets, *J. London Math. Soc.* **11** (1936), 139-141.

[GJ] L. Gillman and M. Jerison, *Rings of continuous functions*, Springer-Verlag.

[GRS] R. Graham, B. Rothschild and J. Spencer, *Ramsey Theory*, Wiley, New York, 1980.

[Ha] G.H. Hardy, *A Mathematician's Apology*, Cambridge University Press, 1990.

[H1] D. Hilbert, Über die Irreduzibilität ganzer rationaler Funktionen mit ganzzahligen Koeffizienten, *J. Math.* **110** (1892), 104-129.

[H2] N. Hindman, Finite sums from sequences within cells of a partition of **N**, *J. Combinatorial Theory* (Series A) **17** (1974) 1-11.

[H3] N. Hindman, Ultrafilters and combinatorial number theory, *Number Theory Carbondale*, M. Nathanson, ed., *Lecture Notes in Math.* **751** (1979) 119-184.

[H4] N. Hindman, Partitions and sums and products of integers, *Trans. AMS* **247** (1979), 227-245.

[H5] N. Hindman, The semigroup β**N** and its applications to number theory, *The Analytical and Topological Theory of Semigroups*, K. H. Hoffman, J. D. Lawson and J. S. Pym, editors, de Gruyter, 1990.

[HJ] A.W. Hales and R.I. Jewett, Regularity and positional games, *Trans. AMS* **106** (1963), 222-229.

[Ho] E. Hopf, *Ergodentheorie*, Chelsea, New York, 1948.

[HR] E. Hewitt and K. Ross, *Abstract Harmonic Analysis I*, Springer-Verlag, 1963.

[K1] A. Y. Khintchine, *Three Pearls of Number Theory*, Graylock Press, Rochester, N.Y., 1952.

[K2] A. Y. Khintchine, Eine Verschärfung des Poincaréschen Wiederkehrsatzes, *Comp. Math.* **1** (1934), 177-179.

[KB] N. Krylov and N. Bogoliouboff, La théorie générale de la measure dans son application à l'étude des systèmes dynamiques de la mécanique non linéare, *Ann. of Math.* **38** (1937), 65-113.

[KN] B. O. Koopman and J. von Neumann, Dynamical systems of continuous spectra, *Proc. Nat. Acad. Sci. U.S.A.* **18** (1932), 255-263.

[Kr] I. Kriz, Large independent sets in shift-invariant graphs. Solution of Bergelson's problem, *Graphs and Combinatorics* **3** (1987), 145-158.

[L1] A. Leibman, Multiple recurrence theorem for nilpotent group actions, *Geom. and Funct. Anal.* **4** (1994), 648-659.

[L2] A. Leibman, Multiple recurrence theorems for nilpotent group actions, Ph.D. dissertation, Technion, Haifa, 1995.

[M] R. McCutcheon, Three results in recurrence, *Ergodic Theory and its Connections with Harmonic Analysis* (edited by K. Petersen and I. Salama), *London Math. Soc. Lecture Notes Series* **205**, Cambridge University Press, 1995, 349-358.

[Mao] Mao Tsetung, *Selected Readings from the Works of Mao Tsetung*, Foreign Language Press, Peking, 1971.

[N] K. Numakura, On bicompact semigroups, *Math. J. of Okayama University* **1** (1952), 99-108.

[Pa] W. Parry, *Topics in Ergodic Theory*, Cambridge Univ. Press, Cambridge, 1981.

[Pe] K. Petersen, *Ergodic Theory*, Cambridge Univ. Press, Cambridge, 1981.

[Ra] R. Rado, Note on combinatorial analysis, *Proc. London Math. Soc.* **48** (1993), 122-160.

[Ro] K. Roth, Sur quelques ensembles d'entiers, *C.R. Acad. Sci. Paris* **234** (1952), 388-390.

[S] A. Sárközy, On difference sets of integers III, *Acta. Math. Acad. Sci. Hungar.*, **31** (1978) 125-149.

[Sz1] E. Szemerédi, On sets of integers containing no four elements in arithmetic progression, *Acta Math. Acad. Sci. Hungar.* **20** (1969), 89-104.

[Sz2] E. Szemerédi, On sets of integers containing no k elements in arithmetic progression, *Acta Arith.* **27** (1975), 199-245.

[W] N. Wiener, *Time Series*, The M.I.T. Press, 1949.

Flows on homogeneous spaces: a review

S.G. Dani

School of Mathematics
Tata Institute of Fundamental Research
Homi Bhabha Road
Bombay 400 005, India

Introduction

Let G be a Lie group and C be a closed subgroup of G. The quotient G/C is also called a homogeneous space. To each $a \in G$ corresponds a translation T_a given by $xC \mapsto axC$. More generally any subgroup A of G yields an action on G/C where each $a \in A$ acts as the translation T_a as above. Similarly, given an automorphism ϕ of G such that $\phi(C) = C$ we can define a transformation $\bar{\phi} : G/C \to G/C$ by $\bar{\phi}(xC) = \phi(x)C$ for all $x \in G$; these transformations are called automorphisms of G/C. The composite transformations of the form $T_a \circ \bar{\phi}$, where T_a is a translation and $\bar{\phi}$ is an automorphism, are called affine automorphisms.

In the sequel we will be interested mainly in the homogeneous spaces G/C which admit a finite Borel measure invariant under the action of G. For a Lie group G we shall denote by $\mathcal{F}(G)$ the class of all closed subgroups C such that G/C admits a finite G-invariant measure. Discrete subgroups from the class $\mathcal{F}(G)$ are called *lattices*.

The transformations and actions as above form a rich class of 'dynamical systems'. The study of the asymptotic behaviour of their orbits is of great significance not only because of their intrinsic appeal but also on account of various applications to varied subjects such as Number theory, Geometry, Lie groups and their subgroups etc. As may be expected the systems have been studied from various angles; most concepts of dynamics have been considered in the context of these systems, with varying degrees of success in understanding them. My aim in this article is to give an exposition of the results concerning mainly the asymptoptic behaviour of the orbits such as their closures, their distribution in the ambient space etc. and discuss their applications, especially to problems of diophantine approximation.

§1 Homogeneous spaces - an overview

Let me begin with some examples which are particular cases of the general class of systems introduced above. I will also recall special features of various classes of homogeneous spaces, which it would be convenient to bear in mind in the sequel.

1.1. Rotations of the circle: Let $G = I\!R$ and $C = \mathbf{Z}$. Then $G/C = I\!R/\mathbf{Z} = I\!T$, the circle group, and the translations in the above sense are the usual rotations of the circle. Also, the usual angle measure is invariant under all rotations. (Alternatively one can think of the same systems setting $G = I\!T$ and C to be the trivial subgroup). The translation corresponding to an irrational number t is called an 'irrational rotation'; it is a rotation by an angle which is an irrational multiple of π.

1.2. Affine transformations of tori: By a torus we mean the quotient $I\!R^n/\mathbf{Z}^n$ or equivalently the cartesian power $I\!T^n$, n being any natural number. Any $A \in GL(n, \mathbf{Z})$, namely an integral $n \times n$ matrix with determinant ± 1 defines an automorphism $\overline{A}$ of $I\!R^n/\mathbf{Z}^n$ by $\overline{A}(v + \mathbf{Z}^n) = (Av) + \mathbf{Z}^n$, A being viewed as a linear transformation via the standard basis of $I\!R^n$. Conversely any automorphism of $I\!R^n/\mathbf{Z}^n$ arises in this way. Composing translations and automorphisms we get what are called affine automorphisms; see [CFS] and [Wa] for details.

1.3. Nilflows and solvflows: Let G be a nilpotent Lie group and let $C \in \mathcal{F}(G)$. Then G/C is called a nilmanifold and the action of any subgroup of G on G/C is called a nilflow. Similarly when G is solvable G/C is called a solvmanifold and the action of any subgroup on G/C is called a solvflow.

A solvmanifold G/C admits a finite G-invariant measure if and only if it is compact (see [R], Theorem 3.1). If G is nilpotent then for any $C \in \mathcal{F}(G)$, the connected component C^0 of the identity in C is a normal subgroup of G (see [R], Corollary 2 of Theorem 2.3); in this case G/C can be viewed as a homogeneous space of the Lie group G/C^0, by the subgroup C/C^0; as the latter is a discrete subgroup, while studying the flows as above for nilpotent Lie groups there is no loss of generality in assuming C to be discrete, namely a lattice.

Let me also recall here that if G is a nilpotent Lie group and C is a lattice in G then $[G, G]C$ is a closed subgroup; see [Mal] and [R] for detailed results on the structure of lattices in nilpotent Lie groups. The factor homogeneous space $G/[G, G]C$ is called the maximal torus factor of G/C. Many properties of flows on G/C turn out to be characterisable in terms of the corresponding (factor) action on the torus factor.

In the year 1960-61 a conference on *Analysis in the large* was held at Yale University and a number of interesting results on nilflows and solvflows came out as a result of the interaction of the participants; see [AGH]. The study was pursued later extensively by L. Auslander and other authors, proving several results on the structure of solvmanifolds and the dynamical behaviour of orbits of solvflows (see [Au1], [Au2]). Through the following sections I shall recall only those of the results which are closely related to the theme of the exposition.

A particular class of solvflows which are not nilflows turns out to be of special interest especially on account of certain results of J. Brezin and C. C. Moore [BM]. A solvable Lie group is called *Euclidean* if it is a covering group of a Lie group of the form $K \cdot V$, semidirect product, where V is a vector group (topologically isomorphic to $\mathbb{R}^n$ for some n) and K is a compact abelian subgroup of $GL(V)$ (acting as automorphisms of V). A compact solvmanifold is called *Euclidean* if it can be written as a homogeneous space of a Euclidean solvable group. A compact solvmanifold is Euclidean if and only if it is finitely covered by a torus. A general compact solvmanifold admits a unique maximal Euclidean quotient; that is, given a solvable Lie group G and a closed subgroup C of G such that G/C is compact there exists a unique minimal closed subgroup L containing C, such that G/L is Euclidean; if M is the largest closed normal subgroup of G contained in L (the intersection of all conjugates of L) then G/M is a Euclidean solvable group; see [BM] for details on these and other properties of Euclidean solvmanifolds. In many ways this Euclidean factor plays the same role for solvmanifolds as the maximal torus factor for nilmanifolds.

1.4. Geodesic and horocycle flows: Let G be the group $PSL(2, \mathbb{R}) = SL(2, \mathbb{R})/\{\pm I\}$, where I is the identity matrix, and Γ be a discrete subgroup of G. Suppose that Γ contains no elements of finite order. Then G/Γ can be realised canonically as the space of unit tangents of the surface $S = \mathbb{H}/\Gamma$, where $\mathbb{H}$ is the upper half plane and G acts as its group of isometries with respect to the Poincaré metric (see [B]). Further, the geodesic flow corresponding to S is given (after suitable identification) by the action of the one-parameter subgroup of G which is the image of the subgroup $D = \{\text{diag}(e^t, e^{-t}) | t \in \mathbb{R}\}$ in $PSL(2, \mathbb{R})$. Conversely any surface of constant negative curvature can be realised as $\mathbb{H}/\Gamma$ for a discrete subgroup Γ of $PSL(2, \mathbb{R})$ with no element of finite order (namely the fundamental group of S) and the associated geodesic flow corresponds to the action on G/Γ of the one-parameter subgroup as above. Thus the geodesic flows belong to the class of flows on homogeneous spaces. Also, Γ is a lattice in G if and only if the surface is of finite Riemannian area.

A particular flow plays a crucial role in the classical study of the geodesic flow of a surface of constant negative curvature, by G.A. Hedlund, E. Hopf and others. It is called the horocycle flow; to be precise there are two horocycle flows associated to each geodesic flow, the so called 'contracting' and 'expanding' horocycle flows, but viewed as flows themselves they are equivalent. In our notation the flow can be given as the action of the (the image in G of the) one-parameter subgroup $\{\begin{pmatrix} 1 & t \\ 0 & 1 \end{pmatrix} \mid t \in I\!R\}$, on G/Γ as above; when G/Γ is the unit tangent bundle of a surface of constant negative curvature this action coincides with the classical horocycle flow defined geometrically (see [He], [Man], [Ve], [Gh]).

Geodesic flows on n-dimensional Riemannian manifolds of constant negative curvature and finite (Riemannian) volume were studied through flows on homogeneous spaces by Gelfand and Fomin [GF]: Let $G = SO(n, 1)$, the special orthogonal group of a quadratic form of signature $(n, 1)$, Γ be a lattice in G and $\{g_t\}$ be a one-parameter subgroup of G whose action on the Lie algebra of G is diagonalisable over $I\!R$ (there exists such a one-parameter subgroup and it is unique upto conjugacy and scaling). Let M be a compact subgroup of G centralised by g_t for all $t \in I\!R$. The flow induced by $\{g_t\}$ on G/Γ factors to a flow on $M\backslash G/\Gamma$. Any geodesic flow of a manifold of constant negative curvature and finite Riemannian volume can be realised as such a flow, with M as the unique maximal compact subgroup of the centraliser of $\{g_t\}$ and Γ a suitable lattice; this may be compared with the one-dimensional case above. In a similar vein the geodesic flows on all locally symmetric spaces were considered by F. Mautner [Mau].

1.5. Modular homogeneous space: One particular homogeneous space and the flows on it are of great significance in many applications to Number theory. This consists simply of taking $G = SL(n, I\!R)$ and $\Gamma = SL(n, \mathbb{Z})$, the subgroup consisting of integral matrices in G. The homogeneous space has a finite G-invariant measure; that is, Γ is a lattice in G (see [R], Corollary 10.5). On the other hand (unlike the lattices in solvable groups) the quotient space is noncompact. It turns out that as far as behaviour of orbits of flows and the proofs of the results are concerned, these examples involve most of the intricacies of the general case.

The following model for the above homogeneous space is one of the main reasons for the interest in it from a number-theoretic point of view. Consider the set $\mathcal{L}_n$ of lattices in $I\!R^n$ with unit discriminant (volume of any fundamental domain for the lattice). The action of $SL(n, I\!R)$ on $I\!R^n$ induces an action on $\mathcal{L}_n$ which can be readily seen to be transitive. Further, $SL(n, \mathbb{Z})$ is precisely the isotropy subgroup for the lattice generated by the standard basis. Hence via the action $\mathcal{L}_n$ may be identified with $SL(n, I\!R)/SL(n, \mathbb{Z})$. On $\mathcal{L}_n$ there is an intrinsically defined topology, in which two lattices are near if and only if they have bases which are near (see [Ca] for details); this topology can be seen to correspond to the quotient topology on $SL(n, I\!R)/SL(n, \mathbb{Z})$ under the identification as above. A well-known criterion due to K. Mahler asserts that a sequence of lattices $\{\Lambda_i\}$ in $\mathcal{L}_n$ is divergent (has no convergent subsequence) if and only if there exists a sequence $\{x_i\}$ in $\mathbb{R}^n - (0)$ with $x_i \in \Lambda_i$ for all i and $x_i \to 0$ as $i \to \infty$ (see [R], Corollary 10.9). The invariant measure on $SL(n, \mathbb{R})/SL(n, \mathbb{Z})$ is also related to the Lebesgue measure on $\mathbb{R}^n$ in an interesting way (see [Si]), which plays an important role in certain applications to diophantine approximation (e.g. Corollary 8.3, below).

1.6. Lattices in real algebraic groups: A subgroup H of $GL(n, \mathbb{R})$, $n \geq 2$, is said to be algebraic if there exists a set $\mathcal{P}$ of polynomials in the n^2 (coordinate) variables such that $H = \{g = (g_{ij}) \in GL(n, \mathbb{R}) \mid P(g_{ij}) = 0 \text{ for all } P \in \mathcal{P}\}$; further H is said to be defined over $\mathbb{Q}$ if $\mathcal{P}$ can be chosen to consist of polynomials with rational coefficients. A theorem of Borel and Harish-Chandra gives a necessary and sufficient condition for $H \cap GL(n, \mathbb{Z})$, where H is an algebraic subgroup of $GL(n, \mathbb{R})$ defined over $\mathbb{Q}$, to be a lattice in H. The condition involves there being no "nontrivial characters defined over $\mathbb{Q}$". Without going into the technical definitions let me only mention that, in particular, if H is an algebraic subgroup defined over $\mathbb{Q}$ such that the unipotent matrices contained in H and a compact subgroup of H together generate H then $H \cap GL(n, \mathbb{Z})$ is a lattice in H. This gives an abundant class of homogeneous spaces. Also, as we shall see later, they are of importance in the study of orbit closures of unipotent one-parameter subgroups, in the case of the modular homogeneous spaces as above.

1.7. Homogeneous spaces of semisimple Lie groups: We shall now look at another class of homogeneous spaces, of which those in Sections 1.4 and 1.5 are particular cases. Let G be a semisimple Lie group and $C \in \mathcal{F}(G)$. Then G has a unique maximal compact connected normal subgroup, say G_0. Clearly G_0C is a closed subgroup of G and G/G_0C is a homogeneous space which is a factor of G/C. In studying flows on G/C, in many ways it is adequate to consider the factor actions on G/G_0C. Thus it is enough to consider homogeneous spaces of groups which have no nontrivial compact factor groups (namely such that any surjective homomorphism on to a compact group is trivial), a condition which holds for the quotient group G/G_0 as above. (By a theorem of H. Weyl the condition is equivalent to there being no nontrivial compact connected normal subgroup in G; it is however customary to express the condition in terms of factors rather than normal subgroups).

Now let G be a semisimple Lie group with no nontrivial compact factors and let $C \in \mathcal{F}(G)$. By Borel's density theorem (see [R]; see also [D6] and [D8] for more general versions of Borel's density theorem) C^0, the connected component of the identity in C is a normal subgroup of G. Hence (as seen in the case of nilpotent Lie groups) there is no loss of genetality in assuming that C is a lattice. Further, Borel's density theorem also shows that any lattice contains a subgroup of finite index in the center of G (see [R], Corollary 5.17). Hence passing to a quotient the center may be assumed to be finite.

Lattices in semisimple Lie groups with no nontrivial compact factors can be 'decomposed' into 'irreducible' lattices: A lattice Γ is said to be *irreducible* if for any closed normal subgroup F of positive dimension $F\Gamma$ is dense in L. The assertion about decomposition is the following: given a semisimple Lie group G with no nontrivial compact factors and a lattice Γ in G there exist closed normal subgroups $G_1, ..., G_k$, for some $k \geq 1$, such that G is locally the direct product of $G_1, ..., G_k$ (that is, $G = G_1 G_2 \cdots G_k$ and the pairwise intersections of G_i's are discrete central subgroups) and $G_i \cap \Gamma$ is an irreducible lattice in G_i for each $i = 1, ..., k$ (see [R], Theorem 5.22); if Γ' is the product of $G_i \cap \Gamma$, $i = 1, ..., k$, then it follows that Γ' is a lattice in G and hence, in particular, of finite index in Γ. Thus, upto a combination of finite coverings and factoring modulo finite central subgroups the homogeneous space G/Γ is a product of the spaces $G_i/G_i \cap \Gamma$, in each of which the lattice involved is irreducible. It may also be noted that for each i, $G/(\Pi_{j \neq i} G_j)\Gamma$ is an irreducible factor of G/Γ.

It is an important fact about lattices in semisimple Lie groups G that if G has trivial center and no compact factors and R-rank at least 2 then any irreducible lattice in G is 'arithmetic'. I will not go into the details of this (see [Z2] and [Mar5] for general theory in this respect); though it plays a role in some proofs, it will not be directly involved in the discussion of the results. Let me only mention that it means that the class of homogeneous spaces corresponding to the lattices as above is closely related to the examples in section 1.6 above.

1.8. General homogeneous spaces: Just as a general Lie group is studied via its special subgroups and factors which are simpler, homogeneous spaces and flows on them are also studied by reducing to simpler cases which I have discussed above. There are various results which make this possible. While it is tempting to recall here some of the results are involved in the sequel, they would probably seem rather technical and devoid of context at this point. I will therefore recall them only as and when necessary. The reader is referred especially to [R], [Au1], [Au2], [BM] and [Mar7] for the general theory.

§2 Ergodicity

Let (X, μ) be a measure space and H be a group which acts (measurably) on X preserving the measure μ. The action is said to be *ergodic* if for any measurable set E such that $\mu(hE \Delta E) = 0$ for all $h \in H$ (namely almost H-invariant) either $\mu(E) = 0$ or $\mu(X - E) = 0$. When X and H are locally compact second countable topological spaces and the action is continous, as will be the case in the sequel, for any Borel measure μ ergodicity is equivalent to either $\mu(E)$ or $\mu(X - E)$ being 0 for every H-invariant Borel set E (namely E such that $hE = E$ for all $h \in H$).

The ergodicity condition is of importance in the study of asymptotic behaviour of orbits in view of the following results.

2.1. Lemma (Hedlund): *Let X be a second countable Hausdorff topological space and let H be a group of homoemorphisms of X. Let μ be a Borel measure on X invariant under the action of H. Suppose that $\mu(\Omega) > 0$ for all nonempty open subsets of X. Then there exists a Borel subset Y such that $\mu(X - Y) = 0$ and for all $y \in Y$ the orbit Hy is dense in X.*

Proof: It is easy to see that if $\{\Omega_i\}_{i=1}^{\infty}$, is a basis for the topology then the assertion holds for the set $Y = \bigcap_i H\Omega_i$.

Remark: By a similar argument one can also see that if H as in the lemma is cyclic, say generated by a homeomorphism ϕ, then there exists a set Z of full measure such that for $z \in Z$, $\{\phi^i(z)|i = 1, 2, ...\}$ is dense in X.

Thus when there is ergodicity with respect to a measure of full support then the orbits of almost all points are dense. A considerably stronger implication of ergodicity to asymptotic behaviour of orbits comes from the following so called individual ergodic theorem of Birkhoff.

2.2. Theorem: *Let (X, μ) be a measure space such that $\mu(X) = 1$. Then for any measurable transformation ϕ preserving the measure μ and any $f \in L^1(X, \mu)$ the sequence of functions*

$$S_k(f)(x) = \frac{1}{k}\Sigma_{i=0}^{k-1} f(\phi^i(x))$$

converges almost everywhere and in L^1.

Similarly, if $\{\phi_t\}$ is a measurable flow on X (that is, $\{\phi_t\}_{t \in \mathbb{R}}$ is a one-parameter group of measurable transformations of X and $(t, x) \rightarrow \phi_t(x)$ is measurable) preserving μ then for any $f \in L^1(X, \mu)$ the functions S_T, $T > 0$, defined by

$$S_T(f)(x) = \frac{1}{T}\int_0^T f(\phi_t(x))\mathrm{d}t$$

converge almost everywhere and in L^1, as $T \rightarrow \infty$.

In either of the cases the limit f^* is invariant under the action (that is, $f^*(\phi(x)) = f^*(x)$ a.e. or $f^*(\phi_t(x)) = f(x)$ a.e. for all $t \in \mathbb{R}$, respectively) and $\int f^* \mathrm{d}\mu = \int f \mathrm{d}\mu$.

Observe that when the action is ergodic with respect to μ then for any f as in the theorem f^* has to be the constant $\int f \mathrm{d}\mu$ a.e.. The convergence of averages as above for a point x signifies how the orbit of x (under the transformation ϕ or the flow $\{\phi_t\}$) is 'distributed' in the ambient space. This will be formulated rigorously later. Here let me discuss only a special case that is typical for orbits of an ergodic transformation (or flow). Let X be a locally compact second countable space. A sequence $\{x_i\}$ in X is said to be *uniformly distributed* with respect to a measure μ on X if for any bounded continuous function f on supp μ (the support of μ)

$$\frac{1}{k}\Sigma_{i=0}^{k-1} f(x_i) \longrightarrow \int_X f \, \mathrm{d}\mu,$$

as $k \to \infty$. Similarly a curve $\{x_t\}_{t\geq0}$ in X is said to be uniformly distributed with respect to a measure μ on X if

$$\frac{1}{T}\int_0^T f(x_t)\, dt \longrightarrow \int_X f \, \mathrm{d}\mu,$$

as $T \to \infty$, for all bounded continuous functions f on supp μ. It can be verified that if a sequence $\{x_i\}$ is uniformly distributed with respect to a measure μ with full support then for any Borel subset E such that the boundary of E in X is of μ measure 0,

$$\frac{\#\{i \,|\, 0 \leq i \leq n - 1 \text{ and } x_i \in E\}}{n} \longrightarrow \mu(E),$$

as $n \to \infty$ (# stands for the cardinality of the set following the symbol). A similar assertion also holds for a uniformly distributed curve $\{x_t\}$, with the cardinality replaced by the Lebesgue measure of the set of t such that $x_t \in E$. The ergodic theorem readily implies the following:

2.3. Corollary: *Let X be a locally compact second countable Hausdorff space and let ϕ be a homeomorphism of X (respectively, let $\{\phi_t\}$ be a continuous one-parameter flow on X). Let μ be a probability measure invariant and ergodic with respect to ϕ (respectively $\{\phi_t\}$). Then there exists a Borel subset Y of X such that $\mu(X - Y) = 0$ and for any $y \in Y$, the sequence $\{\phi^i(y)\}$ (respectively the curve $\{\phi_t(y)\}_{t\geq0}$) is uniformly distributed with respect to the measure μ.*

Let me now come to the question of proving ergodicity. To begin with let us consider a general set up. Let (X, μ) be a measure space with $\mu(X) < \infty$ and let T be an invertible measure-preserving transformation of X. Then one defines a unitary operator U_T on $L^2(X, \mu)$ by setting $(U_T f)(x) = f(T^{-1}x)$ for all $f \in L^2(X, \mu)$ and $x \in X$. It is easy to see that T is ergodic if and only if any $f \in L^2(X, \mu)$ such that $U_T f = f$ a.e. is constant a.e. or, equivalently, 1 is an eigenvalue of U_T with multiplicity one. This observation has been one of the major tools of proving ergodicity for many systems, ever since it was introduced by Koopman in 1931 (see [H] for some historical details). It is easy to apply it to the class of rotations of the circle, using Fourier series, and conclude that a rotation is ergodic (with respect to the angle measure) if and only if it is an irrational rotation. Similarly for affine transformations of tori using Fourier expansions one proves the following ([CFS], [H], [Wa]).

2.4. Theorem: *Let $\phi = T_a \circ \overline{A}$ be an affine transformation of $\mathbb{T}^n$, $n \geq 2$, where $a \in \mathbb{T}^n$ and $A \in GL(n, \mathbb{Z})$. Then ϕ is ergodic with respect to the Haar measure on $\mathbb{T}^n$ if and only if the following holds: no root of unity other than 1 is an eigenvalue of A and the subgroup generated by a and $\{x^{-1}\phi(x) \mid x \in \mathbb{T}^n\}$ is dense in $\mathbb{T}^n$.*

When a is the identity element the condition for ergodicity reduces to no root of unity being an eigenvalue of A.

Theorem 2.4 generalises to nilmanifolds in a simple form as follows (see [Gr] and [P]):

2.5. Theorem: *Let G be a nilpotent Lie group and Γ be a lattice in G. An affine automorphism ϕ of G/Γ is ergodic if and only if its factor $\overline{\phi}$ on $G/[G,G]\Gamma$ is ergodic.*

Let us now consider a general homogeneous space $X = G/C$ where G is a Lie group and $C \in \mathcal{F}(G)$. Through the rest of the section we shall consider only actions of subgroups of G, by translations; results for groups of affine automorphisms can be deduced with more technical work which we shall not go into. Let m denote the G-invariant probability measure on G/C. Then we can define a unitary representation of G over $L^2(X, m)$ by $(U_g f)(x) = f(g^{-1}x)$ for all $g \in G$, $f \in L^2(X, m)$ and $x \in X$. Observe that for $a \in G$, U_a is the unitary operator associated to the translation T_a in the sense as above. Thus to prove T_a to be ergodic it is enough to show that U_a does not fix any nonzero vector orthogonal to all constant functions in $L^2(X, m)$. In doing this one can use the representation $g \mapsto U_g$ of G. A detailed study along these lines has led to satisfactory criteria for ergodicity of subgroup actions. Before going to the general results let me describe some simple observations which enable proving ergodicity for a fairly wide class of actions.

2.6. Lemma (Mautner phenomenon): *Let G be a topological group and $g \mapsto U_g$ be a (strongly continuous) unitary representation of G over a Hilbert space $\mathcal{H}$. Let $a, b \in G$ be such that $a^i b a^{-i}$ converges to the identity element as $i \to \infty$. Then any element of $\mathcal{H}$ which is fixed by U_a is also fixed by U_b; that is, for $p \in \mathcal{H}$, $U_a(p) = p$ implies that $U_b(p) = p$.*

The above lemma was noted first in [AG]; see also [D1]. More generally the following is also true [Mar7]:

2.7. Lemma: *Let G, $\mathcal{H}$ and U be as in Lemma 2.6. Let F be any subgroup of G. Let $x \in G$ be such that for any neighbourhood Ω of the identity in G, x is contained in the closure of $F\Omega F$. Then $U_x(p) = p$ for any $p \in \mathcal{H}$ such that $U_f(p) = p$ for all $f \in F$.*

The lemma follows immediately from the fact that $g \mapsto \| U_g(p) - p \|^2 = 2\|p\|^2 - 2\text{Re} <U_g(p), p>$ is a continuous function on G which is constant on the double cosets of the subgroup F; the latter assertion follows from the last expression and the unitarity of the representation U.

Let G be a Lie group and let e be the identity element in G. For $a \in G$ we denote by $U^+(a)$ and $U^-(a)$ the subgroups defined by

$$U^+(a) = \{g \in G \,|\, a^i g a^{-i} \to e \text{ as } i \to \infty\}$$

and

$$U^-(a) = \{g \in G \,|\, a^{-i} g a^i \to e \text{ as } i \to \infty\};$$

these are 'horospherical subgroups' associated to a (see §3 for more about them). Let A_a be the closed subgroup generated by $U^+(a)$ and $U^-(a)$; we shall call this the *Mautner subgroup* associated to a. It can be shown that A_a is a normal subgroup of G. If $A_a = G$ for an element a then Lemma 2.6 implies that the translation of G/C by a is ergodic for any $C \in \mathcal{F}(G)$; any $f \in L^2$ which is fixed by U_a is also fixed by $U_{a^{-1}}$ and hence by $U^+(a)$ and $U^-(a)$ by Lemma 2.6 and so by the closed subgroup generated by them. One immediate consequence of this is the following.

2.8. Corollary: *Let $G = SL(n, \mathbb{R})$ and $C \in \mathcal{F}(G)$. Let $a \in G$ be a matrix which has an eigenvalue λ (possibly complex) such that $|\lambda| \neq 1$. Then the translation T_a of G/Γ is ergodic. In particular, the geodesic flows of surfaces with constant negative curvature and finite area are ergodic.*

The analogous statement also follows for any irreducible lattice in any semisimple Lie group with no nontrivial compact factors; in this case we demand that the linear transformation $\text{Ad}\, a$ of the Lie algebra have an eigenvalue of absolute value other than 1 (for $SL(n, \mathbb{R})$ this condition is equivalent to the one in the above Corollary).

More generally the Mautner phenomenon yields the following.

2.9. Corollary: *Let G be a Lie group and $C \in \mathcal{F}(G)$. Let $a \in G$ and A_a be the Mautner subgroup associated to a. Let $X = G/C$ and m be the G-invariant probability measure on X. Then we have the following:*

i) if $f \in L^2(X, m)$ and $U_a f = f$ then $U_g f = f$ for all $g \in A_a$.

ii) if $G' = G/A_a$, $C' = \overline{CA_a}/A_a$ and $a' = a\overline{CA_a}$ then the translation T_a of $X = G/C$ is ergodic if and only if the translation $T_{a'}$ of G'/C' by a' is ergodic.

This reduces the question of ergodicity of translations to the special case when the Mautner subgroup is trivial. It can be seen that this condition is equivalent to all the eigenvalues of $\text{Ad}\, a$ being of absolute value 1. Lemma 2.7

can be used to deal with some part of this case also. I will however illustrate this only with the following simple example.

2.10. Proposition: *Let* $G = SL(2, \mathbb{R})$, Γ *be a lattice in* G *and* $a = \begin{pmatrix} 1 & 1 \\ 0 & 1 \end{pmatrix}$. *Then* T_a *is ergodic. In particular, the horocycle flow is ergodic.*

Proof: Let F be the cyclic subgroup generated by a. A simple computation shows that for any rational number α and $\epsilon > 0$ there exists a $\theta \in \mathbb{R}$ such that $0 < |\theta| < \epsilon$ and $\begin{pmatrix} \alpha & 0 \\ \theta & \alpha^{-1} \end{pmatrix} \in F \begin{pmatrix} 1 & 0 \\ \theta & 1 \end{pmatrix} F$. This shows that for any neighbourhood Ω of the identity in G the closure of $F\Omega F$ contains all diagonal matrices. Hence by Lemma 2.7 any F-invariant L^2 function on G/Γ is also invariant under all diagonal matrices and hence constant by Corollary 2.8. This completes the proof.

Using a similar argument and the Jacobson-Morosov Lemma (see [J]) one can prove the following assertion for cyclic subgroups F; the general assertion can then be deduced using Lemma 7.1 of [Mo1].

2.11. Theorem: *Let* G *be a semisimple Lie group with finite center and no compact factors and* Γ *be an irreducible lattice in* G. *Let* F *be a subgroup of* G. *Then the action of* F *on* G/Γ *is ergodic if and only if* $\overline{F}$ *is noncompact.*

This result is essentially due to C. C. Moore [Mo1] though he did not put it in this form; on the other hand the results in [Mo1] are stronger inasmuch as they deal with mixing properties as well. The reader is also referred to [D1], where similar results are proved, in a more general context, along the line of argument indicated above. For a general semisimple group the conditions for ergodicity can now be seen to be the following.

2.12. Theorem: *Let* G *be a connected semisimple Lie group. Let* K *be the maximal compact connected normal subgroup and let* G' *be the smallest normal subgroup such that the quotient is compact. Let* $C \in \mathcal{F}$ *and* $a \in G$. *Then the translation* T_a *of* G/C *is ergodic if and only if* $G'C$ *is dense in* G *and the factor transformation of* T_a *on any irreducible factor of* G/KC *is ergodic.*

Let me now proceed to describe the results in the general case. They are obtained by a closer analysis of the spectrum of the associated unitary operator (see [Au], [Mo1], [D3], [BM], [Z2]).

A subgroup F of a Lie group L is said to be Ad-*compact* if the group of linear transformations $\{\text{Ad}\, x : \mathcal{L} \to \mathcal{L} \,|\, x \in F\}$, where $\mathcal{L}$ is the Lie algebra of L, is contained in a compact subgroup of $GL(\mathcal{L})$. Now let G a be a Lie group. Then for any subgroup F of G there exists a unique minimal normal closed subgroup M_F such that FM_F/M_F is Ad-compact as a subgroup of G/M_F (see [Mo2]); I shall call M_F the Mautner-Moore subgroup associated to F. The general ergodicity criterion is given by the following theorem due to C. C. Moore:

2.13. Theorem (Moore, 1980): *Let G be a Lie group, F be a subgroup of G and let M_F be the Mautner-Moore subgroup associated to F. Then the following holds:*

i) If $g \mapsto U_g$ is a unitary representation of G over a Hilbert space $\mathcal{H}$ and $\mathcal{V}$ is a finite-dimensional subspace of $\mathcal{H}$ invariant under U_f for all $f \in F$ then $U_g(v) = v$ for all $v \in \mathcal{V}$ and $g \in M_F$.

ii) If $C \in \mathcal{F}(G)$ then the action of F on G/C is ergodic if and only if the action of $\overline{FM_F}$ on G/C is ergodic if and only if the action of $\overline{FM_F}/M_F$ on the factor G'/C' is ergodic, where $G' = G/M_F$ and $C' = \overline{CM_F}/M_F$.

iii) If $C \in \mathcal{F}(G)$ and $M_F C$ is dense in G then the action of F on G/C is ergodic.

Assertions (ii) and (iii) may be easily seen to follow from assertion (i). A proof of (i) for one-parameter subgroups may be found in [Mo2]; the case of (infinite) cyclic subgroups F can be deduced from that of one-parameter subgroups by embedding a suitable power of the generating element in a one-parameter subgroup (see [D3], §6 for an idea of this) and the general case may be concluded from the latter using Lemma 7.1 of [Mo1].

One can see that if G is a nilpotent Lie group and $C \in \mathcal{F}(G)$ then for $a \in G$, T_a is not ergodic unless $[G, G] \subset M_a$. Therefore for translations Theorem 2.5 is a special case of the above general result. For a semisimple Lie group the criterion yields Theorems 2.11 and 2.12.

We shall now see a characterisation of ergodicity of translations in terms ergodicity of its factors on simpler homogeneous spaces. It may be recalled that in a connected Lie group G there is a smallest closed normal subgroup R, namely the solvable radical, such that G/R is a semisimple Lie group. On the other hand there also exists a smallest closed normal subgroup S such that G/S is solvable; if we set $G_0 = G$ and inductively define G_{i+1}, $i \geq 0$, to be the closure of $[G_i, G_i]$, then by dimension considerations G_i is the same subgroup for all large i and the common subgroup can be seen to have the desired property.

2.14. Theorem: *Let G be a connected Lie group and $C \in \mathcal{F}(G)$. Let R be the solvable radical of G and S be the smallest closed normal subgroup such that G/S is solvable. Then for $a \in G$, $T_a : G/C \to G/C$ is ergodic if and only if the translations $T_a : G/\overline{RC} \to G/\overline{RC}$ and $T_a : G/\overline{SC} \to G/\overline{SC}$ are ergodic; (the quotient spaces $G/\overline{RC}$ and $G/\overline{SC}$ can be realised as homogeneous spaces of G/R and G/S and when this is done the condition is equivalent to T_{aR} and T_{aS} being ergodic).*

A simple argument for deducing this from Theorem 2.13 is given by A. N. Starkov in [St2]. The criterion was proved earlier in [D3] under an additional condition of 'admissability' of C, that there exist a Lie group F, a lattice Γ in F and a continuous homomorphism $\psi : F \to G$ such that $\overline{\psi(\Gamma)} = C$; the condition holds trivially for a lattice. In [BM], where a detailed analysis of conditions for ergodicity and spectrum of flows was carried out for flows on homogeneous of finite volume, the above assertion was proved under a weaker 'admissability condition' that there exists a solvable Lie subgroup A of G containing the radical such that AC is closed (or equivalently $A/A \cap C$ is compact). Subsequently it was proved independently by D. Witte [Wi1] and A. N. Starkov [St1] that the latter admissability condition in fact holds for all $C \in \mathcal{F}(G)$; see also [Z3] for a more general result in this regard.

The characterisation in terms of factors would be complete with the following theorem of Brezin and Moore [BM]; (actually in [BM] only flows induced by one-parameter subgroups are considered, but the result for translations follows from it via suitable embeddings).

2.15 Theorem: *Let G be a connected solvable Lie group, $C \in \mathcal{F}(G)$ and $a \in G$. Then the translation T_a of G/C is ergodic if and only if the factor on the maximal euclidean quotient is ergodic.*

In view of Theorems 2.11, 2.12, 2.14 and 2.15 to complete the story of ergodicity of the translations it only remains to know criteria in the case of euclidean solvmanifolds. These are indeed completely understood and are described in [BM] (see Corollary 5.3 there). I will however not go into the details, since it would involve introducing more notation which does not seem worthwhile at this stage, but content myself by commenting that the criterion involves 'rational structures' and is comparable to the case of tori except that it is more elaborate.

2.16. Remark: In the following sections we will be largely concerned with actions of subgroups generated by unipotent elements; an element u of a Lie group G is said to be unipotent if $\operatorname{Ad} u$ is a unipotent linear transformation, that is, $(\operatorname{Ad} u - I)^n = 0$ for some n, where I is the identity transformation (equivalently, 1 is the only eigenvalues of $\operatorname{Ad} u$). It may be worthwhile to note some additional properties in this case with regard to ergodicity. Let G be a connected Lie group and $C \in \mathcal{F}(G)$. Let U be a closed subgroup of G generated by unipotent elements (as elements of G). Then there exists a closed normal subgroup L of G containing U such that LC is closed and the action of U on LC/C is ergodic; note that the action on G/C consists of disjoint union of closed invariant sets on each of which the action corresponds to the U action on LC/C. This can be easily deduced from Theorem 2.12. In view of the observation in studying the actions we need to consider only ergodic ones. Now let G, C and U be as above and suppose that the U-action on G/C ergodic. Then the connected component C^0 of the identity in C is a normal subgroup of G (see [Wi1] for the ideas involved; there U is taken to be cyclic but the argument generalises). Since it involves no loss of generality to go modulo a closed normal subgroup, this reduces the study to the case when C is a lattice.

In ergodic theory one also studies various other notions, such as mixing, entropy, Kolmogorov mixing, Bernoullicity etc. and these have also been studied for the systems under consideration. Some of the papers referred to above indeed deal with these properties as well; many papers have results on mixing, which is a property related to the spectrum. I will however not discuss the results here. The reader is referred to the survey articles [D13] and [Mar7] for some details in this regard. Let me however recall for reference in the sequel that a measure-preserving transformation T of a probability space (X, μ) is said to be *weakly mixing* if the constant functions are the only eigenfunctions of U_T and that this condition is equivalent to the cartesian square $T \times T$ of T (acting on $X \times X$) is ergodic.

Let me conclude this section on ergodicity with some remarks about what happens when ergodicity does not hold. While almost all orbits are dense when ergodicity holds, *no* orbit is dense if ergodicity does not hold. This can easily be read off from the criteria as above. In fact when a translation or flow is not ergodic, there exist invariant nonconstant C^∞ functions (see [BM]). It is shown in [St3] that given a flow on a homogeneous space of finite volume, induced by a one-parameter subgroup, the space can be partitioned in to smooth invariant submanifolds such that all of them have smooth measures invariant and ergodic with respect to the flow; further each of the submanifolds is finitely covered by a homogeneous space of a Lie group by a lattice and the restriction of the flow to the submanifolds is the image of a flow on the homogeneous space, induced by a one-parameter subgroup of the Lie group. This enables one to reduce the study of flows on homogeneous spaces as above to the ergodic case.

§3 Dense orbits; some early results

In the previous section we saw the conditions for ergodicity of group actions by translations (and certain affine automorphisms) of homogeneous spaces with finite invariant measure and it was also noted that when ergodicity holds the orbits of almost all points are dense in the homogeneous space and, when the subgroup is either cyclic or a one-parameter subgroup, almost all of them (not necessarily the same set) are also uniformly disributed with respect to the probability measure invariant under the ambient Lie group. We now address the next set of questions arising naturally from this: are all orbits dense? if not, can we describe the closures of the orbits which are not dense?

what can we say about the distribution of an individual orbit if it is not uniformly distributed?

Through the remaining sections the aim will be to discuss a class of systems with a remarkably orderly behaviour in this respect.

I will begin by recalling that for an irrational rotation all orbits are dense and uniformly distributed with respect to the angle measure. This may be deduced by observing that (because of commutativity) any two orbits are translates of each other so once some orbit is dense (respectively, uniformly distributed) the same has to hold for all the orbits; the result can also be found in books on Number Theory, with different proofs. In the same way for a translation of the n-dimensional torus $I\!\!R^n / Z\!\!\!Z^n$ by $(\alpha_1, ..., \alpha_n)$, where $\alpha_1, ..., \alpha_n \in I\!\!R$, all orbits are dense and uniformly distributed in the torus (with respect to the Haar measure) whenever the translation is ergodic, namely when $\alpha_1, ..., \alpha_n$ and 1 are linearly independent over $Q\!\!\!Q$. The density assertion is known as Kronecker's theorem and the uniform distribution assertion is a theorem of H. Weyl (see [CFS], [Wa] for details).

An action of a group G on a locally compact space X is said to be *minimal* if there is no proper closed nonempty G-invariant subset; clearly an action is minimal if and only if orbits of all points are dense in X. A homeomorphism is said to be minimal if the corresponding action of the cyclic group is minimal. Our observation above means that every ergodic translation of a torus is minimal.

If we move from translations to affine automorphisms of tori ergodicity no longer implies all orbits being dense (leave alone uniformly distributed). In fact for any automorphism $\overline{A}$, $A \in GL(n, Z\!\!\!Z)$, the orbit of any point of the form $q + Z\!\!\!Z^n \in I\!\!R^n / Z\!\!\!Z^n$, where q is a vector with rational coordinates, is periodic (finite) and hence can not be dense even if $\overline{A}$ is ergodic. In general, apart from the periodic orbits there are also a whole lot of other orbits which are not dense (see [DGS]). Nevertheless for some affine automorphisms the situations is different:

3.1.   Theorem (Furstenberg, 1961): *Let $A \in GL(n, Z\!\!\!Z)$ be a unipotent matrix and let $a \in I\!\!R^n$ be such that the affine automorphism $T = T_a \circ \overline{A}$ of $I\!\!T^n = I\!\!R^n / Z\!\!\!Z^n$ is ergodic. Then all orbits of $T_a \circ \overline{A}$ are uniformly distributed (and in particular dense) in $I\!\!T^n$.*

The result is proved by showing that the Haar measure is the only probability measure on $I\!\!T^n$ invariant under an affine automorphism T as in the theorem. A homeomorphism T of a locally compact topolological space X is called *uniquely ergodic* if it admits only one invariant probability measure. Thus an affine automorphism as in the above theorem is uniquely ergodic. A straightforward argument shows that if a homeomorphism of a compact space is uniquely ergodic and μ is the the unique invariant probability measure then the orbits of all points of supp μ are uniformly distributed with respect to μ (see [F1] or [CFS], for instance).

Theorem 3.1 yields an ergodic-theoretic proof of the following classical result of H.Weyl on the distribution of the fractional parts of values of a polynomial; see [CFS], Ch.7, Section 2, for a proof.

3.2. Theorem: *Let $P(t) = a_0 t^n + a_1 t^{n-1} + ... + a_n$ be a polynomial of degree $n \geq 1$, with real coefficients. Suppose that at least one of $a_k, 0 \leq k \leq n-1$, is irrational. Then the sequence $\xi_n = <P(n)>$ of fractional parts of $P(n)$ is uniformly distributed in the interval $[0, 1]$.*

In [Gr] L. Green proved that any ergodic nilflow is minimal. An interesting application of this to diophantine approximation was given in [AH], proving the following analogue of one of Weyls results (cf. [We], Theorem 14) on uniform distribution; we recall here that a set of integers is said to be relatively dense if the difference between successive integers in the set is bounded above.

3.3 Theorem: *Let $P_i(t) = \Sigma_{j=1}^n a_{ij} t^j$, where a_{ij} are integers such that $\Sigma_j |a_{ij}| > 0$ for each $i = 1, ...n$. Let $\alpha_1, ..., \alpha_n$ be real numbers which together with 1 form a linearly independent set over $\mathbb{Q}$. Then for any $\epsilon > 0$ and $\theta_1, ..., \theta_n \in \mathbb{R}$ there exists a relatively dense set M of integers such that for each $i = 1, ..., n$ and $m \in M$, $\alpha_i P_i(m) - \theta_i$ differs from an integer by at most ϵ.*

Furstenberg's theorem was generalised by W. Parry to affine automorphisms of nilmanifolds, proving unique ergodicity of $T = T_a \circ \overline{A}$ when T is ergodic and A is unipotent (see [P]).

In [AB] Auslander and Brezin obtained a generalisation of Weyl's criterion for uniform distribution of sequences on tori in the setting of solvmanifolds; their approach is more general than the present one, in that they consider also avarages along more general sequences of sets rather than the intervals in $\mathbb{Z}$ or $\mathbb{R}$ in our discussion. In the course of their study they prove in particular that for a nilpotent Lie group G and any lattice Γ in G the action of any subgroup H of G on G/Γ is uniquely ergodic whenever it is ergodic; for cyclic subgroups and, with a little argument, for connected Lie subgroups this follows also from the result of Parry. As for the case of a general solvable case, while the analogue of Weyl's criterion proved in [AB] would apply in particular to actions of cyclic and one-parameter subgroups, the precise implications to uniform distribution (or density) of orbits do not seem to have been analysed in literature, in terms of the 'position' of the subgroup in the ambient group.

Let us now come to the case of semisimple groups, starting with the simplest case of $SL(2, \mathbb{R})$. In this case results on orbit closures which can be compared to what I have recalled for affine transformations goes back to a classical paper of G. A. Hedlund [He] on the horocycle flow; see also [Man] and [Gh] for exposition and proof.

3.4. Theorem (Hedlund, 1936): *Let $G = SL(2, \mathbb{R})$ and Γ be a lattice in G. For $t \in \mathbb{R}$ let $h_t = \begin{pmatrix} 1 & t \\ 0 & 1 \end{pmatrix}$ and let $H = \{h_t \mid t \in \mathbb{R}\}$. Then every H-orbit on G/Γ is either periodic or dense in G/Γ. If G/Γ is compact then all orbits are dense while if G/Γ is noncompact then the set of points with periodic orbits is nonempty and consists of finitely many immersed cylinders, each of which is dense in G/Γ.*

Thus in the compact quotient case the action is minimal. When the quotient G/Γ is noncompact there are finitely many 'cusps'; these correspond to the cusps of the surface of constant negative curvature, they being finitely many when the area is finite (see [B]). To each cusp there corresponds a one-parameter family of periodic orbits, one periodic orbit with each value of the period, which together form an immersed cylinder in G/Γ.

Hedlund's ideas were followed up in a paper of L. Greenberg [Gre], where the author also noted the following simple *duality principle*, which interrelates results on orbits of flows on finite-volume homogeneous spaces with those for lattice actions on certain 'large' homogeneous spaces and in turn for actions on linear spaces, as we shall see below.

3.5. Proposition: *Let G be a Lie group, C and H be a closed subgroups of G. Then for $g \in G$ the H-orbit of gC is dense in G/C if and only if the C-orbit of $g^{-1}H$ is dense in G/H.*

The proof is immediate, both the statements being equivalent to the set HgC being dense in G. Via this observation the study of subgroup actions on homogeneous spaces can be related to the study orbits of lattices (or more generally finite covolume subgroups) on linear spaces as follows: Let G be a Lie group and $C \in \mathcal{F}(G)$ and consider a linear action of G on a finite-dimensional $\mathbb{R}$-vector space V. One would then like to understand, for $v \in V$, the closure of the orbit Cv. Fix a point $v_0 \in V$ and let $v \in Gv_0$. The Proposition implies in particular that for $g \in G$, Cgv_0 is dense in $Gv = Gv_0$ if the orbit $Hg^{-1}C/C$ of $g^{-1}C$ is dense in G/C, H being the isotropy subgroup $\{x \in G \mid xv_0 = v_0\}$ of v_0. Greenberg also proved the following theorem, which via the duality principle generalises the compact quotient case of Theorem 3.4.; in the sequel I shall follow the terminology that a lattice Γ in a Lie group G is said to be *uniform* if the quotient G/Γ is compact and *nonuniform* otherwise.

3.6. Theorem (Greenberg, 1963): *Let $G = SL(n, \mathbb{R})$ and let Γ be a uniform lattice in G. Then for any $v \in \mathbb{R}^n - \{0\}$ the Γ-orbit Γv is dense in $\mathbb{R}^n$.*

Similarly if $G = Sp(2n, \mathbb{R})$, the symplectic group realised canonically as a subgroup of $SL(2n, \mathbb{R})$, and Γ is a lattice in G then for the natural action of G on $\mathbb{R}^{2n}$ the Γ-orbit of any nonzero vector is dense in $\mathbb{R}^{2n}$.

Theorem 3.6 implies in particular that if $f(x_1, ..., x_n)$ is a real quadratic form in n variables then for any uniform lattice Γ in $SL(n, \mathbb{R})$, the set $\{f(\gamma_{11}, \gamma_{21}, ..., \gamma_{n1}) \,|\, \gamma = (\gamma_{ij}) \in \Gamma\}$ is dense in $\mathbb{R}$. The particular case of this with $n = 2$ and f a positive definite form was proved by K. Mahler [Ma], who conjectured it to be true for indefinite forms as well.

Many other number theoretic applications involve $SL(n, \mathbb{Z})$, which is a nonuniform lattice. Consider first its natural action on $\mathbb{R}^n$. In this case one does not expect all orbits to be dense, since the orbits of points with rational coordinates are in fact discrete. It was proved in [D], following the ideas of the paper of Greenberg that those and their scalar multiples are in fact the only exceptions. It was also proved that, for even n, the orbit of a point v under the subgroup $Sp(n, \mathbb{Z})$, consisting of integral $n \times n$ symplectic matrices, is dense if v is not a scalar multiple of a rational point; see Theorem 3.7 below for a more general result.

The reader would notice that in terms of the actions on homogeneous spaces Greenberg's result involves considering orbits of a rather large subgroup. ¿From the point of view of dynamics this suggests asking what happens for smaller subgroups. One class of subgroups to which the study was extended in the intermediate period, that needs to be mentioned, is the class of horospherical subgroups.

A subgroup U of a Lie group G is said to be *horospherical* if there exists a $a \in G$ such that

$$U = \{g \in G \,|\, a^i g a^{-i} \to e \text{ as } i \to \infty\},$$

e being the identity element in G; specifically U is called the horospherical subgroup corresponding to a. It is not difficult to see that a horospherical subgroup is always a connected (not necessarily closed) Lie subgroup. If G is a semisimple Lie group then a subgroup is horospherical if and only if it is the unipotent radical of a parabolic subgroup.

It may be noticed that the subgroup $\{\begin{pmatrix} 1 & t \\ 0 & 1 \end{pmatrix} \,|\, t \in \mathbb{R}\}$ of $SL(2, \mathbb{R})$, which defines the horocycle flow, is a horospherical subgroup in $SL(2, \mathbb{R})$ in the above sense. In the case of higher-dimensional manifolds of constant negative curvature and finite volume, the classical horospherical foliations associated to the geodesic flow are given by orbits of horospherical subgroups; in the notation as in §1.4, if U is the horospherical subgroup corresponding to g_t, $t > 0$, then the images of U-orbits on G/Γ in $M\backslash SO(n, 1)/\Gamma$ give the horospherical foliation.

Generalising the minimality result of Hedlund for the horocycle flow on compact quotients of $SL(2, I\!R)$ by lattices, it was proved by Veech in [V1] that if G is a connected semisimple Lie group with no nontrivial compact factors and Γ is a uniform lattice in G then the action of any maximal horospherical subgroup of G on G/Γ is minimal. In [V2] he proved another minimality result (see [V2], Theorem 1.3) which shows in particular that for G and Γ as in the preceding assertion the action of a horospherical subgroup of G is minimal whenever it is ergodic. It is noted that the argument uses ideas from Anosov's proof of density of leaves of horospherical foliations (see [An], Theorem 15), which itself is also a generalisation of Hedlund's theorem as above. Using a similar argument it was shown in [D2] that if G is any connected Lie group, Γ is a uniform lattice in G, U is the horospherical subgroup corresponding to a weakly mixing (see § 2) affine automorphism $T = T_a \circ \overline{A}$, where $a \in G$ and A is an automorphism of G such that $A(\Gamma) = \Gamma$, then the U-action on G/Γ is minimal; by the horospherical subgroup corresponding to $T_a \circ \overline{A}$ we mean the subgroup $\{g \in G | (\sigma_a \circ A)^i(g) \to e$ as $i \to \infty\}$, where σ_a is the inner automorphism corresponding to a and e is the identity element. While these results pertain only to minimality, various results were also obtained around the same time on unique ergodicity of horospherical flows, which in particular imply minimality; it would however be convenient to postpone going over them, to § 5 where I will be discussing invariant measures in greater detail.

For a general, not necessarily uniform, lattice Γ the orbit closures of actions of horospherical subgroups on G/Γ were considered in [DR], [D15], [D16] and [St5]; I will take up these results in the next section after introducing a certain perspective. Let me conclude this section with some results on diophantine approximation, in the spirit of Theorem 3.6 but involving nonuniform lattices, obtained in my joint paper [DR] with S. Raghavan.

3.7. Theorem: *Let $\Gamma = SL(n, Z\!\!\!Z)$ and $1 \leq p < n$. Let V be the cartesian product of p copies of $I\!R^n$, equipped with the componentwise action of Γ. Then for $v = (v_1, ..., v_p)$ the Γ-orbit is dense in V if and only if the subspace of $I\!R^n$ spanned by the vectors $v_1, ..., v_p$ does not contain any nonzero rational vector (vector with rational coordinates with respect to the natural basis of $I\!R^n$).*

It was also shown that if n is even and $Sp(n, Z\!\!\!Z)$ is the group of integral symplectic matrices (with respect to a nondegenerate symplectic form) the $Sp(n, Z\!\!\!Z)$-orbit of any symplectic p-tuple, where $1 \leq p \leq n/2$, is dense in the space of symplectic p-tuples; a p-tuple $(v_1, ..., v_p)$ is said to be symplectic if $\omega(v_i, v_j) = 0$ for all $1 \leq i, j \leq p$, ω being the symplectic form defining the symplectic group.

The theorem can be applied to study the values of systems of linear forms, over integral points and also over primitive integral points. We recall that an integral n-tuple is said to be *primitive* if the entries are not all divisible by a positive integer exceeding 1.

3.8.   Corollary: *Let* $\xi_1, ..., \xi_p$ *be real linear forms in* n *variables, where* $1 \leq p < n$. *Suppose that no nontrivial linear combination of* $\xi_1, ..., \xi_p$ *is a rational form (that is, if* $\lambda_1, ..., \lambda_p \in \mathbb{R}$ *are such that* $\sum \lambda_i \xi_i$ *is a form with rational coefficients, then* $\lambda_i = 0$ *for all* i*). Then for any* $a_1, ..., a_p \in \mathbb{R}$ *and* $\epsilon > 0$ *there exists a primitive integral* n-*tuple* $(x_1, ..., x_n)$ *such that*

$$|\xi_i(x_1, ..., x_n) - a_i| < \epsilon \text{ for all } i = 1, ..., p.$$

§ 4 Conjectures of Oppenheim and Raghunathan

One major aspect to be noticed in Hedlund's theorem is that for the systems considered, even when there are both dense as well as non-dense orbits (as in the noncompact quotient case) the closures of all orbits are 'nice' objects geometrically. This is also the case Greenberg's generalisation of Hedlund's theorem and the other similar results which I mentioned above. From the point of view of dynamics this is quite an unusual behaviour. One would naturally wonder how general this phenomenon is. The question got an added impetus from an observation of M. S. Raghunathan that a well-known conjecture going back to a paper of A. Oppenheim from 1929, on values of indefinite quadratic forms at integral points would be settled if an analogue of Hedlund's theorem is proved for the case $G = SL(3, \mathbb{R})$ and $\Gamma = SL(3, \mathbb{Z})$ for the action of the special orthogonal group $SO(2, 1)$ corresponding to a quadratic form of signature $(2, 1)$. Let me recall the conjecture and indicate the connection.

Let $Q(x_1, ..., x_n) = \sum_{ij} a_{ij} x_i x_j$ be a quadratic form with real coefficients. We are interested in the set of values of Q on integral n-tuples and especially in the question whether it is dense in $\mathbb{R}$. Assume for simplicity that Q is nondegenerate. It is clear that if Q is a definite form (either positive definite or negative definite) then the set of values under consideration is discrete. Similarly if the coefficients a_{ij} are all rational multiples of a fixed real number (equivalently if all the ratios are rational) then also the set as above is discrete. The conjecture in question was that if Q is indefinite and not a multiple of a rational quadratic form and $n \geq 3$ then the set of values at integral points is indeed dense; the idea of the conjecture can be traced back to a paper of A. Oppenheim in 1929, though the original statement is somewhat weaker. The problem was attacked by several number theorists, including H. Davenport, H. Heilbronn, B. J. Birch, H. Ridout, A. Oppenheim, G. L. Watson and others, mainly using the circle method and through many papers together by the sixties the conjecture was known to hold for all quadratic forms when $n \geq 21$ and in special cases in lower number of variables. There has also been some work on the problem in the recent years by number theoretic methods by H. Iwaniec, R. C. Baker and

H. P. Schlickewey. I will not go into the details of this and various nuances of
the conjecture but would refer the interested reader to [Le], [Mar7], [Gh] and
[B2]. I may mention however that the condition $n \geq 3$ can not be omitted;
it is easy to see that the quadratic form $x_1(\lambda x_1 - x_2)$ does not take nonzero
values arbitrarily close to 0 over integral x_1, x_2, if λ is a badly approximable
number, namely an irrational number with bounded partial quotients.

If Q is a quadratic form satisfying the conditions of the conjecture then
$Q(\mathbb{Z}^n) = Q(H\Gamma\mathbb{Z}^n)$, where $\Gamma = SL(n, \mathbb{Z})$ and H is the special orthogonal group corresponding to Q, namely $H = \{g \in SL(n, \mathbb{R}) \mid Q(gv) =
Q(v)$ for all $v \in \mathbb{R}^n\}$; we view n-tuples as column vectors and consider the
action of matrices by left multiplication. Now if it is shown that the orbit
$H\Gamma/\Gamma$ of the H action on $SL(n, \mathbb{R})/\Gamma$ is either dense or closed in $SL(n, \mathbb{R})/\Gamma$,
à la Hedlund, then it would follow that $Q(\mathbb{Z}^n)$ is dense in $\mathbb{R}$ unless $H\Gamma$ is
closed. One can see that the latter holds only if Q is a multiple of a form with
rational coefficients, a condition which is disallowed in the hypothesis of the
conjecture. Thus the conjecture can be approached via the study of orbits of
the special orthogonal groups on $SL(n, \mathbb{R})/SL(n, \mathbb{Z})$; while considering the
totality of quadratic forms we fix a special orthogonal group and study the
closures of all orbits.

As a strategy of attacking the problem via study of flows on homogeneous
spaces Raghunathan formulated a more general conjecture. This pertains to
orbit-closures of actions of unipotent elements and unipotent subgroups on
G/Γ, where G is a Lie group and Γ is a lattice; recall that an element $u \in G$
is said to be *unipotent* if $\operatorname{Ad} u$ is a unipotent linear transformation of the
Lie algebra of G; a subgroup U of G is unipotent if all elements of U are
unipotent.

4.1. Conjecture. *Let G be a Lie group and Γ be a lattice in G. Let U be a
unipotent subgroup of G. Then the closure of any U-orbit is a homogeneous
set; that is, for any $x \in G/\Gamma$ there exists a closed subgroup F of G such that
$\overline{Ux} = Fx$.*

(In [D7] where the conjecture first appeared in print the statement was
formulated only for one-parameter subgroups U; the weaker form only reflects
my choice for the write-up and does not fully convey what was meant. The
lack of concern for generality in the formulation there was partly on account of
knowing that the conjecture being true for one-parameter subgroups implies
it being true for all connected unipotent subgroups; see [D14], Theorem 3.8,
for an idea of the proof of this).

The minimality results mentioned above confirm the conjecture in their
respective cases. The result from [DR] involved in Theorem 3.7 implies validity of the conjecture for certain specific horospherical subgroups of $SL(n, \mathbb{R})$
and $Sp(n, \mathbb{R})$ (unipotent radicals of parabolic subgroups of maximum dimen-

sion, in the first case) acting on the quotients of the groups by $SL(n, \mathbb{Z})$ and $Sp(n, \mathbb{Z})$ respectively. In [D15] the conjecture was verified for horospherical flows on not necessarily compact homogeneous spaces of all reductive Lie groups. The case of not necessarily reductive Lie groups was studied in [D12] (Appendix), [D16] and [St5] and certain partial results were obtained towards the conjecture. I will mention a couple of the results; though we will be seeing later some results which are much stronger in spirit, it does not seem easy to read off these results from them. In [D16] it was shown that if G is a connected Lie group, Γ is a lattice in G, R is the radical of G and U is a horospherical subgroup associated to an element of G and if U acts ergodically on G/Γ then an orbit $Ug\Gamma/\Gamma$, $g \in G$, is dense if and only if $Ug\overline{R\Gamma}/\overline{R\Gamma}$ is dense in $G/\overline{R\Gamma}$. Starkov [St5] proved that (under the same notation) if G/Γ is compact then for $x = g\Gamma \in G/\Gamma$, where $g \in G$, $\overline{Ux} = Fx$ for the subgroup F defined as follows: let U^- be the horospherical subgroup opposite to U (corresponding to the inverse element), M be the subgroup generated by U and U^- and set $F = \overline{M(g\Gamma g^{-1})}$ (it may be noted that M is a normal subgroup and hence F is indeed a subgroup).

Oppenheim's conjecture was settled by G. A. Margulis in 1986-87 (see [Mar3], [Mar4]). While it is based on the study of flows on homogeneous spaces, his proof proceeds somewhat differently than the strategy indicated above. He proved that for the action of $SO(2,1)$ on $SL(3, \mathbb{R})/SL(3, \mathbb{Z})$ any relatively compact orbit is actually compact. By the Mahler criterion recalled in § 1.5 this implies that any quadratic form as in Oppenheim's conjecture takes arbitrarily small values. For a form which does not take the value 0 at any integral point, this would mean that 0 is a limit point of the set of values and then a result of Oppenheim implies that the set of values is actually dense. For forms admitting integral zeros Margulis produced a somewhat technical variation of the argument, showing that the conjecture holds in this case also. In [DM1] we proved the following stronger result.

4.2. Theorem: *For the action of* $SO(2,1)$ *on* $SL(3, \mathbb{R})/SL(3, \mathbb{Z})$, *any orbit is either dense or closed.*

(A similar result was also proved there for all lattices satisfying a certain condition and we had mentioned that in fact the condition holds for all lattices. A proof of the latter was given in [DM4] but while the basic idea there is well founded, there are serious presentational errors which make the proof unsatisfactory; a proof of the relevant part may be found in [EMS]; the author also hopes to present it in detail at a suitable place).

The reader may notice that though $SO(2,1)$ is not a unipotent subgroup the conclusion in Theorem 4.2 is similar to that in Conjecture 4.1. The theorem would actually follow if one knew the validity of the conjecture. Nevertheless it is convenient to view both the statements as particular cases of the following conjecture formulated by Margulis in [Mar2] (see also [Mar7]).

4.3. Conjecture : *Let G be a Lie group and $C \in \mathcal{F}(G)$. Let H be a subgroup which is generated by the unipotent elements contained in it. Then the closure of any H-orbit on G/C is a homogeneous set.*

Though the conjecture is stated for all $C \in \mathcal{F}(G)$, using Remark 2.16 and some technical arguments it can be reduced to the case of lattices; in the sequel I shall discuss only the case of lattices. Theorem 4.2 verifies a particular case of the Conjecture 4.3. Evidently the latter is also satisfied in the cases where Conjecture 4.1 holds. We shall see more about the conjecture later.

Observe that Theorem 4.2 yields a proof of Oppenheim's conjecture by the argument indicated earlier. It also yields the following stronger result, in which we consider the values only on *primitive* integral n-tuples; this may be compared with Corollary 3.8 in the case of systems of linear forms.

4.4. Theorem: *Let Q be a nondegenerate indefinite quadratic form on $\mathbb{R}^n$, $n \geq 3$, which is not a multiple of a form with rational coefficients. Let $\mathcal{P}$ denote the set of primitive integral n-tuples. Then $Q(\mathcal{P})$ is dense in $\mathbb{R}$.*

Subsequently in [DM3] we also produced a rather elementary proof for Theorem 4.4, involving only standard material on finite-dimensional vector spaces and topological groups; I should add that Zorn's lemma was used to conclude existence of minimal elements in certain families of compact subsets. In [D17] I constructed a variation of the argument not involving choice of minimal subsets. However I learnt later from A. Katok that (by an observation which he attributed to S. Simpson) the existence of minimal sets as required in the earlier proof could be obtained without the use of Zorn's lemma.

In [DM2] we verified Conjecture 4.1 for the case of $G = SL(3, \mathbb{R})$, Γ any lattice in G and $U = \{u_t\}$ a 'generic' unipotent one-parameter subgroup, namely such that $u_t - I$ has rank 2 for $t \neq 0$, I being the identity matrix. This result has the following consequence.

4.5. Corollary: *Let Q be a nondegenerate indefinite quadratic form on $\mathbb{R}^3$. Let L be a linear form on $\mathbb{R}^3$ such that the plane $\{v \in \mathbb{R}^3 | L(v) = 0\}$ and the double cone $\{w \in \mathbb{R}^3 | Q(w) = 0\}$ intersect in a line and are tangential along the line. Suppose that no linear combination of Q and L^2 is a rational quadratic form. Then for any $a, b \in \mathbb{R}$ and any $\epsilon > 0$ there exists a primitive integral triple x such that*

$$|Q(x) - a| < \epsilon \text{ and } |L(x) - b| < \epsilon.$$

Margulis has pointed out in [Mar6] that Theorem 4.2 can be strengthened to the following.

4.6. Theorem: *If Γ is a lattice in $SL(3, \mathbb{R})$ and Ω is a nonempty open subset $SL(3, \mathbb{R})/\Gamma$ then there are only a finite number of $SO(2,1)$-orbits on $SL(3, \mathbb{R})/\Gamma$ disjoint from Ω and each of them is closed.*

This was applied to the problem of minima of rational quadratic forms. For a nondegenerate quadratic form Q on $\mathbb{R}^n$ let $m(Q)$ denote the infimum of $\{|Q(x)|| \, x \in \mathbb{Z}^n - (0)\}$ and let $\mu(B) = |d(Q)|^{-1/n} m(Q)$, where $d(Q)$ denotes the discriminant of Q. The set M_n of all numbers of the form $\mu(Q)$ where Q is a nondegenerate quadratic form on $\mathbb{R}^n$ is called the Markov spectrum. For $n = 2$ this is realted to the classical Markov numbers. It follows from Markov's work that $M_2 \cap (\frac{4}{9}, \infty)$ is a countable discrete subset of $(\frac{4}{9}, \infty)$; the intersection with the interval $[0, \frac{4}{9}]$ is an uncountable set with a complicated topological structure. From Theorem 4.6 Margulis deduced the following.

4.7. Corollary: *For $n \geq 3$, for any $\epsilon > 0$ the set $M_n \cap (\epsilon, \infty)$ is finite; further, there are only finitely many equivalence classes of quadratic forms Q with $\mu(Q) > \epsilon$.*

(It may be noted that the corollary is essentially about $n = 3$ or 4, since for $n \geq 5$ by Meyer's theorem $M_n = \{0\}$). The result was proved by J. W.S.Cassels and H. F. P. Swinnerton-Dyer earlier under the assumption of the Oppenheim conjecture being true. The reader is referred to [Mar6] and [Gh] for more details.

§5 Invariant measures of unipotent flows

In the last section we saw various results on closures of orbits but nothing was said how they are distributed in space, except for certain assertions about all orbits being uniformly distributed in the whole space. The general question of distribution of orbits involves the study of invariant measures of the flows. For any action the set of all invariant probability measures is a convex set and the ergodic ones (namely those with respect to which the action is ergodic) form its extreme points. Further, any invariant probability measure has an 'ergodic decomposition' as a direct integral of ergodic invariant probability measures (see [Ro], for instance). Therefore to understand the set of all finite invariant measures it is enough to classify the ergodic invariant measures. In this section I will describe the results in this regard. They will be applied in §7 to the problem of distribution of orbits. Let me begin with the following theorem of Furstenberg which may be said to have been instrumental in initiating a full-fledged study of invariant measures and its application to proving Raghunathan's conjecture.

5.1. Theorem (Furstenberg, 1972): *Let $G = SL(2, \mathbb{R})$, Γ a uniform lattice in G and $U = \{\begin{pmatrix} 1 & t \\ 0 & 1 \end{pmatrix} \mid t \in \mathbb{R}\}$. Then the G-invariant probability measure is the only U-invariant probability measure on G/Γ.*

As in the earlier cases this implies that the orbits of U are uniformly distributed with respect to the G-invariant probability measure on G/Γ. The result was generalised by Veech [V2] to actions of maximal horospherical subgroups of semisimple Lie groups on quotients of the latter by uniform lattices. R. Bowen [Bo] and R. Ellis and W. Perrizo [EP] showed that if G is a connected Lie group, Γ is a unifrom lattice in G and $a \in G$ is such that T_a is a weakly mixing (see § 2) translation of G/Γ then the action of the horospherical subgroup U corresponding to a on G/Γ is uniquely ergodic; namely the G-invariant probability measure is the only probability measure invariant under the U-action. Interestingly these proofs and that of Veech in the case of semisimple Lie groups are all substantially different from each other; while Bowen's proof is geometric the other two are analytical. Recall that the horospherical subgroups are nilpotent and hence in particular solvable groups. Since for the action of a solvable group any compact invariant subset supports an invariant measure, these results imply minimality the results recalled in § 3 for the flows under consideration.

Let us now return temporarily to the case of the horocycle flows on G/Γ where $G = SL(2, I\!R)$ and Γ is a lattice in G. If G/Γ is noncompact, namely when Γ is a nonuniform lattice, by Hedlund's theorem we know that there exist continuous families of periodic orbits and thus, in contrast to the above situation, there is a large class of invariant probability measures; each periodic orbit supports an invariant probability measure. It turns out however that together with the G-invariant probability measure these measures account for all ergodic invariant measures. In [D4] I obtained a classification of invariant measures for a class of horospherical flows which in the case of the horocycle flow on $SL(2, I\!R)/\Gamma$, implies the following.

5.2. Theorem : *Let $G = SL(2, I\!R)$, Γ a lattice in G and $U = \{\begin{pmatrix} 1 & t \\ 0 & 1 \end{pmatrix} \mid t \in I\!R\}$. Let μ be an ergodic U-invariant probability measure on G/Γ. Then either μ is G-invariant or there exists a $x \in G/\Gamma$ such that the U-orbit of x is periodic and the support of μ is contained in Ux.*

Let us now consider the question in a more general set up. Let G be a Lie group and Γ be a lattice in G. A canonical class of probability measures on G/Γ arises as follows. Let H be any closed subgroup of G such that $H \cap (g\Gamma g^{-1})$ is a lattice in H for some $g \in G$. Since $H/(H \cap g\Gamma g^{-1})$ can be identified in a natural way with $Hg\Gamma/\Gamma$, the H-orbit of $g\Gamma$, the latter admits a H-invariant probability measure; we shall think of it as a probability measure on G/Γ assigning zero measure to the complement of the orbit. Measures on G/Γ arising in this way are called *algebraic measures;* that is, μ is algebraic if there exists a closed subgroup H of G such that μ is H-invariant and its support is contained in a single orbit of H; such an orbit is always closed (see

[R], Theorem 1.13).

It should be noted that the class of closed subgroups H yielding algebraic measures depends on the lattice Γ. While for some lattices the only algebraic measures other than the G-invariant measure may come from compact subgroups of G (the latter are uninteresting from our point of view and may be considered trivial) for most lattices, including all non-uniform lattices as a matter of fact, there exist algebraic measures arising from proper noncompact subgroups of G. For example, when $G = SL(2, \mathbb{R})$ and Γ is a nonuniform lattice, the measures on periodic orbits of $H = \{ \begin{pmatrix} 1 & t \\ 0 & 1 \end{pmatrix} \mid t \in \mathbb{R} \}$ discussed earlier are algebraic measures of this kind. For $G = SL(n, \mathbb{R})$ and $\Gamma = SL(n, \mathbb{Z})$ there are noncompact closed subgroups of various intermediate dimensions intersecting Γ in a lattice; e.g. the subgroup consisting of all upper triangular unipotent matrices, the subgroup $SL(m, \mathbb{R})$, where $1 \le m < n$, embedded (say) in the upper left corner in $SL(n, \mathbb{R})$ in the usual way, etc.; more generally if H is a real algebraic subgroup defined over $\mathbb{Q}$ and satisfying the condition as in the theorem of Borel and Harish-Chandra (see § 1.6) then we get an algebraic probability measure on the H-orbit of the point Γ (the coset of the identity element). We note that in particular if Q is a nondegenerate indefinite quadratic form on $\mathbb{R}^n$ with rational coefficients (with respect to the standard basis) then $SO(Q)$, the corresponding special orthogonal group, intersects Γ is a lattice and yields an algebraic measure supported on $SO(Q)\Gamma/\Gamma$.

Now let G be any Lie group, Γ be a lattice in G and let U be a subgroup of G. Collecting together algebraic measures arising from subgroups H containing U gives us a class of U-invariant measures. The key conjecture for the classification of invariant measures is the following.

5.3. Conjecture: *Let G be a Lie group, Γ be a lattice in G and U be a subgroup generated by the unipotent elements contained in it. Then any U-invariant ergodic probability measure is algebraic.*

For the case of $G = SL(2, \mathbb{R})$ and U the unipotent one-parameter subgroup corresponding to the horocycle flow, this statement is equivalent to Theorem 5.2. The conjecture was verified in [D7] for maximal horospherical subgroups of reductive Lie groups, where the conjecture was formulated for one-parameter subgroups (the case of one-parameter subgroups implies the corresponding statement for all connected unipotent subgroups).

In what constitutes a landmark development in the area, through a series of papers [R1], [R2], [R3] Marina Ratner proved Conjecture 5.3 for a large class of subgroups U, including all unipotent subgroups. She proved the following.

5.4. Theorem (Ratner, 1991): *Let G be a Lie group and Γ be a discrete subgroup of G. If U is a unipotent subgroup of G then for the U-action on G/Γ any finite ergodic U-invariant measures is algebraic. The same conclusion also holds for the action of any Lie subgroup U of G satisfying the following conditions:*

i) U^0 is generated by the unipotent one-parameter subgroups contained in it and

ii) U/U^0 is finitely generated and each coset of U^0 in U contains a unipotent element.

It may be emphasized that Γ is not assumed to be a lattice but just any discrete subgroup. The general proof is quite intricate and involves several steps. The reader may find it helpful to first go through [R5] and [Gh] where the basic ideas are explained dealing only with the case of the horocycle flows.

Recently Margulis and Tomanov have given another proof of the above theorem in the case of algebraic groups; their proof bears a strong influence of Ratner's original arguments but is substantially different in its approach and methods; see [MT].

Given a Lie group G and two unimodular closed subgroups H and Γ of G there is also a canonical duality between H-invariant measures on G/Γ and Γ-invariant measures on G/H. We note that up to Borel isomorphism G can be viewed as a cartesian product of G/H and H. Using this we can associate with any measure on G/H a measure on G which is invariant under the action of H by multiplication on the right. If the measure on G/H is Γ invariant then the corresponding measure on G is also Γ-invariant, under the left action. This and a corresponding observation for measures on G/Γ yields the one-one correspondence between the two classes of measures as above; see [F2] and [D4]. This correspondence is used in the proofs of some of the results on classification of invariant measures of actions of horopherical subgroups (see [F2], [V2], [D4] and [D7]). On the other hand via the duality Theorem 5.4 yields a description of Γ-invariant measures on G/U (notation as in Theorem 5.4). In turn this may be applied to describe, in the same fashion as for orbit-closures, the measures on vector spaces invariant under actions of certain discrete groups. The strategy in particular yields the following Corollary; I will not go into the details of other more technical results that can be obtained.

5.5. Corollary: *Let $\Gamma = SL(n, \mathbb{Z})$, $1 \le p < n$ and consider the Γ-action on the p-fold cartesian product V of $\mathbb{R}^n$, as in Theorem 3.7. Let λ be a (locally finite) ergodic Γ invariant measure on V. Then the exists a Γ-invariant subspace W and $a \in V$ such that $\Gamma(W+a)$ is closed and λ is the measure supported on $\Gamma(W+a)$ whose restriction to $W+a$ is the translate of a Lebesgue* *measure on W.*

Various applications of the classification of invariant measures will be discussed in later sections. I will conclude the present section by describing a way of 'arranging' the invariant measures of unipotent flows, which is useful in these applications.

Let G and Γ be as before. Let $\mathcal{H}$ be the class of all proper closed subgroups H of G such that $H \cap \Gamma$ is a lattice in H and H contains a unipotent subgroup U of G acting ergodically on $H/H \cap \Gamma$. Then $\mathcal{H}$ is countable (see [R4], Theorem 1; see also Proposition 2.1 of [DM6] which is a variation of the assertion and implies it in view of Corollary 2.13 of [Sh1]). For any subgroups H and U of G let

$$X(H, U) = \{g \in G \mid Ug \subseteq gH\}.$$

Now let U be a subgroup of G generated by the unipotent elements contained in it and acting ergodically on G/Γ. Let μ be any ergodic U-invariant probability measure other than the G-invariant probability measure. By Theorem 5.4 μ is algebraic; thus there exists a closed subgroup L containing U such that μ is the L-invariant measure on a L-orbit, say $Lg\Gamma/\Gamma$. Since μ is not G-invariant L is a proper closed subgroup. Put $H = g^{-1}Lg$. Then $H \cap \Gamma$ is a lattice in H. Also $g^{-1}Ug$ is generated by unipotent elements, contained in H and acts ergodically on $H/H \cap \Gamma$. Hence $H \in \mathcal{H}$. Also, $Ug \subseteq Lg = gH$ and hence $g \in X(H, U)$. Therefore the support of μ, which is the same as $Lg\Gamma/\Gamma = gH\Gamma/\Gamma$, is contained in $X(H, U)\Gamma/\Gamma$. Thus for every ergodic invariant probability measure μ other than the G-invariant measure $\mu(\cup_{H \in \mathcal{H}} X(H, U)\Gamma/\Gamma) = 1$. Since by ergodic decomposition any invariant probability measure π can be expressed as a direct integral of ergodic invariant measures (see [Ro], for instance) this implies that $\pi(\cup_{H \in \mathcal{H}} X(H, U)\Gamma/\Gamma) > 0$, unless π is G-invariant. Since $\mathcal{H}$ is countable this yields the following characterisation of the G-invariant measure on G/Γ.

5.6. Corollary : *Let G, Γ and U be as above. Let π be a U-invariant probability measure on G/Γ such that $\pi(X(H, U)\Gamma/\Gamma) = 0$ for all $H \in \mathcal{H}$. Then π is G-invariant.*

§6 Homecoming of trajectories of unipotent flows

As a part of the strategy involved in one of his proofs of the arithmeticity theorem, Margulis proved that if $\{u_t\}$ is a unipotent one-parameter subgroup of $SL(n, \mathbb{R})$ then for any $x \in SL(n, \mathbb{R})/SL(n, \mathbb{Z})$ there exists a compact subset K of $SL(n, \mathbb{R})/SL(n, \mathbb{Z})$ such that the set $\{t \geq 0 | u_t x \in K\}$ is unbounded; that is, the trajectory of $\{u_t\}$ starting from x keeps returning to K (see [Mar1]). After studying finite invariant measures of maximal horospherical flows in [D4] I realised that Margulis' proof could be strengthened to conclude that for actions of unipotent subgroups all ergodic invariant measures are finite (see [D5] and [D10]); for the case of a one-parameter subgroup $\{u_t\}$ and x as above this involves proving that for a suitable compact subset K the set $\{t \geq 0 | u_t x \in K\}$ has positive upper density. The line of argument later yielded the following result (see [D14], Theorem 3.5; the semisimplicity condition in the statement there is redundant in view of Lemma 9.1 of [D4]; it came to be included in the hypothesis only in the flow of the general development there):

6.1. Theorem: *Let G be a connected Lie group and Γ be a lattice in G. Then for any $\epsilon > 0$ there exists a compact subset K of G/Γ such that for any unipotent one-paramater subgroup $\{u_t\}$ and $g \in G$ at least one of the following conditions holds:*

i) $l(\{t \in [0,T] | u_t g\Gamma \in K\}) \geq (1 - \epsilon)T$ for all large T, or

ii) there exists a proper closed connected subgroup H of G such that $H \cap \Gamma$ is a lattice in H and $\{g^{-1} u_t g\}$ is contained in H.

It may be noticed that if the second alternative holds then the whole $\{u_t\}$-orbit of $g\Gamma$ is contained in the lower-dimensional homogeneous space $gH\Gamma/\Gamma \approx L/L \cap g\Gamma g^{-1}$, where $L = gHg^{-1}$; in view of this in various contexts one can use a suitable induction hypothesis and reduce the problem to considering only the situation as in condition (i). It is clear in particular that for any $\{u_t\}$ and any $g \in G$ as in the hypothesis there exists a compact subset K, depending on the $\{u_t\}$ and g, such that the conclusion as in (i) holds. The compact subset can be chosen to be common for all unipotent one-parameter subgroups and g from a given compact subset; that is, the following holds:

6.2. Theorem: *Let G be a connected Lie group and Γ be a lattice in G. Let a compact subset F of G/Γ and $\epsilon > 0$ be given. Then there exists a compact subset K of G/Γ such that for any unipotent one-parameter subgroup $\{u_t\}$ of G, $x \in F$ and $T \geq 0$ such that $u_s x \in F$ for some $s \in [0,T]$ we have*

$$l(\{t \in [0,T] \mid u_t x \in K\}) \geq (1 - \epsilon)T.$$

This follows from Proposition 2.7 of [D14] together with the arithmeticity theorem; (see [DM6], Theorem 6.1; it may also be noted that it is enough to prove the assertion in the theorem for $x \in F$, which can then be applied to $u_s x$ and the one-parameter subgroups $\{u_t\}$ and $\{u_{-t}\}$ to get the desired result).

I may also mention that the closed subgroup as in the second alternative in Theorem 6.1 can be chosen from a special class of subgroups; for instance if G is semisimple (the general case can be reduced to this via Lemma 9.1 of [D4]) then H can be chosen to be contained in a 'Γ-parabolic' subgroup;

by a Γ-parabolic subgroup we mean a parabolic subgroup whose unipotent radical intersects Γ in a lattice. This follows from Theorem 1 of [DM4]; (as mentioned earlier, though the result is correct the proof in [DM4] involves some errors). A more general result in the direction of Theorem 6.3 has been proved in [EMS] where they consider multiparameter polynomial trajectories; I will however not go into the details.

The above theorems are used in studying the distribution of orbits of unipotent flows, as we shall see in the next section. I will devote the rest of this section to a discussion of various other applications of the theme. Let me begin with the following result derived in [D14] (see Theorem 3.9 there).

6.3. Corollary: *Let G be a connected Lie group and Γ be a lattice in G. There there exists a compact subset C of G such that the following holds: if U is a closed connected unipotent subgroup which is not contained in any proper closed subgroup H such that $H \cap \Gamma$ is a lattice in H then $G = C\Gamma U = U\Gamma C^{-1}$.*

The motivation for such a theorem is the following: if Γ is a uniform lattice in G then there exists a compact subset C of G such that $G = C\Gamma = \Gamma C^{-1}$. While this can not be true for a nonuniform lattice, the theorem means that it is true modulo any unipotent subgroup not contained in a subgroup H as in the hypothesis. This has the following interesting consequence.

6.4. Corollary: *Let G be a connected Lie group and Γ be a lattice in G. Consider a linear action of G on a finite-dimensional vector space V. Let $v \in V$ be such that the following conditions are satisfied:*

i) there exists a sequence $\{g_i\}$ in G such that $g_i v \to 0$, as $i \to \infty$, and

ii) the isotropy subgroup of v (that is, $\{g \in G | gv = v\}$) contains a connected unipotent subgroup U which is not contained in any proper closed subgroup H such that $H \cap \Gamma$ is a lattice in H.

Then there exists a sequence $\{\gamma_i\}$ in Γ such that $\gamma_i v \to 0$.

This may be viewed as diophantine approximation with matrix argument, with elements of lattices replacing integers. It can be applied for instance to various representations of $SL(n, \mathbb{R})$ and we get conditions for orbits of $SL(n, \mathbb{Z})$ to have points arbitrarily close to zero. It also implies (see [D14], Corollary 4.3) that if G is a connected semisimple Lie group with trivial center and Γ is a lattice in G then for any unipotent element u which is not contained in any proper closed subgroup such that $H \cap \Gamma$ is a lattice in H, the closure of the Γ conjugacy class $\{\gamma u \gamma^{-1} | \gamma \in \Gamma\}$ contains the identity element.

By one of the remarks above, Theorem 6.2 implies in particular that for the action of a unipotent one-parameter subgroup on G/Γ, where G is a Lie group and Γ is a lattice, every ergodic invariant (locally finite) measure is finite. It was proved in [D5] and [D10] (in the arithmetic and the general case respectively) that in fact any invariant measure is a countable sum of finite invariant measures. Using this and some more arguments Margulis obtained another proof of the theorem of Borel and Harish-Chandra on the finiteness of the invariant measures of certain homogeneous spaces. More generally he proved the following (see [Mar2]).

6.5. Theorem: *Let G be a connected Lie group and Γ be a lattice in G. Let H be connected Lie subgroup such that if R is the radical of H and N is the subgroup consisting of all unipotent elements of G contained in R then R/N is compact. If μ is a locally finite H-invariant mesure on G/Γ then there exists a sequence $\{X_i\}$ of H-invariant Borel subsets of G/Γ such that $\mu(X_i)$ is finite for all i and $\cup X_i = G/\Gamma$; if μ is ergodic then it is a finite measure. In particular, the H-invariant measure on any closed orbit of H on G/Γ is a finite measure.*

The results of this section are also used (see [DM1]), together with a lemma of Y. Dowker (see Lemma 1.6 in [DM1]), to conclude that for unipotent flows as above all minimal (closed nonempty) sets are compact (this is of course implied by Ratner's theorem verifying Raghunathan's conjecture, (Theorem 7.4 below), but cumulatively that involves a much longer argument). The reader is also referred to [Mar8] for a proof of the corresponding assertion for actions of more general unipotent groups.

§ 7 Distribution and closures of orbits

In this and the following sections I will discuss the distribution of orbits, using Theorem 5.4 on classification of invariant measures and Theorem 6.2. The reader is also referred to [D18] for a similar discussion, presented somewhat differently. I will begin with a cautionary note that when there are more than one invariant probability measures, an orbit need not be uniformly distributed with respect to any of them. Roughly speaking, over time the 'distribution' may oscillate between various invariant measures, possibly uncountably many of them. Rigorously this may be described as follows. Let T be a homeomorphism of a locally compact space X. Let $x \in X$ and for any natural number N let π_N be the probability measure which is the average of the N point masses supported at the points $x, Tx, ..., T^{N-1}x$; that is, for any Borel set E, $\pi_N(E) = N^{-1}\Sigma_{i=0}^{N-1} \chi_E(T^i(x))$, χ_E being the characteristic function of E. The orbit of x being uniformly distributed with respect to a measure μ is equivalent to π_N converging to μ (in the weak topology on the space of measures, with respect to bounded continuous functions) as $N \to \infty$. In considering convergence it is convenient to have the space to be compact and therefore if X is noncompact we view the measures as being on its one-point compactification, say $\tilde{X}$; if X is compact we set $\tilde{X} = X$. Then convergence of $\{\pi_N\}$ to a probability measure on X is equivalent to there being a unique limit in the space of probability measures on $\tilde{X}$ and its assigning zero mass to the point at infinity. It is easy to see that any limit point is a T-invariant probability measure. If X is compact and there is only one invariant probability measure, namely if the transformation is uniquely ergodic, then this immediately implies that the orbit is uniformly distributed. On the other hand if there are several invariant probability measures then different subsequences of $\{\pi_N\}$ may converge to different probability measures. This is what is meant by oscillation of the distribution of the orbit. For a general dynamical system this happens quite frequently (see e.g.[DGS], [DMu]).

Before proceeding further it may be noted that a similar discussion applies to one-parameter flows. If $\{\phi_t\}$ is a one-parameter subgroup of homeomorphisms of a locally compact space X and $x \in X$ then for any $T > 0$ we define a probability measure π_T by setting $\pi_T(E)$, for any Borel set E, to be $T^{-1}l(\{t \in [0,T] \mid \phi_t(x) \in E\})$, where l is the Lebesgue measure on $\mathbb{R}$. The orbit of x under $\{\phi_t\}$ being uniformly distributed with respect to a probability measure μ is equivalent to π_T converging to μ as $T \to \infty$.

In a general sense the task of proving uniform distribution of an orbit can be divided in to two parts; first to understand all invariant measures of the transformation or flow in question and then to identify which of them can occur as limit points of the families of measure associated to any particular orbit, as above.

Using Theorem 5.4 on the classification of invariant measures of the horocycle flow it was shown, in [D9] for $\Gamma = SL(2, \mathbb{Z})$ and in [DS] for a general lattice, that all nonperiodic orbits are uniformly distributed with respect to the invariant volume. That is,

7.1. Theorem: *Let $G = SL(2, \mathbb{R})$, Γ be any lattice in G and $U = \{u_t\}$, where $u_t = \begin{pmatrix} 1 & t \\ 0 & 1 \end{pmatrix}$ for all $t \in \mathbb{R}$. Let $x \in G/\Gamma$ be such that $u_t x \neq x$ for any $t > 0$. Then the $\{u_t\}$-orbit of x is uniformly distributed with respect to the G-invariant probability measure. Also, for any $s \neq 0$, $\{u_s^i x\}_{i=0}^{\infty}$ is uniformly distributed with respect to the G-invariant probability measure.*

Before going over to the general case, let me note the following interesting consequence of the $SL(2, \mathbb{Z})$-case of the above theorem, proved in [D9].

7.2. Theorem: *Let α be an irrational number. For any $T > 0$ let $I(T)$ denote the set of all natural numbers k such that $k \leq T < k\alpha>$ and the integral part of $k\alpha$ is coprime to k. Then*

$$\frac{1}{T}\Sigma_{I(T)} <k\alpha>^{-1} \longrightarrow \frac{1}{\zeta(2)} = \frac{6}{\pi^2},$$

as $T \to \infty$, where ζ is the Riemann zeta function.

The classification of invariant measures of flows made it possible to address the questions of closures and distribution of the orbits formulated earlier. Using Ratner's theorem together with a technique from [DM2] Nimish Shah [Sh1] proved Conjecture 4.1 in the cases when either G is a reductive Lie group of $I\!R$-rank 1 or $G = SL(3, I\!R)$ (he proved Theorem 7.3 in these cases). More complete results in this respect were obtained by Ratner around the same time, by a totally different method, proving the following.

7.3. Theorem (Ratner, 1991): *Let G be a connected Lie group and Γ be a lattice in G. Let $U = \{u_t\}$ be a unipotent one-parameter subgroup of G. Then for any $x \in G/\Gamma$ there exists an algebraic probability measure μ such that the $\{u_t\}$-orbit of x is uniformly distributed with respect to μ.*

She proved also the corresponding assertion for cyclic subgroups generated by unipotent elements. From the theorem she deduced Raghuanthan's conjecture (see Conjecture 4.1) for unipotent subgroups and also the following stronger assertion.

7.4. Theorem (Ratner, 1991): *Let G and U be as in Theorem 5.4. Let Γ be a lattice in G. Then for any $x \in G/\Gamma$, $\overline{Ux}$ is a homogeneous subset with finite invariant measure; that is, there exists a closed subgroup F such that Fx admits a finite F-invariant measure and $\overline{Ux} = Fx$.*

Let G, Γ and U be as in Theorem 7.3. We call a point $x \in G/\Gamma$ *generic* for the U-action if there does not exist any proper closed subgroup F containing U such that Fx admits a finite F-invariant measure; a posteriori, in view of Theorem 7.3 this coincides with the usual notion of generic points when U is either cyclic or a one-parameter subgroup; in the sequel however we shall mean only the condition described here. Consider the orbit of a generic point x. In this case the assertion of Theorem 7.3 is that it is uniformly distributed with respect to the G-invariant probability measure, say m. Let me say a few words about how the assertion of the theorem is proved in this case; (for a non-generic point the theorem would follow if we consider an appropriate homogeneous subspace). For $T > 0$ let π_T be the probability measure on G/Γ given by $\pi_T(E) = T^{-1} l(\{t \in [0, t] | u_t x \in E\})$. We have to show that π_T converge to m as $T \to \infty$. Let X denote G/Γ if the latter is compact and the one-point compactification if it is noncompact. We view π_T as measures on X, in the obvious way. It is then enough to show that if $\{T_i\}_{i=1}^{\infty}$ is a sequence such that $T_i \to \infty$ and the sequence $\{\pi_{T_i}\}$ converges, to say π, in the space of probability measures on X then π is the measure assigning 0 mass to the complement of G/Γ in X and G-invariant on the subset G/Γ. Let such a sequence $\{T_i\}$ be given and let π be the limit of $\{\pi_{T_i}\}$. If G/Γ is noncompact and ∞ is the point at infinity then the desired condition $\pi(\infty) = 0$ follows from Theorm 6.2. Now consider π as a probability measure on G/Γ. Recall that any limit such as π is U-invariant. The G-invariance of π is deduced by proving that $\pi(X(H,U)\Gamma/\Gamma) = 0$ for all $H \in \mathcal{H}$ (carrying this out constitutes the major task in the proof) and applying Corollary 5.5.

For proving that $\pi(X(H,U)\Gamma/\Gamma) = 0$ for $H \in \mathcal{H}$, I will indicate a different approach than in Ratner's proof in [R4]. This approach is involved in many papers by now, including [DM2], [DM5], [Sh1], [Sh2], [MS], [EMS]; the arguments in [DS] also follow the same scheme, though in a simpler context. The main idea in this is that the sets $X(H,U)\Gamma/\Gamma$ are locally like affine submanifolds and a trajectory of $U = \{u_t\}$ corresponds to a polynomial trajectory and hence it does not spend too much time too close to the submanifold. To make sense out of this one associates to each $H \in \mathcal{H}$ a representation $\rho_H : G \to GL(V_H)$ over a finite-dimensional real vector space V_H, and a $p_H \in V_H$ such that the following holds: if $\eta_H : G \to V_H$ denotes the orbit map $g \mapsto \rho_H(g)p_H$ and A_H is the subspace of V_H spanned by $\eta_H(X(H,U))$ then $X(H,U) = \eta^{-1}(A_H) = \{g \in G \mid \eta_H(g) \in A_H\}$; specifically ρ_H may be chosen to be the h-th exterior power of the adjoint representation of G, where $h = \dim H$, and p_H to be a nonzero point in the one-dimensional subspace corresponding to the Lie subalgebra of H; see [DM6] for details (see also [EMS]).

The main ingredients in the proof are as follows. Let $\{T_i\}$ be the sequence as above and for each i let $\sigma_i = \{u_t x \mid 0 \le t \le T_i\}$. Firstly we prove that if A is an algebraic subvariety in a vector space V, then for any compact subset C of A and $\epsilon > 0$ there exists a (larger) compact subset D of A such that for any segment the proportion of time spent near C to that spent near D is at most ϵ; specifically, for any neighbourhood Φ of D there exists a neighbourhood Ψ of C such that for $y \notin \Phi$, any unipotent one-parameter subgroup $\{v_t\}$ of G and $T \ge 0$,

$$l(\{t \in [0,T] \mid v_t y \in \Psi\}) \le \epsilon\, l(\{t \in [0,T] \mid v_t y \in \Phi\}).$$

This depends on certain simple properties of polynomials and the fact that orbits of unipotent one-parameter subgroups in linear spaces are polynomial curves. Now consider V_H and any compact subset C of A_H and let D be the corresponding subset as above. Let $g \in G$ be such that $g\Gamma = x$. We apply the above assertion to the segments $\{u_t g \gamma p_H \mid 0 \le t \le T_i\}$, $\gamma \in \Gamma$. It turns out that there exists a neighbourhood Φ of D such that the points $g\gamma p_H$ are contained in Φ for at most two distinct γ, if we restrict to $g \in G$ such that $g\Gamma$ lies in a compact set disjoint from the 'self-intersection set' of $X(H,U)\Gamma/\Gamma$, namely the union of its proper subsets of the form $(X(H,U) \cap X(H,U)\alpha)\Gamma/\Gamma$, $\alpha \in \Gamma$. Using this and an inductive argument for the points on the self-intersection set we can combine the information about the individual segments and conclude that $\mu(\eta_H^{-1}(C)) < \epsilon$. Varying C and ϵ we get that $\mu(\eta_H^{-1}(A)\Gamma/\Gamma) = 0$, as desired.

It was noted earlier that Raghunathan's conjecture (Theorem 7.4) and even the weaker result Theorem 4.2 implies Oppenheim's conjecture. The former has the following stronger consequence in that direction; the particular case of it for $k = 2$ follows from Theorem 4.2 (see [DM1], Theorem 1).

7.5. Corollary: *Let B be a real nondegenerate symmetric bilinear form on $\mathbb{R}^n$, where $n \geq 3$, which is not a multiple of a rational form. Let $1 \leq k \leq n-1$ and $v_1, ..., v_k \in \mathbb{R}^n$. Then for any $\epsilon > 0$ there exist primitive integral vectors $x_1, ..., x_k$ such that*

$$|B(x_i, x_j) - B(v_i, v_j)| < \epsilon \quad \text{for all } i, j = 1, ..., k.$$

§8 Aftermath of Ratner's work

The classification of invariant measures of unipotent flows has yielded not only to a proof of Raghunathan's orbit-closure conjecture but a host of other interesting results as well. This section will be devoted to describing some of these.

Let me begin with some strengthenings of Theorem 7.3 on uniform distribution of orbits of unipotent one-parameter subgroups. Let G be a Lie group, Γ a lattice and $U = \{u_t\}$ be a unipotent one-parameter subgroup. As noted earlier, the theorem implies in particular that if $x \in G/\Gamma$ is a generic point then its orbit is uniformly distributed with respect to the G-invariant probability measure, say m. Thus, for any bounded continuous function φ on G/Γ

$$\frac{1}{T} \int_0^T \varphi(u_t x) dt \to \int_{G/\Gamma} \varphi dm \quad \text{as } i \to \infty$$

for any generic point x. A natural question is whether the convergence of the averages (for a fixed φ) is uniform over compact subsets of the set of generic points. Similarly one may ask what happens if we vary the unipotent one-parameter subgroup. These questions were considered in [DM6] and the results were applied to obtain lower estimates for the number of solutions in large enough sets, for quadratic inequalities as in Oppenheim's conjecture; see Corollary 8.3 below. One of our first results in this direction is the following:

8.1. Theorem: *Let G be a connected Lie group, Γ be a lattice in G and m be the G-invariant probability measure on G/Γ. Let $\{u_t^{(i)}\}$ be a sequence of unipotent one-parameter subgroups converging to a unipotent one-parameter subgroup $\{u_t\}$; that is $u_t^{(i)} \to u_t$ for all t, as $i \to \infty$. Let $\{x_i\}$ be a sequence in G/Γ converging to a point $x \in G/\Gamma$. Suppose that x is generic for the action of $\{u_t\}$. Let $\{T_i\}$ be a sequence in $\mathbb{R}^+$, $T_i \to \infty$. Then for any bounded continuous function φ on G/Γ,*

$$\frac{1}{T_i} \int_0^{T_i} \varphi(u_t^{(i)} x_i) dt \to \int_{G/\Gamma} \varphi dm \quad \text{as } i \to \infty.$$

(Marina Ratner has mentioned in [R6] that Marc Burger had pointed out to her in December 1990, which happens to be before we started our work on the above questions, that such a strengthening of her theorem can be derived applying her methods).

In proving the theorem we use Theorem 5.4 on classification of invariant measures but not Theorem 7.3 on uniform distribution. Our approach, which was indicated in the previous section in the special case of Theorem 7.3, is quite different from Ratner's approach.

We also proved another such 'uniform version' of uniform distribution in which we consider also the averages as on the left hand side for non-generic points as well, together with those for the generic points. The unipotent one-parameter subgroup is also allowed to vary over compact sets of such subgroups; the class of unipotent one-parameter subgroups of G is considered equipped with the topology of pointwise convergence, when considered as maps from $I\!R$ to G.

Let G and Γ be as above and let $\mathcal{H}$, $X(H, U)$ (for any subgroups H and U) be as defined in §6. Recall that the set of generic points for the actions of a one-parameter subgroup $U = \{u_t\}$ consists of $\bigcup_{H \in \mathcal{H}} X(H, U)\Gamma/\Gamma$. The following result deals simultaneously with averages for generic as well as non-generic points outside certain *compact* subsets from *finitely* many $X(H, U)\Gamma/\Gamma$.

8.2. Theorem: *Let G, Γ and m be as before. Let $\mathcal{U}$ be a compact set of unipotent one-parameter subgroups of G. Let a bounded continuous function φ on G/Γ, a compact subset K of G/Γ and $\epsilon > 0$ be given. Then there exist finitely many subgroups $H_1, ..., H_k \in \mathcal{H}$ and a compact subset C of G such that the following holds: For any $U = \{u_t\} \in \mathcal{U}$ and any compact subset F of $K - \bigcup_{i=1}^{k}(C \cap X(H_i, U))\Gamma/\Gamma$ there exists a $T_0 \geq 0$ such that for all $x \in F$ and $T > T_0$,*

$$\Big| \frac{1}{T} \int_0^T \varphi(u_t x) \, dt - \int_{G/\Gamma} \varphi \, dm \Big| < \epsilon.$$

In [DM6] this was proved for $\mathcal{U}$ consisting of a single one-parameter subgroup; essentially the same proof goes through for a general compact set of one-parameter subgroups.

¿From Theorem 8.2 we deduced the following asymptotic lower estimates for the number of solutions of quadratic inequalities in regions of the form $\{v \in I\!R^n \,|\, \nu(v) \leq r\}$, as $r \to \infty$, where ν is a continuous 'homogeneous' function on $I\!R^n$; we call a function ν homogeneous if $\nu(tv) = t\nu(v)$ for all $t > 0$ and $v \in I\!R^n$. We use the notation $\#$ to indicate cardinality of a set and λ for the Lebesgue measure on $I\!R^n$.

8.3. Corollary: *Let $n \geq 3$, $1 \leq p < n$ and let $\mathcal{Q}(p,n)$ denote the set of all quadratic forms on $\mathbb{R}^n$ with discriminant ± 1 and signature $(p, n - p)$. Let $\mathcal{K}$ be a compact subset of $\mathcal{Q}(p,n)$ (in the topology of pointwise convergence). Let ν be a continuous homogeneous function on $\mathbb{R}^n$ such that $\nu(v) > 0$ for all $v \neq 0$. Then we have the following:*

i) for any interval I in $\mathbb{R}$ and $\theta > 0$ there exists a finite subset $\mathcal{E}$ of $\mathcal{K}$ such that each $Q \in \mathcal{E}$ is a scalar multiple of a rational form and for any $Q \in \mathcal{K} - \mathcal{E}$

$$\#\{z \in \mathbb{Z}^n \mid Q(z) \in I, \nu(z) \leq r\} \geq (1 - \theta)\lambda(\{v \in \mathbb{R}^n \mid Q(v) \in I, \nu(v) \leq r\})$$

for all large r; further, for any compact subset $\mathcal{C}$ of $\mathcal{K} - \mathcal{E}$ there exists $r_0 \geq 0$ such that for all $Q \in \mathcal{C}$ the inequality holds for all $r \geq r_0$.

ii) if $n \geq 5$, then for $\epsilon > 0$ there exist $c > 0$ and $r_0 \geq 0$ such that for all $Q \in \mathcal{K}$ and $r \geq r_0$

$$\#\{z \in \mathbb{Z}^n \mid |Q(z)| < \epsilon, \nu(z) \leq r\} \geq c\lambda(\{v \in \mathbb{R}^n \mid |Q(v)| < \epsilon, \nu(v) \leq r\}).$$

There exist nondegenerate rational quadratic forms in 4 variables with no nontrivial integral solutions. Since for a rational form, for sufficiently small $\epsilon > 0$ the set as on the left hand side consists of integral solutions, it follows that the condition that $n \geq 5$ is necessary for the conclusion in the second assertion to hold. On the other hand for $n \geq 5$ by Meyer's theorem any nondegenerate rational quadratic form in n variables has a nontrivial integral solution (see [Se]); the theorem is used in the proof of the second part of the corollary.

It can be verified that in terms of r the volume terms appearing on the right hand side of the inequalities are of the order of r^{n-2}. A particular case of interest is of course when ν is the euclidean norm, in which case the regions involved are balls of radius r. I may mention here that for a single quadratic form which is not a multiple of a rational form, an estimate as in (i) was obtained, for the case of balls, by S. Mozes and myself, with a (possibly small) positive constant in the place $1 - \theta$ as above (unpublished).

In Theorem 8.1 we considered the trajectories of a sequence of points $\{x_i\}$ which converge to a generic point. Evidently it is not a necessary condition for the conclusion of the theorem to hold. The following theorem describes the general picture of what happens if we omit the condition.

8.4. Theorem: *Let $\{u_t^{(i)}\}$ be a sequence of unipotent one-parameter subgroups such that $u_t^{(i)} \to u_t$ for all t, let $U_i = \{u_t^{(i)}\}$ for all i and $U = \{u_t\}$. Let $\{x_i\}$ be a convergent sequence in G/Γ such that for any compact subset of Φ of G, $\{i \in \mathbb{N} \mid x_i \in (\Phi \cap X(H, U_i))\Gamma/\Gamma\}$ is finite. Let x be the* *limit of* $\{x_i\}$. *Then for any* $H \in \mathcal{H}$ *such that* $x \in X(H,U)\Gamma/\Gamma$ *there exists a sequence* $\{\tau_i\}$ *of positive real numbers such that the following holds: if* $\sigma_i = \{u_t^{(i)}x_i \mid 0 \le t \le T_i\}$, *where* $\{T_i\}$ *is a sequence in* $\mathbb{R}^+$, $T_i \to \infty$ *and the normalised linear measures on the segments* σ_i *converge to* μ *as* $i \to \infty$ *then*

i) if $\limsup (T_i/\tau_i) = \infty$ *then* $\mu(X(H,U)\Gamma/\Gamma) = 0$ *and*

ii) if $\limsup (T_i/\tau_i) < \infty$ *then there exists a curve* $\psi : ([0,1] - D) \to X(H,U)$, *where* D *is a finite subset of* $[0,1]$, *such that* $\eta_H \circ \psi$ *extends to a polynomial curve from* $[0,1]$ *to* V_H, $\operatorname{supp} \mu$ *meets* $\psi(t)N^0(H)\Gamma/\Gamma$ *for all* $t \in [0,1] - D$ *and is contained in their union; here* $N^0(H)$ *denotes the subgroup of the normaliser of* H *in* G *consisting of the elements* g *for which the map* $h \mapsto ghg^{-1}$ *preserves the Haar measure on* H.

An analogue of this result can also be proved for divergent sequences, that is, when $x_i \to \infty$ in X, for ∞ in the place of $X(H,U)\Gamma/\Gamma$.

Roughly speaking the theorem says that given the sequence $\{x_i\}$ if we take long enough segments then any limit of linear measures along the segments is G-invariant. If the segments are rather short (in the particular context of the sequences), though of lengths tending to infinity, then the limit gets distributed over a family of sets of the form $g_t H\Gamma/\Gamma$, for certain $H \in \mathcal{H}$. With some further analysis one can describe an ergodic decomposition of the limit.

The theorem also readily implies that for $\{u_t^{(i)}\}$ and $\{x_i\}$ as in the hypothesis there exists a sequence $\{\tau_i\}$ of positive real numbers such that the conclusion as in Theorem 8.1 holds for any sequence $\{T_i\}$ such that $T_i/\tau_i \to \infty$. On the other hand one can also conclude that the topological limit of sufficiently long orbit-segments contains generic points. Specifically, the following holds (cf. [DM6], Theorem 4).

8.5. Theorem: *Let* $\{u_t^{(i)}\}$ *and* $\{x_i\}$ *be as in Theorem 8.4. Then there exists a sequence* $\{t_i\}$ *in* $\mathbb{R}^+$ *such that* $\{u_{t_i}x_i\}$ *has a subsequence converging to a generic point with respect to the limit one-parameter subgroup. Further,* $\{t_i\}$ *may be chosen from any subset* R *of* $\mathbb{R}^+$ *for which there exists an* $\alpha > 0$ *such that* $l(R \cap [0,T]) \ge \alpha T$ *for all* $T \ge 0$.

The classification of invariant measures and the method sketched in the last section for analysing the limit measures has been used recently in many papers and several interesting results have been obtained: Mozes and Shah [MS] show that the set of probability measures on G/Γ which are invariant and ergodic under the action of some (not a fixed one) unipotent one-parameter subgroup is a closed subset of the space of probability measures. Shah [Sh2] applies the method to extend Ratner's uniform distribution theorem (Theorem 7.3) to polynomial trajectories. His result implies in particular the following assertion, which is in the spirit of H. Weyl's theorem of uniform distribution of polynomial trajectories (see [L]), in a more general context.

8.6. Theorem: *Let G be a closed subgroup of $SL(n, \mathbb{R})$ for some $n \geq 2$ and Γ be a lattice in G. Let $\phi : \mathbb{R} \to SL(n, \mathbb{R})$ be a polynomial map (all coordinate functions are polynomials) such that $\phi(G)$ is contained in G. Then the curve $\{\phi(t)\Gamma/\Gamma\}_{t>0}$ is uniformly distributed in G/Γ with respect to an algebraic measure.*

There are also more technical variations of this and multivariable versions (for polynomial maps from $\mathbb{R}^n$) proved in the paper. Using the multivariable case the author also generalises Theorem 7.3 to actions of higher-dimensional unipotent groups, answering in the affirmative a question raised by Ratner in [R4]. Incidentally, though in discussing distribution of orbits I have restricted to only cyclic or one-parameter subgroups it is possible to consider similar questions for actions of a larger class of groups, including all nilpotent Lie groups. The result of Shah pertains to such a question for actions on homogeneous spaces as above.

The set of ideas was applied by Eskin, Mozes and Shah [EMS] to get some notable results on the growth of the number of lattice points on certain subvarieties of linear spaces, within distance r from the origin, as $r \to \infty$. Their results generalise the results of Duke, Rudnik and Sarnak [DRS] and Eskin and McMullen [EM]; the results of these earlier papers apply essentially only to affine symmetric varieties, a limitation which has been overcome in [EMS]. This is achieved through the study of limits of certain sequences of probability measures on G/Γ, where G is a reductive Lie group and Γ is a lattice in G. Specifically, the sequences are of the form $\{g_i\mu\}$, where $\{g_i\}$ is a sequence in G and μ is an algebraic probability measure corresponding to some closed orbit of a reductive subgroup H (which may not necessarily contain any nontrivial unipotent element). In the case when H is a maximal connected reductive subgroup of G intersecting Γ in a lattice and $\{g_i\}$ is such that $\{g_i H\}$ is a divergent sequence in G/H (namely, has no convergent subsequence) then $\{g_i\mu\}$ converges to the G-invariant probability measure on G/Γ. Appropriate generalisations are also obtained in the case when H is not necessarily maximal, but I will not go into the details. I will also not go into the details of the application to the counting problem as above in its generality, but content myself by describing an interesting particular case; the reader is referred to [BR] for some consequences of the result.

8.7. Theorem: *Let p be a monic polynomial of degree n with integer coefficients and irreducible over $\mathbb{Q}$. Then the number of integral $n \times n$ matrices with p as the characteristic polynomial and Hilbert-Schmidt norm less than T is asymptotic to $c_p T^{n(n-1)/2}$, where c_p is a constant.*

§9 Miscellanea

1. *Other applications :*

In the earlier sections in discussing applications I concentrated largely on diophantine approximation and related questions. The ideas and results have found applications in various in other contexts also. Let me mention here some of the problems, without going in to the details (but giving suitable references wherever possible), to which the results are applied. No attempt is made to be exhaustive in respect of the applications and I am only mentioning the results which have come to my notice, just to give a flavour of the variety of possibilities in this regard.

Classification of transformations and flows upto isomorphism is one of the central problems in ergodic theory. Similarly it is of interest to understand factors of such systems, joinings etc. Ratner's theorem on invariant measures enables one to understand these issues satisfactorily in the context of the unipotent flows (or translations by unipotent elements) on homogeneous spaces of finite volume. On the other hand Ratner's work on the classification of invariant measures and Raghunathan's conjecture draws quite considerably from her earlier work on these problems. The reader is referred to [R2] and [Wi2] for details on the problems and the results on them.

In the course of his work on the structure of lattices in semisimple Lie groups Margulis showed that if G is a semisimple group of $I\!R$-rank at least 2, P is parabolic subgroup of G and Γ is a lattice in G then all measurable factors of the Γ-action on G/P are (up to isomorphism) the actions on G/Q, where Q is a parabolic subgroup containing P (see [Mar5], [Z2]). He raised the question whether a similar assertion is true for topological factors of the action. The question was answered in the affirmative in [D11]; certain partial results were obtained earlier in [Z1] and [Sp]. The arguments are based on results about orbit closures of certain subgroups. A more general result in this direction has been proved by Shah in [Sh3], where he considers, given a connected Lie group L, a closed semisimple subgroup G of L, a lattice Λ in L and a parabolic subgroup P of G the factors of the G-actions on $L/\Lambda \times G/P$ which can be factored further to get the projection factor on to G/P; under certain appropriate additional conditions it is shown that the factor is of the form $L/\Lambda \times G/Q$ for a parabolic subgroup Q of G. In the case when $L = G$ this corresponds to studying the factors of the Γ-action on G/P.

A. N. Starkov has applied Ratner's theorem on invariant measures (Theorem 5.4) to prove a conjecture of B. Marcus that any mixing flow on a finite-volume homogeneous space is mixing of all degrees (see [St6]). The idea of proving mixing via the study of invariant measures was earlier employed by S. Mozes [Moz1], who proved that for a Lie group G such that $Ad\,G$

is closed and the center of G is finite, any mixing action on a Lebesgue space is mixing of all degrees.

R. Zimmer has applied Ratner's invariant measures theorem in [Z5], to obtain interesting information on the fundamental groups of compact manifolds with an action of a semisimple Lie group of $I\!R$-rank greater than one. He also used it in [Z4] to show that for certain homogeneous spaces G/H, where G is a real algebraic group and H is an algebraic subgroup, there are no lattices; that is, there exist no discrete subgroups Γ of G such that $\Gamma \backslash G/H$ is compact. E. Glasner and B. Weiss [GW] have used Ratner's description of joinings of the horocycle flows to give an example of a simple weakly mixing transformation with nonunique prime factors.

2. *Asymptotics of the number of integral solutions*

It is natural to ask whether the expressions for the lower estimates as in Corollary 8.3 are actually assymptotic to the corresponding numbers on the left hand side. It was shown by P. Sarnak that there exist quadratic forms in 3 variables for which this is not true. Recently A. Eskin, G. A. Margulis and S. Mozes have shown the answer to be in the affirmative for all nondegenerate indefinite quadratic forms with signatures different from $(2,1)$ and $(2,2)$, in particular for all nondegenerate indefinite quadratic forms in 5 or more variables; for forms with signatures $(2,1)$ and $(2,2)$ the authors obtain upper estimates which are $\log r$ times larger than the lower estimates as in Corollary 8.3; (at the time of this writing the results are in the process of being written).

3. *Extensions*

The reader would have noticed that in the results on orbit closures and invariant measures there has always been a condition involving unipotence. It should be noted that even in the case of affine automorphisms of tori the dynamical behaviour is quite different when the automorphism component has an eigenvalue of absolute value other than 1 (see [DGS] and [F1]); however if all eigenvalues are of absolute value 1 then the behaviour is similar to the unipotent case (see [DMu]). Ratner's work has been followed up in this respect by A. N. Starkov (see [St4] and [St7]), to get a wider perspective in the case of flows on finite-volume homogeneous spaces, namely on G/C, where G a Lie group and $C \in \mathcal{F}(G)$, defined by one-parameter subgroups $\{g_t\}$. He shows that if $\{g_t\}$ is quasiunipotent, namely for all t all (complex) eigenvalues of $\text{Ad}\, g_t$ are of absolute value 1, then all orbit closures are smooth manifolds and all finite ergodic invariant measures consist of smooth measures on these manifolds; they need not however be homogeneous spaces when the one-parameter subgroup is not unipotent. On the other hand it is shown that if $\{g_t\}$ is not quasiunipotent then there exist orbit-closures which are not smooth manifolds.

In another direction, one can also look for other subgroups, not necessarily generated by unipotent elements, for which the conclusion as in Theorems 5.4 or 7.4 hold. Certain interesting examples of such subgroups have been given by Ratner (see [R6], Theorem 9) and Mozes (see [Moz2]).

Before concluding I would like to mention that analogues of many of the results for actions of unipotent subgroups and in particular the conjecture of Raghunathan have been proved in p-adic and S-arithmetic cases as well (see [MT] and [R7]). The analogue of Oppenheim's conjecture in the p-adic and S-arithmetic cases was proved earlier by A. Borel and Gopal Prasad (see [BP1], [BP2], [B1], and [B2]).

Acknowledgement: The author would like to thank Nimish Shah for his comments on a preliminary version of this article.

References

[An] D. V. Anosov, *Geodesic flows on closed manifolds of negative curvature*, Trudy Mat. Inst. Steklov 90 (1967), (Russian) = Proceedings of the Steklov Inst. of Math. 90 (1967).

[Au1] L. Auslander, *An exposition of the structure of solvmanifolds I, Algebraic Theory*, Bull. Amer. Math. Soc. 79 (1973), 227-261.

[Au2] L. Auslander, *An exposition of the structure of solvmanifolds II, G-induced flows*, Bull. Amer. Math. Soc. 79 (1973), 262-285.

[AB] L. Auslander and J. Brezin, *Uniform distribution on solvmanifolds*, Adv. Math. 7 (1971), 111-144.

[AG] L. Auslander and L. Green, *G-induced flows*, Amer. J. Math. 88 (1966), 43-60.

[AGH] L. Auslander, L. Green and F. Hahn, *Flows on homogeneous spaces*, Annals of Mathematical Studies, No. 53, Princeton University Press, 1963.

[AH] L. Auslander and F. Hahn, *An application of nilflows to diophantine approximations*, In: Flows on homogeneous spaces, Annals of Mathematical Studies, No. 53, Princeton University Press, 1963.

[B] A. F. Beardon, *The Geometry of Discrete Groups*, Springer Verlag, 1983.

[B1] A. Borel, *Values of quadratic forms at S-integral points*, In: Algebraic Groups and Number Theory (Ed.: V. P. Platonov and A. S. Rapinchuk), Usp. Mat. Nauk 47, No. 2 (1992), 117-141, Russian Math. Surveys 47, No.2 (1992), 133-161.

[B2] A. Borel, *Values of indefinite quadratic forms at integral points and flows on spaces of lattices*, Progress in Mathematics lecture, Vancouver, 1993

[BP1] A. Borel and Gopal Prasad, *Valeurs de formes quadratiques aux points entiers*, C. R. Acad. Sci., Paris 307, Serie I (1988), 217-220.

[BP2] A. Borel and Gopal Prasad, *Values of isotropic quadratic forms at S-integral points*, Compositio Math. 83 (1992), 347-372.

[BR] M. Borovoi and Z. Rudnik, *Hardy-Littlewood varieties and semisimple groups*, Invent. Math. 119 (1995), 37-66.

[Bo] R. Bowen, *Weak mixing and unique ergodicity on homogeneous spaces*, Israel J. Math. 23 (1976), 267-273.

[BM] J. Brezin and C. C. Moore, *Flows on homogeneous spaces: a new look*, Amer. J. Math. 103 (1981), 571-613.

[Ca] J. W. S. Cassels, *An Introduction to the Geometry of Numbers*, Springer Verlag, 1959.

[CFS] I. P. Cornfeld, S. V. Fomin and Ya. G. Sinai, *Ergodic Theory*, Springer Verlag, 1982.

[D] J. S. Dani, *Density properties of orbits under discrete groups*, J. Ind. Math. Soc. 39 (1975), 189-218.

[D1] S. G. Dani, *Kolmogorov automorphisms on homogeneous spaces*, Amer. J. Math. 98 (1976), 119-163.

[D2] S. G. Dani, *Bernoullian translations and minimal horospheres on homogeneous spaces*, J. Ind. Math. Soc., 39 (1976), 245-280.

[D3] S. G. Dani, *Spectrum of an affine transformation*, Duke Math. J. 44 (1977), 129-155.

[D4] S. G. Dani, *Invariant measures of horospherical flows on noncompact homogeneous spaces*, Invent. Math. 47 (1978), 101-138.

[D5] S. G. Dani, *On invariant measures, minimal sets and a lemma of Margulis*, Invent. Math. 51 (1979), 239-260.

[D6] S. G. Dani, *A simple proof of Borel's density theorem*, Math. Zeits. 174 (1980), 81-94.

[D7] S. G. Dani, *Invariant measures of horospherical flows on homogeneous spaces*, Invent. Math. 64 (1981), 357-385.

[D8] S. G. Dani, *On ergodic invariant measures of group automorphisms*, Israel J. Math. 43 (1982), 62-74.

[D9] S. G. Dani, *On uniformly distributed orbits of certain horocycle flows*, Ergod. Th. and Dynam. Syst. 2 (1982), 139-158.

[D10] S. G. Dani, *On orbits of unipotent flows on homogeneous spaces*, Ergod. Th. Dynam. Syst. 4 (1984), 25-34.

[D11] S. G. Dani, *Continuous equivariant images of lattice-actions on boundaries*, Ann. Math. 119 (1984), 111-119.

[D12] S. G. Dani, *Divergent trajectories of flows on homogeneous spaces and diophantine approximation*, J. Reine Angew. Math. 359 (1985), 55-89.

[D13] S. G. Dani, *Dynamics of flows on homogeneous spaces: a survey*, In: Colloquio de Sistemas Dinamicos (Ed: J. A. Seade and G. Sienra), (Proceedings of a conf., Guanajuato, 1983), Aportaciones Mat. 1, Soc. Mat. Mexicana, 1985, pp. 1-30.

[D14] S. G. Dani, *On orbits of unipotent flows on homogeneous spaces - II*, Ergod. Th. Dynam. Syst. 6 (1986), 167-182.

[D15] S. G. Dani, *Orbits of horospherical flows*, Duke Math. J. 53 (1986), 177-188.

[D16] S. G. Dani, *Dense orbits of horospherical flows*, In: Dynamical Systems and Ergodic Theory, Banach Center Publications 23, PWN Scientific Publ., Warsaw, 1989.

[D17] S. G. Dani, *A proof of Margulis' theorem on values of quadratic forms, independent of the axiom of choice*, L'enseig. Math. 40 (1994), 49-58.

[D18] S. G. Dani, *Flows on homogeneous spaces*, Proceedings of ICM-1994.

[DM1] S. G. Dani and G. A. Margulis, *Values of quadratic forms at primitive integral points*, Invent. Math. 98 (1989), 405-424.

[DM2] S. G. Dani and G. A. Margulis, *On orbits of generic unipotent flows on homogeneous spaces of $SL(3, I\!R)$*, Math. Ann. 286 (1990), 101-128.

[DM3] S. G. Dani and G. A. Margulis, *Values of quadratic forms at integral points: an elementary approach*, L'enseig. Math. 36 (1990), 143-174.

[DM4] S. G. Dani and G. A. Margulis, *On the limit distributions of orbits of unipotent flows and integral solutions of quadratic inequalities*, C. R. Acad. Sci. Paris, Ser. I, 314 (1992), 698-704.

[DM5] S. G. Dani and G. A. Margulis, *Asymptotic behaviour of trajectories of unipotent flows on homogeneous spaces*, Proc. (Math. Sci.) Ind. Acad. Sci. 101 (1991), 1-17.

[DM6] S. G. Dani and G. A. Margulis, *Limit distributions of orbits of unipotent flows and values of quadratic forms*, Adv. in Sov. Math. (AMS Publ.) 16 (1993), 91-137.

[DMu] S. G. Dani and S. Muralidharan, *On ergodic avarages for affine lattice actions on tori*, Monats. Math. 96 (1983), 17-28.

[DR] S. G. Dani and S. Raghavan, *Orbits of euclidean frames under discrete linear groups*, Israel J. Math. 36 (1980), 300-320.

[DS] S. G. Dani and J. Smillie, *Uniform distribution of horocycle orbits for Fuchsian groups*, Duke Math. J. 51 (1984), 185-194.

[DGS] M. Denker, C. Grillenberger and K. Sigmund, *Ergodic Theory on Compact Spaces*, Lecture Notes in Math. 527, Springer Verlag, 1976.

[DRS] W. Duke, Z. Rudnik and P. Sarnak, *Density of integer points on affine homogeneous varieties*, Duke Math. J. 71 (1993), 143-180.

[EP] R. Ellis and W. Perrizo, *Unique ergodicity of flows on homogeneous spaces*, Israel J. Math. 29 (1978), 276-284.

[EM] A. Eskin and C. McMullen, *Mixing, counting and equidistribution in Lie groups*, Duke Math. J. 71 (1993), 181-209.

[EMS] Alex Eskin, Shahar Mozes and Nimish Shah, *Unipotent flows and counting of lattice points on homogeneous varieties*, (Preprint, 1994).

[F1] H. Furstenberg, *Strict ergodicity and transformations of the torus*, Amer. J. Math. 83 (1961), 573-601.

[F2] H. Furstenberg, *Unique ergodicity of the horocycle flow*, In: Recent Advances in Topological Dynamics, (ed. A. Beck), Springer Verlag, 1972.

[GF] I. M. Gelfand and S. V. Fomin, *Geodesic flows on manifolds of constant negative curvature*, Usp. Mat. Nauk 7 (1952), 118-137.

[Gh] E. Ghys, *Dynamique des flots unipotents sur les espaces homogènes*, Sem. Bourbaki 1992-93, Exp. 747 , Asterisque 206 (1992), 93-136.

[GW] E. Glasner and B. Weiss, *A simple weakly mixing transformation with non-unique prime factors*, Amer. J. Math. 116 (1994), 361-375.

[Gr] L. Green, *Nilflows, measure theory*, In: Flows on Homogeneous Spaces, Ann. of Math. Studies, No. 53, pp. 59-66.

[Gre] L. Greenberg, *Discrete groups with dense orbits*, In: Flows on Homogeneous Spaces, Ann. of Math. Studies, No. 53, pp. 85-103.

[H] P. R. Halmos, *Lectures in Ergodic Theory*, The Mathematical Society of Japan, 1956.

[He] G. A. Hedlund, *Fuchsian groups and transitive horocycles*, Duke Math. J. 2 (1936), 530-542.

[J] N. Jacobson, *Lie Algebras*, Interscience, New York, 1962.

[L] C. G. Lekkerkerker, *Geometry of Numbers*, Walters Nordhoff, Groningen; North Holland, Amsterdam-London, 1969.

[Le] D. J. Lewis, *The distribution of values of real quadratic forms at integer points*, Proceed. of Symp. in Pure Math. XXIV, Amer. Math. Soc., 1973.

[Ma] K. Mahler, *An arithmetic property of groups of linear transformations*, Acta. Arith. 5 (1959), 197-203.

[Mal] A. Malcev, *On a class of homogeneous spaces*, Izv. Akad. Nauk SSSR Ser. Mat. 13 (1949), 9-32 (in Russian)= Amer. Math. Soc. Transl. 39 (1949).

[Man] A. Manning, *Dynamics of geodesic and horocycle flows on surfaces of constant negative curvature*, In: Ergodic Theory, Symbolic Dynamics and Hyperbolic spaces (Ed: T. Bedford, M. Keane and C. Series), Oxford Univ. Press, 1991.

[Mar1] G. A. Margulis, *On the action of unipotent groups in the space of lattices*, In: Proceedings of the Sumer School on Group Representations (Bolyai Janos Math. Soc., Budapest, 1971), Akad. Kiado, Budapest 1975, pp 365-370.

[Mar2] G. A. Margulis, *Lie groups and ergodic theory*, In: Algebra - Some Current Trends, Proceedings (Varna, 1986), pp. 130-146, Springer Verlag, 1988.

[Mar3] G. A. Margulis, *Formes quadratiques indéfinies et flots unipotents sur les espaces homogènes*, C. R. Acad. Sci. Paris, Ser I, 304 (1987), 247-253.

[Mar4] G. A. Margulis, *Discrete subgroups and ergodic theory*, In: Number Theory, Trace Formulas and Discrete Subgroups, pp. 377-398, Academic Press, 1989.

[Mar5] G. A. Margulis, *Discrete Subgroups of Semisimple Lie Groups*, Springer Verlag, 1989.

[Mar6] G. A. Margulis, *Orbits of group actions and values of quadratic forms at integral points*, Festschrift in honour of I. Piatetski Shapiro, Isr. Math. Conf. Proc. 3 (1990), 127-150.

[Mar7] G. A. Margulis, *Dynamical and ergodic properties of subgroup actions on homogeneous spaces with applications to number theory*, In: Proceed. of the Internat. Congress of Math. (Kyoto, 1990), pp. 193-215, Springer, 1991.

[Mar8] G. A. Margulis, *Compactness of minimal closed invariant sets of actions of unipotent subgroups*, Geometry Dedicata 37 (1991), 1-7.

[MT] G. A. Margulis and G. M. Tomanov, *Invariant measures for actions of unipotent groups over local fields on homogeneous spaces*, Invent. Math. 116 (1994), 347-392.

[Mau] F. J. Mautner, *Geodesic flows on symmetric Riemannian spaces*, Ann. Math. 65 (1957, 416-431.

[Mo1] C. C. Moore, *Ergodicity of flows on homogeneous spaces*, Amer. J. Math. 88 (1966), 154-178.

[Mo2] C. C. Moore, *The Mautner phenomenon for general unitary representations*, Pacific J. Math. 86 (1980), 155-169.

[Moz1] S. Mozes, *Mixing of all degrees of Lie group actions*, Invent. Math. 107 (1992), 235-141 and Invent Math. 119 (1995), 399.

[Moz2] S. Mozes, *Epimorphic subgroups and invariant measures*, (Preprint).

[MS]   Shahar Mozes and Nimish Shah, *On the space of ergodic invariant measures of unipotent flows,* Ergod. Th. and Dynam. Syst. 15 (1995), 149-159.

[P]   W. Parry, *Ergodic properties of affine transformations and flows on nilmanifolds,* Amer. J. Math. 91 (1969), 757-771.

[R]   M. S. Raghunathan, *Discrete Subgroups of Lie Groups,* Springer Verlag, 1972.

[R1]   Marina Ratner, *Strict measure rigidity for unipotent subgroups of solvable groups,* Invent. Math. 101 (1990), 449-482.

[R2]   Marina Ratner, *On measure rigidity of unipotent subgroups of semisimple groups,* Acta. Math. 165 (1990), 229-309.

[R3]   Marina Ratner, *On Raghunathan's measure conjecture,* Ann. Math. 134 (1991), 545-607.

[R4]   Marina Ratner, *Raghunathan's topological conjecture and distributions of unipotent flows,* Duke Math. J. 63 (1991), 235-280.

[R5]   Marina Ratner, *Raghunathan's conjectures for $SL(2,\mathbb{R})$,* Israel J. Math. 80 (1992), 1-31.

[R6]   Marina Ratner, *Invariant measures and orbit closures for unipotent actions on homogeneous spaces,* Geom. and Funct. Anal. 4 (1994), 236-257.

[R7]   Marina Ratner, *Raghunathan's conjecture for cartesian products of real and p-adic groups,* Duke Math. J. (to appear).

[Ro]   V. A. Rohlin, *Selected topics from metric theory of dynamical systems,* Usp. Mat. Nauk 4 (No. 2) (1949), 57-128 (Russian) = Amer. Math. Soc. Translations 49 (1966), 171-239.

[Se]   J.-P. Serre, *A Course in Arithmetic,* Springer Verlag, 1973.

[Sh1] Nimish Shah, *Uniformly distributed orbits of certain flows on homogeneous spaces,* Math. Ann. 289 (1991), 315-334.

[Sh2] Nimish Shah, *Limit distributions of polynomial trajectories on homogeneous spaces,* Duke Math. J. 75 (1994), 711-732.

[Sh3] Nimish Shah, *Limit distributions of expanding translates of certain orbits on homogeneous spaces,* (preprint).

[Si]   C. L. Siegel, *A mean value theorem in geometry of numbers,* Ann. of Math. (2) 46 (1945), 340-347.

[Sp]   R. J. Spatzier, *On lattices acting on boundaries of semisimple groups,* Ergod. Th. and Dynam. Sys. 1 (1981), 489-494.

[St1] A. N. Starkov, *Spaces of finite volume,* Vestnik Mosk. Univ., Ser. Mat., 5 (1986), 64-66.

[St2] A. N. Starkov, *On a criterion for the ergodicity of G-induced flows,* Usp. Mat. Nauk 42 (1987), No 3, 197-198 (Russian) = Russian Math. Surveys 42 (1987), 233-234.

[St3] A. N. Starkov, *Ergodic decomposition of flows on homogeneous spaces of finite volume,* Mat. Sbornik 180 (1989), 1614-1633 (Russian) = Math. USSR Sbornik 68 (1991), 483-502.

[St4] A. N. Starkov, *Structure of orbits of homogeneous flows and Raghunathan's conjecture,* Usp. Mat. Nauk 45 (1990), 219-220 (Russian) = Russian Math. Surveys 45 (1990), 227-228.

[St5] A. N. Starkov, *Horospherical flows on spaces of finite volume,* Mat. Sbornik 182 (1991), 774-784 (Russian) = Math. USSR Sbornik 73 (1992), 161-170.

[St6] A. N. Starkov, *On mixing of higher degrees of homogeneous flows,* Dokl. Russ. Akad. Nauk 333 (1993), 28-31.

[St7] A. N. Starkov, *Around Ratner,* (preprint, ICTP, Trieste, 1994).

[V1] W. A. Veech, *Minimality of horospherical flows,* Israel J. Math. 21 (1975), 233-239.

[V2] W. A. Veech, *Unique ergodicity of horocycle flows,* Amer. J. Math. 99 (1977), 827-859.

[Ve] A. Verjovsky, *Arithmetic, geometry and dynamics in the unit tantent bundle of the modular orbifold,* In: Proceedings of a conf. on *Dynamical Systems,* Santiago (1990), pp. 263-298, Pitman Re. Notes Math. Ser. 285, Longman Sci. Tech., Harlow, 1993.

[Wa] P. Walters, *An Introduction to Ergodic Theory,* Springer Verlag, 1982.

[We] H. Weyl, *Über die gleichverteilung von Zahlen mod. Eins,* Math. Ann. 77 (1916), 313-352.

[Wi1] D. Witte, *Zero entropy affine maps on homogeneous spaces,* Amer. J. Math. 109 (1987), 927-961.

[Wi2] D. Witte, *Measurable quotients of unipotent translations on homogeneous spaces,* Trans. Amer. Math. Soc. 345 (1994), 577-594.

[Z1] R. J. Zimmer, *Equivariant images of projective space under the action of $SL(n, \mathbb{Z})$,* Ergod. Th. and Dynam. Sys. 1 (1981), 519-522.

[Z2] R. J. Zimmer, *Ergodic Theory and Semisimple Groups,* Birkhauser, Boston, 1984.

[Z3] R. J. Zimmer, *Amenable actions and dense subgroups of Lie groups*, J. Funct. Anal. 72 (1987), 58-64.

[Z4] R. J. Zimmer, *Superrigidity, Ratner's theorem and fundamental groups*, Israel J. Math. 74 (1991), 199-207.

[Z5] R. J. Zimmer, *Discrete groups and non-Riemannian homogeneous spaces*, J. Amer. Math. Soc. 7 (1994), 159-168.

THE VARIATIONAL PRINCIPLE
FOR HAUSDORFF DIMENSION:
A SURVEY

DIMITRIOS GATZOURAS AND YUVAL PERES

University of Crete and University of California, Berkeley

1. Introduction

The variational principle for entropy highlights the significance of measures of maximal entropy. Indeed, these are widely considered to be the most important invariant measures from a purely dynamical perspective. However, other invariant measures can be better adapted to the geometry. This is illustrated by the self-map f of $[0,1)$, defined by $f(x) = 3x/2$ for $x \in [0, 2/3)$ and $f(x) = 3x - 2$ for $x \in [2/3, 1)$. Since f is conjugate to the 2-shift, its topological entropy is $\log 2$; Lebesgue measure is invariant under f, but does not have maximal entropy.

More generally, given an expanding self-map f of a compact manifold (which by a result of Shub must then be a torus), it is generally accepted that the measure with the greatest geometric significance among f-invariant measures is the one equivalent to Lebesgue measure. (See Krzyzewski and Szlenk (1969) for the existence of such measures.) If instead we consider an expanding map $f : K \to K$ of a compact set K of zero Lebesgue measure, the closest analogue to such a measure is an *f-invariant measure of full Hausdorff dimension*. This survey addresses the existence question for such measures, which is mostly open, and discusses known partial results. We sketch some of the proofs, and give references for the complete arguments (which are somewhat technical).

Consider a smooth map $f : U \to M$, where $U \subset M$ is open and M is a Riemannian manifold. (Throughout the survey the term manifold will refer to a C^1 connected manifold.)

Y. Peres' work was partially supported by NSF grant DMS-9404391.

Typeset by $\mathcal{A}\mathcal{M}\mathcal{S}$-TEX

Problem 1. *Let $K \subset U$ be a compact f-invariant set (i.e., $f(K) \subset K$). Assuming that f is expanding on K, i.e.,*

$$\| Df^n(x)(u) \| \geq C\lambda^n \| u \| \qquad \forall\, u \in T_x M, \; x \in K, \; n \geq 1$$

for some constants $C > 0$ and $\lambda > 1$, under what conditions is it true that K supports an f-invariant measure of full Hausdorff dimension?

This problem was raised in Gatzouras and Lalley (1992), and in Kenyon and Peres (1995a).

By an f-invariant measure we mean an invariant Borel probability measure. The Hausdorff dimension of a Borel probability measure μ on a metric space is defined by

$$\dim_H(\mu) = \inf \{ \, \dim_H(B) : B \text{ a Borel set with } \mu(B) = 1 \, \} \; ;$$

Naturally, we refer to a measure μ with $\dim_H(\mu) = \dim_H(K)$ as a *measure of full dimension* for K.

We remark that Problem 1 can be formulated in an arbitrary metric space setting, but assuming K lies in a manifold covers most applications.

To motivate the assumption that f is expanding, we recall a result of Manning and McClusky (1983): *There exist Axiom-A diffeomorphisms f, with corresponding basic sets $\Omega(f)$, whose dimensions cannot even be approximated by the dimensions of the ergodic measures they support, i.e., the following strict inequality holds:*

$$\dim_H(\Omega(f)) > \sup \{ \, \dim_H(\mu) : \mu(\Omega(f)) = 1, \; \mu \text{ ergodic } \} \, .$$

2. The conformal case

Call a differentiable mapping f defined on an open set in a manifold *conformal*, if the derivative $Df(x)$ at each point is a scalar multiple of an isometry. (Denote the magnitude of this scalar by $|Df(x)|$.) Invariant measures of full Hausdorff dimension *do* exist for arbitrary compact sets invariant under an expanding conformal map (see Theorem 1 below).

Consider first the map $f(x) = bx(\mod 1)$ of the circle, where $b > 1$ is an integer. By applying the Shannon-McMillan-Breiman theorem, Billingsley (1960) showed that

$$(2.1) \qquad \dim_H(\mu) = \frac{h_\mu(f)}{\log b}$$

for any ergodic f-invariant measure μ, where h_μ is the measure-theoretic entropy. Furstenberg (1967) showed that

$$(2.2) \qquad \dim_H(K) = \frac{h_{\text{top}}(f|_K)}{\log b}$$

for any invariant set K in the circle, where h_{top} is the topological entropy. Combining these two formulas, the variational principle for entropy implies a variational principle for dimension:
(2.3)
$$\dim_H(K) = \max \{ \dim_H(\mu) : \mu(K) = 1, \ \mu \text{ ergodic and } f - \text{invariant} \} .$$

(The maximum is attained since for expansive maps there always exists an ergodic measure of maximal entropy).

The same proof works for conformal expanding linear endomorphisms of the torus. Subsequent development of the thermodynamic formalism enabled an extension of this argument to nonlinear expanding maps. The following is an easy consequence of the variational principle for topological pressure:

Theorem 1. *Let $f : U \to M$ be a C^1 map, where $U \subset M$ is open and M is a Riemannian manifold. If f is conformal, then any compact f-invariant set $K \subset U$, on which f is expanding, carries an f-invariant ergodic measure μ with $\dim_H(\mu) = \dim_H(K)$.*

For $f \in C^{1+\eta}$ (for $\eta > 0$) and K a repeller (i.e., $f(K) = K$ and $f^{-1}(K) \cap V = K$ for some open neighbourhood of K), which is topologically transitive, this result is implicit in Ruelle (1982), and, in a different context, in Bowen (1979). Other versions of this are in works of Bedford, Hofbauer, Raith and Urbanski; the version cited is from Gatzouras and Peres (1995).

Proof-sketch for Theorem 1: The ingredients of the proof are the variational principle for pressure and the following well known extension of (2.1): *If μ is an ergodic measure supported by a compact set on which $\inf |Df(x)| > 1$ then*

$$(2.4) \qquad\qquad \dim_H(\mu) = \frac{h_\mu(f)}{\int \log |Df| d\mu} .$$

By taking a power of f if necessary, we may assume that $\inf_K |Df(x)| > 1$. Define $\varphi(x) := \log |Df(x)|$, for $x \in K$, and let θ be the unique nonnegative real number satisfying $P(-\theta\varphi, f|_K) = 0$, where P denotes topological pressure. By the variational principle for pressure, and expansiveness of f, there exists an ergodic f-invariant measure μ, which is an equilibrium state for the function $-\theta\varphi$ on K. (See the remarks at the end of Walters (1976).) By the choice of θ and formula (2.4), the dimension of this measure is θ. To prove Theorem 1, it only remains to verify that $\dim_H(K) \le \theta$; this is straightforward using the non-variational definition of pressure (see Walters (1976, 1982)).

3. Nonconformal maps

The question of existence of invariant measures of full dimension is mostly open. Besides the case where f is conformal, there are only few other cases where an answer is known. (Maps f such that for each x, all the eigenvalues of the derivative $Df(x)$ have the same modulus, can often be handled like conformal maps; we exclude them from the discussion in this section.)

Nonconformal maps expand at different rates in different directions, and there is no general analogue of (2.4). In some cases, e.g., when f is a smooth expanding self-mapping of a surface, which has globally defined strong and weak unstable foliations, one can express the dimension of an ergodic measure μ as a positive combination of entropies. Denote by f_* the action of f on the leaves of the strong unstable foliation, and by $\pi_*\mu$ the projection of μ to the space of leaves. Then

$$(3.1) \qquad \dim_H(\mu) = \frac{h_\mu(f)}{\lambda_\mu^{(1)}(f)} + \left(\frac{1}{\lambda_\mu^{(2)}(f)} - \frac{1}{\lambda_\mu^{(1)}(f)} \right) h_{\pi_*\mu}(f_*),$$

where $0 < \lambda_\mu^{(2)}(f) < \lambda_\mu^{(1)}(f)$ are the Lyapunov exponents of the system (f, μ). This follows from a general formula of Ledrappier and Young (1985), as explained in Ledrappier (1991). The general definition of weak and strong unstable foliations depends on Oseledec's theorem (see Ruelle (1979), Ledrappier and Young (1985)). Two points u, v are on the same leaf of the strong unstable foliation if there exist u_n, v_n such that

$$f^n(u_n) = u, \quad f^n(v_n) = v \quad \text{and} \quad \lim_n \frac{1}{n} \log \left(\mathrm{dist}(u_n, v_n) \right) = -\lambda_\mu^{(1)}(f).$$

The weak unstable foliation is defined analogously using the smaller Lyapunov exponent $\lambda_\mu^{(2)}(f)$. For simplicity, we will concentrate below mostly on the case where the two foliations are given by the horizontal and vertical lines, respectively. Even in this case Problem 1 is open; a major challenge is to use Lyapunov charts to transfer results from this to more general situations.

We now describe some nonconformal maps for which Problem 1 has been solved.

3.1 Diagonal linear maps: Fix integers $1 < m \leq n$ and let D be a subset of

$\{0, \ldots, n-1\} \times \{0, \ldots, m-1\}$. Consider the set $K \subset [0,1]^2$ defined by

$$(3.2) \qquad K = \left\{ \sum_{k=1}^\infty \begin{pmatrix} n^{-k} & 0 \\ 0 & m^{-k} \end{pmatrix} d_k \; : \; d_k \in D, \, \forall \, k \in \mathbf{N} \right\}.$$

Viewed as a subset of the torus $[0,1)^2$, the compact set K is invariant under the toral endomorphism $x \longmapsto \begin{pmatrix} n & 0 \\ 0 & m \end{pmatrix} x$, acting modulo 1. We refer to

K as a *self-affine carpet*, because

$$K = \bigcup_{d \in D} \left\{ \begin{pmatrix} n^{-1} & 0 \\ 0 & m^{-1} \end{pmatrix} (K + d) \right\}.$$

The Hausdorff dimension of these sets was found by McMullen (1984) (who called them "general Sierpinski carpets"), and independently by Bedford (1984). In the course of their (different) proofs, they establish the following.

Theorem 2. [McMullen (1984), Bedford (1984)]. *The set K defined in (3.2) supports an invariant Bernoulli probability measure μ, which is of full Hausdorff dimension.*

The Ledrappier-Young formula (3.1) implies that μ is the *unique* ergodic measure of full dimension on K (see Kenyon and Peres (1995a)).

The self-affine carpet K may be obtained by applying the representation map

$$(3.3) \qquad R(\{d_k\}_{k \in \mathbb{N}}) = \sum_{k=1}^{\infty} \begin{pmatrix} n^{-k} & 0 \\ 0 & m^{-k} \end{pmatrix} d_k$$

to the full shift $D^{\mathbb{N}}$. The next theorem generalizes the result of McMullen and Bedford in two ways: higher dimensions and arbitrary invariant sets.

Theorem 3. [Kenyon and Peres (1995a)]. *Let f be a (linear) expanding endomorphism of the d-dimensional torus, which has integer eigenvalues, or more generally eigenvalues with an integer power λ^k. Then any f-invariant subset of the torus supports an f-invariant ergodic measure of full Hausdorff dimension.*

3.2 Diagonal piecewise-linear maps: Gatzouras and Lalley (1992) extend the results on self-affine carpets in a different direction. Fix numbers $0 < a_{ij} < b_i < 1$ and $0 < c_{ij}, d_i < 1$, for $j = 1, \ldots, n_i$ and $i = 1, \ldots, m$, such that the rectangles $Q_{ij} := (c_{ij}, c_{ij} + a_{ij}) \times (d_i, d_i + b_i)$ are pairwise disjoint subsets of $(0,1)^2$. Set $D = \{(i,j) : 1 \le j \le n_i \text{ and } 1 \le i \le m\}$. Denote $T_{ij}(x) = \begin{pmatrix} a_{ij} & 0 \\ 0 & b_i \end{pmatrix} x + \begin{pmatrix} c_{ij} \\ d_i \end{pmatrix}$, so that T_{ij} maps the open unit square onto Q_{ij}. These maps are contractions, so by Hutchinson (1981) there exists a unique compact set K satisfying

$$K = \bigcup_{(i,j) \in D} T_{ij}(K).$$

Let f be any mapping satisfying $f|_{Q_{ij}} = (T_{ij}|_{Q_{ij}})^{-1}$. Then K is a compact f-invariant subset of $[0,1]^2$ and K corresponds to the full shift $D^{\mathbb{N}}$

again, this time under the representation map

$$R_f(\{(i_k, j_k)\}_{k \in \mathbb{N}}) = \begin{pmatrix} c_{i_1 j_1} \\ d_{i_1} \end{pmatrix}$$

$$+ \sum_{k=1}^{\infty} \begin{pmatrix} a_{i_1 j_1} & 0 \\ 0 & b_{i_1} \end{pmatrix} \cdots \begin{pmatrix} a_{i_k j_k} & 0 \\ 0 & b_{i_k} \end{pmatrix} \begin{pmatrix} c_{i_{k+1} j_{k+1}} \\ d_{i_{k+1}} \end{pmatrix}.$$

Theorem 4. [Gatzouras and Lalley (1992)]. *The invariant set K supports an f-invariant ergodic measure of full Hausdorff dimension. The measure of full dimension may be chosen to be Bernoulli.*

(In this theorem, it is not known if the ergodic measure of full dimension is necessarily unique.)

Other diagonally self-affine sets were analyzed earlier by Przytycki and Urbanski (1989), who related them to a problem of Erdős on absolute continuity of infinitely convolved Bernoulli measures; major progress on the latter problem has been recently achieved by Solomyak (1995), who used an idea of Pollicott and Simon (1995).

3.3 Piecewise-linear maps with rotations: The papers by Falconer (1988), and Hueter and Lalley (1995) find the dimension for certain classes of self-affine sets. Each of these sets is defined as the attractor for a finite collection of contracting affine maps; it is invariant under an expanding map f obtained by patching together local inverses of the contractions. The key difference from the cases discussed in the previous subsections is that the linear parts of the contractions do not commute. The measures of full dimension constructed in these papers are f-invariant and ergodic. The expanding maps f obtained in this manner are quite complicated, but only invariant sets with simple symbolic dynamics (corresponding to full-shifts) have been analyzed. We refer to Edgar (1992) for an insightful review of Falconer's theorem and related results.

3.4 Products of nonlinear conformal maps: For $i = 1, 2$, let U_i be an open subset of a Riemannian manifold M_i and $f_i : U_i \to M_i$ a conformal C^1 mapping. Denote $f(x, y) = (f_1(x), f_2(y))$. We stress that an f-invariant set $K \subset U_1 \times U_2$ need *not* be of the form $K = K_1 \times K_2$.

A compact set Λ_i is called an *invariant repeller* for f_i if:
- f_i is expanding on Λ_i.
- $f_i(\Lambda_i) = \Lambda_i$.
- $f_i^{-1}(\Lambda_i) \cap V_i = \Lambda_i$ for some open neighbourhood V_i of Λ_i.

If f_i has an invariant repeller Λ_i then $f_i|_{\Lambda_i}$ has Markov partitions of arbitrarily small diameter. Hence, there exist subshifts of finite type, Ω_1 and Ω_2, and a representation map $R_f : \Omega_1 \times \Omega_2 \to \Lambda_1 \times \Lambda_2$ which is continuous, surjective, shift-commuting and finite-to-one.

We need one more definition. Let $\mathcal{L}$ be a finite alphabet and let X be a subshift of $\mathcal{L}^{\mathbb{N}}$. X is said to satisfy **specification** if there is an $m \in \mathbb{N}$ with the following property: Given any two blocks ξ_1 and ξ_2 that are legal in X (i.e., may be extended to sequences in X), there is a block η of length m, such that the block $\xi_1 \eta \xi_2$ is legal in X. (It follows by induction that any finite number of legal blocks in X can be glued together to a legal block in X by inserting appropriate blocks of length m between them.) See Denker, Grillinberger and Siegmund (1976) for a proof that shifts of finite type and (more generally) sofic shifts satisfy specification.

Theorem 5. [Gatzouras and Peres (1995)]. *For $i = 1, 2$ let U_i be an open subset of a Riemannian manifold M_i and let $f_i : U_i \to M_i$ be a conformal C^1 mapping with invariant repeller Λ_i. Set $f(x, y) = (f_1(x), f_2(y))$. If*

$$\min_{x \in \Lambda_1} |Df_1(x)| \geq \max_{y \in \Lambda_2} |Df_2(y)|,$$

then for any f-invariant subset $K \subset \Lambda_1 \times \Lambda_2$, for which the subshift $R_f^{-1}(K)$ satisfies specification, we have

$$\dim_H(K) = \sup_{\mu \in \mathrm{ERG}_f(K)} \dim_H(\mu).$$

Here $\mathrm{ERG}_f(K)$ is the set of f-invariant ergodic (Borel probability) measures supported by K.

3.5 Proof outlines. *Proof-sketch for the McMullen-Bedford result (Theorem 2):* Any probability vector $\{p(d) : d \in D\} = \boldsymbol{p}$ defines a probability measure μ_p on K which is the image of the product measure $\boldsymbol{p}^{\mathbb{N}}$ under the representation map R in (3.3). We restrict attention to probability vectors $\boldsymbol{p}$ such that $p(d)$ depends only on the second coordinate of d.

For $d \in D$, denote by $a(d)$ the number of elements of D that have the same second coordinate as d, and denote $\alpha = \frac{\log m}{\log n}$. By applying the strong Law of large numbers, McMullen essentially showed that

$$(3.4) \qquad \dim_H(\mu_p) = \frac{1}{\log m} \sum_{d \in D} p(d)[\log \frac{1}{p(d)} + \log(a(d)^{\alpha - 1})].$$

(He did not use this terminology, however.) This yields a lower bound on $\dim_H(K)$. Bedford derived the same lower bound by combining the law of large numbers with a theorem of Marstrand. Note that (3.4) can now be viewed as a special case of the Ledrappier-Young formula (3.1).

An easy and well known calculation shows that the right-hand-side of (3.4) is maximized for $p(d) = \frac{1}{Z} a(d)^{\alpha - 1}$ where $Z = \sum_{d \in D} a(d)^{\alpha - 1}$. Note that for this choice of $\boldsymbol{p}$, $\dim_H(\mu_p) = \log_m(Z)$, so this is also a lower bound

for $\dim_H(K)$. The most surprising part of McMullen's proof shows that it is also an upper bound, by employing the following observation:
Let $\{S_k\}$ be a sequence with bounded increments. For any $0 < \alpha < 1$ we have

$$\limsup_{k \to \infty} \left[\frac{S_k}{k} - \frac{S_{\alpha k}}{\alpha k} \right] \geq 0.$$

Bedford proves the sharp upper bound on $\dim_H(K)$ by decomposing K into subsets that are "typical" for different measures of the type μ_p.

Remark on the proof Theorem 4: Broadly speaking, Gatzouras and Lalley (1992) use McMullen's approach for the lower bound, and Bedford's approach for the upper bound; the details are considerably more involved, however.

Proof-sketch for Theorems 3 and 5: To simplify the exposition we limit the generality and consider the case where $M_1 = M_2 = \mathbb{R}$. Furthermore, for $i = 1, 2$ we suppose that A_i is a union of two disjoint closed subintervals of $[0, 1]$ and f_i maps each of these intervals onto $[0, 1]$. We assume that the derivatives satisfy $f_1'(x) > f_2'(y) > 1$ for all $x \in A_1$ and $y \in A_2$.

(The simplest examples are the linear maps $f_1(x) = nx \,(\mathrm{mod}\,1)$ and $f_2(y) = my \,(\mathrm{mod}\,1)$, with $1 < m < n$. For nonlinear examples, consider the quadratic maps $f_1(x) = \alpha x(1-x)$ and $f_2(y) = \beta y(1-y)$, where $\beta > 2 + \sqrt{5}$ and $\alpha > 2 + \sqrt{4 + \beta^2}$, and take A_i to be the preimage of $[0, 1]$ under f_i.)

Set $\Lambda_i = \cap_{k \geq 0} f_i^{-k}(A_i)$ for $i = 1, 2$. Then the mapping $f(x, y) = (f_1(x), f_2(y))$ satisfies the hypotheses of Theorem 5. Here $\Omega_i = \{0, 1\}^{\mathbb{N}_0}$, for both $i = 1, 2$ and the representation map R_f is $1 - 1$. The set $\cap_{l=0}^{k-1} f_1^{-l}(A_1) \times \cap_{l=0}^{k-1} f_2^{-l}(A_2)$ consists of 2^k rectangles which can be coded by the finite sequences $\omega \in (\{0, 1\} \times \{0, 1\})^k$; we call each such rectangle $Q(\omega)$.

Fix an arbitrary f-invariant set $K \subset \Lambda_1 \times \Lambda_2$. The idea is to approximate K by self-affine carpets. Now K differs from a self-affine carpet in two ways:
(a) It corresponds to a subshift rather than a full shift;
(b) The map f with respect to which K is invariant may be nonlinear.

Thus the approximation must deal with both of these differences. Let D_l consist of those blocks ω in $(\{0, 1\} \times \{0, 1\})^l$ that are legal in $R_f^{-1}(K)$, i.e., $\omega \in D_l$ iff $Q(\omega) \cap K \neq \emptyset$. The first step is to approximate K by $\widetilde{K}_l := R_f(\Omega^{(l)})$, where $\Omega^{(l)}$ is the set consisting of all infinite sequences which are concatenations of elements of D_l (i.e., $\Omega^{(l)}$ is isomorphic to the full shift $D_l^{\mathbb{N}}$). This takes care of (a). To deal with (b) we approximate $\widetilde{K}_l$ by the self-affine carpet K_l which is constructed from the rectangles $\{Q(\omega) : \omega \in D_l\}$ as in subsection 3.2. By Theorem 4 there exists a measures μ_l on K_l with $\dim_H(\mu_l) = \dim_H(K_l)$. Furthermore, there exists a natural mapping $g_l : \widetilde{K}_l \to K_l$, which is $1 - 1$ and onto. Both g_l and g_l^{-1} are almost Lipshitz: in particular g_l and its inverse are both Hölder continuous with exponents tending to 1 as $l \to \infty$. This implies that g_l does not change dimensions by much.

Transfer μ_l, through g_l^{-1}, to a measure $\widehat{\mu}_l$ on $\widetilde{K}_l$. This new measure will by construction be f^l-invariant. Setting $\widetilde{\mu}_l = \frac{1}{l}\sum_{i=0}^{l-1}\widehat{\mu}_l \circ f^{-i}$ yields an f-invariant ergodic measure. The next step is to take a weak* subsequential limit μ of the $\widetilde{\mu}_l$. It is not hard to verify that μ is supported by K; however, μ *is not necessarily ergodic.* Now

$$\dim_H(\widetilde{\mu}_l) = \dim_H(\widehat{\mu}_l) \approx \dim_H(\mu_l) = \dim_H(K_l) \approx \dim_H(\widetilde{K}_l) \geq \dim_H(K)$$

(since $K \subset \widetilde{K}_l$ for all l) and μ would be an f-invariant measure of full Hausdorff dimension if we knew that:

(3.5)   the function $\mu \longmapsto \dim_H(\mu)$ is upper semi-continuous.

The linear case: completing Theorem 3. Assume for concreteness that $f_1(x) = nx \,(\mathrm{mod}\,1)$ and $f_2(y) = my \,(\mathrm{mod}\,1)$. Formula (3.1) for the dimension of an ergodic measure ν reduces to

$$\dim_H(\nu) = \frac{1}{n}h_\nu(f) + \left(\frac{1}{m} - \frac{1}{n}\right)h_{\nu_y}(f_2),$$

where ν_y is projection of ν onto the y-axis. Since μ is a weak*-limit point of the μ_l,

$$\limsup_{l \to \infty}\dim_H(\mu_l) \leq \frac{1}{n}h_\mu(f) + \left(\frac{1}{m} - \frac{1}{n}\right)h_{\mu_y}(f_2)$$

by the upper semi-continuity of entropy for expansive maps. Using the ergodic decomposition of μ, it follows that

(3.6) $$\dim_H(\mu) \geq \frac{1}{n}h_\mu(f) + \left(\frac{1}{m} - \frac{1}{n}\right)h_{\mu_y}(f_2)$$

Therefore some ergodic component of μ is an invariant ergodic measure of full dimension.

The nonlinear case: completing Theorem 5. In the example we are considering formula (3.1) becomes

(3.7) $$\dim_H(\nu) = \frac{h_\nu(f)}{\int |Df_1|d\nu_x} + \left(\frac{1}{\int |Df_2|d\nu_y} - \frac{1}{\int |Df_1|d\nu_x}\right)h_{\nu_y}(f_2)$$

where ν_x and ν_y are the projections of ν onto the x- and y- axes again. While the inequality
(3.8)

$$\limsup_{l \to \infty}\dim_H(\mu_l) \leq \frac{h_\mu(f)}{\int |Df_1|d\mu_x} + \left(\frac{1}{\int |Df_2|d\mu_y} - \frac{1}{\int |Df_1|d\mu_x}\right)h_{\mu_y}(f_2)$$

still holds, we cannot show an inequality like (3.6) for a non-ergodic invariant measure by appealing to ergodic decomposition, because of the dependence of the denominators in (3.7) on ν.

We now outline briefly the rest of the proof of Theorem 5. Recall that K is an f-invariant set approximated by the self-affine carpets K_l. Let m be as in the definition of specification for $R_f^{-1}(K)$. Define a mapping $\Psi_l : K \to K_l$ as follows. For $(x, y) \in K$ let $\omega = R_f^{-1}(x, y)$; let ω' be the infinite sequence obtained from ω by keeping its first l digits, deleting its next m digits, then keeping the next l digits of ω, deleting the m following, and so on. Finally, define $\Psi_l(x, y)$ to be the element of K_l corresponding to the sequence ω'. The mapping Ψ_l is surjective, by specification and compactness. Furthermore, since m is constant, the maps Ψ_l are almost Lipshitz; more precisely, they are Hölder continuous with exponents tending to 1 as $l \to \infty$. This implies that Ψ_l does not change dimensions by much. Now pull the measures μ_l from K_l to measures μ_l^* on K, through Ψ_l. Then

$$\dim_H(\mu_l^*) \approx \dim_H(\mu_l) = \dim_H K_l \geq \dim_H K$$

and we have a sequence of measures *supported on K*, whose dimensions converge to the dimension of K.

4. Concluding remarks

By using Markov partitions and formula (3.1), we can translate Problem 1 for the maps considered in subsection 3.4 into a question in symbolic dynamics:

Problem 2. *Let X and Y be subshifts, where Y a factor of X with factor map $\pi : X \to Y$. Let $\phi : X \to \mathbb{R}$ and $\psi : Y \to \mathbb{R}$ be positive continuous functions, with $\phi(x) \geq \psi \circ \pi(x)$ for all $x \in X$ and denote by σ_X and σ_Y the (left) shifts on X and Y respectively. Is the maximum of*

$$\frac{h_\mu(\sigma_X)}{\int \phi d\mu} + \left(\frac{1}{\int \psi \circ \pi d\mu} - \frac{1}{\int \phi d\mu} \right) h_{\mu \circ \pi^{-1}}(\sigma_Y)$$

over all σ_X-invariant measures, attained on the set $ERG_{\sigma_X}(X)$ of ergodic measures on X?

Even in the cases where we know invariant measures of full dimension exist there are still some interesting open problems. For example, Kenyon and Peres (1995a) show that the measure found by McMullen is the *unique* ergodic measure of full Hausdorff dimension for the self-affine carpets of subsection 3.1. However, uniqueness is not known for more general invariant subsets $R(\Omega)$ (where R is defined in (3.3)), even if Ω is a transitive subshift of finite type. Furthermore, we would like to know whether measures of full dimension have good mixing properties. This can also be formulated as a problem in symbolic dynamics:

Problem 3. *Let X be a transitive subshift of finite type and let Y be a symbolic factor of X (so that Y is a sofic shift). Denote by π the factor map and by σ_X, σ_Y the shift maps on X and Y respectively. Fix $\alpha > 0$ and consider all the ergodic shift-invariant measures μ on X which maximize the expression*

$$h_\mu(\sigma_X) + \alpha h_{\mu \circ \pi^{-1}}(\sigma_Y).$$

Are all these measures mixing? Bernoulli? Is there in fact a unique maximizing measure?

Remark. The preprints by Hu (1994) and Zhang (1994), prove interesting inequalities for the Hausdorff dimension of sets and measures invariant under more general nonconformal maps than those considered in Section 3. These papers do not directly address the variational principle for dimension, but it is quite possible that combining their methods with those described in this survey will yield a substantial extension of Theorem 5.

REFERENCES

T. Bedford, *Crinkly curves, Markov partitions and box dimension self-similar sets*, Ph.D. Thesis, University of Warwick (1984).

T. Bedford and M. Urbanski, *The box and Hausdorff dimension of self-affine sets*, Ergod. Th. & Dynam. Sys. **10** (1990), 627-644.

P. Billingsley, *Hausdorff dimension in probability theory*, Illinois J. of Math. **4** (1960), 187-209.

R. Bowen, *Some systems with unique equilibrium states*, Math. Sys. Theory **8** (1974), 193-202.

_____, *Hausdorff dimension of quasi-circles*, Publ. Math. I.H.E.S. **50** (1979), 11-26.

M. Denker, C. Grillenberger and K. Sigmund, *Ergodic theory on compact spaces*, Lec. Notes in Math., No. 527, Springer Verlag, 1976.

G. A. Edgar, *Fractal dimension of self-affine sets: some examples*, Supplemento ai Rendiconti del Cirrolo Matematico di Palermo, Serie II **28** (1992), 341-358.

K.J. Falconer, *The Hausdorff dimension of self-affine fractals*, Math. Proc. Camb. Phil. Soc. **103** (1988), 339-350..

_____, *Fractal Geometry-Mathematical foundations and applications*, Wiley, New York, 1990.

H. Furstenberg, *Disjointness in ergodic theory, minimal sets, and a problem in Diophantine approximation*, Math. Sys. Theory **1** (1967), 1-49.

D. Gatzouras and S. Lalley, *Hausdorff and box dimensions of certain self-affine fractals*, Indiana Univ. Math. J. **41** (1992), 533-568.

D. Gatzouras and Y. Peres, *Invariant measures of full Hausdorff dimension for some expanding maps*, preprint, to appear in Ergod. Th. & Dynam. Sys. (1995).

N.T.A. Haydn and D. Ruelle, *Equivalence of Gibbs and equilibrium states for homeomorphisms satisfying expansiveness and specification*, Comm. Math. Phys. **148** (1992), 155-167.

F. Hofbauer and M. Urbanski, *Fractal properties of invariant subsets for piecewise monotonic maps of the interval*, Trans. Amer. Math. Soc. **343** (1994), 659-671.

H. Hu, *Dimensions of invariant sets of expanding maps*, preprint (1994).

124   *ERGODIC THEORY OF* $\mathbf{Z}^d$ *ACTIONS*

I. Hueter and S. Lalley, *Falconer's formula for the Hausdorff dimension of a self-affine set in* $\mathbf{R}^2$, Ergod. Th. & Dynam. Sys. **15** (1995), 77-97.

J.E. Hutchinson, *Fractals and self-similarity*, Indiana University Math. J. **30** (1981), 271-280.

K. Krzyzewski and W. Szlenk, *On invariant measures for expanding differentiable mappings*, Studia Mathematica **T, XXXIII** (1969), 83-92.

R. Kenyon and Y. Peres, *Measures of full dimension on affine-invariant sets*, preprint, to appear in Ergod. Th. & Dynam. Sys. (1995a).

———, *Hausdorff dimensions of sofic affine-invariant sets*, preprint, to appear in Israel J. of Math. (1995b).

F. Ledrappier, *On the dimension of some graphs*, Symbolic Dynamics and its Applications, Contemp. Math. (P. Walters, ed.), vol. **135**, 1991, pp. 285-294.

F. Ledrappier and M. Misiurewicz, *Dimension of invariant measures for maps with exponent zero*, Ergod. Th. & Dynam. Sys. **5** (1985), 595-610.

F. Ledrappier and L.S. Young, *The metric entropy of diffeomorphisms II*, Ann. Math. **122** (1985), 540-574.

A. Manning and H. McCluskey, *Hausdorff dimension for horseshoes*, Ergod. Th. & Dynam. Sys. **3** (1983), 251-260.

K. McMullen, *The Hausdorff dimension of general Sierpinski carpets*, Nagoya Math. J. **96** (1984), 1-9.

Y. Peres, *The self–affine carpets of McMullen and Bedford have infinite Hausdorff measure*, Math. Proc. Camb. Phil. Soc. **116** (1994), 513-526.

M. Pollicott and K. Simon, *The Hausdorff dimension of* $\lambda-$*expansions with deleted digits*, Trans. Amer. Math. Soc. **347** (1995), 967-983.

F. Przytycki and M. Urbanski, *On the Hausdorff dimension of some fractal sets*, Studia Math. **93** (1989), 155-186.

P. Raith, *Hausdorff dimension for piecewise monotonic maps*, Studia Mathematica **T. XCIV** (1989), 17-33.

D. Ruelle, *Statistical mechanics on a compact set with* $\mathbf{Z}^\nu$ *action satisfying expansiveness and specification*, Trans. Amer. Math. Soc. **185** (1973), 237-251.

———, *Thermodynamic formalism*, Encyclopedia of Mathematics and its Applications, Vol. 5 (Gian-Carlo Rota, ed.), Addison-Wesley, Reading, Mass., 1978.

———, *Ergodic Theory of differentiable dynamical systems*, Publ. Math. I.H.E.S. **50** (1979), 27-58.

———, *Repellers for real analytic maps*, Ergod. Th. & Dynam. Sys. **2** (1982), 99-107.

B. Solomyak, *On the random series* $\sum \pm \lambda^n$ *(an Erdős problem)*, preprint, to appear in Ann. Math. **142** (1995).

S. Takahashi, *A variational formula for spectra of linear cellular automata*, J. Analyse Math. **64** (1994), 1-51.

P. Walters, *A variational principle for the pressure of continuous transformations*, Amer. J. Math. **97** (1976), 937-971.

———, *An Introduction to ergodic theory*, Springer-Verlag, Berlin, 1982.

B. Weiss, *Subshifts of finite type and sofic systems*, Monatsh. Math. **77** (1973), 462-474.

L.S. Young, *Dimension, entropy and Lyapunov exponents*, Ergod. Th. & Dynam. Sys. **2** (1982), 109-124.

Y. Zhang, *A relation between Hausdorff dimension and topological pressure with applications*, preprint (1994).

DEPARTMENT OF MATHEMATICS, UNIVERSITY OF CRETE, 714 09 IRAKLION, CRETE, GREECE

E-mail address: gatzoura@edu.uch.gr

DEPARTMENT OF STATISTICS, UNIVERSITY OF CALIFORNIA, BERKELEY, CA 94720, U.S.A.
E-mail address: peres@stat.berkeley.edu

BOUNDARIES OF INVARIANT MARKOV OPERATORS :
THE IDENTIFICATION PROBLEM

Vadim A. Kaimanovich

CNRS URA-305
Institut de Recherche Mathématique de Rennes

Initially this paper was conceived as a short appendix to the recent article of Olshanetsky [Ol2] on the Martin boundary of symmetric spaces (which appeared nearly 25 years after the first research announcement [Ol1]), and was supposed to provide the reader with a background information about what was going on in the area during all that time. A preliminary version was circulated in 1993 under the title "An introduction to boundary theory of invariant Markov operators". Alas, soon it outgrew any reasonable limits for such an appendix, so that instead I decided to make of it a separate survey of the development of the boundary theory of invariant Markov operators on groups and homogeneous spaces during the last 2-3 decades (preserving, however, a special section devoted to boundaries of symmetric spaces, semi-simple Lie groups and their discrete subgroups). Still, trying to keep the survey as brief as possible, I had to omit a (rather large) number of topics closely connected with the boundary theory (ergodic properties and singularity of the harmonic measure with respect to other natural boundary measures; connections with such numerical characteristics as the spectral radius, growth, the rate of escape and the Hausdorff dimension of the harmonic measure; harmonic invariant measures of the geodesic flow, etc.). My intention was to concentrate on general ideas and methods used for describing the Martin boundary and its probabilistic counterpart – the Poisson boundary. These methods mostly bear a geometrical nature and in a sense are complementary to the approach used by Olshanetsky (direct estimate of the Green kernel by means of the harmonic analysis).

For a more extensive bibliography on random walks and Markov operators on algebraic and geometrical structures the reader is referred to [KV] and [An2], and to recent surveys [Gu6], [Ly] and [Wo3]. I am grateful to Martine Babillot for numerous helpful remarks and suggestions.

Typeset by $\mathcal{A}\mathcal{M}\mathcal{S}$-TEX

0. Markov operators and harmonic functions

0.1. General Markov operators.

A *Markov operator* on a measure space (X, m) is a linear operator $P : L^\infty(X, m) \hookleftarrow$ such that

1) P preserves the cone of non-negative functions;
2) $P\mathbf{1} = \mathbf{1}$ for the constant function $\mathbf{1}(x) \equiv 1 \; \forall \, x \in X$;
3) $Pf_n \downarrow \mathbf{0}$ a.e. whenever $f_n \downarrow \mathbf{0}$.

If P is an *integral operator*, i.e., $Pf(x) = \int p(x, y) f(y) dm(y)$, then the kernel functions $p(x, \cdot)$ are called its *transition densities*, and the probability measures $\pi_x = p(x, \cdot)m$ – its *transition probabilities*. If P is a *Lebesgue measure space* (i.e., its non-atomic part is isomorphic to the unit interval with the Lebesgue measure on it; e.g., see [CFS]), then the transition probabilities π_x (generally speaking, singular with respect to m), can be uniquely (mod 0) defined for a.e. point $x \in X$ by using existence of *regular conditional probabilities* in Lebesgue spaces, and the operator P can be presented as $Pf(x) = \int f(y) \, d\pi_x(y)$ [Ka9]. A measurable function f on X is called *P-harmonic* if $f = Pf$, i.e., if f satisfies the following *mean value property*: $f(x) = \int f(y) \, d\pi_x(y)$ for a.e. $x \in X$.

The adjoint operator of a Markov operator P acts in the space of measures θ absolutely continuous with respect to m, so that $\langle \theta P, f \rangle = \langle \theta, Pf \rangle$ (a standard convention in the theory of Markov operators is that one puts P on the right to denote the action of its adjoint operator on measures). For an arbitrary *initial distribution* $\theta \prec m$ let $\mathbf{P}_\theta$ be the corresponding *Markov measure* in the *path space* $X^{\mathbf{Z}_+} = \{\overline{x} = (x_0, x_1, \ldots)\}$ of the *Markov chain* determined by the operator P. The *one-dimensional distribution* of the measure $\mathbf{P}_\theta$ at time n is θP^n. See [Dy2], [Fo2], [Kr], [Re] for a general theory of Markov operators.

0.2. Examples of Markov operators.

If the state space X is countable (and m is the counting measure on X), then a Markov operator P is determined by the set of its transition probabilities $p(x, y)$. If X is a *locally finite graph*, then the transition probabilities $p(x, y) = 1/d_x$, where y is a neighbour of x and d_x is the total number of such neighbours, determine the Markov operator of the *simple random walk* on X [Wo3].

Another important class of Markov operators arises from *diffusion processes* on smooth manifolds X. In this case the time 1 transition operator $P = P^1$ can be included in the *semigroup of operators* $P^t = e^{t\mathcal{D}}$, where $\mathcal{D}$

is the *generating operator* of the diffusion. The transition densities of the operators P^t are fundamental solutions of the corresponding *heat equation* $\partial u/\partial t = \mathcal{D}u$. All P-harmonic functions are smooth, and the "global harmonicity" with respect to the operator P is equivalent to the usual "local harmonicity" ($\mathcal{D}f \equiv 0$) determined by the operator $\mathcal{D}$. If $\mathcal{D}$ is the Laplacian of a Riemannian metric on X, then this diffusion process is called the *Brownian motion* on X [An2].

For both these classes of Markov operators their transition probabilities are absolutely continuous (with respect to the counting measure in the first case and with respect to the smooth measure class in the second case). There is also an interesting class of Markov operators such that their transition probabilities, although being singular "globally" (with respect to the measure m on the state space X), can be considered as absolutely continuous in a "local sense". Namely, suppose that the space (X, m) is endowed with a *measured equivalence relation* R (e.g., see [HK]). Then X splits into equivalence classes (often called *leaves*) $[x] \subset X$, and a.e. leaf $[x]$ carries a measure class denoted $\overline{m}_{[x]}$. In the non-trivial case when the set of leaves is uncountable, the measure classes $\overline{m}_{[x]}$ are a.e. singular with respect to m. In this situation any measurable family of probability measures $\pi_x \prec \overline{m}_{[x]}$, $x \in X$ concentrated on classes $[x]$ determines a *global* Markov operator $P : L^\infty(X, m) \hookleftarrow$ whose transition probabilities π_x are *singular* with respect to m. On the other hand, sample paths of the associated Markov chain are confined to single leaves, and the transition probabilities of the corresponding *local* Markov operators $P_{[x]} : L^\infty([x], \overline{m}_{[x]}) \hookleftarrow$ are *absolutely continuous* with respect to $\overline{m}_{[x]}$.

There are two basic examples of measured equivalence relations: *countable equivalence relations* and *measured Riemannian foliations*, which correspond to two previously described classes of Markov operators (on countable sets and on smooth manifolds). In the first case the equivalence classes are countable [Pau], [Ka16]; in the second case the equivalence classes have a Riemannian structure, and the local Markov operators correspond to diffusion processes on equivalence classes [Ga], [Ka4].

0.3. Invariant Markov operators.

If the measure space (X, m) is endowed with a measure type preserving (left) action of a locally compact group G, then a Markov operator $P : L^\infty(X, m) \hookleftarrow$ is called *invariant* if it commutes with the group action by translations in the space $L^\infty(X, m)$. If $X = G$ is the group itself, and m is the left Haar measure on G, then a Markov operator $P : L^\infty(X, m) \hookleftarrow$ is G-invariant if and only if it has the form $P_\mu f(x) = \int f(xg) \, d\mu(g)$, where μ is a Borel probability measure on G. The corresponding Markov chain is called the (right) *random walk* on G determined by the measure μ. The position of the random walk at time n is $x_n = x_0 h_1 h_2 \cdots h_n$, where h_i are independent μ-distributed random variables (*increments* of the random walk). The operator P_μ acts on measures on G as the *convolution* $\nu P_\mu = \nu \mu$. In particular, if the initial distribution of the random walk is concentrated at the identity e, then the one-dimensional distribution at time n of the corresponding measure $\mathbf{P} = \mathbf{P}_e$ in the path space is the *n-fold convolution* μ_n.

Another example of invariant Markov operators is provided by *covering Markov operators*, in which case G is a countable group, and its action on the space (X, m) is *completely dissipative* (so that X is the union of pairwise disjoint translations gX_0 of a "fundamental domain" X_0). Both random walks on countable groups and diffusion processes on covering manifolds belong to this class [LS], [Ka13]. Yet another example is given by so called *geodesic random walks* on covering metric spaces [KM].

0.4. Quotients of Markov operators.

Let α be a *measurable partition* (e.g., see [CFS]) of the state space (X, m) of a Markov operator P. The operator P determines a *quotient Markov operator* P_α on the *quotient space* (X_α, m_α) if and only if P preserves the space of α-measurable functions on X. In particular, let $X = G$ be a countable group, α – the partition of G into right H-classes gH of a subgroup $H \subset G$, and $P = P_\mu$ – the operator of the right random walk on G determined by a measure μ. Then the projection from G onto G/H determines a G-invariant Markov operator on G/H if and only if

$$\mu(hgH) = \mu(gH) \qquad \forall h \in H, g \in G ,$$

and the transition probabilities of the quotient operator are completely determined by the values $\mu(HgH)$, $\gamma \in H$. Conversely, any G-invariant Markov operator on G/H can be obtained in this way for a measure μ satisfying the above condition. In other words, the space of G-invariant Markov operators on G/H is isomorphic to the measure algebra of the *double coset hypergroup* $H\backslash G/H$, see [KW2]. For a general locally compact group G one can describe in a similar way all invariant Markov operators on the space G/H whenever the partition of G into the right H-classes is measurable. In particular, if G is a semi-simple Lie group with finite center and K – its maximal compact subgroup, then there is a one-to-one correspondence between G-invariant Markov operators on the *symmetric space* G/K and probability measures μ on G, which are bi-invariant with respect to K (cf. Section 3).

0.5. Boundary theory of Markov operators.

There are two approaches to the boundary theory of Markov operators, i.e., to describing the behaviour of the corresponding Markov chains at infinity. One is based on potential theory and leads to the *Martin boundary* and *Martin compactification*. The other one uses methods and language of measure theory and ergodic theory and leads to the *Poisson boundary*. These two boundaries are closely connected; namely, the Martin boundary is responsible for integral representation of *positive harmonic functions*, whereas the Poisson boundary via the *Poisson formula* gives an integral representation of *bounded harmonic functions*. Considered as a measure space with the representing measure of the constant harmonic function, the Martin boundary is isomorphic to the Poisson boundary.

However, there is a principal difference between these two boundaries. The Martin boundary is a *bona fide* topological space and requires for its definition a topology on the state space of the Markov operator (it is defined as the pointwise closure of the Green functions in the projective space of functions on the state space). On the other hand, the Poisson boundary is a purely measure theoretical object and is defined as the space of ergodic components of the time shift in the path space (i.e., as the quotient of the path space with respect a certain measurable partition). The latter definition uses the following fundamental property of *Lebesgue measure spaces*: a one-to-one correspondence between their *measurable partitions*, complete σ-algebras, and their *homomorphic images* (in the category of measure spaces mod 0), e.g., see [CFS]. Thus, the Poisson boundary (unlike the Martin boundary) can be defined for an *arbitrary* Markov operator on a Lebesgue measure space (if the state space of a Markov operator is a Lebesgue space, then the corresponding path space is also a Lebesgue space). Note that attempts to treat the Poisson boundary as a topological space by introducing a topology on it often shroud its true nature and lead to unnecessary technical complications.

Traditionally, the Martin boundary is a more popular object (or, at least, term) than the Poisson boundary, although from a probabilistic point of view it would be more natural to look first at measure-theoretical objects, and only then at topological ones. [Actually, the definition of the Poisson boundary can be traced back to the papers of Blackwell [Bl] and Feller [Fe] which had appeared before the works of Doob [Doo] and Hunt [Hu] on the Martin boundary.] Following this tradition we begin our exposition with the Martin boundary.

1. The Martin boundary

1.1. The Martin compactification.

Suppose first that the state space X of a Markov operator P is countable. The *Green kernel* of the operator P is defined as

$$\mathcal{G}(x,y) = \sum_{n=0}^{\infty} p_n(x,y),$$

where p_n are the *n-step transition probabilities* (transition probabilities of the operator P^n). The Green kernel satisfies the relation $\mathcal{G}(x,y) = \mathcal{F}(x,y)\mathcal{G}(y,y)$, where $\mathcal{F}(x,y)$ is the probability to ever visit the point y starting from the point x. If the operator P is *irreducible* in the sense that $\mathcal{F}(x,y) > 0 \; \forall\, x,y \in X$ (i.e., any two states *communicate*), then either $\mathcal{F}(x,y) = 1, \mathcal{G}(x,y) = \infty \; \forall\, x,y \in X$, or $\mathcal{F}(x,y) < 1, \mathcal{G}(x,y) < \infty \; \forall\, x,y \in X$. In the first case the operator P (and the corresponding Markov chain) is called *recurrent*, and in the second case – *transient* [KSK].

If the operator P is transient, then for any reference point $o \in X$ one can define the *Martin kernel*

$$\mathcal{K}_o(x,y) = \frac{\mathcal{G}(x,y)}{\mathcal{G}(o,y)} = \frac{\mathcal{F}(x,y)}{\mathcal{F}(o,y)}.$$

The map $y \mapsto \mathcal{K}_o(\cdot,y)$ is an embedding of the space X into the space of positive functions on X satisfying the normalizing condition $\mathcal{K}_o(o,y) = 1$. Since $\mathcal{F}(x,y) \geq \mathcal{F}(x,o)\mathcal{F}(o,y)$, one has the uniform bounds

$$\mathcal{F}(x,o) \leq \mathcal{K}_o(x,y) \leq \frac{1}{\mathcal{F}(o,x)},$$

so that the family of functions $\mathcal{K}_o(\cdot,y)$, $y \in X$ is relatively compact in the topology of pointwise convergence.

The pointwise closure of the family $\{\mathcal{K}_o(\cdot,y) : y \in X\}$ and its boundary $\partial_\mathcal{M} X$ are called the *Martin compactification* and the *Martin boundary* of the space X determined by the operator P, respectively. Thus, points γ of the Martin boundary are represented as positive functions $\mathcal{K}_o(\cdot,\gamma)$ on X. Choosing another reference point $o' \in X$ means that the Green kernels $\mathcal{G}(\cdot,y)$ have to be normalized at the point o' instead of the point o, so that the corresponding Martin compactifications are homeomorphic, and

(1.1.1) $$\mathcal{K}_{o'}(x,\gamma) = \frac{\mathcal{K}_o(x,\gamma)}{\mathcal{K}_o(o',\gamma)}.$$

In more invariant terms what we have just said means that the Green kernel is used for embedding the state space X into the compact *projective space* $P\mathbb{R}_+^X$ of non-negative functions on X, after which the Martin compactification is obtained by taking the closure of this embedding.

1.2. Representation of positive harmonic functions.

A positive harmonic function f on X is called *minimal* if any positive harmonic function f' dominated by f is a multiple of f. All minimal harmonic functions belong to the Martin boundary (with proper normalization), and the (Borel) subset of the Martin boundary consisting of minimal harmonic functions is often called the *minimal Martin boundary* (in general, it is not even dense in the whole Martin boundary). Any positive harmonic function f of the operator P can be uniquely represented as an integral of minimal harmonic functions

$$f(x) = \int \mathcal{K}_o(x, \gamma) \, d\nu_{o,f}(\gamma) \,,$$

where $\nu_{o,f}$ is a positive measure on the minimal Martin boundary with the mass $f(o)$. Since this representation is unique, formula (1.1.1) implies that the representing measures of the same function corresponding to different reference points o and o' are connected by the relation

(1.2.1) $$\frac{d\nu_{o',f}}{d\nu_{o,f}}(\gamma) = \mathcal{K}_o(o', \gamma) \,.$$

The representing measure $\nu_o = \nu_{o,1}$ of the constant harmonic function $\mathbf{1}$ is called the *harmonic measure* of the reference point $o \in X$. The measure ν_o has the following interpretation: almost all sample paths of the Markov chain corresponding to the operator P converge in the Martin topology, and the distribution of the corresponding limit points is ν_o provided the starting point is o [Doo]. Formula (1.2.1) can be then rewritten as

(1.2.2) $$\frac{d\nu_x}{d\nu_o}(\gamma) = \mathcal{K}_o(x, \gamma) \,.$$

1.3. Martin boundary in more general situations.

In the same way one can consider the embedding of X into $P\mathbb{R}_+^X$ determined by the λ-*Green kernel* $\mathcal{G}_\lambda(x, y) = \sum \lambda^{-n} p_n(x, y)$ for a parameter $\lambda \neq 1$. Provided that $\mathcal{G}_\lambda(x, y) < \infty \; \forall \, x, y \in X$, the corresponding λ-*Martin boundary* is responsible for integral representation of positive λ-*harmonic functions*, i.e., such functions f that $Pf = \lambda f$.

The Martin boundary can be also defined for more general Markov operators. In the case of diffusion processes on manifolds the construction above can be reproduced almost literally. Note that in this situation one can consider two Green operators: the discrete time operator $\sum_{n=0}^{\infty} P^n$, and the continuous time operator $\int_0^{\infty} P^t \, dt$, which both give the same Martin boundary. See [AC], [An2], [Br], [Del], [Dy2], [Re] for a discussion of the Martin boundary in this and other situations.

One should note here that the Martin boundary does not have good functorial properties in the category of Markov operators; see [Mol1], [Fr], [Ta3], [PW3] for a discussion of problems connected with the Martin boundary for a product of two operators, and [CSa] for an example when the Martin boundary of the operator $P^\mu = \sum \mu(n) P^n$ with P being the Markov operator of the equidistributed random walk on the free group is different from the Martin boundary of P.

1.4. Martin boundary of abelian and nilpotent groups.

Let now P be the Markov operator of the right random walk on a countable group G determined by a probability measure μ. Then the Green kernel is G-invariant, so that $\mathcal{G}(x,y) = \mathcal{G}(x^{-1}y) \; \forall x, y \in G$, and $\mathcal{F}(x,y) > 0 \; \forall x, y \in G$ if and only if the measure μ satisfies the following *non-degeneracy condition*: the semigroup generated by its support coincides with the whole group G (in other words, any two points from the group communicate with respect to the corresponding random walk). The action of the group G extends by continuity to the Martin boundary, and the harmonic measure of an arbitrary point $g \in G$ is $\nu_g = g\nu$, where $\nu = \nu_e$ is the harmonic measure of the identity e.

All minimal harmonic functions on *abelian* and, more generally, *nilpotent groups* are $\mathbb{R}_+$-valued *multiplicative characters* of these groups. For the abelian groups this result is known as the Choquet–Deny theorem [CD]. The idea of the proof is very simple. If f is a harmonic function, then $f(g) \geq f(gx)\mu(x) \; \forall g, x \in G$. Since G is abelian, the translation $T^x f(g) = f(gx) = f(xg)$ is also a μ-harmonic function, and by the above it is dominated by f if $\mu(x) > 0$. The function f being minimal, it implies that the functions $T^x f$ and f are proportional. Thus, if the measure μ is non-degenerate, the function f must be a multiplicative character. As the constant function $\mathbf{1}$ can not be decomposed as a non-trivial integral of multiplicative characters, it is minimal (so that the harmonic measure ν is concentrated on a single point). For nilpotent groups one has to apply the same argument to the center of the group G and to use induction on degrees of nilpotency [Mar1]. The key ingredient of this proof is a *Harnack inequality* for positive harmonic functions, so that it also works for covering Markov operators with a nilpotent deck group [LS], [LP] or for nilpotent Lie groups.

Describing the *Martin compactification* and not just the minimal Martin boundary requires a much harder analysis. It seems, nothing is known about the Martin boundary of nilpotent groups. As for abelian groups, a description of the Martin boundary was obtained for a class of random walks with an exponential moment condition using harmonic analysis methods [NS]. In this case the Martin boundary as a set coincides with the minimal Martin boundary. An analogous result for random walks on compact extensions of $\mathbb{R}^n$ is proven in [Bab1]. On the other hand, there is an example of a transient random walk on $\mathbb{Z}$ with non-minimal points in the Martin boundary [CSa].

1.5. Martin boundary of free groups.

The first example of describing non-trivial Martin boundary of a random walk on a discrete group was that of the random walk on a *free group* F_d with d generators $\{a_{\pm i} : i = 1, \ldots, d\}$ determined by the measure μ equidistributed on the set of generators: $\mu(a_i) = 1/2d$. This random walk coincides with the *simple random walk* on the *homogeneous tree* of degree $2d$ (the *Cayley graph* of the group F_d). Dynkin and Malyutov [DMal] explicitly calculated the Green kernel of this random walk and showed that the Martin compactification coincides with the *end compactification* of the Cayley graph (see also [DYu], [Car]). Thus, the Martin boundary coincides with the *space of ends*, or, in other terms, with the space of *infinite irreducible words* in the alphabet $\{a_i\}$. The harmonic measure ν in this case is equidistributed on the space of ends.

Later, Levit and Molchanov [LM] extended this result to an arbitrary probability measure μ supported by the set of generators $\{a_i\}$ also by calculating the Martin kernel. The crucial idea here is that in order to visit a word, say $g = a_{i_1} a_{i_2} \ldots a_{i_k}$, starting from the group identity e, one has to visit all intermediate points $g_1 = a_{i_1}, g_2 = a_{i_1} a_{i_2}, \ldots$. Thus,
(1.5.1)
$$\mathcal{F}(e, g) = \mathcal{F}(e, g_1) \mathcal{F}(g_1, g_2) \ldots \mathcal{F}(g_{k-1}, g) = \mathcal{F}(e, a_{i_1}) \mathcal{F}(e, a_{i_2}) \ldots \mathcal{F}(e, a_{i_k}) .$$

This method does not use group invariance of the random walk at all and is applicable to any transient nearest neighbour Markov chain on a tree.

For non-nearest neighbour random walks on the free group an explicit calculation of the Green kernel in this way is no longer possible. Nonetheless, a slight modification of the above method still allows one to identify the Martin boundary with the space of ends. One should use the fact that if the lengths of one-step "jumps" are uniformly bounded (the corresponding operator has *bounded range*), then in order to visit a certain point one has to pass through a family of finite *barrier sets* between the origin and this point. Thus, (1.5.1) can be replaced with the formula

$$\mathcal{F}(e, B_k) = \delta_e S^{B_1} S^{B_2} \ldots S^{B_k} \, ,$$

where $\mathcal{F}(e, B_k)$ is the *first hitting distribution* on B_k, and S^{B_i} are the *balayage* operators assigning to an initial distribution λ the first hitting distribution λS^{B_i} on B_i. The size of the barriers B_i is uniformly bounded, and one can show that the corresponding operators S^{B_i} are uniformly contracting in a projective metric, which leads to identification of the Martin boundary with the space of infinite irreducible words for all random walks on the free group determined by a finitely supported measure μ [De2]. With some technical modifications this approach was applied to finitely supported random walks on Fuchsian groups [Se], on groups with infinitely many ends [Wo1], and to finite range random walks on arbitrary trees [PW1], [PW2], [CSW].

1.6. Martin boundary of Gromov hyperbolic spaces.

In the continuous setup Anderson and Schoen [AS] by using a potential theory approach identified the Martin compactification with the *visibility compactification* for the (semigroup of) Markov operator(s) corresponding to the Brownian motion on *Cartan–Hadamard manifolds* with pinched sectional curvatures. In the framework of stochastic differential equations earlier results in this direction were obtained by Kifer [Ki1] and Prat [Pr] under the condition that the curvature varies slowly at infinity; a probabilistic proof of the result of Anderson and Schoen in full generality is given in [Ki2]; see [Ki4] for a survey of the probabilistic aspects of this problem.

A remarkable result of Ancona [An1], [An2] is a far reaching generalization of the approaches used for free groups and for Cartan–Hadamard manifolds. For a given increasing function $\Phi : \mathbb{R}_+ \to \mathbb{R}_+$ with $\Phi(0) = C > 1$ and $\Phi(\infty) = \infty$ he introduces the following geometrical Φ-*chain condition* on a metric space X. A family of open sets $U_1 \supset U_2 \supset \cdots \supset U_m$ together with a sequence of points $x_1, x_2, \ldots, x_m \in X$ is a Φ-chain if

1) $1/C \leq \operatorname{dist}(x_i, x_{i+1}) \leq C$;
2) $\operatorname{dist}(x_i, \partial U_i) \leq C/4$;
3) $\operatorname{dist}(x, U_{i+1}) \geq \Phi(\operatorname{dist}(x, x_i)) \, \forall x \in \partial U_i$.

Now, if P is a Markov operator on X such that its spectral radius in $L^2(X, m)$ is strictly less than 1, and P is "local" (P is the time 1 operator of a diffusion process on a smooth manifold X, or, P is a bounded range operator on a locally finite graph X), then the Green function is *almost multiplicative* along the Φ-chain:

$$\frac{1}{K} \leq \frac{\mathcal{G}(x_m, x_1)}{\mathcal{G}(x_m, x_k)\mathcal{G}(x_k, x_1)} \leq K \qquad \forall\, 1 < k < m$$

for a constant $K = K(\Phi, P)$.

If X is a *hyperbolic space* in the sense of Gromov [Gro2], then there is a Φ-chain along any *infinite geodesic ray* in X, so that *the Green function is almost multiplicative along geodesics*. The latter property alone is sufficient to prove that the Martin compactification coincides with the *hyperbolic compactification* of X [Ka12]. The class of hyperbolic spaces contains both discrete hyperbolic objects (e.g., trees, fundamental groups of compact negatively curved manifolds, and, more generally, convex cocompact groups) and Cartan–Hadamard manifolds with pinched negative curvature. Thus, Ancona's theory is applicable to all these situations provided the Markov operator is local and its spectral radius is less than 1 (note that no group invariance is required).

For Cartan–Hadamard manifolds hyperbolicity is equivalent to the *uniform visibility property*, and the hyperbolic boundary coincides with the *visibility boundary*. Recently, Kifer [Ki3] and Cao [Cao] showed that for Cartan–Hadamard manifolds hyperbolicity implies that the spectral radius of the Markov operator of the Brownian motion is strictly less than 1 (i.e., 0 does not belong to the L^2-spectrum of the Laplacian). Thus, the Martin compactification coincides with the visibility compactification for all Cartan–Hadamard manifolds with uniform visibility property. An analogous result for the Brownian motion on polygonal complexes is announced in [BK].

However, the "local character" of the Markov chain (i.e., the "finite range" condition in discrete situations) seems to be essential for a successful identification of the Martin boundary with the hyperbolic boundary (for example, it is unclear whether Ancona's technique works for hyperbolic graphs with even a "very fast" decay of transition probabilities instead of the finite range condition; cf. the exponential moment condition used in describing the Martin boundary on abelian groups **1.4**). A recent example of Ballmann and Ledrappier [BL2] shows that there is a probability measure with a finite first logarithmic moment on a free group such that the Martin boundary of the corresponding random walk is homeomorphic to the circle and not to the space of ends (although from the measure theoretical point of view the *Poisson boundary* is still the space of ends). This example uses a *discretization* of the Brownian motion on the hyperbolic plane – see **2.6**.

2. THE POISSON BOUNDARY

2.1. The Poisson boundary and the Poisson formula.

Let P be a Markov operator on a *Lebesgue measure space* (X, m). We shall say that two paths $\bar{x}$ and $\bar{x}'$ in the path space $(X^{\mathbf{Z}_+}, \mathbf{P}_m)$ are $\sim$-equivalent if there exist $n, n' \geq 0$ such that $T^n \bar{x} = T^{n'} \bar{x}'$, where $T : (x_n) \mapsto (x_{n+1})$ is the *time shift* in the path space (in other words, $\sim$ is the *trajectory equivalence relation* of the shift T). Let $\mathcal{S}$ be the σ-algebra of all measurable unions of $\sim$-classes (mod 0). Since $(X^{\mathbf{Z}_+}, \mathbf{P}_m)$ is a Lebesgue space, there is a (unique up to an isomorphism) measurable space Γ and a map $\mathbf{bnd} : X^{\mathbf{Z}_+} \to \Gamma$ such that the σ-algebra $\mathcal{S}$ coincides with the σ-algebra of $\mathbf{bnd}$-preimages of measurable subsets of Γ (i.e., Γ is the *space of ergodic components* of the shift T). The space Γ is called the *Poisson boundary* of the operator P. The measure type $[\nu]$ on Γ which is the image of the type of the measure $\mathbf{P}_m$ is called the *harmonic measure type*. For any initial probability distribution $\theta \prec m$ on X the measure $\nu_\theta = \mathbf{bnd}(\mathbf{P}_\theta) \prec [\nu]$ is called the *harmonic measure* corresponding to θ.

Let $H^\infty(X, m, P) \subset L^\infty(X, m)$ be the space of bounded harmonic functions of the operator P. If $f \in H^\infty(X, m, P)$, then as it follows from the *martingale convergence theorem*, for a.e. path $\bar{x}$ there exists the limit $\widehat{f}(\bar{x}) = \lim f(x_n)$ which is clearly $\mathcal{S}$-measurable (i.e., $\widehat{f}$ can be considered as a function on Γ). Moreover, $\langle f, \theta \rangle = \langle \widehat{f}, \nu_\theta \rangle$ for any initial distribution $\theta \prec m$. Conversely, for any $\widehat{f} \in L^\infty(\Gamma, [\nu])$ there is a corresponding function $f \in H^\infty(X, m, P)$. Thus, the spaces $H^\infty(X, m, P)$ and $L^\infty(\Gamma, [\nu])$ are isometric. In particular, triviality of the Poisson boundary is equivalent to absence of non-constant bounded harmonic functions for the operator P (the *Liouville property*). If the operator P has absolutely continuous transition probabilities, then the harmonic measures $\nu_x \prec [\nu]$ can be defined for a.e. point $x \in X$, and this isometry takes the form of the *Poisson formula* $f(x) = \langle \widehat{f}, \nu_x \rangle$ [Ka9].

2.2. Conditional Markov operators.

If A is a measurable subset of the Poisson boundary, then by the Markov property for any cylinder set $C = C_{x_0, x_1, \ldots, x_n}$ in the path space

$$\mathbf{P}_{x_0}(C \cap \mathbf{bnd}^{-1}A) = \mathbf{P}_{x_0}(C)\mathbf{P}_{x_n}(\mathbf{bnd}^{-1}A) = \mathbf{P}_{x_0}(C)\nu_{x_n}(A)$$

(for simplicity we assume that the state space is countable; the general case is treated along the same lines modulo some incantations about Lebesgue spaces, measurable partitions and conditional measures [Ka9]). In other words,

$$\mathbf{P}_{x_0}(C|\mathbf{bnd}^{-1}A) = \frac{\mathbf{P}_{x_0}(C)\nu_{x_n}(A)}{\mathbf{P}_{x_0}(\mathbf{bnd}^{-1}A)} = \mathbf{P}_{x_0}(C)\frac{\nu_{x_n}(A)}{\nu_{x_0}(A)} ,$$

which means that conditioning by A (more rigorously, by $\mathbf{bnd}^{-1}A$) gives rise to the Markov operator $P_A f = P(f\varphi_A)/\varphi_A$, where $\varphi_A(x) = \nu_x(A) = \mathbf{P}_x(\mathbf{bnd}^{-1}A)$ is the harmonic function corresponding to the indicator function $\mathbf{1}_A$ (the operator P_A is called the *Doob transform* of the operator P determined by the harmonic function φ_A). Thus, for a.e. point $\gamma \in \Gamma$ conditioning the measure $\mathbf{P}_{x_0}$ by γ gives the *conditional measure*

$$(2.2.1) \qquad\qquad \mathbf{P}_{x_0}(C|\gamma) = \mathbf{P}_{x_0}(C)\frac{d\nu_{x_n}}{d\nu_{x_0}}(\gamma) .$$

so that a.e. point $\gamma \in \Gamma$ determines the *conditional Markov operator* P_γ which is the Doob transform of P determined by the harmonic function

$$(2.2.2) \qquad\qquad \varphi_\gamma(x) = \frac{d\nu_x}{d\nu_o}(\gamma)$$

(as the Doob transform depends only on the projective class of the function φ_γ, this definition does not depend on the choice of the reference point $o \in X$). The measures $\mathbf{P}_x(\cdot|\gamma)$ are conditional measures on ergodic components of the time shift, so that they are ergodic and the operators P_γ a.e. have trivial Poisson boundary, which means that the functions φ_γ are a.e. minimal. Since the conditional measures in the path space corresponding to different points $\gamma \in \Gamma$ are pairwise singular, different points $\gamma \in \Gamma$ determine different functions φ_γ, or, in other words, *the Radon–Nikodym derivatives* $d\nu_x/d\nu_o$ *separate points from* Γ.

In the situation when one can construct the Martin compactification of the state space X corresponding to the operator P, formula (2.2.2) coincides with formula (1.2.2), so that the Poisson boundary admits a realization on the minimal Martin boundary with the family of harmonic measures ν_x.

2.3. The tail boundary.

Another measure-theoretic boundary associated with a Markov operator is the *tail boundary*. Its definition is analogous to the definition of the Poisson boundary with the equivalence relation $\sim$ replaced by the equivalence relation $\approx$ such that $\overline{x} \approx \overline{x}'$ if $T^n\overline{x} = T^n\overline{x}'$ for a certain n. The *tail σ-algebra* $\mathcal{A}^\infty$ of all measurable unions of $\approx$-classes is the limit of the decreasing sequence of σ-algebras $\mathcal{A}_n^\infty$ determined by the positions of sample paths at times $\geq n$. In other words, the tail boundary is the quotient of the path space with respect to the *tail partition* η which is the measurable intersection $\bigwedge \eta_n$ of the decreasing sequence of measurable partitions η_n corresponding to σ-algebras $\mathcal{A}_n^\infty$ (note that this definition automatically implies that the tail boundary of the product of two Markov operators is the product of their tail boundaries). One can say that the tail boundary completely describes the *stochastically significant behaviour* of the Markov chain at infinity.

The tail boundary is the Poisson boundary for the *space-time* operator $\tilde{P}f(\cdot, n) = Pf(\cdot, n+1)$ on $X \times \mathbb{Z}$, so that it gives integral representation of *bounded harmonic sequences* $f_n = Pf_{n+1}$ on X (which are counterparts of so-called *parabolic harmonic functions* in the classical setting). The tail boundary is endowed with a natural action of the time shift T induced by the time shift in the path space, and the Poisson boundary is the space of *ergodic components* of the tail boundary with respect to T.

The Poisson and the tail boundaries are sometimes confused, and, indeed, they do coincide for "most common" Markov operators (such operators are called *steady* in [Ka9]).

General criteria of triviality of these boundaries and of their coincidence are provided by *0-2 laws*, see [De3], [Ka9]. In particular, for a given initial distribution θ the tail boundary of a Markov operator P is trivial $\mathbf{P}_\theta$ – mod 0 if and only if

$$\lim_{n \to \infty} \|\theta P^{n+k} - \nu P^n\| = 0 \qquad \forall\, k \geq 0,\ \nu \prec \theta P^k$$

(i.e., if the one-dimensional distribution νP^n of the chain at time $n + k$ is asymptotically the same independently of the initial distribution at time k). The tail and the Poisson boundaries coincide $\mathbf{P}_\theta$ – mod 0 iff

$$(2.3.1) \quad \lim_{n \to \infty} \|\nu P^n - \nu P^{n+d}\| = 0 \qquad \forall\, k \geq 0, d > 0,\ \nu \prec \theta P^k \wedge \theta P^{k+d}.$$

Finally, the Poisson boundary is trivial $\mathbf{P}_\theta$ – mod 0 iff

$$(2.3.2) \qquad \lim_{n \to \infty} \frac{1}{n+1} \left\| \sum_{k=0}^{n} (\theta - \nu) P^k \right\| = 0 \qquad \forall\, \nu \prec \sum_{k=0}^{\infty} 2^{-k} \theta P^k.$$

2.4. The identification problem.

Suppose that one has a measurable map π from the path space $(X^{\mathbb{Z}_+}, \mathbf{P}_m)$ to a space Z such that $\pi(\overline{x}) = \pi(\overline{x}')$ whenever $\overline{x} \sim \overline{x}'$, and let $\lambda_\theta = \pi(\mathbf{P}_\theta)$ be the corresponding measures on Z (for example, if for a.e. path $\overline{x}$ there exists a limit $\pi(\overline{x}) = \lim x_n \in Z$, then the map π satisfies this condition). Then by the definition of the Poisson boundary the space Z is a quotient of the Poisson boundary Γ, and the space $L^\infty(B, [\lambda])$ is isometrically isomorphic to a subspace in $H^\infty(X, m, P)$. Thus, describing the Poisson boundary in terms of geometrical or combinatorial structures associated with the state space X would consist of two steps: first, one has to exhibit a space Z with the above property, and, then, one has to show that in fact Z is isomorphic to the *whole* Poisson boundary, i.e., the projection $\Gamma \to Z$ is an *isomorphism* of measure spaces. In other words, first one has to exhibit a certain system of invariants, and then to show completeness of this system (e.g., see [Bi], [MP] for a description of Euclidean domains for which the Poisson boundary can be identified with the topological boundary). A particular case of the problem of describing the Poisson boundary is proving its triviality (i.e., showing that the one-point space is the only $\sim$-measurable quotient of the path space).

2.5. The Poisson boundary of random walks on groups.

For random walks on groups the Poisson boundary is endowed with a natural group action (induced from the group action by translations in the path space), and $\nu_{g\theta} = g\nu_\theta \ \forall g \in G, \theta \prec m$. This action is ergodic because any G-invariant harmonic function on G is obviously constant. We assume that the measure μ is *spread-out*, i.e., there is $n \geq 1$ such that the convolution μ_n is non-singular with respect to the Haar measure. In this case the measure $\nu = \nu_e$ on the Poisson boundary corresponding to the initial distribution δ_e is absolutely continuous with respect to any ν_θ with $\nu_\theta \sim m$. If, in addition, the measure μ is non-degenerate, then $\nu \sim \nu_\theta$ for any $\nu_\theta \sim m$ (hence, ν is quasi-invariant). The measure μ is *μ-stationary* in the sense that $\nu = \mu\nu$. Below we shall always consider the Poisson boundary of random walks on groups as a measure space with the measure ν corresponding to the initial distribution δ_e. The Poisson boundary (Γ, ν) is either trivial or purely non-atomic. See [Ka13] for general ergodic properties of the group action on the Poisson boundary and on its quotients.

By a theorem of Foguel [Fo1] the sequence of n-fold convolutions μ_n of a probability measure μ on a locally compact group has the property that for any $d > 0$ either $\|\mu_n - \mu_{n+d}\| \to 0$, or the measures μ_n and μ_{n+d} are pairwise singular for any n. Thus, for random walks on groups the tail and the Poisson boundaries coincide (mod 0) with respect to the initial distribution δ_e (for, if μ_k and μ_{k+d} are pairwise singular for any k, then there are no measures ν in (2.3.1) absolutely continuous with respect to $\delta_e P^k = \mu_k$ and $\delta_e P^{k+d} = \mu_{k+d}$ simultaneously). This is a key ingredient of the entropy theory of random walks (**2.10**). Interrelations between the tail and the Poisson boundaries for an arbitrary initial distribution are discussed in the recent paper [J].

The Poisson boundary is trivial for all non-degenerate measures on abelian and nilpotent groups (because in this case the constant function 1 is minimal – see the argument in **1.4**). Note that although this result is commonly referred to as the Choquet–Deny theorem, its first proof for countable abelian groups was obtained by Blackwell who used direct estimates for proving asymptotic invariance of convolution powers [Bl]. For general abelian groups it can be also deduced from triviality of the *exchangeable* σ-algebra of the sequence of increments of the random walk (the *Hewitt–Savage 0-1 law* – e.g., see [Me]). Yet another proof (valid for any corecurrent covering Markov operator with a nilpotent deck group) can be found in [Ka13]. Apparently, the total number of known different proofs of the Choquet–Deny theorem should be somewhere between 10 and 15 (we shall try to count them elsewhere).

A group G is called *amenable* if there exists a sequence of probability measures λ_n on G with the property $\|\lambda_n - g\lambda_n\| \to 0$ $\forall g \in G$ (such sequence is called *asymptotically invariant*). This is just one from a long list of equivalent definitions of amenability; e.g., see [Pat]. As it follows from the 0-2 law (2.3.2), for a non-degenerate measure μ the Poisson boundary is trivial iff the sequence of Cesaro averages of the n-fold convolutions μ_n is asymptotically invariant. Moreover, if the measure μ is in addition *aperiodic*, i.e., if the measures μ_n and μ_{n+1} are pairwise non-singular for a certain n, then the Poisson boundary is trivial iff the sequence μ_n is asymptotically invariant. Thus, the Poisson boundary is non-trivial for all non-degenerate measures on a non-amenable group (see [KV], [Ka9] and references therein).

Conversely, if the group G is *amenable*, then using asymptotically invariant sequences one can always construct a measure μ with trivial Poisson boundary [KV], [Ro] (but there may also be measures with a non-trivial boundary; see examples below).

For a symmetric spread-out measure μ on an amenable Lie group the Poisson boundary is always trivial [BR] (see below **2.9**). On the contrary, the example from **2.7** shows that the Poisson boundary may well be non-trivial for all finitely supported symmetric probability measures on discrete amenable (in particular, solvable) groups. However, the constant function 1 always belongs to the Martin boundary for a symmetric measure μ on a discrete amenable group G [Nor] (so that, if the Poisson boundary is non-trivial, then the Martin boundary contains a non-minimal point).

Note that triviality of the Poisson boundary (absence of bounded harmonic functions) by no means implies absence of positive harmonic functions. See [BE] for general results on existence of positive harmonic functions on discrete solvable groups.

The formula $x_n = gh_1h_2 \cdots h_n$ states an isomorphism of the shift T in the path space of the random walk (G, μ) and the *skew product* $(g, \overline{h}) \mapsto (gh_1, \mathbf{B}\overline{h})$ over the Bernoulli shift $\mathbf{B}$ in the space of increments $\overline{h} = (h_n)$. Thus, the Poisson boundary can be also defined as the *Mackey range* of the G-valued *cocycle* $g \mapsto gh_1$ of the (unilateral) Bernoulli shift [Zi]. In particular, the action of G on the Poisson boundary is always *amenable* (even if the group G itself is not amenable). A direct proof of this property can be also obtained by using the fact that the Poisson boundaries of the conditional random walks determined by points from the Poisson boundary are trivial (by definition of the Poisson boundary), so that the 0-2 law (2.3.2) implies that Cesaro averages of one-dimensional distributions of conditional random walks have the asymptotic invariance property equivalent to amenability of the action (cf. the argument above). Conversely, any measure type preserving amenable ergodic action of a locally compact group G is isomorphic to the Poisson boundary of a certain so called matrix-valued random walk on G (which need not be homogeneous in time) [CW], [EG], [AEG]. However, clearly there are amenable G-spaces (e.g., with a finite invariant measure) which are not isomorphic to any "usual" Poisson boundary $\Gamma(G, \mu)$.

See [Wi] for a description of the Poisson boundary in terms of the *group algebra* of G. Yet another (in a sense dual) definition of the Poisson boundary can be given in terms of topological dynamics [DE], [Dok].

2.6. Discretization of covering Markov operators.

Let $\widetilde{P} : L^\infty(\widetilde{X}, \widetilde{m})$ be a covering Markov operator with deck group G (see **0.3** for a definition). A natural condition to impose on the operator $\widetilde{P}$ in order to connect its Poisson boundary with the Poisson boundary of an appropriate random walk on the deck group is *corecurrence*, i.e., recurrence (in the sense of Harris) of the quotient operator P on the quotient space $(X, m) = (\widetilde{X}, \widetilde{m})/G$ (otherwise there is no reason to expect that the behaviour of the sample paths of the corresponding Markov chain on $\widetilde{X}$ at infinity could be described just in terms of a single orbit in $\widetilde{X}$). The idea of the following discretization procedures goes back to Furstenberg [Fu4], [Fu5] and consists in presenting for an arbitrary point $x \in \widetilde{X}$ its harmonic measure ν_x on the Poisson boundary of the operator $\widetilde{P}$ as a convex combination of the harmonic measures $\nu_{go} = g\nu_o$, $g \in G$ for a fixed reference point $o \in \widetilde{X}$.

Suppose that the operator $\widetilde{P}$ satisfies the following weak form of the *Harnack inequality*: for any two points $x, y \in \widetilde{X}$ there is $\varepsilon > 0$ and a probability measure θ on $\widetilde{X}$ such that

$$(2.6.1) \qquad\qquad \nu_x = \varepsilon\nu_y + (1 - \varepsilon)\nu_\theta$$

For any $x \in \widetilde{X}$ let $F(x)$ be the supremum of all $\varepsilon > 0$ for which there exist a probability measure $\varkappa$ on G and a probability measure θ on $\widetilde{X}$ such that

$$(2.6.2) \qquad\qquad \nu_x = \varepsilon\sum \varkappa(g)\nu_{go} + (1 - \varepsilon)\nu_\theta .$$

By (2.6.1) the function F is strictly positive; further, it is easily seen to be G-invariant and superharmonic with respect to the operator $\widetilde{P}$, so that it defines a bounded superharmonic function of the quotient operator P, which must be constant as P is recurrent. Denote by $\overline{F} > 0$ the value of F, then for any $x \in \widetilde{X}$ we can present the harmonic measure ν_x as the sum (2.6.2) with, say, $\varepsilon = \overline{F}/2$. Starting from the point $x = o$ and repeating this procedure for points from the support of the measure θ, and so on, we obtain that finally the whole measure ν_x will be replaced by a convex combination of measures $\nu_{go}, g \in G$, which means that

$$\nu_o = \sum \mu(g)g\nu_o = \sum \mu(g)\nu_{go} ,$$

i.e., ν_o is μ-stationary for the resulting measure μ on G.

By the Poisson formula, this implies that the restriction of any bounded $\widetilde{P}$-harmonic function to the orbit Go is μ-harmonic. On the other hand, μ-stationarity of the measure ν_o alone does *not* necessarily imply that, conversely, any bounded μ-harmonic function can be uniquely extended to a $\widetilde{P}$ harmonic function (although this may be the case in some special situations, e.g., if the group G is word hyperbolic). Example [Ka3]: if $\widetilde{P}$ is the Markov operator corresponding to the Brownian motion on a cover of a compact Riemannian manifold with a polycyclic fundamental group G, then the Poisson boundary of $\widetilde{P}$ is trivial, so that the harmonic measure ν_o is trivially μ-stationary for any measure μ on G, whereas there exist measures on G with non-trivial Poisson boundary (hence, with non-trivial bounded harmonic functions).

By making the decomposition (2.6.2) more specific, one can obtain a probability measure μ on G such that the random walk (G, μ) is naturally connected with the operator $\widetilde{P}$. First recall that if T is a Markov stopping time for the Markov chain (x_n) on $\widetilde{X}$ determined by the operator $\widetilde{P}$ then the distribution λ_T of x_T has the property that the harmonic measure ν_{λ_T} on the Poisson boundary of $\widetilde{P}$ coincides with the harmonic measure of the initial distribution at time 0. Now suppose that the decomposition (2.6.2) can be chosen in such a way that $(1 - \varepsilon)\theta < \lambda_T$ for a certain Markov stopping time T (provided the starting point is x). Then one can make the above discretization procedure defined entirely in terms of Markov stopping times (without using group invariance!) [LS], which allows one to prove that the Poisson boundary of $\widetilde{P}$ coincides with the Poisson boundary of the random walk on G determined by the resulting measure μ [K8] (see also more detailed expositions in [Ka13] and [KM]). Namely, for any bounded $\widetilde{P}$-harmonic function its restriction to the orbit Go is μ-harmonic, and, conversely, any bounded μ-harmonic function on $G \cong Go$ uniquely extends to a bounded $\widetilde{P}$-harmonic function.

Imposing some additional conditions on the decomposition (2.6.2) allows one to obtain the measure μ such that the random walk (G, μ) has the same Green function (hence, the same Martin boundary) as the original operator $\widetilde{P}$; moreover, if $\widetilde{P}$ is *reversible*, then the measure μ can be chosen symmetric [BL2]. Another discretization procedure, giving a measure μ with the same *positive* harmonic functions as the operator $\widetilde{P}$, is described in [An2].

The Harnack inequality is satisfied for diffusion processes, hence, the Poisson boundary of (the Brownian motion on) a corecurrent covering Riemannian manifold coincides with the Poisson boundary of the deck group G with an appropriate measure μ on it. Another class of Markov operators for which one can check the Harnack inequality is provided by *geodesic random walks* and similar discrete time chains with uniformly bounded jumps, so that, for example, by using this discretization procedure and a description of the Poisson boundary of the mapping class group one can show that the Poisson boundary of geodesic random walks on *Teichmüller space* coincides with the Thurston boundary [KM].

2.7. μ-Boundaries.

Any G-space which is a $\sim$-measurable image of the path space is the quotient (Γ_ξ, ν_ξ) of the Poisson boundary with respect to a G-invariant measurable partition ξ. Such quotients are called *μ-boundaries*. By definition, the Poisson boundary is the maximal μ-boundary. The measure ν_ξ itself and almost all conditional measures in the fibers of the projection $\Gamma \to \Gamma_\xi$ are purely non-atomic (unless $\Gamma_\xi = \{\cdot\}$ or $\Gamma_\xi = \Gamma$, respectively) [Ka13]. Another way of characterizing a μ-boundary is to say that it is a G-space with a μ-stationary measure λ such that $x_n \lambda$ weakly converges to a δ-measure for a.e. path (x_n) of the random walk (G, μ) [Fu4].

Thus, the problem of describing the Poisson boundary of (G, μ) consists of two parts (cf. **2.4**) :

(1) To find (in geometric or combinatorial terms) a μ-boundary (B, λ);

(2) To show that this μ-boundary is maximal.

If a certain compactification of the group G has the property that sample paths of the random walks on G converge a.e. in this compactification (so that it is a μ-boundary), and this μ-boundary is in fact isomorphic to the Poisson boundary of (G, μ), then it means that this compactification is indeed *maximal in a measure theoretical sense*, i.e., there is no way (up to measure 0) of splitting further the boundary points of this compactification. Note that this property has nothing to do with solvability of the *Dirichlet problem* with respect to this compactification. For example, the Dirichlet problem is trivially solvable for the one-point compactification; on the other hand, even if the boundary of a certain group compactification can be identified with the Poisson boundary, it does *not* imply in general that the Dirichlet problem is solvable (or even that the support of the harmonic measure is the whole topological boundary); see [KW1], [Wo4] for a discussion of related questions.

For finding a μ-boundary one can apply various direct methods of describing non-trivial behaviour of sample paths at infinity. In the case of Lie groups they usually amount to proving convergence in appropriate homogeneous spaces of the group [Fu2], [Az], [Rau], see the example in **2.9** below.

For discrete groups the variety of situations is much wider [Ka1], [KV]. We shall describe here an example of a finitely generated solvable group G such that any finitely supported probability measure μ on G has a non-trivial μ-boundary (note that, however, such an example does not exist within the class of *polycyclic groups* – see **2.10**).

Denote by $\mathrm{fun}(\mathbf{Z}^k, \mathbf{Z}_2)$ the additive group of finitely supported $\{0, 1\}$-valued configurations on the integer lattice $\mathbf{Z}^k$, and let the group of *dynamical configurations* $G_k = \mathbf{Z}^k \ltimes \mathrm{fun}(\mathbf{Z}^k, \mathbf{Z}_2)$ be the semi-direct product determined by the action T of $\mathbf{Z}^k$ on $\mathrm{fun}(\mathbf{Z}^k, \mathbf{Z}_2)$ by translations. The group product in G_k has the form $(z_1, f_1)(z_2, f_2) = (z_1 + z_2, f_1 + T^{z_1} f_2)$. Thus, if the projection of the random walk (G_k, μ) to $\mathbf{Z}^k$ is transient, and the measure μ is finitely supported (in fact, finiteness of the first moment of μ is sufficient [Ka8]), then for a.e. sample path (x_n, φ_n) the configurations φ_n converge pointwise to a certain (not finitely supported) configuration φ_∞, so that the space $\mathrm{Fun}(\mathbf{Z}^k, \mathbf{Z}_2)$ of all configurations on $\mathbf{Z}^k$ with the resulting measure on it is a non-trivial μ-boundary. Since for $k \geq 3$ *any* non-degenerate random walk on $\mathbb{Z}^k$ is transient, we obtain that all non-degenerate measures with a finite first moment (in particular, finitely supported) on the groups G_k, $k \geq 3$ have a non-trivial Poisson boundary. On the other hand, as the groups G_k are amenable, there always exist probability measures on G_k with trivial Poisson boundary (see **2.5**); however, the argument above shows that they can not be chosen to be finitely supported or even have a finite first moment. The strip criterion (**2.12**) allows one to prove that if a measure μ has a finite first moment and the drift of its projection to $\mathbb{Z}^k$ is non-zero, then the space of limit configurations φ_∞ is indeed the Poisson boundary. However, for random walks with zero drift (which are all transient for $k \geq 3$) the question is still open.

The problem of describing the Poisson boundary for the groups G_k is closely connected with the problem of describing the *exchangeable σ-algebra* of a transient Markov chain (x_n) on a countable set X, i.e., the σ-algebra of all events in the path space invariant with respect to finite permutations of the parameter set $\mathbb{Z}_+$. The relation with the groups G_k becomes clear if one takes into account the fact that the exchangeable σ-algebra coincides with the tail σ-algebra of the *extended* chain (x_n, φ_n) on $X \times \mathrm{fun}(X, \mathbb{Z}_+)$, where $\varphi_n = \delta_{x_0} + \delta_{x_1} + \cdots + \delta_{x_{n-1}} \in \mathrm{fun}(X, \mathbb{Z}_+)$. Transience means that the functions φ_n a.e. converge pointwise to a function $\varphi_\infty \in \mathrm{Fun}(X, \mathbb{Z}_+)$. Clearly, φ_∞ is measurable with respect to the exchangeable σ-algebra. Do these functions generate the whole exchangeable σ-algebra? For random walks on $\mathbb{Z}^k$ with finite first moment and non-zero drift it follows from the strip criterion [Ka15]. James and Peres [JP] noticed that the answer is always positive if for a.e. sample path (x_n) there exists an infinite number of times n such that $p(x_i, x_j) = 0$ for any two $i < n, j > n$. Using this idea they proved that the exchangeable σ-algebra is generated by the final occupation times φ_∞ for any random walk on $\mathbb{Z}^k$ determined by a finitely supported measure. However, their approach apparently does not apply to the groups of dynamical configurations G_k.

Other examples of non-trivial μ-boundaries obtained from "elementary" probabilistic and combinatorial considerations include random walks on the *infinite symmetric group* and on some *locally finite solvable groups* [Ka1]. Whether these μ-boundaries are maximal is still an open question.

The following very useful idea of Furstenberg [Fu4] gives a general method for constructing μ-boundaries. Let B be a separable compact G-space; its compactness implies that there exists a μ-stationary probability measure λ on B. Now, the martingale convergence theorem implies that for a.e. sample path $\overline{x} = (x_n)$ the sequence of translations $x_n \lambda$ converges weakly to a measure $\lambda(\overline{x})$, and

$$(2.7.1) \qquad\qquad \lambda = \int \lambda(\overline{x}) \, d\mathbf{P}(\overline{x}) \, .$$

Thus, the map $\overline{x} \mapsto \lambda(\overline{x})$ allows one to consider the space of probability measures on B as a μ-boundary. This map plays an important role in Margulis' rigidity theory [Mar2].

Further, the measure λ is necessarily purely non-atomic provided that the group generated by supp μ does not fix a finite subset of B. If the action of G on B has the property that for any non-atomic measure λ all weak limit points of the translations $g\lambda$ as g tends to infinity in G are δ-measures, then almost all measures $\lambda(\overline{x})$ are δ-measures, so that (B, λ) is a μ-boundary [Fu5]

This property is satisfied for all group compactifications $\overline{G} = G \cup \partial G$ such that if $g_n \to \gamma_+ \in \partial G$ and $g_n^{-1} \to \gamma_- \in \partial G$, then $g_n \xi \to \gamma_+$ uniformly outside of every neighbourhood of γ_- in $\overline{G}$ [Wo2] (see also the earlier paper [CSo] devoted to groups of tree automorphisms). Another sufficient condition is existence of a G-equivariant map S assigning to pairs of distinct points (γ_-, γ_+) from ∂G non-empty subsets ("strips") $S(\gamma_-, \gamma_+) \subset G$ such that for any distinct $\gamma_0, \gamma_1, \gamma_2 \in \partial G$ there exist neighbourhoods $\mathcal{O}_o \subset \overline{G}$ and $\mathcal{O}_1, \mathcal{O}_2 \subset \partial G$ with $S(\gamma_-, \gamma_+) \cap \mathcal{O}_o = \varnothing$ for all points $\gamma_- \in \mathcal{O}_1, g_+ \in \mathcal{O}_2$ [Ka15]. Moreover, under these conditions convergence of translations $x_n \lambda$ implies convergence of x_n in $\overline{G}$, and the corresponding hitting distribution is the *unique* μ-stationary measure on ∂G (uniqueness follows from the decomposition (2.7.1) and the fact that the limit of $x_n \lambda$ depends on (x_n) only).

The *hyperbolic compactification* of word hyperbolic groups and the *end compactification* of finitely generated groups satisfy both these conditions. A more involved analysis allows one to prove convergence in the *Furstenberg compactification* of the corresponding symmetric space and uniqueness of μ-stationary measure on the *Furstenberg boundary* for probability measures μ on a *semi-simple Lie group* G under natural irreducibility conditions [GR1], in particular, if the group generated by supp μ is Zariski dense [GM] (see Section 3). Another example is that of convergence to the visibility boundary and uniqueness of μ-stationary measures for random walks on *cocompact lattices in rank one Cartan–Hadamard manifolds* [Bal]. A new class of examples was recently considered in [KM]: random walks on the *mapping class group* converge in the *Thurston compactification* of Teichmüller space, and the corresponding limit distribution is the unique μ-stationary measure on the *Thurston boundary* (in fact, it is concentrated on the subset of the Thurston boundary consisting of *uniquely ergodic projective measured foliations*).

2.8. The Poisson boundary of Lie groups.

There are two general ideas which are especially helpful for identification of the Poisson boundary of Lie groups. The first one is used for finding out group elements $g \in G$ such that their action on the Poisson boundary is trivial, i.e., such that $f(gx) = f(x) \ \forall f \in H^\infty(G, \mu), x \in G$ (these group elements are called *μ-periods*). If the sequence $(x_n^{-1} g x_n)$ has a limit point in G for a.e. path (x_n), then g is a μ-period [Az], [Gu1]. In particular, all elements from the center of G are μ-periods. This gives another proof of triviality of the Poisson boundary for abelian groups.

If the group G is compactly generated, for any compact generating subset K of G containing a neighbourhood of the identity one can define the *length function* on G as $\delta_K(g) = \min\{n : g \in K^n\}$, and the corresponding (left-invariant) *word distance* as $d_K(g_1, g_2) = \delta_K(g_1^{-1} g_2)$. Then $g \in G$ is a μ-period if a.e. $\liminf d_K(x_n, g x_n) < \infty$.

The other idea is used for proving maximality of a given μ-boundary Z. Denote by π the projection from the Poisson boundary Γ onto Z. Suppose that a subgroup $H \subset G$ acts simply transitively on Z. If for a point $z \in Z$ one has a non-constant bounded function φ_z on the fiber $\Gamma_z = \pi^{-1}(z)$, then using the action of H one can extend φ_z to a non-constant bounded H-invariant function φ on Γ corresponding to a non-constant bounded H-invariant harmonic function. Thus, if one knows that there are no non-constant H-invariant bounded harmonic functions on G, then Z in fact coincides with the Poisson boundary.

In the case when G is a non-compact semi-simple Lie group with finite center, these ideas allowed Furstenberg [Fu1] to identify the Poisson boundary for an absolutely continuous non-degenerate measure μ with the *Furstenberg boundary* B of the corresponding symmetric space (see below Section 3). He also showed that the harmonic measure in this case is absolutely continuous with respect to the Haar measure on B.

By a skillful development of these ideas (and with a heavy use of the structure theory of Lie groups) Azencott [Az], Guivarc'h [Gu1] and, finally, Raugi [Rau] obtained a description of the Poisson boundary for an arbitrary spread out probability measure with a finite first moment (with respect to a length function δ_K) on a connected Lie group G as a G-space determined by a family of cocycles associated with the measure μ.

This approach is also applicable in discrete situations when the group of automorphisms of the Markov operator is big enough to act transitively on a would-be Poisson boundary. For example, for random walks on *buildings of reductive split groups over local fields* it leads to a description of the Poisson boundary analogous to the Furstenberg boundary of symmetric spaces [Ge1], [Ge2].

2.9. Example: solvable Lie groups and polycyclic groups.

To give the reader a general impression of how this technique works let us consider the simplest solvable group – the *affine group* $G = \text{Aff}(\mathbb{R}) = \{x \mapsto ax+b, a \in \mathbb{R}_+, b \in \mathbb{R}\}$ of the real line. Finiteness of the first moment of a probability measure μ on G is equivalent to $\int \log^+(|a(g)| + |b(g)|)\, d\mu(g) < \infty$. Let $x_n = (a_n, b_n)$ be a sample path of the random walk (G, μ), and $g = (1, b)$. Then $x_n^{-1} g x_n = (1, a_n^{-1}b)$. Thus, if $\alpha = \int \log a(g)\, d\mu(g) \geq 0$, then $H = \{(1, b)\}$ is a subgroup of the group of μ-periods, so that any bounded μ-harmonic function on G depends on the component $a(g)$ only. Since the abelian group $\{(a, 0)\}$ does not have bounded harmonic functions, the Poisson boundary of the random walk (G, μ) is trivial.

In the *contracting* case $\alpha < 0$ the situation is different, and looking at the formula for the group product in G one can immediately see that for a.e. path (x_n) there exists the limit $b_\infty = \lim_{n \to \infty} b_n \in \mathbb{R}$. Since the subgroup H acts on $\mathbb{R}$ simply transitively, and there are no H-invariant bounded harmonic functions, $\mathbb{R}$ with the corresponding family of harmonic measures is isomorphic to the Poisson boundary of (G, μ).

For this group a description of the minimal Martin boundary was obtained by Elie [El]. What is interesting is that even for a symmetric measure μ (when the Poisson boundary is trivial) there exist non-trivial positive harmonic functions given by Radon–Nikodym derivatives of a (unique) σ-finite μ-stationary measure on $\mathbb{R}$, so that the Martin boundary is non-trivial (see also [BBE]). Describing the Martin compactification seems to be much more difficult and is an open problem (except for the case of some invariant diffusion processes on $\text{Aff}(\mathbb{R})$ [Mol2], see also **3.3**).

Note that finiteness of the first moment is used in this proof only to establish the convergence $b_n \to b_\infty$, so that the given proof of triviality of the Poisson boundary works for *any* symmetric spread out measure μ on $\text{Aff}(\mathbb{R})$. Developing this idea one can prove that the Poisson boundary is trivial for an arbitrary symmetric spread out measure on an amenable Lie group without any moment conditions whatsoever [BR] (in sharp contrast with the situation for discrete solvable groups – cf. **2.7**).

Consider now a general semi-direct product $G = A \ltimes \mathcal{N}$ determined by an action T of an abelian group $A \cong \mathbb{R}^n$ on a simply connected nilpotent Lie group $\mathcal{N}$ (see [DH]). For any $\alpha \in A$ the corresponding tangent automorphism of the Lie algebra $\mathfrak{N}$ determines a decomposition of $\mathfrak{N}$ into *contracting*, *neutral*, and *expanding* subspaces $\mathfrak{N} = \mathfrak{N}_- \oplus \mathfrak{N}_0 \oplus \mathfrak{N}_+$, which gives rise to a decomposition of $\mathcal{N}$ into a product of the corresponding groups $\mathcal{N}_-, \mathcal{N}_0, \mathcal{N}_+$. For $G = \mathrm{Aff}(\mathbb{R}) = \mathbb{R} \wedge \mathbb{R}$ this decomposition is trivial, and $\mathbb{R} = \mathcal{N}$ coincides with one of the groups $\mathcal{N}_-, \mathcal{N}_0, \mathcal{N}_+$ according to whether α is negative, zero, or positive. Let μ be a probability measure on G with a finite first moment, and $\alpha = \alpha(\mu) \in A$ be the barycenter of the projection of μ to A, then one can show that in the decomposition $\mathcal{N} = \mathcal{N}_- \mathcal{N}_0 \mathcal{N}_+$ determined by α the $\mathcal{N}_-$-component converges for a.e. sample path of the random walk, so that $\mathcal{N}_- \cong G/A\mathcal{N}_0\mathcal{N}_+$ with the resulting measure is a μ-boundary, and if the measure μ is spread out, then $\mathcal{N}_-$ is the Poisson boundary. This example is a key ingredient of the description of the Poisson boundary for spread out measures with finite first moment on general Lie groups.

Note that the proof of convergence does not require any conditions on absolute continuity of the measure μ. Thus, it also works for discrete subgroups of solvable Lie groups, or, slightly more generally, for all *polycyclic groups* (these are solvable groups with a normal series with cyclic quotients; they can be characterized as solvable groups with finitely generated abelian subgroups or as solvable groups of integer matrices). Roughly speaking (up to a *semi-simple splitting*), polycyclic groups are just semi-direct products $G = A \wedge N$ of finitely generated free abelian and torsion free nilpotent groups. If μ is a probability measure on $G = A \wedge N$ with a finite first moment, then the barycenter α of the projection of the measure μ onto A determines a decomposition of the Lie hull $\mathcal{N}$ of the group N into contracting $\mathcal{N}_-$, neutral $\mathcal{N}_0$ and expanding $\mathcal{N}_+$ subgroups, and one can show (basically in the same way as in the case of solvable Lie groups) that the $\mathcal{N}_-$-components converge along a.e. sample path of the random walk, which gives a non-trivial μ-boundary whenever $\alpha \neq 0$ [Ka15]. In the *centered case* $\alpha = 0$ one has to use the entropy theory for establishing triviality of the Poisson boundary for polycyclic groups [Ka8], see **2.10**.

2.10. Entropy and triviality of the Poisson boundary.

A countable group G can not act transitively on a non-trivial Poisson boundary. Also, in the discrete situation usually $d(x_n, gx_n) \to \infty$ a.e. (unless g belongs to the center of G), so that the approach described in **2.8** does not work for countable groups. The notion of *entropy* in explicit [Av1], [Av3], [KV], [De4], [De5] or implicit form (via differential entropy [Fu4], asymptotic growth [Gu2], Hausdorff dimension [Le1], [Le2], [BL1]) turned out to be much more efficient for dealing with the Poisson boundary of random walks on discrete groups. A method based on estimating the *entropy of conditional random walks* incorporates all these approaches [Ka2], [Ka11], [Ka14], [Ka15]. Instead of using structure theory this method relies upon volume estimates for random walks and it is applicable both to discrete and continuous groups. For the sake of simplicity we shall first describe it for the case of random walks on a countable group G (the exposition in **2.10**, **2.11** is based on the papers [KV], [Ka2]).

Let μ be a probability measure on G with finite *entropy*

$$H(\mu) = \sum_{g \in G} -\mu(g) \log \mu(g) .$$

If G is a finitely generated group, and the measure μ has a finite first moment in G, then its entropy is also finite [De5]. The limit

$$h(G, \mu) = \lim_{n \to \infty} \frac{H(\mu_n)}{n}$$

is called the *entropy of the random walk* (G, μ) [Av1], [KV], [De4]. Existence of the limit $h(G, \mu)$ follows from the fact that the sequence $H(\mu_n)$ is subadditive, because for any $m, n > 0$ the $(n+m)$-fold convolution μ_{n+m} is the image of the product of the n-fold and m-fold convolutions μ_n and μ_m under the map $(g_1, g_2) \mapsto g_1 g_2$.

The Poisson boundary of (G, μ) is trivial (with respect to the measure **P** in the path space corresponding to the initial distribution δ_e) if and only if $h(G, \mu) = 0$, and the entropy $h(G, \mu)$ has the following Shannon type *equidistribution property*:

$$\frac{1}{n} \log \mu_n(x_n) \to -h(G, \mu)$$

a.e. and in the space $L^1(\mathbf{P})$ [KV], [De4].

Indeed, let α_k (resp., η_k) be the partitions of the path space $(G^{\mathbf{Z}+}, \mathbf{P})$ such that two paths (x_n) and (x'_n) belong to the same class of α_k (resp., η_k) iff $x_i = x'_i$ $\forall i \leq k$ (resp., $x_i = x'_i$ $\forall i \geq k$). Then the entropy of α_k is $H(\alpha_k) = kH(\mu)$, whereas the *conditional entropy* of α_k with respect to η_n (provided $k < n$) obtained by integrating logarithms of the corresponding conditional probabilities is

$$H(\alpha_k | \eta_n) = kH(\mu) + H(\mu_{n-k}) - H(\mu_n) .$$

As the partitions η_n monotonously decrease to the tail partition η, we have

$$H(\alpha_k | \eta_n) \uparrow H(\alpha_h | \eta) = k\big[H(\mu) - h(G, \mu)\big] .$$

Since triviality of the partition η is equivalent to its independence of all α_k, i.e., to $H(\alpha_k|\eta) = H(\alpha_k)\ \forall\, k > 0$, the tail partition η is trivial (with respect to the measure $\mathbf{P}$) iff $h(G,\mu) = 0$. Coincidence of the tail and Poisson boundaries with respect to the measure $\mathbf{P}$ (see **2.3**) finishes the proof.

The Bernoulli shift $\mathbf{B}$ in the space of increments of the random walk determines the measure preserving transformation $(U\overline{x})_n = x_1^{-1}x_{n+1}$ of the path space $(G^{\mathbb{Z}_+}, \mathbf{P})$. Since

$$\mu_{n+m}(x_{n+m}) \geq \mu_n(x_n)\mu_m(x_n^{-1}x_{n+m}) \, ,$$

and $x_n^{-1}x_{n+m} = (U^n\overline{x})_m$, the equidistribution property immediately follows from Kingman's *subadditive ergodic theorem* [De4].

Thus, the Poisson boundary is trivial iff there exist $\varepsilon > 0$ and a sequence of sets A_n such that $\mu_n(A_n) > \varepsilon$ and $\log|A_n| = o(n)$ (for, the maximal value for the entropy of a probability distribution on A_n is $\log|A_n|$). This implies triviality of the Poisson boundary for all random walks with a finite first moment on groups of *subexponential growth*, cf. with the original paper by Avez [Av2]. [Recall that the class of finitely generated groups of *polynomial growth* coincides with the class of finite extensions of nilpotent groups [Gro1], whereas there are groups with growth intermediate between polynomial and exponential [Gri].]

More generally, if the group G is finitely generated, and μ has a finite first moment, then (again by Kingman's subadditive ergodic theorem) for any length function δ on G there exists a number $l = l(G,\mu,\delta)$ (the *linear rate of escape*) such that $\delta(x_n)/n \to l$ a.e. and in $L^1(G^{\mathbb{Z}_+}, \mathbf{P})$. Then the entropy criterion implies that the Poisson boundary is trivial if $l = 0$. In the case when the measure μ is symmetric and finitely supported, these two conditions are in fact equivalent [Va1]. Whether for any finitely generated group G of exponential growth there exists a probability measure μ with a finite first moment (or just finitely supported) with a non-trivial Poisson boundary is an open question. Another open question is whether for any finitely generated group the Poisson boundaries of all finitely supported symmetric non-degenerate measures are trivial or non-trivial simultaneously. For all known examples the answer to both these question s is "yes".

An interesting consequence of the entropy criterion is that if a measure μ on a countable group G has finite entropy $H(\mu)$, then the Poisson boundaries of the measure μ and of the *reflected measure* $\breve{\mu}(g) = \mu(g^{-1})$ are trivial or non-trivial *simultaneously* (for, the n-fold convolutions μ_n and $\breve{\mu}_n$ have the same entropy, and thereby $h(G,\mu) = h(G,\breve{\mu})$). Surprisingly, this is *not* true in general for measures μ with *infinite* entropy $H(\mu)$; see [Ka1] for an example. Note that for *non-unimodular* locally compact groups the situation is different because of the presence of the modular function in the expression for the density of the reflected measure, so that the differential entropies of the measures μ and $\check{\mu}$ do not have to coincide (cf. **2.13**). This fact can be used for evaluating the entropy $h(G, \mu)$ for random walks on some non-unimodular groups [De6].

2.11. Kullback–Leibler deviation and entropy of conditional walks.

By formula (2.2.1), $\mathbf{P}(C_{e,g}|\gamma) = \mu(g)\frac{dg\nu}{d\nu}(\gamma)$ for any $g \in G$ and a.e. $\gamma \in \Gamma$, so that

$$H(\alpha_1|\eta) = -\int \left[\sum_g \mu(g)\frac{dg\nu}{d\nu}(\gamma)\left[\log \mu(g) + \log \frac{dg\nu}{d\nu}(\gamma)\right]\right] d\nu(\gamma)$$

$$= H(\mu) - \sum_g \mu(g)\, I(\nu|g\nu)\,,$$

where

$$I(\nu|g\nu) = \int \log \frac{dg\nu}{d\nu}(\gamma)\, dg\nu(\gamma)$$

is the *Kullback–Leibler deviation* of the measure ν from the measure $g\nu$. Thus, we have the formula

$$h(G, \mu) = \sum_g \mu(g)\, I(\nu|g\nu)$$

which connects the entropy $h(G, \mu)$ (initially defined in terms of convolutions of the measure μ) with differential entropy of translations of the harmonic measure ν on the Poisson boundary.

Let now (Γ_ξ, ν_ξ) be a quotient of the Poisson boundary with respect to a G-invariant measurable partition ξ. Denote by $\mathbf{bnd}_\xi$ the corresponding map from the path space to Γ_ξ such that $\nu_\xi = \mathbf{bnd}_\xi(\mathbf{P})$. The Kullback–Leibler deviation does not increase under homomorphisms of measure spaces, and equality holds iff the homomorphism preserves the Radon–Nikodym derivatives. Since the Radon–Nikodym derivatives $dg\nu/d\nu$ separate points from Γ (see **2.2**), we have that the *differential entropy*

$$E(\Gamma_\xi, \nu_\xi) = \sum \mu(g)I(\nu_\xi|g\nu_\xi) = \int \log \frac{dx_1\nu_\xi}{d\nu_\xi}(\mathbf{bnd}_\xi \overline{x})\, d\mathbf{P}(\overline{x})$$

does not exceed $h(G, \mu)$, and equality holds iff ξ is the point partition of Γ, i.e., iff the μ-boundary (Γ_ξ, ν_ξ) coincides with the Poisson boundary (Γ, ν).

One can deduce from formula (2.2.1) that the one-dimensional distributions of the *conditional operators* corresponding to points $\gamma_\xi \in \Gamma_\xi$ have the form

$$p_n^\gamma(g) = \frac{dg\nu_\xi}{d\nu_\xi}(\gamma_\xi)\mu_n(g) \ .$$

As it follows from the ergodic theorem applied to the measure preserving transformation U in the path space induced by the Bernoulli shift in the space of increments,

$$\frac{1}{n}\log\frac{dx_n\nu_\xi}{d\nu_\xi}(\mathbf{bnd}_\xi\overline{x}) \to E(\Gamma_\xi, \nu_\xi)$$

a.e. and in the space $L^1(\mathbf{P})$. On the other hand, $1/n\log\mu_n(x_n) \to -h(G,\mu)$, so that

$$(2.11.1) \qquad \frac{1}{n}\log p_n^\gamma(x_n) \to E(\Gamma_\xi, \nu_\xi) - h(G,\mu) \ , \qquad \gamma = \mathbf{bnd}_\xi(\overline{x})$$

a.e. and in $L^1(\mathbf{P})$. Thus, Γ_ξ is the Poisson boundary if and only if there is $\varepsilon > 0$ such that for ν_ξ-a.e. $\gamma_\xi \in \Gamma_\xi$ there is a sequence of sets $A_n = A_n(\gamma)$ such that $p_n^\gamma(A_n) > \varepsilon$ and $\log|A_n| = o(n)$ [Ka2] (if Γ_ξ is the one-point space, then this condition coincides with the condition above for triviality of the Poisson boundary).

2.12. Geometric criteria of maximality.

Now we can formulate two simple geometric criteria of maximality of a μ-boundary for a measure μ with finite entropy. Both require an approximation of the sample paths of the random walk by a certain family of subsets of the group determined by the limit behaviour of the sample path [Ka2], [Ka11], [Ka15]. For simplicity we shall formulate them under the assumption that G is finitely generated. Denote by $d(g_1, g_2) = \delta(g_1^{-1}g_2)$ the left-invariant metric corresponding to a length function δ, and by B_n the ball of radius n in G centered at the identity.

The first criterion says that if there is a family of maps $\pi_n : \Gamma_\xi \to G$ such that a.e.

$$d(x_n, \pi_n(\mathbf{bnd}_\xi\overline{x})) = o(n) \ ,$$

then Γ_ξ is the Poisson boundary ("*ray approximation*").

The other criterion applies simultaneously to a μ-boundary Γ_ξ and to a $\check{\mu}$-boundary $\check{\Gamma}_\zeta$ (where $\check{\mu}(g) = \mu(g)^{-1}$ is the *reflected measure* of μ). If there is a G-equivariant measurable map S assigning to pairs $(\gamma_-, \gamma_+) \in \check{\Gamma}_\zeta \times \Gamma_\xi$ a non-empty subset $S(\gamma_-, \gamma_+) \subset G$ such that for a.e. $(\gamma_-, \gamma_+) \in \check{\Gamma}_\zeta \times \Gamma_\xi$

(2.12.1)
$$\frac{1}{n} |S(\gamma_-, \gamma_+) \cap B_{\delta(x_n)}| \to 0$$

in probability with respect to the measure $\mathbf{P}$ in the space of the sample paths of the random walk (G, μ), then Γ_ξ and $\check{\Gamma}_\zeta$ are the Poisson boundaries of the measures μ and $\check{\mu}$, respectively (*"strip approximation"*).

We shall indicate here a proof of the "strip criterion", the "ray criterion" being an immediate corollary of formula (2.11.1). First note that the formula $x_n = x_{n-1} h_n$ considered for all $n \in \mathbf{Z}$ (and not just for positive n) determines an isomorphism between the space of *bilateral* sequences of independent μ-distributed increments $(h_n), n \in \mathbf{Z}$ and the space $(G^{\mathbf{Z}}, \widetilde{\mathbf{P}})$ of *bilateral* paths $(x_n), n \in \mathbf{Z}$, the latter space being the product of spaces $(G^{\mathbf{Z}_+}, \mathbf{P})$ and $(G^{\mathbf{Z}_+}, \check{\mathbf{P}})$ of unilateral paths $(x_n), n \in \mathbf{Z}_+$ and $(\check{x}_n) = (x_{-n}), n \in \mathbf{Z}_+$ of random walks (G, μ) and $(G, \check{\mu})$, respectively. The bilateral Bernoulli shift in the space of increments induces then an ergodic measure preserving transformation $\widetilde{U}$ of the bilateral path space $(G^{\mathbf{Z}}, \widetilde{\mathbf{P}})$. Denote by $\mathbf{bnd}_+ \overline{x}$ (resp., $\mathbf{bnd}_- \overline{x}$) the point from the boundary Γ_ξ (resp., from $\check{\Gamma}_\zeta$) determined by the "positive" $(x_n), n \in \mathbf{Z}_+$ (resp., "negative" $(\check{x}_n) = (x_{-n}), n \in \mathbf{Z}_+$) part of bilateral path $(x_n), n \in \mathbf{Z}$. Then $\mathbf{bnd}_\pm (\widetilde{U}^n \overline{x}) = x_n^{-1} \mathbf{bnd}_\pm \overline{x}$, so that by equivariance of the strip map S for any $n \in \mathbf{Z}$

$$\widetilde{\mathbf{P}} \big[x_n \in S(\mathbf{bnd}_- \overline{x}, \mathbf{bnd}_+ \overline{x}) \big] = \widetilde{\mathbf{P}} \big[e \in S(\mathbf{bnd}_- \overline{x}, \mathbf{bnd}_+ \overline{x}) \big] = p \,.$$

Since the strips $S(\gamma_-, \gamma_+)$ are a.e. non-empty, ergodicity of $\widetilde{U}$ implies that $p > 0$, so that sample paths of the conditional walk conditioned by $\gamma_+ \in \Gamma_\xi$ belong to $S(\gamma_-, \gamma_+)$ with probability p. [Note that ergodicity of $\widetilde{U}$ implies that the group G acts ergodically on the product of the Poisson boundaries of the measures μ and $\check{\mu}$, see [Ka10], [Ka13].] Now, if the growth of the strips is controlled by (2.12.1), then the entropy of the conditional chain must be zero, so that Γ_ξ is the Poisson boundary of (G, μ). The same argument applied to the random walk $(G, \check{\mu})$ conditioned by $\gamma_- \in \check{\Gamma}_\zeta$ yields maximality of $\check{\Gamma}_\zeta$.

If the strips $S(\gamma_-, \gamma_+)$ grow *subexponentially*, i.e., $\log |S(\gamma_-, \gamma_+) \cap B_n| = o(n)$, then the condition (2.12.1) is satisfied for any probability measure μ with finite first moment (which automatically implies that the entropy $H(\mu)$ is also finite), and if the strips grow *polynomially* then (2.12.1) is satisfied for any measure μ with a finite *first logarithmic moment* $\sum \log_+ \delta(g) \mu(g)$.

The "ray criterion" provides more information than the "strip criterion" about the behaviour of sample paths of the random walk (which can be also useful for other issues than just identification of the Poisson boundary; e.g., see [Ka17]). On the other hand, for checking the ray criterion one often needs rather elaborate estimates, whereas existence of strips is usually almost evident, and estimates of their growth are not very hard.

Let us return to the examples of μ-boundaries constructed in **2.7** and look at how the ray and strip approximation criteria can be used in these situations for proving maximality of these μ-boundaries.

For *word hyperbolic groups* the ray criterion (for measures μ with a finite first moment) amounts to proving that for any sequence (x_n) in a Gromov hyperbolic space such that $d(x_0, x_n)/n \to l > 0$ and $d(x_n, x_{n+1}) = o(n)$ there exists a geodesic γ such that $d(x_n, \gamma(ln)) = o(n)$, whereas the strip corresponding to a pair of points (ξ_-, ξ_+) from the hyperbolic boundary is naturally defined as the union of all geodesics with endpoints ξ_-, ξ_+ [Ka11]. In the case of *groups with infinitely many ends* for any two distinct two ends ξ_-, ξ_+ there exists a ball of minimal radius R separating ξ_- and ξ_+. Then the strip $S(\xi_-, \xi_+)$ is defined as the union of all such R-balls. For *cocompact lattices in rank one Cartan–Hadamard manifolds* for applying the strip criterion one takes geodesics joining pairs of points from the visibility boundary [BL1], [Ka15]. In the case of the *mapping class group* the strips are defined by using Teichmüller geodesic lines associated with any two distinct uniquely ergodic projective measured foliation [KM] (note that in the last three situations the ray approximation either fails or is unknown). In all these cases the strips have linear growth, so that the strip criterion allows one to identify the Poisson boundary with the corresponding geometric boundary for *all measures with finite entropy and finite first logarithmic moment*. However, the problem of proving maximality of the constructed μ-boundaries of these groups (or, just for the free group, to take the simplest case) for an *arbitrary* measure μ is still open.

See Section 3 for a description of the situation with *discrete subgroups of semi-simple Lie groups*.

In the case of *polycyclic groups* (see **2.9**) the ray approximation follows from the *global law of large numbers* for solvable Lie groups, which allows one to approximate (in the enveloping solvable Lie group) a.e. sample path (x_n) by the sequence of powers g^n [Ka8]. The strip criterion requires considering simultaneously with the measure μ its reflected measure $\check{\mu}$. In the same way as for the measure μ we obtain a $\check{\mu}$-boundary $\mathcal{N}_+$ (because the contracting subgroup for the reflected measure $\check{\mu}$ is precisely the expanding subgroup corresponding to μ). Since the points from $\mathcal{N}_-$ (resp., $\mathcal{N}_+$) are identified with cosets of the subgroup $A\mathcal{N}_0\mathcal{N}_+$ (resp., $A\mathcal{N}_0\mathcal{N}_-$) in $A \ltimes \mathcal{N}$, any pair of points from $\mathcal{N}_-$ and $\mathcal{N}_+$ determines (as intersection of the corresponding cosets) a coset of $A\mathcal{N}_0$. After a modification this gives an equivariant map assigning to pairs of points from $\mathcal{N}_- \times \mathcal{N}_+$ subsets in G. Now, according to the strip criterion, for proving maximality we have to show that these strips are "thin enough", which boils down to an easy estimate of the growth of the neutral component along sample paths of the random walk. The general situation is intermediate between two extreme cases: when $\mathcal{N}_0 = \{e\}$, and when $\mathcal{N}_0 = \mathcal{N}$. In the first case the strips are just cosets of A, so that the required estimate is trivial; whereas in the second case the strips coincide with G, and proving maximality of $\mathcal{N}_- = \{e\}$ is equivalent to proving triviality of the Poisson boundary of (G, μ) [Ka15].

2.13. Entropy for continuous groups and other generalizations.

The entropy approach (with some technical modifications) can be also used for finding out when the Poisson boundary is trivial for random walks with absolutely continuous measure μ on general locally compact groups (in particular, Lie groups) [Av3], [Gu3], [De5], [Va2], [Al]. Here one should replace the entropy $H(\mu)$ with the *differential entropy*

$$H_{diff}(\mu) = -\int \log \frac{d\mu}{dm}(g)\, d\mu(g)\,,$$

where m is the left Haar measure on G. For a measure μ with a non-trivial singular part the differential entropy is infinite. However, the Poisson boundaries of measures $\mu = \mu' + \mu''$ and $\widetilde{\mu} = (\delta_e + \mu' + \mu_2' + \ldots)\mu''$ are the same [Wi]. Thus, in the case when the measure μ is spread out, one can choose μ'' to be absolutely continuous with bounded density and apply the entropy theory to the measure $\widetilde{\mu}$. If the measure μ has a finite first moment, then $\widetilde{\mu}$ also has a finite first moment and finite differential entropy. Formula (2.11.1) can be proved in this setup with convergence in the space L^1, which is sufficient for formulating geometric criteria of boundary maximality analogous to those from **2.12** [Ka14].

Recall that the problem with the discrete groups was that the group was "too small" to act on the Poisson boundary transitively, whereas for groups which are "larger" than Lie groups (e.g., the *affine groups* Aff(T_2) *of a homogeneous tree* of degree 3 which consists of all tree automorphisms which preserve a given end and contains Aff($\mathbb{Q}_2$) as a subgroup) the group is "too big" to have a subgroup with a free action on the Poisson boundary. However, the entropy approach can deal with all these situations in a uniform way. For example, for the *real affine group* Aff($\mathbb{R}$) taking for strips the sets $A_\gamma = \{(a, b) : b = \gamma\}$ shows at once that the Poisson boundary coincides with $\mathbb{R}$ in the contracting case $\alpha < 0$ and is trivial in the expanding case $\alpha > 0$. The same idea works both for the *dyadic affine group* Aff($\mathbb{Z}[\frac{1}{2}]$) [Ka8], [Ka15] and for the groups Aff(T_d) [CKW].

The ideas described in **2.10–2.12** can be applied not only to random walks on groups but also to other invariant Markov operators. One example is provided by covering Markov operators [Ka13], in particular, operators corresponding to Brownian motion on regular covers of compact Riemannian manifolds [Ka3]. The entropy methods can be applied here either directly or by using *discretization procedures* connecting the Poisson boundary of a covering operator with the Poisson boundary of an appropriate random walk on the deck group (**2.6**). Without going into details we mention here absence of bounded harmonic functions on polycyclic covers of compact Riemannian manifolds proved in [Ka3] by entropy methods (cf. with triviality of the Poisson boundary for polycyclic groups with centered measures, **2.9**); expositions of this proof are given in [An2] and [Ly]. This result can be used for proving absence of Kählerian structures on certain compact complex manifolds. Namely, suppose that the fundamental group of such manifold X is polycyclic, and its universal covering space admits a bounded holomorphic function. Since polycyclic covers do not admit bounded harmonic functions, there are no Riemannian metrics on X for which holomorphic functions are harmonic, in particular, no Kählerian metrics [Ka6]. See [Li] for criteria of absence of bounded holomorphic functions on covers of compact complex manifolds.

There is another class of Markov operators which are "sufficiently homogeneous", although generally speaking they are not endowed with any group action. These are Markov operators on equivalence relations, see **0.2**. A finite stationary measure of the global Markov operator (such measures were called "harmonic" in the case of Riemannian foliations in [Ga]) gives rise to a shift invariant measure in the path space. The general theory of Markov operators with finite stationary measure immediately implies that in this situation *the global Markov operator P is ergodic* (i.e., the shift in its path space is ergodic, or, equivalently, there are no global bounded P-harmonic functions) *iff the equivalence relation is ergodic* in the sense that there are no non-trivial measurable unions of equivalence classes; e.g., see [Ka9]. Thus, by the ergodic theorem the behaviour of a. e. sample path is "statistically homogeneous" globally (although every path is confined to a single leaf). However, the local structure of a single leaf can be very complicated; in particular, the local leafwise operators may well have a non-trivial Poisson boundary.

The entropy methods can be used in this situation for identifying (or proving triviality of) the Poisson boundary of a single leaf. Among the examples are the Brownian motion on Riemannian foliations [Ka4], random walks in random environment on groups (i.e., essentially, random walks along orbits of group actions) [Ka7], and, more generally, random walks on arbitrary discrete equivalence relations [Ka16]. Although any discrete equivalence relation is generated by a measure type preserving action of a certain countable group [FM], there are also numerous "natural" equivalence relations whose origin has nothing to do with group actions (e.g., the transversal equivalence relation of a foliation, or the trajectory equivalence relation of a non-invertible transformation). In the first case one is often able to construct stationary measures of global Markov operators by using holonomy invariant measures, and in the second case by using Ruelle's Perron–Frobenius theorem. Another example is the equivalence relation in the space of rooted trees (more generally, graphs) with trivial automorphism group, which is obtained by taking different roots of the same tree; then the equivalence classes are obviously given a tree structure, and any *Galton–Watson* branching process determines in a natural way a stationary measure for the simple random walk along the equivalence classes [LPP] (see also [Ka17] for other examples of stationary measures in this situation).

Note that the fact that triviality of the leafwise Poisson boundaries (in complete analogy with the group case) implies *amenability of the foliation* (or of the equivalence relation) was mentioned already in [CFW]; another more constructive proof follows from the 0-2 laws (cf. **2.5**). Thus, the entropy technique can be also used for establishing amenability of equivalence relation.

3. SEMI-SIMPLE LIE GROUPS AND SYMMETRIC SPACES

3.1. Asymptotic geometry and compactifications of symmetric spaces.

Let G be a non-compact real semi-simple Lie group with finite center, and K – its maximal compact subgroup (example: $G = \mathrm{SL}(n, \mathbb{R})$ and $K = \mathrm{SO}(n)$). By $S = G/K$ we shall denote the corresponding *symmetric space*, which is a simply connected Riemannian manifold with non-positive sectional curvatures (a *Cartan–Hadamard manifold*).

The Riemannian manifold S being a Cartan–Hadamard manifold, its geometrically most natural compactification is the *visibility compactification* [EO]: a sequence of points x_n converges in this compactification if and only if directions of the geodesic rays $[o, x_n]$ converge for a certain ($\equiv$ any) reference point $o \in S$. In other words, the boundary of S in this compactification ∂S (the *sphere at infinity* of S) can be identified with the unit sphere of the tangent space at the point $o = K \in S$. Another interpretation of the sphere at infinity is that it is the space of *asymptotic classes* of geodesic rays in S (two rays are *asymptotic* if they are within bounded distance one from the other). For any $\gamma \in \partial S$ and $x \in S$ there exists a unique geodesic ray which emanates from x and belongs to the class γ.

Let $\mathfrak{a}_+^1$ (resp., $\bar{\mathfrak{a}}_+^1$) be the set of unit vectors in a dominant Weyl chamber $\mathfrak{a}_+$ of the Lie algebra $\mathfrak{a}$ of a fixed Cartan subgroup A (resp., in the closure $\bar{\mathfrak{a}}_+$). In the case $G = SL(n, \mathbb{R})$ the group A consists of all diagonal matrices with positive entries, and $\mathfrak{a}_+$ is the set of all vectors $(\lambda_1, \lambda_2, \ldots, \lambda_n)$ with $\lambda_1 > \lambda_2 > \cdots > \lambda_n$ and $\lambda_1 + \lambda_2 + \cdots + \lambda_n = 0$. Any geodesic ray in S has the form $\xi(t) = g \exp(ta)o$ with $g \in G$ and a uniquely determined vector $a = a(\xi) \in \bar{\mathfrak{a}}_+^1$, and if the origin of ξ is o, then $g \in K$. Moreover, any two asymptotic geodesic rays ξ_1, ξ_2 have the same vectors $a(\xi_1) = a(\xi_2)$. Hence, the orbits of the group G in ∂S are in one-to-one correspondence with vectors from $\bar{\mathfrak{a}}_+^1$. Stabilizers $\mathrm{Stab}_\gamma \subset G$ of points $\gamma \in \partial S$ are *parabolic subgroups* of G, and Stab_γ is a minimal parabolic subgroup iff γ belongs to the orbit ∂S_a corresponding to a vector $a \in \mathfrak{a}_+^1$, see [Kar]. In particular, for the geodesic ray $\xi(t) = \exp(ta)o$, $a \in \mathfrak{a}_+^1$ the stabilizer of the corresponding point from ∂S is the *standard minimal parabolic subgroup* $P = MAN$ determined by the *Iwasawa decomposition* $G = KAN$, where M is the centralizer of A in K (if $G = SL(n, \mathbb{R})$, then N is the nilpotent group of upper triangular matrices with 1's on the diagonal).

Thus, G-orbits S_a in ∂S corresponding to non-degenerate vectors a are isomorphic to the homogeneous space $B = G/P$, which is called the *Furstenberg boundary* of the symmetric space S (or, of the group G). Orbits corresponding to degenerate vectors are quotients of G by non-minimal parabolic subgroups, i.e., quotients of B. For $G = SL(n, \mathbb{R})$ the boundary B is the space of full flags in $\mathbb{R}^n$, and its quotients are spaces of partial flags. In the rank one case the sphere at infinity consists of a single G-orbit and coincides with the Furstenberg boundary.

Another interpretation of the Furstenberg boundary can be given in terms of *Weyl chambers* in maximal totally geodesic flat subspaces of S (*flats*) which for general symmetric spaces play the same role as geodesic rays and infinite geodesic in the rank one case [Mos], [Im].

A geodesic ξ in S is contained in a uniquely determined flat iff $a(\xi) \in \mathfrak{a}_+^1$. For a given flat f and a basepoint $x \in f$ the connected components of the set obtained by removing from f all geodesics $\xi \subset f$ with $a(\xi) \in \bar{\mathfrak{a}}_+^1 \setminus \mathfrak{a}_+^1$ passing through x are called the *Weyl chambers* of f based at x. Fix a vector $a \in \mathfrak{a}_+^1$, then one can associate with a Weyl chamber c based at a point x the uniquely determined geodesic ray $\xi = \xi(c) \subset c$ with $a(\xi) = a$ starting at x, and, conversely, any geodesic ray with $a(\xi) = a$ uniquely determines a Weyl chamber. Two Weyl chambers are called *asymptotic* if they are within a bounded distance one from the other. Equivalently, two Weyl chambers c_1, c_2 are asymptotic if and only if the corresponding geodesic rays $\xi(c_1)$ and

$\xi(c_2)$ are asymptotic. Thus, the set of *asymptotic classes of Weyl chambers* can be identified with any orbit ∂S_a, $a \in \mathfrak{a}_+^1$, i.e., with the Furstenberg boundary B.

A flat with a distinguished class of asymptotic Weyl chambers is called an *oriented flat*. Thus, any oriented flat $\overline{f}$ determines a point $\pi_+(\overline{f}) \in B$. We say that two orientations of the same flat $f \in F$ are *opposite* if the corresponding Weyl chambers in f based at the same point $x \in f$ are opposite (i.e., the intersection of their closures consists of the point x only). Denote by $-\overline{f}$ the flat opposite to $\overline{f}$, and let $\pi_-(\overline{f}) = \pi_+(-\overline{f})$. The points $\pi_-(\overline{f}), \pi_+(\overline{f}) \in B$ can be considered as "endpoints" of the oriented flat $\overline{f}$ (recall that in the rank one case flats are just geodesics).

The *Bruhat decomposition* of G into the union of double cosets of the group P parameterized by the *Weyl group* determines a stratification of the product $B \times B$ into G-orbits (again parameterized by the Weyl group), such that the only orbit Ω of maximal dimension consists exactly of those pairs of points (b_-, b_+) from B for which there exists a unique oriented flat $\overline{f}$ with $b_\pm = \pi_\pm(\overline{f})$. In particular, for any point $b \in B$ the set $B_b = \{b' : (b, b') \in \Omega\}$ is open and dense in B. In the rank one case B_b is just the complement of the point b in B.

Denote by m the unique K-invariant probability measure on B, then $g_1 m = g_2 m$ for $g_1, g_2 \in G$ iff $g_2 = g_1 k$ for a $k \in K$, so that the map $go \mapsto gm$ determines an embedding of the symmetric space S into the space of Borel probability measures on B. The closure of the set $\{gm\}, g \in G$ in the weak topology gives the *Furstenberg compactification* of the space S [Fu1]. The boundary of the Furstenberg compactification is *larger* than the Furstenberg boundary B (except for the rank one case), as the limit measures on B are not necessarily δ-measures.

The Furstenberg compactification coincides with (the maximal) one of the compactifications introduced earlier by Satake [Sa], so that it is often called the *Furstenberg-Satake compactification* [Moo]. Namely, the Furstenberg compactification can be obtained by taking a realization of S by Hermitian positively defined matrices determined by an irreducible faithful representation of G with dominant weight in the interior of a Weyl chamber (in the case $G = \mathrm{SL}(n, \mathbb{R})$ such a realization is given just by the formula $go \mapsto (gg^t)^{1/2}$), and taking then the closure in the projective space of matrices. Roughly speaking, the difference between the visibility and the Furstenberg compactifications is in using projectivization before or after applying the exponential map. The visibility and Furstenberg compactifications are incomparable (unless S has rank one, in which case they coincide). For example, for $G = \mathrm{SL}(3, \mathbb{R})$ the sequences of diagonal matrices $\exp(n + \sqrt{n}, n - \sqrt{n}, -2n)$ and $\exp(n, n, -2n)$ converge to the same point in the visibility compactification, but to different points in the Fursten-berg compactification; on the other hand, the sequences $\exp(n, 0, -n)$ and $\exp(2n, n, -3n)$ converge to the same point in the Furstenberg compactifi-cation, and to different points in the visibility compactification.

Yet another compactification of S was introduced by Karpelevich [Kar]. One can associate a symmetric space S_γ with rank strictly less than the rank of S and a projection $\pi_\gamma : S \to S_\gamma$ with any point γ from the visibility boundary ∂S (the space S_γ consists of a single point iff the vector $a = a(\gamma)$ is in the interior of the Weyl chamber). Then the Karpelevich compactifi-cation is defined inductively (we do not describe here its topology): in the rank 1 case it coincides with the visibility compactification, whereas in a higher rank case the closure of S is the disjoint union of S and Karpelevich compactifications of all symmetric spaces $S_\gamma, \gamma \in \partial S$. If the rank of S is greater than one, then the Karpelevich compactification is stronger than both the visibility and the Furstenberg compactifications.

See [Ta5] for a description of these compactifications restricted to a Cartan subgroup A (i.e., to a flat in S).

3.2. Convergence in symmetric spaces.

We begin with convergence in the Furstenberg compactification. Con-traction properties of the action of G on B allow one to use the idea of Furstenberg described in **2.7** and to prove the following result [GR1].

A measure λ on the Furstenberg boundary B is called *irreducible* if $\lambda(B \setminus B_b) = 0 \ \forall b \in B$. Let μ be a probability measure on G. If the closed semigroup $\mathrm{sgr}(\mu)$ in G generated by $\mathrm{supp}\,\mu$ satisfies certain natural irreducibility conditions (these conditions are satisfied if $\mathrm{sgr}(\mu)$ is Zariski dense in G [GM], although they are weaker than Zariski density [GR2]), then there exists a measurable map $\bar{x} \mapsto \pi(\bar{x})$ from the path space of the random walk (G, μ) to the Furstenberg boundary B such that for any ir-reducible probability measure λ on B and a.e. sample path $\bar{x} = (x_n)$ the sequence of measures $x_n \lambda$ converges weakly to the δ-measure at the point $\pi(\bar{x})$; the distribution of $\pi(\bar{x})$ is the unique μ-stationary measure on B, and it is irreducible. In particular, the sequence $x_n o$ converges to $\pi(\bar{x})$ in the Furstenberg compactification. Now, polar or Iwasawa decompositions give rise to natural transitive actions on B, which implies (see **2.8**) that the Poisson boundary can be identified with the Furstenberg boundary for spread out measures on G (if one considers arbitrary initial distributions on G rather than δ_e, then the Poisson boundary is one of finite covers of B described in [Fu1]).

Now let us look at the convergence in the visibility compactification.

For a point $x \in S$ denote by $a = a(x)$ the uniquely determined vector from $\bar{\mathfrak{a}}_+$ such that $x = k \exp(a)o$ for $k \in K$, then $d(o,x) = \|a\|$. We shall say that a sequence of points z_n in S is *regular* if there exists a geodesic ray ξ and a constant $l > 0$ such that $d(z_n, \xi(ln)) = o(n)$, i.e., if the sequence z_n asymptotically follows a geodesic. Clearly, any regular sequence z_n converges in the visibility compactification, and have the properties:

(i) $d(z_n, z_n + 1) = o(n)$;
(ii) $\exists\, a = \lim a(z_n)/n \in \bar{\mathfrak{a}}_+$.

It turns out that, conversely, any sequence in S satisfying the conditions (i) and (ii) is regular [Ka5].

Let μ be a probability measure with a finite first moment on G, i.e. $\int d(o, go)\, d\mu(g) < \infty$, and let $z_n = x_n o$, where x_n is a sample path of the random walk (G, μ), i.e., $x_n = h_1 \cdots h_n$ is the product of n independent μ-distributed increments h_i. Applying Kingman's multiplicative ergodic theorem to finitely dimensional irreducible representations of G one can prove existence of the limit (ii), which is called the *Lyapunov vector* of the measure μ. Since $d(z_n, z_{n+1}) = d(o, h_{n+1}o)$, condition (i) is also satisfied because the measure μ has finite first moment. Thus, for a.e. sample path (x_n) the sequence $x_n o$ is regular, and, in particular, converges in the visibility compactification (provided the Lyapunov vector is non-zero). Under the same irreducibility conditions as for convergence in the Furstenberg compactification the Lyapunov vector belongs to $\mathfrak{a}_+$ [GR2].

The notion of regular sequences in symmetric spaces is inspired by the classical notion of Lyapunov regularity of matrices. A sequence of matrices $A_n \in \mathrm{SL}(n, \mathbb{R})$ is *Lyapunov regular* if and only if the sequence $z_n = (A_n^t A_n)^{1/2}$ is regular in the symmetric space corresponding to the group $\mathrm{SL}(n, \mathbb{R})$ [Ka5]. The above criterion of regularity in symmetric spaces is in a sense similar to Raghunathan's lemma [Rag] (see also [Ru], [GM]) which gives an estimate for the rate of convergence of eigenspaces of positively definite matrices. As well as Raghunathan's lemma, this criterion can be used for proving the *Oseledec multiplicative ergodic theorem* [Os].

The difference between using convergence in the Furstenberg and visibility compactification for identifying the Poisson boundary is in a trade-off between the moment and irreducibility conditions. Depending on situation, one can use either of these compactifications for describing the Poisson boundary by applying the corresponding geometric criterion (**2.12**). If μ is a non-degenerate measure on a *Zariski dense* discrete subgroup of G with finite first logarithmic moment $\sum \log^+ d(o, go)\mu(g)$ and finite entropy, then irreducibility of the harmonic measures of μ and $\check{\mu}$ on the Furstenberg boundary B allows one to assign to a.e. (with respect to the product of these harmonic measures) pair of points in B a uniquely determined flat in S; since flats have polynomial growth, by using the strip criterion we obtain that B is the Poisson boundary of the measure μ [Ka15]. For an *arbitrary* discrete subgroup of G provided the measure μ has finite first moment $\sum d(o, go)\mu(g)$ one can use convergence in the visibility compactification and the ray criterion for identifying the Poisson boundary with an orbit ∂S_a (if the Lyapunov vector is zero, then the Poisson boundary is trivial) [Ka2] (see also [Le2]). For example, using an embedding of a *polycyclic group* into $\mathrm{SL}(n, \mathbb{Z})$, one can obtain another description of the Poisson boundary for measures with finite first moment on polycyclic groups as a set of flags (not necessarily full ones) in $\mathbb{R}^n$ (cf. **2.12**).

3.3. Brownian motion on symmetric spaces.

The Laplace–Beltrami operator Δ of the Riemannian metric on S determines a diffusion process (Brownian motion) on S. Since the group G acts on S by isometries, the transition probabilities of the Brownian motion are G-invariant. Hence, the one-dimensional distribution π_t at time t of the Brownian motion starting at the origin o is K-invariant, and there exists a uniquely determined convolution semigroup μ_t on G such that the measures μ_t are bi-invariant with respect to K and π_t is the image of μ_t under the map $g \mapsto go$. By K-invariance, the integral of the measures gm on the Furstenberg boundary B with respect to μ_t coincides with the (unique) K-invariant measure m. Thus, for any function $\widehat{f} \in C(B)$ the Poisson integral $\langle \widehat{f}, gm \rangle$ determines a bounded harmonic function on S, so that by the (continuous time) martingale convergence theorem a.e. sample path of the Brownian motion converges in the Furstenberg compactification (cf. **2.7**). Using the fact that the measures μ_t charge the whole group G, one can check that the limit points of the Brownian motion a.e. belong to the Furstenberg boundary B (i.e., the corresponding limit measures are δ-measures, cf. **3.2**); then by K-invariance the harmonic measure of a point $go \in S$ must be gm.

The fact that B is the whole Poisson boundary can be proved by considering the action of the groups K or N on the boundary B [Fu1] (see **2.8**). Note that another proof is provided by the entropy technique (cf. **2.12**, **2.13**).

Thus, the *Poisson formula for symmetric spaces* has the form $f(gK) = \langle \widehat{f}, gm \rangle$ with the *Poisson kernel* being given by the derivatives dgm/dm. For the hyperbolic plane which is the simplest symmetric space this is precisely the classical Poisson formula for bounded harmonic functions in the (Poincaré) disc.

A representation of all minimal harmonic functions on S corresponding to a real eigenvalue λ was obtained by Karpelevich [Kar] using the Martin approach and later by Guivarc'h [Gu4] (see also [Ta1]) by more geometrical methods. A survey of harmonic functions on symmetric spaces and their boundary behaviour can be found in [Ko1], [Ko2], [Ko3].

Recall that if the rank of S is greater than one, the boundary of the Furstenberg compactification is larger than the Furstenberg boundary, i.e., there exist sequences $g_n \in G$ such that the limit of the harmonic measures $g_n m$ is not a δ-measure. This fact leads to the following property of harmonic functions on higher rank symmetric spaces [Fu3]:

> (∗) There exists $\varepsilon > 0$ such that for any two harmonic functions f_1, f_2 on S with $0 \le f_i(x) \le 1$ and $f_i(o) \ge \frac{1}{2} - \varepsilon$ the minimum of f_1 and f_2 does *not* tend to zero at infinity.

Using a discretization of the Brownian motion one can show that for any lattice Γ in G there is a probability measure μ on Γ such that μ-harmonic functions on Γ have the property (∗) (this application was the reason why Furstenberg considered the discretization problem). On the other hand, discrete subgroups of rank one semi-simple Lie groups (or, more generally, discrete groups of isometries of Gromov hyperbolic spaces) can not have the property (∗) as it follows from the description of their μ-boundaries constructed in **2.7**. Thus, lattices in higher rank semi-simple Lie groups can not be realized as discrete subgroups of rank one semi-simple groups [Fu3], which was one of the first results of rigidity theory. Applying this idea to the *mapping class group* allows one to prove that neither the mapping class group itself nor its non-elementary subgroups can be lattices in higher rank semi-simple Lie groups [KM]. See [Gu5] for other applications of the boundary theory to algebraic properties of subgroups of semi-simple groups.

In the visibility compactification trajectories of the Brownian motion on S converge a.e. to the orbit of the vector $\rho/\|\rho\|$, where ρ is the half-sum of positive roots in $\mathfrak{a}$ taken with their multiplicities (the vector ρ belongs to the interior of the Weyl chamber so that this orbit is, of course, B). This fact was discovered and rediscovered several times in various contexts [Fu2], [Os], [Tu], [Vi],[Or], [MM], [NRW], [Ta4].

Apparently, the simplest way to prove convergence in the visibility compactification (i.e., in *polar coordinates*) is to identify the symmetric space S with the solvable group NA using the Iwasawa decomposition (i.e., to introduce *horospheric coordinates* on S), and to deduce then convergence in polar coordinates from convergence in horospheric coordinates. Let $\mathfrak{n}$ and $\mathfrak{a}$ be the Lie algebras of the groups N and A, respectively, so that $T_o S = \mathfrak{n} \oplus \mathfrak{a}$. Then the Laplacian Δ of the symmetric space splits as the sum $\Delta = \Delta_{\mathfrak{n}} \oplus \Delta_{\mathfrak{a}}$ of differential operators on N and on A, and $\Delta_{\mathfrak{a}} = \Delta_0 - 2\rho$, where Δ_0 is the Euclidean Laplacian on $\mathfrak{a} \approx \mathbb{R}^d$ (here d is the rank of S). Thus, the

Brownian motion on S in the horospheric NA coordinates is a diffusion process with "negative" drift -2ρ, so that it converges in horospheric (cf. with the group $\mathrm{Aff}(\mathbb{R})$ in **2.9**) and polar coordinates [Ta2], [Bab4]. Yet another proof can be obtained by using only the fact that for the Brownian paths $x_t = k_t \exp(a_t)o$ there exists the limit $\lim a_t/t = 2\rho$ and that a.e. $\sup\{d(x_t, x_{t+\tau}) : 0 \le \tau \le 1\} = o(t)$. These two properties alone guarantee that x_t converges both in polar and in horospheric coordinates and $\delta(x_t, k_\infty \exp(2\rho t)o) = o(t)$ for the limit geodesic $\xi(t) = k_\infty \exp(\rho/\|\rho\|t)o$ [Ka5].

A description of the Martin boundary was obtained by Dynkin [Dy1] for the group $\mathrm{SL}(n, \mathbb{C})$, by Nolde [Nol] for a complex semi-simple Lie group, and by Olshanetsky [Ol1], [Ol2] in the general case. Olshanetsky's approach uses harmonic analysis on G to obtain an asymptotic of the Green kernel (from an asymptotic for the zonal spherical functions) and gives also (the same) description of λ-Martin boundaries corresponding to eigenvalues $\lambda > \lambda_0$ strictly larger than the top of the spectrum $\lambda_0 = -\|\rho\|^2$ of the Laplacian on S. Note that a similar approach was used by Bougerol [Bo1], [Bo2] to obtain an asymptotic of the Green kernel in the interior of the Weyl chamber for a general random walk on G determined by a bi-invariant (with respect to K) measure μ.

As it is explained in [GJT], the description of Olshanetsky means that *the Martin compactification is the minimal compactification dominating both the visibility and the Furstenberg compactifications.* The Martin compactification is weaker than the Karpelevich compactification (there exists a continuous equivariant map from the latter to the former), and they coincide iff the rank of S is not greater than 2 (in the rank one case all compactifications coincide).

For $\lambda = \lambda_0$ asymptotics of Olshanetsky show that the Martin compactification coincides with the Furstenberg–Satake compactification. Another, more geometrical proof of the latter fact was obtained by Guivarc'h and Taylor by geometrical methods [GT1], [GT2].

In the particular case when S is a product of two (or several) hyperbolic spaces $\mathbb{H}_n$ (not necessarily of the same dimension n) an asymptotic of the Green kernel (and a description of the Martin boundary) was also obtained by Giulini and Woess [GW] by integrating the asymptotics of the heat kernel on hyperbolic spaces due to Davies and Mandouvalos [DMan]. In the discrete situation the Martin boundary for the simple random walk on a product of two homogeneous trees was described in [PW4] by using the asymptotics of Lalley [La1]. In the same way, by using the asymptotics from [La2], one can obtain a description of the Martin boundary, say, for a random walk determined by a finitely supported measure on a product of two free groups.

A recent paper by Guivarc'h [Gu7] contains a detailed exposition of the boundary theory for semi-simple Lie groups (both over the real and local fields). In particular, using estimates of the Green kernel due to Anker and Ji [AJ] and Harnack's inequality, Guivarc'h obtains a description of the Martin boundary of symmetric spaces by more geometrical methods than in [Ol2]. As it is pointed out in [Gu7], explicit formulas for the spherical functions on semi-simple groups over local fields given in [Mac] lead to a description of the Martin boundary in that case (analogous to the real situation).

Since the Brownian motion on S can be interpreted as an invariant diffusion process on the solvable group NA, the results of Olshanetsky give a description of the Martin boundary for this process (see [Bab3] for an instructing survey of this approach).

Note that so far the only way of identifying the Martin boundary for a *solvable Lie group* R is realizing it as a *homogeneous Cartan–Hadamard manifold* M and considering on R the invariant diffusion process corresponding to the Brownian motion on M (see [Bab2] for asymptotic properties of such processes). The only other example is a class of invariant diffusion processes on the affine group, for which the Martin boundary was described in [Mol2] by analytic means. Geometrically these processes correspond to the Brownian motion on the upper half-plane with a vertical drift.

If M is a rank one symmetric space, or if M is a non-symmetric harmonic space corresponding to an H-*type* group [DR], then the Martin compactification is homeomorphic to the sphere. Olshanetsky's results give an example of solvable groups with "complicated" Martin boundary. However, in contrast with the Poisson boundary, the situation with the Martin boundary (especially, with describing its topology rather than just identifying minimal harmonic functions) for all other random walks on solvable groups remains almost completely unclear (e.g., cf. conjectures from [LS]).

References

[AC]   R. Azencott, P. Cartier, *Martin boundaries of random walks on locally compact groups*, Proc. 6th Berkeley Symp. Math. Stat. & Prob., vol. 3, Univ. Calif. Press, Berkeley, 1972, pp. 87–129.

[AEG]   S. Adams, G. A. Elliott, T. Giordano, *Amenable actions of groups*, Trans. Amer. Math. Soc. **344** (1994), 803–822.

[AJ]   J.-P. Anker, L. Ji, *Heat kernel and Green function estimates for semi-simple non-compact Lie groups*, preprint (1995).

[Al]   G. Alexopoulos, *On the mean distance of random walks on groups*, Bull. Sci. Math. **111** (1987), 189–199.

[An1] A. Ancona, *Negatively curved manifolds, elliptic operators and the Martin boundary*, Ann. of Math. **125** (1987), 495–536.

[An2] ———, *Théorie du potentiel sur les graphes et les variétés*, Springer Lecture Notes in Math. **1427** (1990), 4–112.

[AS] M. T. Anderson, R. Schoen, *Positive harmonic functions on complete manifolds of negative curvature*, Ann. of Math. **121** (1985), 429–461.

[Av1] A. Avez, *Entropie des groupes de type fini*, C. R. Acad. Sci. Paris, Sér. A **275** (1972), 1363–1366.

[Av2] ———, *Théorème de Choquet–Deny pour les groupes à croissance non exponentielle*, C. R. Acad. Sci. Paris, Sér. A **279** (1974), 25–28.

[Av3] ———, *Harmonic functions on groups*, Differential Geometry and Relativity, Reidel, Dordrecht-Holland, 1976, pp. 27–32.

[Az] R. Azencott, *Espaces de Poisson des groupes localements compacts*, Springer Lecture Notes in Math., vol. 148, Springer, Berlin, 1970.

[Bab1] M. Babillot, *Théorie du renouvellement pour des chaînes semi-markoviennes transientes*, Ann. Inst. H. Poincaré Probab. Statist. **24** (1988), 507–569.

[Bab2] ———, *Comportement asymptotique de mouvement brownien sur une variété homogène à courbure négative ou nulle*, Ann. Inst. H. Poincaré Probab. Statist. **27** (1991), 61–90.

[Bab3] ———, *Potential at infinity on symmetric spaces and Martin boundary*, Proceedings of the Conference on Harmonic Analysis and Discrete Potential Theory (Frascati, 1991) (M. A. Picardello, ed.), Plenum, New York, 1992, pp. 23–46.

[Bab4] ———, *A probabilistic approach to heat difffusion on symmetric spaces*, J. Theoret. Probab. **7** (1994), 599–607.

[Bal] W. Ballmann, *On the Dirichlet problem at infinity for manifolds of nonpositive curvature*, Forum Math. **1** (1989), 201–213.

[BBE] M. Babillot, P. Bougerol, L. Elie, *The random difference equation $X_n = A_n X_{n-1} + B_n$ in the critical case*, preprint (1995).

[BE] P. Bougerol, L. Elie, *Existence of positive harmonic functions on groups and on covering manifolds*, Ann. Inst. H. Poincaré Probab. Statist. **31** (1995), 59–80.

[Bi] C. J. Bishop, *A characterization of Poissonian domains*, Ark. Mat. **29** (1991), 1–24.

[BK] M. Brin, Yu. Kifer, *Brownian motion and harmonic functions on polygonal complexes*, Proc. Symp. Pure Math. **57** (1995), 227–237.

[Bl] D. Blackwell, *On transient Markov processes with a countable number of states and stationary transition probabilities*, Ann. Math. Stat. **26** (1955), 654–658.

[BL1] W. Ballmann, F. Ledrappier, *The Poisson boundary for rank one manifolds and their cocompact lattices*, Forum Math. **6** (1994), 301–313.

[BL2] ———, *Discretization of positive harmonic functions on Riemannian manifolds and Martin boundary*, preprint (1992).

[Bo1] P. Bougerol, *Comportement asymptotique des puissances de convolution sur un espace symétrique*, Astérisque **74** (1980), 29–45.

[Bo2] ———, *Comportement à l'infini du noyau potentiel sur un espace symétrique*, Springer Lecture Notes in Math. **1096** (1984), 90–115.

[Br] M. Brelot, *On topologies and boundaries in potential theory*, Springer Lecture Notes in Math., vol. 175, Springer, Berlin, 1971.

[BR] L. Birge, A. Raugi, *Fonctions harmoniques sur les groupes moyennables*, C. R. Ac. Sci. Paris, Sér. A **278** (1974), 1287–1289.

[Cao] J. Cao, *A new isoperimetric estimate and applications to the Martin boundary*, preprint (1993).

[Car]   P. Cartier, *Fonctions harmoniques sur un arbre*, Sympos. Math., vol. 9, Academic Press, New York, 1972, pp. 203–270.

[CD]   G. Choquet, J. Deny, *Sur l'equation de convolution $\mu = \mu * \sigma$*, C. R. Ac. Sci. Paris, Sér. A **250** (1960), 799–801.

[CFS]   I. P. Cornfeld, S. V. Fomin, Ya. G. Sinai, *Ergodic theory*, Springer, New York, 1982.

[CFW]   A. Connes, J. Feldman, B. Weiss, *An amenable equivalence relation is generated by a single transformation*, Ergod. Th. & Dynam. Sys. **1** (1981), 431–450.

[CKW]   D. I. Cartwright, V. A. Kaimanovich, W. Woess, *Random walks on the affine group of a homogeneous tree*, Ann. Inst. Fourier (Grenoble) **44** (1994), 1243–1288.

[CSa]   D. I. Cartwright, S. Sawyer, *The Martin boundary for general isotropic random walks in a tree*, J. Theor. Prob. **4** (1991), 111-136.

[CSo]   D. I. Cartwright, P. M. Soardi, *Convergence to ends for random walks on the automorphism group of a tree*, Proc. Amer. Math. Soc. **107** (1989), 817–823.

[CSW]   D. I. Cartwright, P. M. Soardi, W. Woess, *Martin and end compactifications for nonlocally finite graphs*, Trans. Amer. Math. Soc. **338** (1993), 679–693.

[CW]   A. Connes, E. J. Woods, *Hyperfinite von Neumann algebras and Poisson boundaries of time dependent random walks*, Pacific J. Math. **137** (1989), 225–243.

[De1]   Y. Derriennic, *Sur la frontière de Martin des processus de Markov à temps discret*, Ann. Inst. H. Poincaré Séct. B **9** (1973), 233–258.

[De2]   ———, *Marche aléatoire sur le groupe libre et frontière de Martin*, Z. Wahrscheinlichkeitsth. Verw. Geb. **32** (1975), 261–276.

[De3]   ———, *Lois "zéro ou deux" pour les processus de Markov, applications aux marches aléatoires*, Ann. Inst. H. Poincaré Séct. B **12** (1976), 111–129.

[De4]   ———, *Quelques applications du théorème ergodique sous-additif*, Astérisque **74** (1980), 183–201.

[De5]   ———, *Entropie, théorèmes limites et marches aléatoires*, Springer Lecture Notes in Math. **1210** (1986), 241–284.

[De6]   ———, *Entropy and boundary for random walks on locally compact groups – the example of the affine group*, Transactions of the 10th Prague Conference on Information Theory, Statistical Decision Functions, Random Processes (Prague, 1986), vol. A, Reidel, Dordrecht, 1988, pp. 269–275.

[DE]   D. P. Dokken, R. Ellis, *The Poisson flow associated with a measure*, Pacific J. Math **141** (1990), 79–103.

[DH]   E. Damek, A. Hulanicki, *Boundaries for left-invariant subelliptic operators on semi-direct products of nilpotent and abelian groups*, J. reine und ang. Math. **411** (1990), 1–38.

[DMal]   E. B. Dynkin, M. B. Malyutov, *Random walks on groups with a finite number of generators*, Soviet Math. Dokl. **2** (1961), 399–402.

[DMan]   E. B. Davies, N. Mandouvalos, *Heat kernel bounds on hyperbolic spaces and Kleinian groups*, Proc. London Math. Soc. **57** (1988), 182–208.

[Dok]   D. P. Dokken, *μ-Harmonic functions on locally compact groups*, J. d'Analyse Math. **52** (1989), 1–25.

[Doo]   J. Doob, *Discrete potential theory and boundaries*, J. Math. Mech. **8** (1959), 433–458.

[DR]   E. Damek, F. Ricci, *Harmonic analysis on solvable extensions of H-type groups*, J. Geom. Anal. **2** (1992), 213–248.

[Dy1]   E. B. Dynkin, *Non-negative eigenfunctions of the Laplace–Beltrami operator and Brownian motion on certain symmetric spaces*, Amer. Math. Soc. Translations

72 (1968), 203–228.

[Dy2] ———, *Markov Processes and Related Problems of Analysis*, London Math. Soc. Lecture Note Series, vol. 54, Cambridge Univ. Press, Cambridge, 1982.

[DYu] E. B. Dynkin, A. A. Yushkevich, *Markov processes: Theorems and problems*, Plenum, New York, 1969.

[EG] G. A. Elliott, T. Giordano, *Amanable actions of discrete groups*, Ergod. Th. & Dynam. Sys. **13** (1993), 289–318.

[El] L. Elie, *Noyaux potentiels associés aux marches aléatoires sur les espaces homogènes. Quelques exemples clefs dont le groupe affine*, Springer Lecture Notes in Math. **1096** (1984), 223–260.

[EO] P. Eberlein, B. O'Neill, *Visibility manifolds*, Pacific J. Math. **46** (1973), 45–109.

[Fe] W. Feller, *Boundaries induced by non-negative matrices*, Trans. Amer. Math. Soc. **83** (1956), 19–54.

[FM] J. Feldman, C. C. Moore, *Ergodic equivalence relations, cohomology, and von Neumann algebras. I*, Trans. Amer. Math. Soc. **234** (1977), 289–324.

[Fo1] S. R. Foguel, *Iterates of a convolution on a non-abelain group*, Ann. Inst. H. Poincaré Séct. B **11** (1975), 199–202.

[Fo2] S. R. Foguel, *Harris operators*, Israel J. Math. **33** (1979), 281–309.

[Fr] A. Freire, *On the Martin boundary of Riemannian products*, J. Diff. Geom. **33** (1991), 215–232.

[Fu1] H. Furstenberg, *A Poisson formula for semi-simple Lie groups*, Ann. of Math. **77** (1963), 335–386.

[Fu2] ———, *Non-commuting random products*, Trans. Amer. Math. Soc. **108** (1963), 377-428.

[Fu3] ———, *Poisson boundaries and envelopes of discrete groups*, Bull. Amer. Math. Soc. **73** (1967), 350–356.

[Fu4] ———, *Random walks and discrete subgroups of Lie groups*, Advances in Probability and Related Topics, vol. 1, Dekker, New York, 1971, pp. 3–63.

[Fu5] ———, *Boundary theory and stochastic processes on homogeneous spaces*, Proc. Symp. Pure Math., vol. 26, AMS, Providence R. I., 1973, pp. 193–229.

[Ga] L. Garnett, *Foliations, the ergodic theorem and Brownian motion*, J. Funct. Anal. **51** (1983), 285-311.

[Ge1] P. Gérardin, *On harmonic functions on symmetric spaces and buildings*, Canadian Math. Soc. Conference Proceedings **1** (1981), 79–92.

[Ge2] ———, *Harmonic functions on buildings of reductive split groups*, Operator Algebras and Group Representations, vol. 1, Pitman, Boston, 1984, pp. 208–221.

[GJT] Y. Guivarc'h, L. Ji, J. Taylor, *Compactifications of symmetric spaces*, C. R. Ac. Sci. Paris, Sér. I **317** (1993), 1103–1108.

[GM] I. Ya. Goldsheid, G. A. Margulis, *Lyapunov exponents of a product of random matrices*, Russian Math. Surveys **44:5** (1989), 11–71.

[GR1] Y. Guivarc'h, A. Raugi, *Frontière de Furstenberg, propriétés de contraction et théorèmes de convergence*, Z. Wahrscheinlichkeitsth. Verw. Geb. **69** (1985), 187–242.

[GR2] ———, *Propriétés de contraction d'un semi-groupe de matrices inversibles. Coefficients de Liapunoff d'un produit de matrices aléatoires indépendantes*, Israel J. Math. **65** (1989), 165–196.

[Gri] R. I. Grigorchuk, *The growth degrees of finitely generated groups and the theory of invariant means*, Math. USSR Izv. **48** (1985), 259–300.

[Gro1] M. Gromov, *Groups of polynomial growth and expanding maps*, Publ. Math.

IHES **53** (1981), 53–73.

[Gro2] _____, *Hyperbolic groups*, Essays in Group theory (S. M. Gersten, ed.), MSRI Publ., vol. 8, Springer, New York, 1987, pp. 75–263.

[GT1] Y. Guivarc'h, J. C. Taylor, *The Martin compactification of the polydisc at the bottom of the positive spectrum*, Colloq. Math. **60/61** (1990), 537–547.

[GT2] _____, *The Martin compactification of a symmetric space at the bottom of the positive spectrum*, preprint (1993).

[Gu1] Y. Guivarc'h, *Croissance polynomiale et périodes des fonctions harmoniques*, Bull. Soc. Math. France **101** (1973), 333–379.

[Gu2] _____, *Quelques propriétés asymptotiques des produits de matrices aléatoires*, Springer Lecture Notes in Math. **774** (1980), 177–250.

[Gu3] _____, *Sur la loi des grands nombres et le rayon spectral d'une marche alèatoire*, Astérisque **74** (1980), 47–98.

[Gu4] _____, *Sur la représentation intégrale des fonctions harmoniques et des fonctions propres positives dans un espace Riemannien symétrique*, Bull. Sci. Math. **108** (1984), 373–392.

[Gu5] _____, *Produits de matrices aléatoires et applications aux propriétés géometriques des sous-groupes de groupe linéaire*, Ergod. Th. & Dynam. Sys. **10** (1990), 483–512.

[Gu6] _____, *Marches aléatoires sur les groupes et problèmes connexes*, preprint (1992).

[Gu7] _____, *Compactifications of symmetric spaces and positive eigenfunctions of the Laplacian*, preprint (1995).

[GW] S. Giulini, W. Woess, *The Martin compactification of the Cartesian product of two hyperbolic spaces*, J. reine angew. Math. **444** (1993), 17–28.

[HK] S. Hurder, A. Katok, *Ergodic theory and Weil measures for foliations*, Ann. of Math. **126** (1987), 221–275.

[Hu] G. A. Hunt, *Markoff chains and Martin boundaries*, Illinois J. Math. **4** (1960), 313–340.

[Im] H.-C. Im Hof, *An Anosov action on the bunddle of Weyl chambers*, Ergod. Th. & Dynam. Sys. **5** (1985), 587–593.

[J] W. Jaworski, *On the asymptotic and invariant σ-algebras of random walks on locally compact groups*, Probab. Theory Relat. Fields **101** (1995), 147–171.

[JP] N. James, Y. Peres, *Cutpoints and exchangeable events for random walks*, preprint (1995).

[Ka1] V. A. Kaimanovich, *Examples of non-commutative groups with non-trivial exit boundary*, J. Soviet Math. **28** (1985), 579–591.

[Ka2] _____, *An entropy criterion for maximality of the boundary of random walks on discrete groups*, Soviet Math. Dokl. **31** (1985), 193–197.

[Ka3] _____, *Brownian motion and harmonic functions on covering manifolds. An entropy approach*, Soviet Math. Dokl. **33** (1986), 812–816.

[Ka4] _____, *Brownian motion on foliations: entropy, invariant measures, mixing*, Funct. Anal. Appl. **22** (1988), 326–328.

[Ka5] _____, *Lyapunov exponents, symmetric spaces and multiplicative ergodic theorem for semi-simple Lie groups*, J. Soviet Math. **47** (1989), 2387–2398.

[Ka6] _____, *Harmonic and holomorphic functions on covers of complex manifolds*, Mat. Zametki **46** (1989), 94–96.

[Ka7] _____, *Boundary and entropy of random walks in random environment*, Proceedings of the 5-th International Vilnius Conference on Probability and Mathematical Statistics (Vilnius, 1989), vol. 1, Mokslas, Vilnius, 1990, pp. 573–579.

[Ka8] _____, *Poisson boundaries of random walks on discrete solvable groups*, Proceedings of Conference on Probability Measures on Groups X (Oberwolfach, 1990) (H. Heyer, ed.), Plenum, New York, 1991, pp. 205–238.

[Ka9] _____, *Measure-theoretic boundaries of Markov chains, 0-2 laws and entropy*, Proceedings of the Conference on Harmonic Analysis and Discrete Potential Theory (Frascati, 1991) (M. A. Picardello, ed.), Plenum, New York, 1992, pp. 145–180.

[Ka10] _____, *Bi-harmonic functions on groups*, C.R. Ac. Sci. Paris **314** (1992), 259–264.

[Ka11] _____, *The Poisson boundary of hyperbolic groups*, C. R. Ac. Sci. Paris, Sér. I **318** (1994), 59–64.

[Ka12] _____, *Ergodicity of harmonic invariant measures for the geodesic flow on hyperbolic spaces*, J. reine angew. Math. **455** (1994), 57–103.

[Ka13] _____, *The Poisson boundary of covering Markov operators*, Israel J. Math **89** (1995), 77–134.

[Ka14] _____, *Boundary and entropy of locally compact groups revisited*, in preparation.

[Ka15] _____, *The Poisson formula for groups with hyperbolic properties*, in preparation.

[Ka16] _____, *Markov operators on equivalence relations*, in preparation.

[Ka17] _____, *The Hausdorff dimension of the harmonic measure on trees*, in preparation.

[Kar] F. I. Karpelevich, *The geometry of geodesics and the eigenfunctions of the Beltrami–Laplace operator on symmetric spaces*, Trans. Moscow Math. Soc. **14** (1965), 51–199.

[Ki1] Y. Kifer, *Brownian motion and harmonic functions on manifolds of negative curvature*, Theor. Prob. Appl. **21** (1976), 81–95.

[Ki2] _____, *Brownian motion and positive harmonic functions on complete manifolds of non-positive curvature*, Pitman Research Notes in Math., vol. 150, Longman, Harlow, 1986, pp. 187–232.

[Ki3] _____, *Spectrum, harmonic functions, and hyperbolic metric spaces*, Israel J. Math **89** (1995), 377–428.

[Ki4] _____, *Harmonic functions on Riemannian manifolds: a probabilistic approach*, Progr. Probab., vol. 34, Birkhäuser, Boston, 1994, pp. 199–207.

[KM] V. A. Kaimanovich, H. Masur, *The Poisson boundary of the mapping class group and of Teichmüller space*, preprint (1994).

[Ko1] A. Koranyi, *Harmonic functions on symmetric spaces*, Symmetric Spaces, Dekker, New York, 1972, pp. 379–412.

[Ko2] _____, *Compactifications of symmetric spaces and harmonic functions*, Springer Lecture Notes Math. **739** (1979), 341-366.

[Ko3] _____, *A survey of harmonic functions on symmetric spaces*, Proc. Symp. Pure Math., vol. 35, AMS, Providence R. I., 1979, pp. 323–344.

[Kr] U. Krengel, *Ergodic Theorems*, de Gruyter, Berlin, 1985.

[KSK] J. G. Kemeny, J. L. Snell, A. W. Knapp, *Denumerable Markov chains*, Springer, New York, 1976.

[KV] V. A. Kaimanovich, A. M. Vershik, *Random walks on discrete groups: boundary and entropy*, Ann. Prob. **11** (1983), 457–490.

[KW1] V. A. Kaimanovich, W. Woess, *The Dirichlet problem at infinity for random walks on graphs with a strong isoperimetric inequality*, Probab. Theory Relat. Fields **91** (1992), 445–466.

[KW2]   ———, *Construction of discrete, non-unimodular hypergroups*, preprint (1994).

[La1]   S. Lalley, *Saddle-point approximations and space-time Martin boundary for nearest-neighbour random walk on a homogeneous tree*, J. Theoret. Probab. **4** (1991), 701–723.

[La2]   ———, *Finite range random walk on free groups and homogeneous trees*, Ann. Probab. **21** (1993), 2087–2130.

[Le1]   F. Ledrappier, *Une relation entre entropie, dimension et exposant pour certaines marches aléatoires*, C. R. Acad. Sci. Paris, Sér. I **296** (1983), 369–372.

[Le2]   ———, *Poisson boundaries of discrete groups of matrices*, Israel J. Math. **50** (1985), 319–336.

[Li]   V. Ya. Lin, *Liouville coverings of complex spaces and amenable groups*, Math. USSR – Sbornik **132** (1987), 197–216.

[LM]   B. Ya. Levit, S. A. Molchanov, *Invariant Markov chains on a free group with a finite number of generators*, in Russian (translated in English in Moscow Univ. Math. Bull.), Vestnik Moscow Univ. **26**:4 (1971), 80–88.

[LP]   Lin V. Y., Pinchover Y., *Manifolds with group actions and elliptic operators*, Memoirs Amer. Math. Soc. **112**:540 (1994), 1–78.

[LPP]   R. Lyons, R. Pemantle, Y. Peres, *Ergodic theory on Galton–Watson trees: speed of random walk and dimension of harmonic measure*, Ergod. Th. & Dynam. Sys. **15** (1995), 593–619.

[LS]   T. Lyons, D. Sullivan, *Function theory, random paths and covering spaces*, J. Diff. Geom. **19** (1984), 299–323.

[Ly]   T. Lyons, *Random thoughts on reversible potential theory*, Summer School on Potential Theory (Joensuu 1990) (Ilpo Laine, ed.), Joensuu Publ. in Sciences, vol. 26, 1992, pp. 71–114.

[Mac]   I. G. Macdonald, *Spherical functions on a group of p-adic type*, Publ. Ramanujan Inst. **2** (1971).

[Mar1]   G. A. Margulis, *Positive harmonic functions on nilpotent groups*, Soviet Math. Dokl. **166** (1966), 241–244.

[Mar2]   ———, *Discrete Subgroups of Semisimple Lie Groups*, Springer-Verlag, Berlin, 1991.

[Me]   P.-A. Meyer, *Probabilités et Potentiel*, Hermann, Paris, 1966.

[MM]   M. P. Malliavin, P. Malliavin, *Factorisation et lois limites de la diffusion horizontale au dessus d'un espace riemannien symétrique*, Springer Lecture Notes Math. **404** (1974), 164–217.

[Mol1]   S. A. Molchanov, *On Martin boundaries for the direct product of Markov chains*, Theor. Prob. Appl. **12** (1967), 307–310.

[Mol2]   ———, *Martin boundaries for invariant Markov processes on a solvable group*, Theor. Prob. Appl. **12** (1967), 310–314.

[Moo]   C. C. Moore, *Compactifications of symmetric spaces. I, II*, Amer. J. Math. **86** (1964), 201–218, 358–378.

[Mos]   G. D. Mostow, *Strong Rigidity of Locally Symmetric Spaces*, Ann. of Math. Studies, vol. 78, Princeton Univ. Press, Princeton New Jersey, 1973.

[MP]   T. S. Mountford, S. C. Port, *Representations of bounded harmonic functions*, Ark. Mat. **29** (1991), 107–126.

[Nol]   E. L. Nolde, *Non-negative eigenfunctions of the Laplace–Beltrami operator on symmetric spaces with a complex group of motions*, in Russian, Uspekhi Mat. Nauk **21**:5 (1966), 260–261.

[Nor]   S. Northshield, *Amenability and superharmonic functions*, Proc. Amer. Math. Soc. **119** (1993), 561–566.

[NRW]   J. R. Norris, L. C. G. Rogers, D. Williams, *Brownian motion on ellipsoids*, Trans. Amer. Math. Soc. **294** (1986), 757–765.

[NS]   P. Ney, F. Spitzer, *The Martin boundary for random walk*, Trans. Amer. Math. Soc. **121** (1966), 116–132.

[Ol1]   M. A. Olshanetsky, *The Martin boundaries of symmetric spaces with nonpositive curvature*, in Russian, Uspekhi Mat. Nauk **24**:6 (1969), 189–190.

[Ol2]   ———, *Martin boundaries for real semi-simple Lie groups*, J. Funct. Anal. **126** (1994), 169–216.

[Or]   A. Orihara, *On random ellipsoids*, J. Fac. Sci. Univ. Tokyo Sect. 1A Math. **17** (1970), 73–85.

[Os]   V. I. Oseledec, *A multiplicative ergodic theorem. Lyapunov characteristic numbers for dynamical systems*, Trans. Moscow Math. Soc. **19** (1969), 197–231.

[Pat]   A. L. T. Paterson, *Amenability*, AMS, Providence, R. I., 1988.

[Pau]   F. Paulin, *Analyse harmonique des relations d'équivalences mesurées discrètes*, preprint (1994).

[Pr]   J.-J. Prat, *Étude de la frontière de Martin d'une variété à courbure négative. Existence de fonctions harmoniques*, C. R. Ac. Sci. Paris, Sér. A **284** (1977), 687–690.

[PW1]   M. A. Picardello, W. Woess, *Martin boundaries of random walks: ends of trees and groups*, Trans. Amer. Math. Soc. **302** (1987), 185–205.

[PW2]   ———, *Harmonic functions and ends of graphs*, Proc. Edinburgh Math. Soc. **31** (1988), 457–461.

[PW3]   ———, *Martin boundaries of Cartesian products of Markov chains*, Nagoya Math. J. **128** (1992), 153–169.

[PW4]   ———, *The full Martin boundary of the bi-tree*, preprint (1992).

[Rag]   M. S. Raghunathan, *A proof of Oseledec' multiplicative ergodic theorem*, Israel J. Math **32** (1979), 356–362.

[Rau]   A. Raugi, *Fonctions harmoniques sur les groupes localement compacts à base dénombrable*, Bull. Soc. Math. France. Mémoire **54** (1977), 5–118.

[Re]   D. Revuz, *Markov Chains*, 2nd revised ed., North-Holland, Amsterdam, 1984.

[Ro]   J. Rosenblatt, *Ergodic and mixing random walks on locally compact groups*, Math. Ann. **257** (1981), 31–42.

[Ru]   D. Ruelle, *Characteristic exponents and invariant manifolds in Hilbert space*, Ann. of Math. **115** (1982), 243–290.

[Sa]   I. Satake, *On representations and compactifications of symmetric Riemannian spaces*, Ann. of Math. **71** (1960), 77–110.

[Se]   C. Series, *Martin boundaries of random walks on Fuchsian groups*, Israel J. Math. **44** (1983), 221–242.

[Ta1]   J. C. Taylor, *Minimal functions, martingales, and Brownian motion on a noncompact symmetric space*, Proc. Amer. Math. Soc. **100** (1987), 725–730.

[Ta2]   ———, *Brownian motion in non-compact symmetric spaces, asymptotic behaviour in Iwasawa coordinates*, Contemp. Math. **73** (1988), 303–332.

[Ta3]   ———, *The product of minimal functions is minimal*, Bull. London Math. Soc **22** (1990), 499-504; *Erratum*, ibid **24** (1992), 379–380.

[Ta4]   ———, *Brownian motion in non-compact symmetric spaces, asymptotic behaviour in polar coordinates*, Canadian J. Math. **43** (1991), 1065–1085.

[Ta5]   ———, *Compactifications determined by a polyhedral cone decomposition of* $\mathbb{R}^n$, Proceedings of the Conference on Harmonic Analysis and Discrete Potential Theory (Frascati, 1991) (M. A. Picardello, ed.), Plenum, New York, 1992, pp. 1–14.

[Tu]   V. N. Tutubalin, *On limit theorems for the product of random matrices*, Theor. Prob. Appl. **10** (1965), 15–27.

[Va1]   N. Th. Varopoulos, *Long range estimates for Markov chains*, Bull. Sc. Math. **109** (1985), 225–252.

[Va2]   ———, *Information theory and harmonic functions*, Bull. Sci. Math. **110** (1986), 347–389.

[Vi]   A. D. Virtser, *Central limit theorem for semi-simple Lie groups*, Theor. Prob. Appl. **15** (1970), 667–687.

[Wi]   G. Willis, *Probability measures on groups and some related ideals in group algebras*, J. Funct. Anal. **92** (1990), 202–263.

[Wo1]   W. Woess, *Boundaries of random walks on graphs and groups with infinitely many ends*, Israel J. Math. **68** (1989), 271–301.

[Wo2]   W. Woess, *Fixed sets and free subgroups of groups acting on metric spaces*, Math. Z. **214** (1993), 425–440.

[Wo3]   ———, *Random walks on infinite graphs and groups – a survey on selected topics*, Bull. London Math. Soc. **26** (1994), 1–60.

[Wo4]   ———, *Dirichlet problem at infinity for harmonic functions on graphs*, preprint (1994).

[Zi]   R. J. Zimmer, *Amenable ergodic group actions and an application to Poisson boundaries of random walks*, J. Funct. Anal. **27** (1978), 350–372.

VADIM A. KAIMANOVICH, IRMAR, UNIVERSITÉ RENNES-1, CAMPUS BEAULIEU, RENNES 35042, FRANCE

E-mail address: kaimanov@univ-rennes1.fr

SQUARING AND CUBING THE CIRCLE -RUDOLPH'S THEOREM

William Parry

This is an exposition of Rudolph's theorem whose value, we hope, lies in its brevity. Most of the basic ideas can be found in [3], but here, for once, we avoid the machinery of symbolic dynamics as, it seems to us, a direct approach is more illuminating, although we concede Rudolph's theorem suggests a number of problems in the realm of $\mathbb{Z}^2$ lattice dynamics and of higher rank hyperbolic action. The latter topic is the subject of Katok and Spatzier's paper [2], where an extensive programme generalising Rudolph's theorem to Anosov actions of $\mathbb{R}^k$, $\mathbb{Z}^k$ and $\mathbb{Z}^k_+$ may be found together with substantial contributions to this programme. In the same work an outline of other work in this area is given.

In a rather different direction, one should also mention Host's paper [1] in which methods of a Fourier analytical character are utilised to extend Rudolph's theorem.

As the referee has pointed out, there seems to be a number of alternative accounts and generalisations in currency. The present work began as a rewriting of Rudolph's theorem for my own benefit. However, a number of colleagues expressed an interest so, perhaps, it may be of service to others.

The title, of course, does not indicate an attempt to achieve the impossible. We are concerned with the measure-theoretic problem of invariant probabilities for the maps of the circle given by squaring and cubing or, more generally, for the maps

$$Sx = px \bmod 1, \quad Tx = qx \bmod 1$$

where p, q are relatively prime.

A word of caution is in order. Equations are to be interpreted modulo sets of measure zero. This needs special emphasis as we shall be looking at fibred spaces with atomic distributions on these fibres.

§1. Generalities

If S is a countable to one measure-preserving transformation of the Lebesgue probability space $(X, \mathcal{B}, m)$ then there exists $Y \in \mathcal{B}$, $m(Y) = 1$, such that $SY = Y$ and S is positively measurable and positively non-singular on Y. (Walters [4].) This enables one to define unambiguously the Radon-Nikodym derivative $S'(x) = \dfrac{dm\ S}{dm}(x)$. If T is a second such transformation commuting with S then

$$
\begin{aligned}
(ST)'(x) &= S'(Tx)T'(x) \\
&= T'(Sx)S'(x).
\end{aligned}
$$

We write
$$
\begin{aligned}
f(x) &= \log S'(x) \geq 0 \\
g(x) &= \log T'(x) \geq 0
\end{aligned}
$$

and we have

$$
f(Tx) - f(x) = g(Sx) - g(x). \tag{1.1}
$$

Alternatively, the information functions $I(\mathcal{B} \mid S^{-1}\mathcal{B})$ are well defined and in fact

$$
f(x) = I(\mathcal{B} \mid S^{-1}\mathcal{B}), \quad g(x) = I(\mathcal{B} \mid T^{-1}\mathcal{B}).
$$

Moreover we can define Perron-Frobenius operators on appropriate spaces by

$$
(L_{-f}w)(x) = \sum_{Sy\ =\ x} e^{-f(y)}w(y), \quad (L_{-g}w)(x) = \sum_{Ty\ =\ x} e^{-g(y)}w(y)
$$

and in view of (1.1) we have

$$
L_{-f}L_{-g} = L_{-g}L_{-f}.
$$

These are Markov operators in that they are positive and $L_{-f}1 = 1$, $L_{-g}1 = 1$.

Moreover, on $L^1(X)$, we have (writing $(Sw)(x) = w(Sx)$, $(Tw)(x) = w(Tx)$)

$$SL_{-f} = E(\,|\,S^{-1}\mathcal{B}), \quad TL_{-g} = E(\,|\,T^{-1}\mathcal{B})$$

and $L_{-f}S = \text{id}$, $L_{-g}T = \text{id}$.

Finally, on $L^2(X)$, it is easy to check that L_{-f}, L_{-g} are the duals of S, T respectively.

Lemma 1.   Suppose that for almost all x we have $T : S^{-1}x \to S^{-1}Tx$ bijectively then $fT = f$ and $gS = g$.

Proof.   In fact $L_{-f}(e^g) = e^g$ since

$$\sum_{Sy = x} e^{-f(y)+g(y)} = \sum_{Sy = x} e^{-f(Ty)+g(Sy)}$$

$$= e^{g(x)} \sum_{Sy = x} e^{-f(Ty)} = e^{g(x)}.$$

Hence $E(e^g\,|\,S^{-1}\mathcal{B}) = e^{g(S)}$, from which one concludes that $e^{g(S)} = e^g$ and therefore $g \circ S = g$, $f \circ T = f$.

§2. Endomorphisms of the circle

We apply the forgoing to the endomorphisms of the circle

$$Sx = px \bmod 1, \quad Tx = qx \bmod 1$$

and suppose that S, T preserve a Borel probability m on the circle with respect to which the action of the semi-group $\{S^mT^n : m, n \in \mathbb{Z}^+\}$ is ergodic. Furthermore we suppose $(p, q) = 1$. We emphasise that it is *not* assumed that either of S, T is ergodic.

Since $(p, q) = 1$ it is clear that the hypothesis of lemma 1 is fulfilled and therefore $f \circ T = f$, $g \circ S = g$.

Associated with each $x \in [0, 1)$ there is a sequence of distributions $d(x, n)$ on the points $S^{-n}S^n x$ which we can ascribe to $\{\,j/p^n : j = 0, 1, ..., p^n-1\}$ since $S^{-n}S^n x =$

$\{x + j/p^n : j = 0, 1, ..., p^n-1\}$. The 'weights' of this distribution are $e^{-f^n(y)}$ for each y such that $S^n y = S^n x$, where $f^n(y) = f(y) + f(Sy) + ... f(S^{n-1}y)$. More exactly, we assign

$e^{-f^n(x + j/p^n)}$ to j/p^n.

Since

$$e^{-f^{n-1}(y)} = \sum_{Sz = y} e^{-f^n(z)}$$

it is clear that $d(x, n-1)$ can be expressed in terms of certain $d(, n)$. In particular if $\mathcal{H}_n$ is the smallest σ-algebra with respect to which all of the distributional functions $d(x, 1)$, $d(x, 2), ... d(x, n)$ are measurable then

$$\mathcal{H}_1 \subset \mathcal{H}_2 \subset ... \mathcal{H}_n.$$

Furthermore $f + f^{n-1} \circ S = f^n$ so that $S^{-1}\mathcal{H}_{n-1} \subset \mathcal{H}_n$. Finally we note that $f^n \circ T = f^n$ so that

$$d(n, x)\left(\frac{j}{p^n}\right) = d(n \; Tx)\left(\frac{qj}{p^n}\right) \quad \text{and}$$

$T^{-1}\mathcal{H}_n = \mathcal{H}_n$. Moreover $q^{\varphi(p^n)} \equiv 1 \mod p^n$ (φ, Euler's function) since $(p^n, q) = 1$ so that $d(n, x)(j/p^n) = d(n, T^{\varphi(p^n)}x)(j/p^n)$, and $T^{-\varphi(p^n)}B = B$ for all $B \in \mathcal{H}_n$.

§3. Rudolph's comparative entropy lemma

Let γ be the partition of $[0, 1)$ into contiguous intervals of length $1/pq$ and choose two sequences of positive integers $\{r_n\}, \{s_n\}$ so that, for a given $\delta > 0$,

$$\left| s_n - r_n \frac{\log p}{\log q} \right| < \frac{\delta}{\log q}$$

i.e. $\dfrac{1}{2} < \dfrac{p^{r_n}}{q^{s_n}} < 2$ with $\delta = \log 2$.

Then each element of $\displaystyle\bigvee_{i=0}^{r_n} S^{-i}\gamma$ is contained in at most 3 elements of $\displaystyle\bigvee_{i=0}^{s_n} T^{-i}\gamma$ and vice-versa – a simple exercise.

This guarantees that for any sub-σ-algebra $\mathcal{A}$

$$\left| H\left(\bigvee_{i=0}^{r_n} S^{-i}\gamma \,\middle|\, \bigvee_{i=0}^{s_n} T^{-i}\gamma \vee \mathcal{A} \right) - H\left(\bigvee_{i=0}^{s_n} T^{-i}\gamma \,\middle|\, \bigvee_{i=0}^{r_n} S^{-i}\gamma \vee \mathcal{A} \right) \right|$$

is bounded, and therefore

$$\lim_{n \to \infty} \left(\frac{1}{s_n} H\left(\bigvee_{i=0}^{s_n} T^{-i}\gamma \,\middle|\, \mathcal{A} \right) - \left(\frac{\log p}{\log q} \right) \frac{1}{r_n} H\left(\bigvee_{i=0}^{r_n} S^{-i}\gamma \,\middle|\, \mathcal{A} \right) \right) = 0.$$

When $T^{-1}\mathcal{A} = \mathcal{A}$ the individual limits exist, which we denote by $h(T \mid \mathcal{A})$ and $h(S \mid \mathcal{A})$, respectively, and we have

Lemma 2.

$$h(T \mid \mathcal{A}) = \frac{\log q}{\log p}\, h(S \mid \mathcal{A}).$$

When $\mathcal{A} = \mathcal{H} = \displaystyle\bigvee_{n=1}^{\infty} \mathcal{H}_n$ we have $T^{-1}\mathcal{H} = \mathcal{H}$ and

$$h(T \mid \mathcal{H}) = H\left(\gamma \,\middle|\, \bigvee_{i=1}^{\infty} T^{-i}\gamma \vee \mathcal{H} \right)$$

$$= H(\mathcal{B} \mid T^{-1}\mathcal{B} \vee \mathcal{H})$$

$$= H(\mathcal{B} \mid T^{-1}\mathcal{B}) = h(T).$$

So by the lemma we have

$$h(S) = h(S \mid \mathcal{H}).$$

Hence $H(\mathcal{H} \mid S^{-1}\mathcal{B}) = H(\mathcal{B} \vee \mathcal{H} \mid S^{-1}\mathcal{B}) - H(\mathcal{B} \mid \mathcal{H} \vee S^{-1}\mathcal{B})$

$$\leq h(S) - h(S \mid \mathcal{H}) = 0,$$

the inequality arising from

$$h(S \mid \mathcal{H}) = \lim_{n \to \infty} \frac{1}{n} H (\gamma \vee \ldots \vee S^{-n}\gamma \mid \mathcal{H})$$

$$= \lim_{n \to \infty} \frac{1}{n} [H(\gamma \mid \overset{n}{\underset{1}{\vee}} S^{-i}\gamma \vee \mathcal{H}) + H(\overset{n}{\underset{1}{\vee}} S^{-i}\gamma \mid \mathcal{H})]$$

$$\leq \lim_{n \to \infty} \frac{1}{n} [H(\gamma \mid \overset{n}{\underset{1}{\vee}} S^{-i}\gamma \vee \mathcal{H}) + H(\overset{n}{\underset{1}{\vee}} S^{-i}\gamma \mid \mathcal{H})]$$

$$\text{(using } S^{-1}\mathcal{H} \subset \mathcal{H})$$

$$\leq \lim_{n \to \infty} \frac{1}{n} (H(\gamma \mid \overset{n}{\underset{1}{\vee}} S^{-i}\gamma \vee \mathcal{H}) + \ldots + H(\gamma \mid \mathcal{H}))$$

$$= H(\gamma \mid \overset{\infty}{\underset{1}{\vee}} S^{-i}\gamma \vee \mathcal{H})$$

$$= H(\mathcal{B} \mid S^{-1}\mathcal{B} \vee \mathcal{H}).$$

In short, we have

Lemma 3.

$$H(\mathcal{H} \mid S^{-1}\mathcal{B}) = 0 \text{ i.e. } \mathcal{H} \subset S^{-1}\mathcal{B}.$$

In fact a similar argument shows $\mathcal{H} \subset \overset{\infty}{\underset{n=0}{\cap}} S^{-n}\mathcal{B}$.

Thus for each pair n, N = 1,2,... $d(x, n)$ is $S^{-N}\mathcal{B}$ measurable. This ensures that $d(x, n) = d(y, n)$ if $S^n x = S^n y$. (Here one has to argue in terms of points of the 'essential' part of the space.) Consequently $d(x, n)$ is, almost surely, symmetric with respect to a

subgroup G_x^n of $\{j/p^n : j = 0, 1, \ldots, p^n-1\}$.

These subgroups are almost surely trivial if and only if $h(S) = 0$ (in which case, by

Lemma 2, $h(T) = 0$) and otherwise there is a non-trivial subgroup G of

$\{j/p : j = 0,1,...,...p-1\}$ such that $E = \{x : G_x^1 = G\}$ has positive measure. Assuming the

latter case we see that E is T invariant and thus for almost all x, $T^n x \in E$ infinitely often

i.e. $f^n(x) \to \infty$, since the joint action of S, T is ergodic. However $1/\text{card } G_x^n = e^{-f^n}(x)$

so that card $G_x^n \to \infty$ almost surely.

For each $k = \pm 1, \pm 2, ...$ we have

$$E(e_k \mid S^{-n}\mathcal{B})(x) = \frac{\displaystyle\sum_{g \in G_x^n} e^{2\pi i k(x + g)}}{\text{Card } G_x^n} = 0$$

when $f^n(x)$ is large enough, where $e_k(x) = e^{2\pi i k x}$, from which we conclude that $\int e_k \, dm = 0$. Hence, when $h(S) > 0$, m is Lebesgue measure. We have therefore proved:

Rudolph's Theorem. If $Sx = px \mod 1$ and $Tx = qx \mod 1$ (where $(p, q) = 1$) preserve the Borel probability m with respect to which $\{S^i T^j : i, j = 0, 1, 2, ...\}$ is ergodic then either $h(S) = h(T) = 0$ or m is Lebesgue measure.

References

[1] B. Host, Nombres normaux, entropie, translations. (Preprint).

[2] A. Katok and R.J. Spatzier, Invariant measures for higher rank hyperbolic abelian actions. (Preprint).

[3] D.J. Rudolph, X2 and X3 invariant measures and entropy. Ergod. Th. & Dynam. Syst. (1990), 10, 395-406.

[4] P. Walters, Roots of n : 1 measure-preserving transformations. J. Lond. Math. Soc. 44 (1969), 7-14.

A Survey of Recent K-Theoretic Invariants for Dynamical Systems

Ian F. Putnam[1]

Department of Mathematics and Statistics,
University of Victoria,
Victoria, B.C.,
Canada.

[1]Supported in part by an NSERC Canada Operating Grant.

185

Section 1:   Introduction

This paper is an attempt to survey some recent results in the theory of topological dynamics which have been obtained using C*-algebras, K-theory and tools which are native to those branches of mathematics. For certain topological dynamical systems, one may construct a C*-algebra and consider its K-theory. This K-theory is then interpreted in purely dynamical terms. The results we have in mind, show how this K-theoretic invariant may be used to classify the system up to orbit equivalence, or other natural equivalences in dynamics.

In addition to surveying recent results, we will also present an introduction to the construction of the C*-algebras involved. It is aimed at readers with a background in dynamics rather than C*-algebras. The pace is quite pedestrian and the intent is to convey the basic ideas. It is not meant to be thorough; we quickly evade technical issues which would challenge these aims. The hope is to give readers in dynamics the notion of these C*algebras as geometric/topological/dynamic objects.

The idea of associating a C*-algebra to a topological dynamical system of some sort goes back to Murray and von Neumann. The actual construction is due to many people (depending on the sort of system involved): Zeller-Mayer, Effros and Hahn, Krieger, ... The most general construction is due to Renault [Ren]. We will consider two sorts of dynamical systems: an equivalence relation satisfying some conditions on a locally compact Hausdorff space X and a countable group Γ acting as homeomorphisms of such a space X. These have substantial overlap - when the action of Γ is free then the equivalence relation whose equivalence classes are the Γ-orbits satisfies the necessary conditions.

A major influence in C*-algebra algebra theory in the late 1970's and throughout the 1980's was the introduction and development of K-theory. To any C*-algebra A, one may associate a pair of abelian groups $K_0(A)$ and $K_1(A)$. The former carries a natural pre-order structure as well. The combination of these two constructions gives a way of associating to a topological dynamical system (as above), a pair of abelian groups, one of which is pre-ordered. One can view this as a "dynamical (co)homology theory". Some obvious questions arise. First, can one give a purely dynamic description of these groups? The answer in the most general sense is probably "no", but in many specific situations this is possible. Secondly, what sort of information is contained in these invariants? what are they good for? This paper is an attempt to survey some of the answers to these questions. Let us begin with some historical perspective.

One area where these ideas had great success was the work of Cuntz and Krieger on subshifts of finite type [Kr2, CuKr]. In addition to constructing an important class of C*-algebras, they also obtained, via K-theory, Williams' invariant for shift equivalence and some of the Bowen-Franks invariants. Unfortunately, the C*-algebra constructions in this case are slightly more complex than we want to consider here, so we will have nothing more to say about this work.

K-theory for C*-algebras has had many successes; two of them concern the irrational rotation C*-algebras and AF-algebras. There has been a great deal of interest in the C*-algebras one constructs from the action of the group of integers on the circle generated by an irrational rotation. The work of Powers, Rieffel [Ri] and Pimsner and Voiculescu [PV] showed that the K-theoretic invariants contained the rotation numbers of the homeomorphism. This was extremely important work from a C*-algebra point of view, but of course the rotation number was a known invariant in dynamics and had been defined without using C*-algebras or K-theory.

Another success of K-theory was Elliott's classification theorem for AF (or approximately

finite-dimensional) C*-algebras [Ell]. Basically, it states that the K_0-group, with its pre-order,

is a complete isomorphism invariant for this class of C*-algebras. These C*-algebras were shown

to be constructed from a certain class of equivalence relations, which we call AF equivalence

relations and describe in section 2. (See Stratila-Voiculescu [Str-V], Krieger [Kr1], Renault

[Ren].) Here was a case where a class of systems could be classified by their K-theoretic

invariants (see section 9). We will describe this result completely in section 10. Although tail

equivalence for a one-sided shift of finite type is an example of such a relation, these systems

in their full generality were not usually considered in dynamics.

The case of a minimal action of the group of integers on a Cantor set was considered in

[Pu2]. Now in this case, one obtains via K-theory invariants which are really new. Actually

this is not strictly accurate - one obtains a group which is not new, but the pre-order structure

is. Not only that, but the results of [Pu2] show that there is a close connection between the

dynamics of a minimal homeomorphism of a Cantor set and those of AF-equivalence relations.

This idea began before [Pu2]; it is certainly present in the work of Pimsner and Voiculescu and

Pimsner [Pi]. Even more so, it is suggested by the work of Veršik [V1, V2] which was a major

influence in later developments. Thus, it is important for us to expand our class of dynamical

systems beyond group actions and consider equivalence relations as well. This is not really

surprising in view of the situation in ergodic theory. We are seeing a program which runs

parallel to that initiated by Dye [D1].

The paper is organized as follows. Sections 2 and 3 describe the equivalence relations

which we consider. Section 4 deals with the construction of the C*-algebra. All of this material

is essentially taken from Renault [Ren]. Here, we have tried to present this in a manner accessible to readers in dynamics, helped by our restriction to considering the case of equivalence relations. We also briefly describe crossed product C*-algebras for group actions on spaces. We also refer the reader to [Sch].

Section 5 is a very brief discussion of K-theory for C*-algebras and section 6 deals with invariant measures for the dynamics and their interpretations in a K-theory context.

In section 7, we describe the calculation of the K-theory invariants for AF-equivalence relations. Section 8 deals with the case of an action of the group of integers on a space where the principal tool is the Pimsner-Voiculescu exact sequence. We also examine the special cases of irrational rotations of the circle and of homeomorphisms of compact totally disconnected spaces.

Section 9 deals with the results on orbit equivalence and in section 10 we discuss "factor equivalences" and "sub-equivalences" as well as the notion of "weak orbit equivalence".

Throughout the paper, we do not make an effort to describe the proofs of the results, except to the extent the main results depend on each other. This then tends to obscure the issue to what extent the dynamical results actually rely on C*-algebra theory for their proofs. Obviously one would like to give dynamical proofs of dynamical facts in the hope that they would be more enlightening. It is hoped that our presentation will help convince the reader that these C*-algebras are indeed geometric/topological/dynamic entities.

I am indebted to many people (over a span of many years) for enlightening conversations on various issues touched on in this paper. It is a pleasure to thank Jack Spielberg, Terry Loring, Marc Rieffel, Anatole Veršik, Sergei Kerov, Rob Bures, David Handelman, Mike Boyle, Eli Glasner, Benjamin Weiss and especially Thierry Giordano, Richard Herman and Christian Skau. I would also like to thank some of the above for sharing early versions of their results with me. Finally I would like to thank Klaus Schmidt, Bill Parry and Peter Walters and the staff at the University of Warwick for their hospitality during my visit in the summer of 1994.

Section 2: Topological Equivalence Relations

The following definitions are due to Renault [Ren] although the ideas of many others are involved. In the language of [Ren], we study principal, r-discrete groupoids with counting measure as a Haar system. This terminology is part of a much more general framework. We discuss the ideas using more conventional terms.

Let X denote a locally compact Hausdorff space. For convenience, let us assume that the one-point compactification of X is metrizable. Let $G \subseteq X \times X$ denote an equivalence relation on X. For x in X, $[x]_G$ (or simply $[x]$) denotes the G-equivalence class of x. We may regard G as an algebraic object (namely a groupoid) by considering the partially defined product:

$$(x, y)(y', z) = (x, z), \quad \text{if } y = y'$$

and inverse

$$(x, y)^{-1} = (y, x),$$

for (x, y), (y', z) in G. We use r and s (for range and source) denote the two canonical projections from G to X; i.e. $r(x, y) = x$, $s(x, y) = y$.

The key hypothesis on G is that it must be equipped with its own topology satisfying some conditions. First G must be σ-compact in this topology. Secondly, $\Delta = \{(x,x) \in G \mid x \in X\}$ should be an open subset of G. Finally, we require that, for any (x,y) in G, there is a neighbourhood U of (x,y) in G such that $r|U$ and $s|U$ are homeomorphisms from U to their respective images in X. This, of course, links the topology

of G to that of X. In particular, since Δ is open, the natural identification of Δ with X

is a homeomorphism. Also, G is locally compact and Hausdorff.

Let us mention some other consequences of our hypotheses, all of which can be proved

quite easily. If $K \subseteq G$ is compact and x is in X, then $r^{-1}\{x\} \cap K$ and $s^{-1}\{x\} \cap K$

are both finite. As G is σ-compact, $r^{-1}\{x\}$ and $s^{-1}\{x\}$ and hence $[x]$ are countable,

for any x in X. If x and y are in X with (x, y) in G, then there are

neighbourhoods V_x and V_y of x and y in X, respectively, and a homeomorphism

$\gamma: V_x \to V_y$ such that $\gamma(x) = y$ and $\{((z, \gamma(z)) \mid z \in V_x\} \subseteq G$ and is pre-compact.

Moreover, the γ above is unique. (In fact, $\gamma = s \circ r^{-1}$.) This last item is important: it

means that G is made up of the graphs of "local homeomorphisms" of X. In fact, this can

even be made precise at a topological level.

Some words of warning are in order. First, the topology of G is rarely the relative

topology of $X \times X$ in interesting examples (see example 3(e)). Secondly, certain equivalence

relations may possess no such topology (see example 3(d)) and for relations which do admit one,

it may not be unique. The constructions we undertake in the next sections depend on this

topology. This is a sublety not present in the measurable or Borel situations. The Borel

structure on G is the same as that from $G \subseteq X \times X$.

Section 3: Examples

(a) Let X be countable (and discrete) and $G = X \times X$, the "trivial" equivalence relation with the discrete topology.

(b) Let X be arbitrary and $G = \Delta$, the "co-trivial" equivalence relation.

(c) Let G and H be equivalence relations on X and Y, respectively. Then $G \times H$ determines an equivalence relation on $X \times Y$. Also, $G \cup H$ is an equivalence relation on $X \cup Y$. (We use $\cup$ to denote disjoint union.)

(d) Let $X = [0, 1] \cup [2, 3] \subseteq \mathbf{R}$ and $G = \Delta \cup \{(x, x+2), (x+2, x) \mid 0 \leq x < 1\}$ with the relative topology from $X \times X$. The point here is that G satisfies the conditions of section 2, while $G' = \Delta \cup \{(1, 3), (3, 1)\}$ does not.

(And now the more serious examples.)

(e) Let Γ denote a countable discrete group acting freely on X (as homeomorphisms). Let

$$G = \{(x, \gamma(x)) \mid x \in X, \gamma \in \Gamma\}.$$

i.e. the G-equivalence classes are simply the Γ-orbits. To topologize G, notice that the map from $X \times \Gamma$ to G sending (x, γ) to $(x, \gamma(x))$ is a bijection since Γ acts freely. Use the product topology on $X \times \Gamma$ shifted over by this map. That is, the sequence $\{(x_n, \gamma_n(x_n))\}$ converges to $(x, \gamma(x))$ in G if and only if $\{x_n\}$ converges to

x in X and $\gamma_n = \gamma$ for all but finitely many n. Here, the local homeomorphisms of

section 2 are just the elements of Γ.

In case $\Gamma = \mathbf{Z}$, the integers, we denote by φ the homeomorphism 1 in Γ. In this

case, we denote G by G_φ and we say G is "singly generated".

(f) AF-equivalence relations: see [Br], [Str-V], [Kr1], [Ren]. Begin with a Bratteli diagram.

This is an infinite graph as shown.

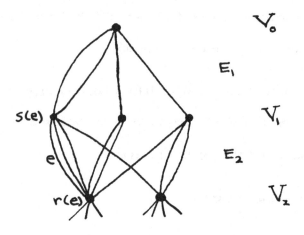

The vertex set is partitioned into levels $V_0, V_1, V_2, \cdots$, as is the edge set $E_1, E_2, E_3, \cdots$.

Each V_n and E_n are finite and an edge e in E_n goes from a vertex in V_{n-1} to one

in V_n, which we denote by $s(e)$ and $r(e)$, respectively. (It is convenient to think of edges as directed.) We insist that there are no sinks. For convenience, we will also assume here that $V_0 = \{v_0\}$ and that there are no sources other than v_0. (For the general situation, see [Pu1].) We let X denote the space of infinite paths in the diagram - beginning at v_0 and directed downward. Then X may be viewed as a closed subset of the infinite product space $\prod_{n \geq 1} E_n$ - a path is given as an edge list. With the relative

topology, X is compact, metrizable and totally disconnected. The equivalence relation G is cofinal or tail equivalence: two paths are equivalent if they agree from some level on. We let G_n denote all pairs of paths which agree after V_n, for $n = 1, 2, 3, \cdots$. That is,

$$G_n = \left\{ \left((e_k)_1^\infty, (f_k) \right)_1^\infty \mid e_k = f_k, \text{ for all } k > n \right\}. \text{ Endow } G_n \text{ with the relative topology of}$$

$X \times X$. Then each G_n is a (compact) open subset of G_{n+1}; $G = \bigcup_n G_n$ is given the

inductive limit topology. That is, a sequence $\{(x_n, y_n)\}$ in G converges to (x, y) in G if and only if $\{x_n\}$ converges to x, $\{y_n\}$ converges to y (in X) and, for some N, (x_n, y_n) is in G_N, for all but finitely many n. The local homeomorphisms γ may be described as follows. Fix two paths $(e_1, e_2, \cdots)$, $(f_1, f_2, \cdots)$ in X (i.e. $e_n, f_n \in E_n$). Suppose, for some fixed N, $e_n = f_n$ for all $n \geq N$. Let

$$V = \left\{ (e'_1, e'_2, e'_3, \cdots) \in X \mid e'_n = e_n, n \leq N \right\}$$
$$W = \left\{ (e'_1, e'_2, e'_3, \cdots) \in X \mid e'_n = f_n, n \leq N \right\}$$

which are both clopen sets in X. Define

$$\gamma(e_1, e_2, \cdots, e_N, e'_{N+1}, e'_{N+2} \cdots) = (f_1, f_2, \cdots, f_N, e'_{N+1}, e'_{N+2}, \cdots).$$

Section 4: C*-algebras

Let us begin with a topological equivalence relation G as in section 2. We want to construct a C*-algebra [KR, P] from G. Begin with the complex linear space of compactly supported, continuous, complex valued functions on G, which we denote $C_c(G)$. Define a product and involution by

$$(fg)(x,y) = \sum_{z \in [x]} f(x,z)g(z,y)$$

$$f^*(x,y) = \overline{f(y,x)},$$

for f, g in $C_c(G)$ and (x,y) in G. As we remarked earlier, for any x in X and compact set $K \subseteq G$, $r^{-1}\{x\} \cap K$ is finite and so the sum used in the product is actually finite. One must also verify that fg and f^* are again in $C_c(G)$. This product and involution should remind one of matrix multiplication and the adjoint of a complex matrix. Indeed, if X is finite and $G = X \times X$ then, as a *-algebra, $C_c(G) \approx M_n$, the $n \times n$ complex matrices, where $n = |X|$.

Let us look at a general property of $C_c(G)$. The space, $C_0(X)$, of continuous complex functions vanishing at infinity on X is a commutative C*-algebra: the product is taken pointwise and the norm is the sup-norm. There is a map $\alpha: C_c(X) \rightarrow C_c(G)$ defined by

$$\alpha(f)(x,y) = \begin{cases} f(x) & \text{if } x = y \\ 0 & \text{otherwise} \end{cases}$$

and α is a *-homomorphism. In the case X is compact, $C(X)$ is a subalgebra of $C_c(G)$.

One views $C(X)$ in $C_c(G)$ as an analogue of the diagonal matrices in M_n. The image

under α of the constant function 1, is a unit for the algebra $C_c(G)$.

To obtain a C*-algebra, we require a norm on $C_c(G)$. We discuss this briefly. The

first option is to consider all *-homomorphisms $\pi : C_c(G) \to B(H)$, where H is a Hilbert

space and $B(H)$ denotes the C*-algebra of all bounded linear operators on H. One defines

a norm on $C_c(G)$ be setting $|f|$ equal to the supremum, taken over all such π, H, of

$|\pi(f)|$ (the operator norm). Of course, one must see this supremum is finite [Ren]. Now,

$C_c(G)$ is not usually complete in this norm; its completion, denoted $C^*(G)$, is a C*-algebra.

A second option is to restrict attention to a smaller class of representations. Fix a point

x in X. Consider $H = \ell^2([x])$ and define $\lambda_x : C_c(G) \to B(H)$ by

$$(\lambda_x(f)\xi)(y) = \sum_{z \in [x]} f(y,z)\,\xi(z)$$

for f in $C_c(G)$, ξ in $\ell^2[x]$, y in $[x]$. Checking that $\lambda_x(f)$ is a well-defined bounded

linear operator and that λ_x is a *-homomorphism are fairly easy. We may then define a

(different) norm on $C_c(G)$ by $|f| = \sup_{x \in X} |\lambda_x(f)\|$. Again, the completion of $C_c(G)$ in this

norm, denoted $C_r^*(G)$, is a C*-algebra called the <u>reduced</u> C*-algebra of G. In general, this

is a quotient of $C^*(G)$. However, in many instances they agree. For example, if Γ in example (e) is amenable, then $C_r^*(G) \approx C^*(G)$. The same conclusion is true for all AF-

equivalence relations. Generally, $C_r^*(G)$ is probably more manageable and more interesting.

For example, if G is minimal in the sense that every equivalence class is dense in X, then

$C_r^*(G)$ is simple (no non-trivial closed two-sided ideals).

Let us examine the C*-algebras of our examples of section 3.

(a) As already noted,

$$C^*(G) \approx C_r^*(G) \approx M_n,$$

the $n \times n$ complex matrices if $|X| = n$ is finite. If $|X|$ is infinite then

$$C^*(G) \approx C_r^*(G) \approx K(l^2(X)),$$

where K denote the C*-algebra of compact operators.

(b) In this case $G = \Delta$ which may be identified with X in the obvious way. Thus

$C_c(G) \approx C_c(X)$ and the product is simply pointwise multiplication of functions. Thus,

$$C^*(G) \approx C_r^*(G) \approx C_0(X),$$

the continuous $\mathbb{C}$-functions vanishing at infinity.

(c) In this situation, it can be shown that $C^*(G \times H) \approx C^*(G) \otimes C^*(H)$, the tensor product

[KR] and $C^*(G \cup H) \approx C^*(G) \oplus C^*(H)$, the direct sum [KR].

(d) For f in $C_c(G)$, define

$\alpha(f)$: $[0, 1] \to M_2$ by

$$\alpha(f)(t) = \begin{cases} \begin{bmatrix} f(t,t) & f(t,t+2) \\ f(t+2,t) & f(t+2,t+2) \end{bmatrix} & 0 \le t < 1 \\[12pt] \begin{bmatrix} f(1,1) & 0 \\ 0 & f(3,3) \end{bmatrix} & t = 1. \end{cases}$$

Then α defines a *-isomorphism from $C^*(G)$ to

$$\{f: [0, 1] \to M_2 \mid f \text{ continuous, } f(1) \text{ diagonal}\}$$

(e) $C^*(G) \approx C_0(X) \times \Gamma$, the crossed-product C*-algebra [Ped, Tom]. We will elaborate a

little, assuming for simplicity that X is compact. For each γ in Γ, let

$$E_\gamma = \{((x, \gamma(x)) \mid x \in X\},$$

which is compact and open in G. Let u_γ denote the characteristic function of E_γ which

is in $C_c(G)$. It is easy to verify that

Additionally, if we let α: $C(X) \to C^*(G)$ be as described earlier, then for any f in

$$u_{\gamma_1} u_{\gamma_2} = u_{\gamma_1 \gamma_2}, \quad \gamma_1, \gamma_2 \in \Gamma$$

$$u_\gamma u_\gamma^* = 1 = u_\gamma^* u_\gamma, \quad \gamma \in \Gamma.$$

$C(X)$ and γ in Γ we have

$$u_\gamma \, \alpha(f) u_\gamma^{-1} = \alpha(f^\gamma),$$

where $f^\gamma(x) = f(\gamma^{-1}(x))$, for all x in X. It can also easily be seen that $\alpha(C(X))$ and

the u_γ's generate $C^*(G)$ as a C*-algebra. That is, (α, u) forms a covariant

representation of (X, Γ) as in [Tom]. The idea is that we have a C*-algebra, $C(X)$ and

a group of automorphisms, $f \to f^\gamma$. We pass to a larger C*-algebra $C(X) \times \Gamma$, containing

$C(X)$ (via α) where the given automorphisms become inner.

Let us also mention that there is a generalization of the construction used here to the case

when Γ does not act freely. The resulting C*-algebra, which we denote by $C(X) \times \Gamma$,

can be described as the universal C*-algebra containing $C(X)$ as a subalgebra and a set of

unitaries $\{u_\gamma \mid \gamma \in \Gamma\}$ satisfying conditions as above. (This can be further generalized to

the case of non-compact X, but we will not discuss this here.) Such a C*-algebra is called

a crossed-product or transformation group C*-algebra. (See [Ped] or [Tom].)

(f) In this case, $C^*(G)$ is called an AF-algebra, for "approximately finite dimensional". We

will see why shortly. Recall first that $G = \bigcup_n G_n$. As each G_n is compact and open in

G, we have inclusions of algebras $C_c(G_0) \subseteq C_c(G_1) \subseteq \cdots \subseteq C_c(G)$ and, moreover,

$\underset{n}{\cup} \, C_c(G_n) = C_c(G)$ which is dense in $C^*(G)$. Let us fix an n and describe $C^*(G_n)$.

Let v be in V_n and let X_v denote the set of all paths in X which pass through v.
We may write

$$X = \underset{v \in V_n}{\cup} \, X_v.$$

Note that for x in X_v, y in X_w with v, w in V_n and $v \neq w$, we have

$(x, y) \notin G_n$. Let X_v^{init} and X_v^{final} denote the set of paths from v_0 to v and from v

onwards, respectively. We note that

$$X_v \approx X_v^{\text{init}} \times X_v^{\text{final}}$$

in an obvious way and the restriction of G_n to X_v is the trivial relation on X_v^{init}, product

with the co-trivial relation on X_v^{final}. Now, using examples (a), (b) and (c), we have

$$C^*(G_n) \approx \bigoplus_{v \in V_n} \left(M_{k(v)} \otimes C\left(X_v^{\text{final}}\right) \right),$$

where $k(v) = \left| X_v^{\text{init}} \right|$. If we let A_n denote the subalgebra of $C^*(G_n)$ which is identified with

$$\bigoplus_v \left(M_{k(v)} \otimes 1 \right)$$

by the isomorphism above, then A_n is finite dimensional. It can also be shown that $\bigcup_n A_n$ is dense in $C^*(G)$. Hence, $C^*(G)$ is called approximately finite dimensional.

The construction here can also be carried out for equivalence relations with uncountable equivalence classes. Again, there are topological considerations. The main new ingredient is a Haar system: a collection of measures, one on each equivalence class. This allows one to define a convolution product on $C_c(G)$. The details are in [Ren].

Section 5: K-Theory

An important tool in the study of C*-algebras is K-theory. We will give only very brief descriptions of the definitions and a few important examples. More thorough treatments may be found in [Bla] and [W-O].

To any C*-algebra A, one associates a pair of abelian groups $K_0(A)$ and $K_1(A)$. Let us assume for simplicity that A has a unit. In the case $A = C^*(G)$, we write $K^i(G)$ rather than $K_i(C^*(G))$. For $K_0(A)$, one begins by considering all self-adjoint idempotents p in all matrix algebras over A. We let $M_n(A)$ denote the n by n matrices over A, which is again a C*-algebra. There is an obvious way of taking direct sums; for p in $M_n(A)$, q in $M_m(A)$, $p \oplus q$ is in $M_{m+n}(A)$. We first identify p and $p \oplus 0_m$ (for all m) and then define an equivalence relation: $p \sim q$ if there is p_0 such that $(p \oplus p_0) = v(q \oplus p_0)v^{-1}$, for some invertible v in a suitable matrix algebra. The set of $\sim$ equivalence classes of self-adjoint idempotents forms a semi-group ($\oplus$ is addition) with unit (0_m) and the cancellation property (built into the definition of equivalence). The Grothendieck group of this semi-group, which consists of formal differences $[p] - [q]$, is $K_0(A)$. This group also contains a natural pre-order; the positive cone $K_0(A)^+$ is the set $\{[p] - [0]\}$. There is also a distinguished element of $K_0(A)^+$, namely $[1]$.

Let us consider a simple example; $A = M_n$. Here, we have $K_0(A) = \mathbf{Z}$. Under this isomorphism $K_0(A)^+$ is identified with $\{0, 1, 2, \cdots\}$ and $[1]$ is identified with n. The

isomorphism takes an idempotent p in A to its rank or trace. Integers greater than n are

in the image of the map by considering idempotents in matrices over A. Negative integers are

the image of formal differences $[0] - [p]$. In a similar way, $K_0(K) = Z$ (although K is

not unital).

Let us mention one more example. Let X be a totally disconnected compact metrizable

space. If p is an idempotent in $M_n(C(X))$, then applying the trace pointwise yields a

continuous integer valued function on X. This map induces a well-defined isomorphism from

$K_0(C(X))$ to $C(X, Z)$, the abelian group of continuous integer valued functions on X. The

positive cone in $K_0(C(X))$ is identified with the non-negative valued functions. The injectivity

of the map above relies on X being totally disconnected.

The group $K_1(A)$ is obtained by considering the group of invertible elements in

$M_n(A)$, for each n. First one identifies an invertible a in $M_n(A)$ with $a \oplus 1_k$, for any

$k \geq 1$. Finally, one declares two invertiles to be equivalent if they are connected by a path of

invertibles in some $M_n(A)$. The result is $K_1(A)$. As examples: $K_1(M_n) = 0$, which reflects

the fact that every invertible $n \times n$ matrix is homotopic to the identity. The same kind of

argument can be used to show $K_1(C(X)) = 0$ if X is totally disconnected. For a more

interesting example, $K_1(C(S^1)) = Z$, where S^1 denotes the unit circle. The map is given as

follows: if a is an invertible in $M_n(C(S^1))$, then its determinant, taken pointwise, is a continuous function from S^1 to $\mathbb{C}$ - $\{0\}$. The isomorphism above assigns to the equivalence

class of a, the winding number of this function.

There is of course a K-theory for topological spaces, $K^*(X)$. Swan's Theorem [Bla]

asserts that there is a natural isomorphism between $K^i(X)$ and $K_i(C(X))$, for $i = 0, 1$. In

what follows, readers who are not familiar with topological K-theory should feel free to take the

conclusion of Swan's Thm, $K^i(X) \approx K_i(C(X))$, as the definition of $K^i(X)$.

Section 6: Invariant Measures

Let X be a locally compact space with an equivalence relation G. As in [Ren], one can define G-invariant measures on X. A regular Borel measure μ on X is G-invariant if $\mu(E) = \mu(\gamma(E))$, for each local homeomorphism γ such that $\text{graph}(\gamma)$ is a precompact, open set in G and E is any Borel set contained in the domain of γ. We will only consider finite measures for the present.

For such a measure μ, one can associate a linear functional τ_μ on $C^*(G)$ (and $C_r^*(G)$). For f in $C_c(G)$, we have

$$\tau_\mu(f) = \int_{x \in X} f(x, x) \, d\mu(x).$$

In example (a) $(G = X \times X)$ when X is finite, we have seen that $C^*(G) \approx M_n$. Letting μ be counting measure on X (normalized so $\mu(X) = 1$), then τ_μ is just the usual trace on M_n (normalized so $\tau_\mu(1) = 1$).

If the measure μ is positive, then τ_μ is a positive functional in the sense that $\tau_\mu(a^*a) \geq 0$ for all a in $C^*(G)$. Also, τ_μ satisfies the trace property:

$$\tau_\mu(ab) = \tau_\mu(ba), \text{ for } a, b \text{ in } C^*(G).$$

This requires the invariance of μ. As an illustrative example, if a is in $C_c(G)$ and its support is contained in the graph of a local homeomorphism γ and $b = a^*$, then the formula

above follows from the invariance of μ under γ.

Suppose τ is a trace on a unital C^*-algebra A; normalized so $\tau(1) = 1$. Extend τ

to $M_n(A)$ by using the usual trace on M_n: $\tau((a_{ij})) = \sum_i \tau(a_{ii})$. This extended map also

has the trace property and is positive. The trace property ensures that equivalent idempotents

have the same trace. The positivity of τ implies that for a self-adjoint idempotent p,

$\tau(p) = \tau(pp) = \tau(p^*p) \geq 0$. We conclude that τ induces a well-defined positive group

homomorphism $\hat{\tau}$ from $K_0(A)$ to $\mathbf{R}$. For any ordered group (G, G^+), a positive group

homomorphism ω from G to $\mathbf{R}$ (positive means $\omega(G^+) \subset [0, \infty)$) is called a state.

Usually, it is also required that ω is normalized to map the order unit of G to one.

Section 7: K-Theory of AF-Equivalence Relations

We will give a brief description of the K-theory of AF-algebras. Let G be an AF-equivalence relation based on a fixed Bratteli diagram (V, E). As described in section 4, we have a sequence of C^*-subalgebras

$$A_0 \subseteq A_1 \subseteq A_2 \subseteq \cdots$$

contained in $C^*(G)$, whose union is dense in $C^*(G)$. Each A_n is finite dimensional:

$$A_n \approx \bigoplus_{v \in V_n} M_{k(v)}$$

where $k(v)$ is the number of paths in the diagram from v_0 to v. In particular,

$$A_0 \approx \mathbb{C}.$$

At the level of K-theory, we have

$$K_0(A_n) \approx K_0\left(\bigoplus_v M_{k(v)}\right)$$

$$\approx \bigoplus_v K_0(M_{k(v)})$$

$$\approx \bigoplus_v \mathbb{Z}$$

$$\approx F(V_n),$$

the free abelian group on V_n. Under this isomorphism, $K_0(A_n)^+$ is identified with the sub semi-group generated by V_n. This is usually referred to as the standard or simplicial order on

$F(V_n)$. The inclusion of A_n in A_{n+1} induces a map from $K_0(A_n)$ to $K_0(A_{n+1})$. Under the

identifications $K_0(A_n) \approx F(V_n)$, $K_0(A_{n+1}) \approx F(V_{n+1})$, this sends vertex v in $V_n \subseteq F(V_n)$ to

$$\sum_{s(e)=v} r(e) \in F(V_{n+1})$$

Another way of saying this, $K_0(A_n) \approx \mathbf{Z}^{i_n}$, where i_n is the cardinality of V_n, and the map

from $\mathbf{Z}^{i_n}$ to $\mathbf{Z}^{i_{n+1}}$ is given by multiplication by the incidence matrix of the edge set E_{n+1}.

(The i, j entry of the matrix is the number of edges from the jth vertex in V_n to the ith

vertex in V_{n+1}.) Thus, we obtain an inductive system of ordered abelian groups

$$F(V_0) = \mathbf{Z} \to F(V_1) \to F(V_2) \to \cdots$$

It follows from the basic properties of K-theory that $K_0(C^*(G))$ is isomorphic to the direct

limit of this system. The limit is taken in the category of ordered abelian groups and this is an

order isomorphism. The distinguish positive element is identified with the image of v_0 in the

direct limit.

Let us give a couple of explicit examples (drawing Bratteli diagrams horizontally).

<u>ex 1.</u>

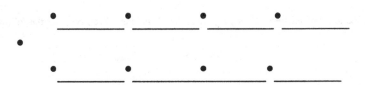

Note that the associated equivalence relation is as follows. The space X is homeomorphic to

$\{1, 2, 3, ..., \infty\}$ and there are two equivalence classes, $\{1, 2, 3, ...\}$ and $\{\infty\}$. The inductive

system of groups is

$$Z \underline{\hspace{2cm}} Z^2 \underline{\hspace{2cm}} Z^2 \underline{\hspace{2cm}} Z^2 \underline{\hspace{2cm}}$$

$$\begin{bmatrix} 1 \\ 1 \end{bmatrix} \qquad \begin{bmatrix} 1 & 1 \\ 0 & 1 \end{bmatrix} \qquad \begin{bmatrix} 1 & 1 \\ 0 & 1 \end{bmatrix}$$

As $\det \begin{bmatrix} 1 & 1 \\ 0 & 1 \end{bmatrix} = 1$, each map (except the first) is a group isomorphism. Hence the limit group

G is isomorphic with the first Z^2 in the sequence. The question is then, what is G^+?

Identifying G with the first Z^2, G^+ consists the elements of G, whose images in the

sequence above eventually lie in $(Z^2)^+$ (simplicial order); i.e.

$$G \approx Z^2$$

$$G^+ \approx \left\{ (a, b) \in Z^2 \mid \begin{bmatrix} 1 & 1 \\ 0 & 1 \end{bmatrix}^k \begin{bmatrix} a \\ b \end{bmatrix} \geq 0 \text{ for some } k \right\}$$

It is easy to check that this is lexicographic order on Z^2.

ex 2. Fix a sequence of integers $a_1,\ a_2,\ a_3,\ \cdots$, each at least one. Then consider the following

Bratteli diagram

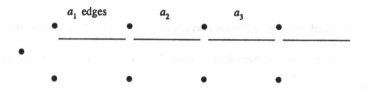

As in the last example, each incidence matrix has determinant one so $G = Z^2$ as an abelian

group. Arguing as in example 1, the order turns out to be

$$G^+ = \{(m,n) \in Z^2 \mid m + \theta n \geq 0\}$$

where θ is the irrational number with continued fraction expansion $[a_1,\ a_2,\ a_3,\ \cdots,]$ i.e.

$G = Z + Z\theta \subseteq \mathbf{R}$, with the relative order from $\mathbf{R}$. In this example, the corresponding AF-

equivalence relation, G, has a unique invariant measure μ. Moreover,

$$\hat{\tau}_\mu \colon\ K^0(G) \to \mathbf{R}$$

induces an order isomorphism from $K^0(G)$ to $Z + Z\theta$.

Ordered abelian groups arising in such a way are called dimension groups. The

fundamental result in the subject is the theorem of Effros, Handelman and Shen [EHS, G] which

gives an abstract characterization of such groups. Specifically, a countable ordered group

(G, G^+) is a dimension group if and only if it is unperforated (if $g \in G$ and $ng \in G^+$ for some $n \geq 1$, then $g \in G^+$) and satisfies the Riesz interpolation property (if $g_1, g_2 \leq g_3, g_4$, then there is g in G such that $g_1, g_2 \leq g \leq g_3, g_4$).

As for K_1, we have $K_1(A_n) = 0$, since matrix algebras have trivial K_1. As is the case for K_0, $K_1(C^*(G))$ is the inductive limit of the system

$$K_1(A_0) \to K_1(A_1) \to K_1(A_2) \to \cdots$$

and hence is the zero group.

<u>Section 8</u>:   <u>Singly Generated Equivalence Relations and the Pimsner-Voiculescu Sequence</u>

In this section, we let X be a compact space and φ be a homeomorphism of X. We want to discuss $K^1(G_\varphi) \approx K_1(C^*(G_\varphi))$, in the case φ has no periodic points. However, the results apply equally well to the crossed product C*-algebra $C(X) \times_\varphi \mathbf{Z}$, whether φ has periodic orbits or not. For convenience, we denote $K_j(C(X) \times_\varphi \mathbf{Z})$ by $K^j(X, \varphi)$.

As described in section 4, there is an inclusion $\alpha: C(X) \to C(X) \times_\varphi \mathbf{Z}$ which induces a map on associated K-groups; i.e.

$$\alpha_*: K^j(X) \to K^j(X, \varphi), \quad j = 0, 1.$$

Similarly, the homeomorphism $\varphi: X \to X$ induces automorphisms $\varphi_*: K^j(X) \to K^j(X)$, $j = 0, 1$.

<u>Theorem 8.1</u> (Pimsner-Voiculescu [PV, Bla])

There is a six-term exact sequence of K-groups

$$
\begin{array}{ccccc}
& id - \varphi_* & & \alpha_* & \\
K^0(X) & \longrightarrow & K^0(X) & \longrightarrow & K^0(X, \varphi) \\
\big\uparrow & & & & \big\downarrow \\
K^1(X, \varphi) & \longrightarrow & K^1(X) & \longrightarrow & K^1(X) \\
& \alpha_* & & id - \varphi_* &
\end{array}
$$

<u>Case</u>:   <u>Irrational rotations of the circle</u>

Consider the case $X = S^1$ and φ is rotation through angle $2\pi\theta$, with θ irrational. In this case, the crossed product C*-algebra is often denoted A_θ and called the irrational rotation C*-algebra. We have

$$K^0(S^1) \approx Z$$
$$K^1(S^1) \approx Z \quad \text{(see section 5)}.$$

The map φ acts trivially on both K^0 and K^1; i.e. $\varphi_* = id$. Then the Pimsner-Voiculescu sequence easily gives

$$K^0(S^1, \varphi) \approx Z^2$$
$$K^1(S^1, \varphi) \approx Z^2.$$

This does not reveal anything about the order on K^0. However, Lebesgue measure on S^1 is φ-invariant, we obtain a state $\hat{\tau}_\lambda \colon K^0(S^1, \varphi) \to \mathbf{R}$. Predating the Pimsner-Voiculescu result, Rieffel [Ri] had explicitly constructed idempotents in A_θ and computed their traces. Using the Pimsner-Voiculescu sequence and Rieffel's construction and computation, one can show that

$$\hat{\tau}_\lambda \colon K^0(S^1, \varphi) \to \{m + n\theta \mid m, n \in Z\} \subseteq \mathbf{R}$$

is an order isomorphism. The distinguished element in $K^0(S^1, \varphi)$ is mapped to 1. (The image $Z + \theta Z$ has the relative order from $\mathbf{R}$.) Notice that it is the <u>ordered</u> group which detects the relevant dynamical information. In particular, one concludes that A_θ and $A_{\theta'}$ are *-isomorphic if and only if $\theta = \theta'$ or $1 - \theta'$ $(0 < \theta, \theta' < 1)$.

Let us mention that the case of rational rotations is considered in [HKS]; the Denjoy examples are done in [PSS]. Finally, we remark that Exel [Ex1] used these ideas to define a notion of rotation number for homeomorphisms of more general spaces.

Case: Zero dimensional systems

Here we let X be a compact totally disconnected metrizable space and φ be a homeomorphism of X. As noted in section 5, we have

$$K^0(X) \approx C(X, \mathbf{Z}), \quad K^1(X) = 0.$$

Under the first isomorphism the action of φ on $C(X, \mathbf{Z})$ becomes the obvious one: $\varphi_*(f) = f \circ \varphi^{-1}$. The Pimsner-Voiculescu sequence yields

$$K^1(X, \varphi) \approx \{f \in C(X, \mathbf{Z}) \mid f = f \circ \varphi^{-1}\}$$
$$K^0(X, \varphi) \approx C(X, \mathbf{Z})/B(X, \varphi),$$

where $B(X, \varphi) = \{f - f \circ \varphi^{-1} \mid f \in C(X, \mathbf{Z})\}$, the so-called coboundaries. Note, in particular that if X contains no non-trivial clopen φ-invariant set, then $K^1(X, \varphi) \approx \mathbf{Z}$. More generally, $K^1(X, \varphi) \neq 0$ since the constant functions are φ-invariant.

Let us consider the order on $K^0(X, \varphi)$. Since the map $\alpha_*: K^0(X) \to K^0(X, \varphi)$ is induced by a C*-algebra homomorphism, it maps the positive cone in $K^0(X)$ into the positive

cone in $K^0(X, \varphi)$. We conclude that $K^0(X, \varphi)^+$ contains the image of

$\{f \in C(X, Z) \mid f \geq 0\}$ in the quotient. The reverse inclusion was shown by Boyle and

Handelman [BH1].

The distinguished positive element corresponds to the element of $C(X, Z)/B(X, Z)$

represented by $f = 1$.

Let us quickly mention the rôle of the order unit in the group $K^0(G)$ as it is especially

relevant for zero-dimensional systems. Start with a minimal Cantor system (X, φ). Let

$E \subseteq X$ be a non-empty clopen set and ψ be the first return map of φ on E which is well-

defined and a minimal homeomorphism of E. The inclusion of $C(E, Z)$ into $C(X, Z)$

(extending a function to be zero on $X - E$) induces an isomorphism of pre-ordered abelian

groups

$$K^0(E, \psi) \rightarrow K^0(X, \varphi)$$

which maps the order unit of the first group to the class of χ_E in the second. Kakutani

equivalent minimal Cantor systems (or flow equivalent minimal Cantor systems) have naturally

order-isomorphic K^0-groups. The order-isomorphism does not preserve the order units. The

order unit keeps track of the size of the space, in some sense.

It is natural to ask about the specific properties of the pre-ordered group $K^0(X, \varphi)$ and

how they relate to the dynamics. We will give a brief (and incomplete) survey of some results.

Poon [Po] showed that if (X, φ) is topologically transitive (i.e. there is a dense orbit) then the pre-order is an order; i.e.

$$K^0(X, \varphi) \cap (-K^0(X, \varphi)) = \{0\}.$$

Boyle and Handelman [BH1] showed that the same conclusion holds if every point of x is chain recurrent under φ. In fact [BH1] shows much more. Let

$$J = K^0(X, \varphi) \cap (-K^0(X, \varphi)).$$

Then $K^0(X, \varphi)/J$ is naturally isomorphic to $K^0(\mathrm{ch}(\varphi), \varphi|\mathrm{ch}(\varphi))$, where $\mathrm{ch}(\varphi) \subseteq X$ denotes the set of chain recurrent points of (X, φ).

In [Pu2], it is shown (as a consequence of 3.3, 4.1 and 5.6) that when (X, φ) is minimal, $K^0(X, \varphi)$ is a simple dimension group. (Here, simple means that there are no non-trivial order ideals in G; i.e. if H is a subgroup of G such that h in H, g in G and $0 \leq g \leq h$ implies g in H, then H is either 0 or G.) We will discuss this more completely in section 9. There is a version of this also for systems which are not minimal, but have a unique minimal set, given in [HPS]. Also in [HPS], it is shown that every simple dimension group arises in this way. This result relies heavily on the fundamental ideas of A. Veršik [V1, V2].

Boyle and Handelman [BH1] showed that chain recurrence in (X, φ) implies $K^0(X, \varphi)$ is unperforated (see section 7). They also discuss the structure of order ideals in $K^0(X, \varphi)$. In contrast to the situation for minimal systems, $K^0(X, \varphi)$ does not have the Riesz interpolation property if (X, φ) is a mixing shift of finite type [BH2].

Section 9: Orbit Equivalence

Let G_1 and G_2 be equivalence relations (with topologies) on X_1 and X_2, respectively. There are two natural notions of equivalence between G_1 and G_2. We say they are isomorphic if there is a homeomorphism $h: X_1 \to X_2$ such that $h \times h \,|\, G_1$ is a homeomorphism from G_1 to G_2. We say that they are (orbit) equivalent if there is a homeomorphism $h: X_1 \to X_2$ such that $h \times h(G_1) = G_2$. We will see examples shortly of equivalence relations which are equivalent but not isomorphic.

Isomorphism is the most usual notion from a C^*-algebraic viewpoint. However, within dynamical systems, orbit equivalence is probably more natural. It is important to note that since the construction of $C^*(G)$ and $K_*(C^*(G))$ depend on the topology of G, these are isomorphism invariants but not orbit equivalence invariants. On the other hand, it is not difficult to show that an orbit equivalence will induce a bijection between the invariant measures.

Let us first discuss singly generated equivalence relations. Let (X, φ), (Y, ψ) be any systems without periodic orbits. Let h be a homeomorphism which induces an orbit equivalence between G_φ and G_ψ. That is, h carries each φ-orbit to a ψ-orbit, so we may define functions $m, n: X \to Z$ by

$$\psi \circ h(x) = h \circ \varphi^{m(x)}(x)$$

$$h \circ \varphi(x) = \psi^{n(x)} \circ h(x),$$

for all x in X. The map h is an isomorphism between G_φ and G_ψ if and only if m

and n are continuous.

Theorem 9.1 (M. Boyle) Let (X, φ) and (Y, ψ) be topologically transitive systems with

no periodic points. Then G_φ and G_ψ are isomorphic if and only if (X, φ) and (Y, ψ)

are flip conjugate; i.e. (X, φ) is topologically conjugate to (Y, ψ) or (Y, ψ^{-1}).

A proof can be found in [GPS], although Boyle's original result dealt with the case when

φ and ψ have periodic orbits, which is much harder.

Let us turn to the question of orbit equivalence of G_φ and G_ψ when X and Y are

connected. The following is a minor modification of a result of Kupka. A result of Sierpinski

states that if a compact connected space X is written as a countable disjoint union of closed

sets $F_1, F_2, \cdots$, then, for some k, $F_k = X$ and $F_i = \phi$, for $i \neq k$. If $h: X \to Y$ is an

orbit equivalence between G_φ and G_ψ, then $F_i = \{x \in X \mid h \circ \varphi(x) = \psi^i \circ h(x)\}$ is closed

for all i in $\mathbf{Z}$. By Sierpinski's result, $F_k = X$, $F_i = \phi$ for $i \neq k$, for some fixed k. It

is easy to check that k must be 1 or -1 and so h is a flip conjugacy.

Connected spaces being a dead end, let us now restrict out attention to relations on

Cantor sets, both AF and singly generated ones. The first fundamental result is the following.

Theorem 9.2 Let (X, G), (Y, H) be AF-equivalence relations, with X, Y compact. Then G and H are isomorphic if and only if

$$K^*(G) \approx K^*(H)$$

as ordered abelian groups with distinguished order units.

The result is originally due to Krieger [Kr1] although proofs can also be found in [Ren] and [GPS], the last using the Bratteli diagram terminology.

In the measure theoretic situation, Dye [D1] showed that every relation generated by a single measure preserving ergodic transformation is orbit equivalent to a hyperfinite relation. Prompted by this result, let us consider whether every relation on the Cantor set generated by a minimal homeomorphism is isomorphic to or orbit equivalent to an AF-equivalence relation. The isomorphism question can be answered immediately by recalling K^1 of an AF-relation is zero, while $K^1(G_\varphi)$ is not, as noted before. We postpone the orbit equivalence question for the moment. We will note the following result which can be regarded as a topological version of Dye's theorem - especially as the proof relies on a kind of Rohlin analylsis.

Theorem 9.3 [Pu2] Let (X, φ) be a minimal Cantor system. Let Y be any closed non-empty subset of X such that $\varphi^n(Y) \cap Y = \phi$ for $n \neq 0$; i.e. Y meets each φ-orbit at most once. Let

$$G_Y = G_\varphi - \{(\varphi^k(y), \varphi^\ell(y)), (\varphi^\ell(y), \varphi^k(y)) \mid k \leq 0, \ell \geq 1 \text{ and } y \in Y\}.$$

(That is, the G_Y-equivalence classes are the φ-orbits, if they don't meet Y and the half-orbits $\{\varphi^n(y) \mid n \geq 1\}$, $\{\varphi^n(y) \mid n \leq 0\}$, for each y in Y.) Then G_Y with the relative topology of G is an AF-equivalence relation. Moreover, there is an exact sequence

$$0 \to \mathbf{Z} \to C(Y, \mathbf{Z}) \to K^0(G_Y) \to K^0(G_\varphi) \to 0.$$

(In the case Y is a single point, the map $K^0(G_Y)$ to $K^0(G_\varphi)$ is an order-isomorphism, hence $K^0(G_\varphi)$ is a dimension group.)

Boyle and Handelman [BH2], Glasner and Weiss [GW] and Host [H] (independently) gave the following dynamical description of $K^0(G_Y)$. This is more or less equivalent to the exact sequence above.

<u>Theorem 9.4</u> $K^0(G_Y) = C(X, \mathbf{Z})/\{f - f \circ \varphi^{-1} \mid f \mid Y = 0\}$.

Poon [Po] showed that the hypotheses may be relaxed. Insteady of requiring φ to be minimal, Y should be such that, for any open set $U \supseteq Y$, we have

$$\bigcup_{n \geq 1} \varphi^n(U) = X.$$

(Boyle and Handelman refer to such a Y as a hitting set.) As a consequence, let X be a non-compact totally disconnected metric space and φ be a minimal homeomorphism of X. The analysis above may be applied to the extension of φ to the one point compactification of X to see that G_φ (on X) itself is an AF-equivalence.

Returning to Theorem 9.3, although it has not appeared in the literature, it would seem that the same proof would apply to a minimal homeomorphism of a compact space X which is not necessarily totally disconnected. The conclusion in this case would be that G_Y is the countable union of equivalence relations where each equivalence class is finite. In analogy with the situation in measure theory, it seems reasonable to call such an equivalence relation hyperfinite. Analogues of such results for actions of groups other than Z are noticeably absent, however.

Theorem 9.5 [GPS] Let (X_1, φ_1), (X_2, φ_2) be two minimal Cantor systems. They are orbit equivalent by a map h such that the associated cocycles m, n each have at most one point of discontinuity if and only if

$$K^0(X_1, \varphi_1) \approx K^0(X_2, \varphi_2)$$

as ordered abelian groups with distinguished order units.

Let us first remark that an h as in this theorem is called a strong orbit equivalence. We give a sketch of the "if" part of the result. The idea is to let Y_i be a single point in X_i and apply theorem 9.3, for $i = 1, 2$. Then we have

$$K^0(G_{Y_i}) \approx K^0(G_{\varphi_i}).$$

By hypothesis and Krieger's result on AF-equivalence relations, G_{Y_1} and G_{Y_2} are isomorphic.

In fact, the isomorphism may be chosen to carry Y_1 to Y_2 and $\varphi_1(Y_1)$ to $\varphi_2(Y_2)$, and then

this map can be shown to be a strong orbit equivalence.

To proceed further, we must discuss the notion of an infinitesimal. In an ordered group

G, an element g is called an infinitesimal if, for any order unit u and integer n, $ng \leq u$.

We let $\mathrm{Inf}(G)$ denote the set of infinitesimals, which is a subgroup of G. In a dimension

group, g is infinitesimal if and only if $\tau(g) = 0$, for all states τ on $G[G]$. (Recall, a

state is a positive group homomorphism $\tau: G \to \mathbf{R}$.)

Now for a minimal Cantor system (X, φ), the states on $K^0(X, \varphi)$ correspond to the

φ-invariant probability measures on X. Let

$$J = \left\{ f \in C(X, \mathbf{Z}) \mid \int f \, d\mu = 0 \text{ for all } \varphi\text{-invariant probability measures } \mu \right\}.$$

Then we have

$$J \supseteq B(X, \varphi)$$

$$\mathrm{Inf}\, K^0(X, \varphi) = J/B(X, \varphi)$$

$$K^0(X, \varphi)/\mathrm{Inf}\, K^0(X, \varphi) = C(X, \mathbf{Z})/J$$

In a natural way, $G/\mathrm{Inf}\, G$ is an ordered group with distinguished order unit, for any dimension

group G.

<u>Theorem 9.6 [GPS]</u> Let G and H be two minimal equivalence relations, each of which is either

    (i) an AF-relation

or   (ii) generated by a single homeomorphism of the Cantor set.

    Then G and H are orbit equivalent if and only if

$$K^0(G)/\text{Inf } K^0(G) \approx K^0(H)/\text{Inf } K^0(H)$$

as ordered groups with distinguished order units.

    There is also a version of this result for "weak orbit equivalence" by Glasner and Weiss which we discuss in the next section.

    Finally, let us mention a result of Boyle and Handelman who investigated the nature of the K-zero invariant in more general systems and especially for subshifts of finite type.

    For any system (X, φ), let $Z(\varphi)$ denote the set of all zeta functions of homeomorphisms flow equivalent to (X, φ) [BH1].

<u>Theorem 9.7 [BH1]</u> Suppose (X, φ) and (Y, ψ) are irreducible shifts of finite type. Then the following are equivalent.

    (a) $(X, \varphi), (Y, \psi)$ are flow equivalent

    (b) $K^0(X, \varphi) \approx K^0(Y, \psi)$ as ordered abelian groups

    (c) $Z(\varphi) = Z(\psi)$.

If (X, φ) and (Y, ψ) are orbit equivalent then they are flow equivalent.

The theorem is proved by showing (a) $\Rightarrow$ (b) $\Rightarrow$ (c) $\Rightarrow$ (a). The proof of (a) $\Rightarrow$ (b) proceeds as follows. Let

$$S(X, \varphi) = [0, 1] \times X/(1, x) \sim (0, \varphi(x))$$

be the suspension of X by φ. First, one shows that $H^1(S(X, \varphi))$ (the first cohomology group) is naturally isomorphic to $K^0(X, \varphi)$. Moreover, the natural orientation on $S(X, \varphi)$ allows one to describe an order on H^1 which agrees with the order on K^0 under the isomorphism above.

The proof that (b) $\Rightarrow$ (c) relies on the following observation. If x in X is periodic period n then the map sending f in $C(X, \mathbf{Z})$ to $\sum_1^n f \circ \varphi^k(x)$ induces a positive group homomorphism from $K^0(X, \varphi)$ to $\mathbf{Z}$, sending the order unit to n. In fact, every such homomorphism occurs this way; i.e. the points of period n may be detected from $K^0(X, \varphi)$.

Section 10: Factors and Sub-Equivalence Relations

In this section, we will first discuss some work of Glasner and Weiss [GW] (which might equally well have been included in the section on orbit equivalence).

The first result deals with the situation of Cantor minimal factors.

Theorem 10.1 [GW] Suppose (X, φ) and (Y, ψ) are minimal Cantor systems and $\pi: X \to Y$ is a continuous map such that $\pi \circ \varphi = \psi \circ \pi$ (i.e. (Y, ψ) is a factor of (X, φ)). Then the map sending f in $C(Y, \mathbb{Z})$ to $f \circ \pi$ in $C(X, \mathbb{Z})$ induces an injective positive group homomorphism

$$\pi^*: K^0(Y, \psi) \to K^0(X, \varphi).$$

Most of the proof is straightforward. The most difficult part is the injectivity of π^*.

Following the terminology of Dye, Glasner and Weiss define the full group of φ, denoted $FG(\varphi)$, to be the set of all homeomorphisms α of X whose graph is contained in G_φ; i.e.

$$\alpha(x) = \varphi^{n(x)}(x),$$

for all $x \in X$, where n is some integer valued function on X. Then two systems (X, φ) and (Y, ψ) are weakly orbit equivalent if there are α in $FG(\varphi)$ and β in $FG(\psi)$ and continuous surjections $\pi: X \to Y$ and $\sigma: Y \to X$ such that $\pi \circ \alpha = \psi \circ \pi$ and

$\sigma \circ \beta = \varphi \circ \sigma$. (This implies, at the level of equivalence relations that $\pi \times \pi (G_\varphi) \supseteq G_\psi$

and $\sigma \times \sigma (G_\psi) \supseteq G_\varphi$.) They also defined two ordered groups H_1 and H_2 to be weakly

equivalent if there are positivity-preserving, order-unit preserving injective group

homomorphisms

$$\alpha_1 \colon \; H_1 \to H_2, \quad \alpha_2 \colon \; H_2 \to H_1.$$

Theorem 10.2 [GW] Let (X, φ) and (Y, ψ) be minimal Cantor systems. They are weakly

orbit equivalent if and only if $K^0(X, \varphi)/\text{Inf } K^0(X, \varphi)$ and $K^0(Y, \psi)/\text{Inf } K^0(Y, \psi)$ are weakly

isomorphic.

If there was a Schroder-Bernstein theorem for either weak orbit equivalence or weak

isomorphism, this would be the main result of [GPS] as in section 9. In fact [GW] show (using

an example of D. Handelman) that such a result fails for weak isomorphism of dimension groups

and hence also for weak orbit equivalence.

Finally in this section, we present a short discussion of the results of [Pu1]. We do not

assume our spaces are zero-dimensional.

First, suppose we have equivalence relations (with topologies) G and H on spaces

X and Y. Also suppose we have two continuous injections $i_0, i_1 \colon Y \to X$ with disjoint

images. We also suppose that each map sends each H-equivalence class onto a G-equivalence

class. This means that i_0 and i_1 induce inclusions of H in G which we also assume are continuous. Finally, we assume that the quotient space X' obtained from X by identifying

$i_0(y)$ and $i_1(y)$, for all y in Y, is Hausdorff and that the quotient map π from X to

X' is proper. Then X' has a natural equivalence relation G' ($= \pi \times \pi(G)$) which carries

the quotient topology from G. This topology has all the correct properties.

Let us consider an example: Let (X, d) be any compact metric space and let φ be

a homeomorphism of X. Suppose x and y are in X satisfying

$$\lim_{n \to \infty} d\big(\varphi^n(x), \varphi^n(y)\big) = 0.$$

This means that X', the space obtained identifying $\varphi^n(x)$ and $\varphi^n(y)$, $n \in \mathbf{Z}$, is Hausdorff.

Moreover, X' has an obvious homeomorphism φ' which is a factor of φ. In this case, let

$G = G_\varphi$, $Y = \mathbf{Z}$, $H = \mathbf{Z} \times \mathbf{Z}$ and

$$i_0(k) = \varphi^k(x), \quad i_1(k) = \varphi^k(y), \quad k \in \mathbf{Z}.$$

Then we have $G' = G_{\varphi'}$.

Theorem 10.3 [Pu1] With G, G', H as above, there is a six-term exact sequence.

$$
\begin{array}{ccccc}
K_1\big(C_r^*(H)\big) & \longrightarrow & K_0\big(C_r^*(G')\big) & \longrightarrow & K_0\big(C_r^*(G)\big) \\[2pt]
\big\uparrow & & & & \big\downarrow \\[2pt]
K_1\big(C_r^*(G)\big) & \longleftarrow & K_1\big(C_r^*(G')\big) & \longleftarrow & K_0\big(C_r^*(H)\big)
\end{array}
$$

In the example above, $C_r^*(H) \cong K$, and $K_0\big(C_r^*(H)\big) \cong \mathbf{Z}$, $K_1\big(C_r^*(H)\big) \cong 0$.

The result above is a kind of equivariant version of a result which is fairly routine in algebraic topology. Let us now turn to a situation which is more intrinsically an equivalence relation phenomena. Suppose G is an equivalence relation (with topology) on X. Suppose $L \subseteq G$ satisfies:

(i) L is closed

(ii) $r(L) \cap s(L) = \phi$

(iii) $G' = G - L - L^{-1}$ is an equivalence relation.

The simplest example is: $X = \{1, \cdots, n\}$, $G = X \times X$,

$L = \{(i, j) \in G \mid 1 \le i \le k, k < j \le n\}$, for some fixed n, $1 \le k < n$. A more interesting example is the following. Let X be a compact metric space and let φ be a homeomorphism of X with no periodic points. Let Y be a non-empty closed subset of X which meets each φ-orbit at most once; i.e. $\varphi^n(Y) \cap Y = \phi$ for $n \ne 0$. Let G_φ be as before and let

$$L = \{(\varphi^k(y), \varphi^\ell(y)) \mid y \in Y, \ k \le 0, \ \ell \ge 1\}.$$

Then in G', the equivalence classes are just the φ-orbits, if the orbit does not meet Y, and the backward half-orbit of y and the forward half-orbit of $\varphi(y)$, for all y in Y; i.e. $G' = G_Y$ as in Theorem 9.3.

Given this set-up, we define

$$H_0 = L^{-1}L$$

$$H_1 = LL^{-1}$$

$$H' = H_0 \cup H_1$$

$$H = H_0 \cup H_1 \cup L \cup L^{-1}.$$

Basically H is just the reduction of G to the set $r(L) \cup s(L) \subseteq X$. Now the relative topology of G on H is not satisfactory. We introduce a new topology which basically is the quotient topology associated with the map

$$\{((x,y),\, (y,z)) \in L^{-1} \times L\} \rightarrow (x,z) \in H_0$$

$$\{((x,y),\, (y,z)) \in L \times L^{-1}\} \rightarrow (x,z) \in H_1$$

and the relative topology on L and L^{-1}. With this topology H is a topological equivalence relation.

Theorem 10.4 [Pul] For G, G', H as above, there is a six-term exact sequence

$$K_0\big(C_r^*(H)\big) \quad \text{———} \quad K_0\big(C_r^*(G')\big) \quad \text{———} \quad K_0\big(C_r^*(G)\big)$$

$$\Big|\hspace{10.5cm}\Big|$$

$$K_1\big(C_r^*(G)\big) \quad \text{———} \quad K_1\big(C_r^*(G')\big) \quad \text{———} \quad K_1\big(C_r^*(H)\big)$$

In the example above, it turns out the space of the equivalence relation H is $Y \times Z$ with H-equivalence classes equal to $\{y\} \times Z$, $y \in Y$. This new topology described above is the product topology on $Y \times Z$. We have

$$K_*\big(C_r^*(H)\big) \approx K_*(C(Y) \otimes K) \approx K^*(Y).$$

In the case that X is totally disconnected and φ is minimal,

$$K^0(Y) \cong C(Y, \mathbf{Z}), \quad \text{since} \quad \dim Y = 0$$

$$K^1(Y) = 0$$

$$K^1(G_\varphi) \cong \mathbf{Z}$$

$$K^1(G_Y) \cong 0 \quad \text{since} \quad G_Y \text{ is AF}$$

and the exact sequence above reduces to the one in 9.3.

References

[Bla] B. Blackadar, *K-theory for Operator Algebras*, MSRI Publ. 5, Springer-Verlag, Berlin-Heidelberg-New York, 1986.

[BH1] M. Boyle and D. Handelman, Orbit equivalence, flow equivalence and ordered cohomology, preprint.

[BH2] M. Boyle and D. Handelman, in preparation.

[BF] R. Bowen and J. Franks, Homology for zero dimensional non-wandering sets, Ann. of Math. 106 (1977), 73-92.

[Br] O. Bratteli, Inductive limits of finite dimensional C*-algebras, Trans. AMS 171 (1972), 195-234.

[CuKr] J. Cuntz and W. Krieger, A class of C*-algebras and topological markov chains, Invent. Math. 56 (1980), 239-250.

[D1] H. Dye, On groups on measure preserving transformations, I; II. Amer. J. Math. 81 (1959), 119-159; 85 (1963), 551-576.

[Ef] E.G. Effros, *Dimensions and C*-algebras*, CMS Lecture Notes 46, AMS, Providence, R.I., 1981.

[EHS] E.G. Effros, D. Handelman and C.L. Shen, Dimension groups and their affine representations, Amer. J. Math. 102 (1980), 385-407.

[Ell] G.A. Elliott, On the classification of inductive limits of sequences of semisimple finite dimensional algebras, J. Algebra 38 (1976), 29-44.

[Ex] R. Exel, Rotation numbers for automorphisms of C*-algebras, Pacific J. Math. 127, No. 1 (1987), 31-89.

[GPS] T. Giordano, I.F. Putnam and C.F. Skau, Topological orbit equivalence and C*-crossed products, preprint.

[GW] E. Glasner and B. Weiss, Weak orbit equivalence of Cantor minimal systems, preprint.

[G] K.R. Goodearl, *Partially ordered abelian groups with interpolation*, Mathematical surveys and monographs 20, AMS, Providence, R.I., 1986.

[H] B. Host, unpublished lecture notes.

[HPS] R.H. Herman, I.F. Putnam and C.F. Skau, Ordered Bratteli diagrams, dimension groups and topological dynamics, International J. Math. 3 (1992), 827-864.

234   ERGODIC THEORY OF $\mathbf{Z}^d$ ACTIONS

[HKS]   R. Hoegh-Krohn and T. Skjelbred, Classification of C*-algebras admitting ergodic
        actions of the two-dimensional torus, J. Reine Ange. Math. 328 (1981), 1-8.

[KR]   R.V. Kadison and J.R. Ringrose, *Fundamentals of the Theory of Operator Algebras,
        I, II*, Academic Press, Orlando, 1983, 1986.

[Kr1]   W. Krieger, On a dimension for a class of homeomorphism groups, Math. Ann. 252
        (1980), 87-95.

[Kr2]   W. Krieger, On dimension functions and topological Markov chains, Invent. Math. 56
        (1980), 239-250.

[GKP]   G.K. Pedersen, *C*-algebras and their automorphism groups*, London Math. Soc.
        Monographs 14, Academic Press, London, 1979.

[Pi]   M. Pimsner, Embedding some transformation group C*-algebras into AF-algebras,
        Ergod. Th. and Dynam. Sys. 3 (1983), 613-626.

[PV]   M. Pimsner and D. Voiculescu, Exact sequences for K-groups and Ext-groups of
        certain crossed-product C*-algebras, J. Operator Th. 4 (1980), 93-118.

[Po]   Y.T. Poon, A K-theoretic invariant for dynamical systems, Trans. AMS 311 (1989),
        515-533.

[Pu1]   I.F. Putnam, On the K-theory of C*-algebras of principal groupoids, preprint.

[Pu2]   I.F. Putnam, The C*-algebras associated with minimal homeomorphisms of the Cantor
        set, Pacific J. Math. 136 (1989), 329-353.

[PSS]   I.F. Putnam, K. Schmidt and C.F. Skau, C*algebras associated with Denjoy
        homeomorphisms of the circle, J. Operator Th. 16 (1986), 99-126.

[Ren]   J. Renault, *A groupoid approach to C*algebras*, Lecture Notes in Math. 793, Springer-
        Verlag, Berlin-Heidelberg-New York, 1980.

[Ri]   M.A. Rieffel, C*-algebras associated with irrational rotations, Pacific J. Math. 93
        (1981), 415-429.

[Sch]   K. Schmidt, Algebraic ideas in ergodic theory, CBMS Lecture Notes 94, AMS,
        Providence, RI, 1990.

[Str-V] S. Stratila and D. Voiculescu, *Representations of AF-algebras and of the group* $U(\infty)$,
        Lecture Notes in Math. 486, Springer-Verlag, Berlin-Heidelberg-New York, 1975.

[Tom] J. Tomiyama, *Invitation to C*-algebras and topological dynamics*, World Scientific Advanced Series in Dynamical Systems 3, World Scientific, Singapore, 1987.

[V1] A. Veršik, Uniform algebraic approximation of shift and multiplication operators, Soviet Math. Dokl. (1) 24 1981), 97-100.

[V2] A. Veršik, A theorem on periodical markov approximation in ergodic theory, Ergodic Theory and Related Topics, Math. Res. 12 Akademie-Verlag, Berlin, 1981, (Vitte, 1981), 195-206.

[WO] N.E. Wegge-Olsen, *K-theory and C*-algebras; a friendly approach,*

Miles of Tiles*

Charles Radin
Mathematics Department
University of Texas
Austin, TX 78712
radin@math.utexas.edu

Introduction. This article corresponds closely to four lectures given in conjunction with the July 1994 workshop at Warwick; the informal format of those lectures is maintained.

Our main goal is to discuss the generalization of subshifts with $\mathbf{Z}^d$ actions, especially those of finite-type, to tiling systems with $\mathbf{R}^d$ actions. Our justification for this generalization is the natural action of the Euclidean group, in particular rotations, on tiling systems. The new "statistical" form of rotational symmetries which appear are a significant addition to the mathematics of subshifts, and of decided value in applications, which we discuss.

There are four sections to this article. We begin slowly, warming up within the familiar context of subshifts. The two themes in section I) are first the contrast between subshifts of finite-type and substitution subshifts, and second the entropy of uniquely ergodic subshifts of finite-type. Sections II) and III) generalize the subject in two ways. In II) we make the leap from subshifts to tiling systems, and discuss further the connection between finite-type and substitution systems. The main examples used for illustration are the pinwheel and Penrose tilings of the plane; see Figures 1 and 2.

Section III) is devoted to the use of intuition from physics (statistical mechanics) to finite-type systems, and in particular leads into the subject of rotational symmetries in such systems. All the previous discussion becomes focused in section IV) with a discussion of statistical symmetries and their use in ergodic theory and in the analysis of patterns in space. Again the main illustrations are the pinwheel and Penrose tilings of the plane.

I. The Wang/Berger phenomenon

First some notation. Throughout this section $\mathcal{A}$ will denote a fixed, finite alphabet of cardinality at least 2, T^t will denote translation by $t \in \mathbf{Z}^d$ on $\mathcal{A}^{\mathbf{Z}^d}$ (compact in the product topology), and X will be a subshift: that

* Research supported in part by NSF Grant No. DMS-9304269 and Texas ARP Grant 003658-113

FIGURE 1.   A PINWHEEL TILING

is, a closed, translation invariant (nonempty) subset of $\mathcal{A}^{\mathbf{Z}^d}$. An element of $\mathcal{A}$ will be called a letter, and an element of X will be called a configuration.

We will be concerned with two types of subshifts, substitution and finite-type, and we consider them first in the classical regime, $d = 1$.

a) $d = 1$.

We begin with substitution subshifts. We define $\mathcal{B}$, the set of "words", as $\bigcup_{K \geq 1} \mathcal{A}^K$. To define a substitution subshift one must have a substitution function, $F : \mathcal{A} \longrightarrow \mathcal{B}$. One then extends F to be a map from $\mathcal{B}$ to itself. (If $b = a_1 a_2 \cdots a_K$, $F(b) \equiv F(a_1)F(a_2) \cdots F(a_K)$.) $\mathcal{B}_F$ is then defined by

Figure 4

$$\mathcal{B}_F \equiv \{b \in \mathcal{B} : b = F^k(a), \text{ for some } a \in \mathcal{A}\} \qquad 1)$$

If $b = F^k(a)$ we call b a letter of "level k". Finally, we define X_F, the substitution subshift associated with F, by

$$X_F \equiv \{x \in \mathcal{A}^{\mathbf{Z}} : \text{ for each } t, j \in \mathbf{Z}, \, (x_{t+1}, x_{t+2}, \cdots x_{t+j}) \subset b$$
$$\text{for some } b \in \mathcal{B}_F\} \qquad 2)$$

We now record some simple facts about substitution subshifts.

Lemma 1. Each $x \in X_F$ "decomposes" into adjacent letters of level 1, and therefore also into adjacent letters of level k, for any fixed k.

FIGURE 2.   A PENROSE TILING

(The proof is a simple exercise using compactness.)

Lemma 2. Assume there exists p such that for all $a \in \mathcal{A}$, $F^p(a)$ contains all $a' \in \mathcal{A}$. Then X_F is uniquely ergodic – that is, there is one and only one translation invariant Borel probability measure on X_F.

(We will sketch the proof in the last section.)

Lemma 3. Assume the decomposition of Lemma 1 is unique for each $x \in X_F$. Then for all $x \in X_F$, $T^t x = x$ implies $t = 0$.

(We will give the proof in the last section.)

To clarify the uniqueness assumption of Lemma 3 we leave it as an exercise to see that there is not uniqueness for the substitution:

$$\mathcal{A} = \{0,1\}, \ F(0) = 101, \ F(1) = 010 \qquad\qquad 3)$$

We now consider finite-type systems. For fixed positive integer K and nonempty subset $\phi \subset \mathcal{A}^K$ we define the finite-type subshift X_ϕ associated with ϕ by

$$X_\phi \equiv \{x \in \mathcal{A}^{\mathbf{Z}} : \text{ for all } t \in \mathbf{Z}, \ (x_{t+1}, x_{t+2}, \cdots x_{t+K}) \in \phi\} \qquad 4)$$

(Note the difference from substitution systems for which restrictions are made for subwords of all sizes.)

Lemma 4. If X_ϕ is nonempty there exists $\tilde{x} \in X_\phi$ and $\tilde{t} \neq 0$ such that $T^{\tilde{t}}\tilde{x} = \tilde{x}$.

Note the difference between Lemmas 3 and 4. Since the assumptions in these lemmas are very weak, we make the intuitive conclusion that these two classes of subshifts, the substitutions and the finite-type subshifts, are essentially disjoint.

b) $d \geq 2$.

We now define $\mathcal{B}$, the set of "words", as $\bigcup_{K \subset \mathbf{Z}^d} \mathcal{A}^K$. And again, to define a substitution subshift one must have a substitution function, $F : \mathcal{A} \longrightarrow \mathcal{B}$. But now there can be a problem extending F to a function from $\mathcal{B}$ to itself. This problem will naturally disappear when we extend our format to tilings. (We will see then that basically substitutions are representations of similarity transformations of Euclidean space, and the discreteness of $\mathbf{Z}^d$ creates artificial difficulties such as this extension problem.) As an example consider the following.

Let $d = 2$, $\mathcal{A} = \{a, b, c, d\}$, and define the substitution function F by

$$F(a) = \begin{matrix} a & c \\ b & d \end{matrix}; \quad F(b) = \begin{matrix} b & d \\ a & c \end{matrix}; \quad F(c) = \begin{matrix} c & a \\ d & b \end{matrix}; \quad F(d) = \begin{matrix} d & b \\ c & a \end{matrix} \qquad 5)$$

This F has a natural extension from $\mathcal{B}$ to itself. And then we can define the subshift associated with F as for $d = 1$. It is easy to check that Lemmas 1, 2 and 3 are still true in this more general setting.

For subshifts of finite-type there is less of a problem; we just fix some bounded $K \subset \mathbf{Z}^d$ and nonempty $\phi \subset \mathcal{A}^K$, and define the subshift X_ϕ associated with ϕ as before. The big difference is that **Lemma 4 is not true for** $d \geq 2$. Specifically, in 1966 Robert Berger [Ber] answered a question of his thesis advisor, the philosopher Hao Wang, by giving an example with $d = 2$ and an alphabet $\mathcal{A}$ with over 20,000 letters, of a subshift X_ϕ of finite-type such that for all $x \in X_\phi$, $T^t x = x$ implies $t = 0$. After much intermediate research, for example by Raphael Robinson [Rob], this situation was clarified by Shahar Mozes in 1989 [Moz].

Theorem 1. Given a substitution subshift $X_F \subset \mathcal{A}^{\mathbf{Z}^2}$ which is uniquely ergodic with unique measure μ_F, (and satisfies some weak conditions we do not list), there exists an alphabet $\tilde{\mathcal{A}}$ and some $\phi \subset \tilde{\mathcal{A}}^K$ such that

i) X_ϕ is uniquely ergodic, with unique measure μ_ϕ;

ii) $(X_\phi, \mathbf{Z}^2, \mu_\phi)$ and $(X_F, \mathbf{Z}^2, \mu_F)$ are metrically conjugate.

Considering the correspondence of this result with Lemmas 3 and 4 we conclude that although for $d = 1$ the class of substitution subshifts and the class of subshifts of finite-type are basically disjoint, for $d \geq 2$ the former is basically a subset of the latter! We call this the Wang/Berger phenomenon, the starting point for research in what is now called aperiodic tilings, or forced tilings. We will follow up this path in later sections, but first we want to add one fundamental fact about subshifts of finite-type, a result by Jacek Miękisz and the author [Rad1].

Theorem 2. If the subshift of finite-type $(X_\phi, \mathbf{Z}^d)$ is uniquely ergodic then it has zero entropy.

sketch of the proof. The case of $d = 1$ is immediate from Lemma 4. Assume for simplicity of notation that $d = 2$ and K is a 2×2 subset of $\mathbf{Z}^2$. Let S_N be the square set of N lattice points of $\mathbf{Z}^2$ centered at the origin. Consider the cylinder sets based on S_N, that is, sets of configurations with fixed values in S_N. Think of the letters of the alphabet as colors, and the fixed values of such configurations as (colored) "pictures". We will assume the entropy is $\alpha > 0$, and obtain a contradiction. From Birkhoff's theorem, as a function of N there are roughly $e^{\alpha N}$ pictures on S_N which appear with positive frequency in all configurations. Consider a picture frame of unit thickness around S_N; there are at most $e^{\beta \sqrt{N}}$ different ways to color in the frame. So for large N there is some colored frame D "compatible with" more than N^2 different pictures on S_N, that is, such that each of those pictures and the colored frame D together appear with positive frequency in all configurations. Fix some configuration x and large N with some picture C on S_N inside the colored frame D. Consider S_{9N}. It "contains" at most roughly N different pictures of shape $\sqrt{N} \times \sqrt{N}$, so there is some picture C', compatible with D, which doesn't appear in S_{9N} in x. If necessary we ignore some appearances of C inside D in S_{9N} so the remaining ones do not overlap. The remainder still occur with positive frequency. Change the configuration x to a new configuration x' by replacing each of these remaining appearances of C by C'. It is easy to see that x' still belongs to our subshift X_ϕ. On the other hand it is also clear that the frequency of C' in x' is higher than it is in x, and this is incompatible with the assumption that X_ϕ is uniquely ergodic. This contradiction proves that the entropy is zero. ∎

We conclude this section with some comments on the assumption of unique ergodicity. This assumption is unusual when considering subshifts of finite-type, because in the classical regime of $d = 1$ it follows from Lemma 4 that only trivial subshifts would have the property. Now the classical regime of $d = 1$ corresponds to, or originated from, considerations of time evolution, wherein configurations can correspond to very different conditions, and be very different one from another. We assert that the regime $d \geq 2$ corresponds to a different paradigm, namely space rather than time translation. And there the assumption of unique ergodicity is much more natural, corresponding as it does to "essentially" only one global pattern

– that is, locally unique if not globally. So the study of subshifts of finite-type with $\mathbf{Z}^d$ action is the study of spatial patterns, and the assumption of unique ergodicity is a natural one of nondegeneracy [Rad3].

II. The pinwheel and Penrose tilings

One of the key ideas in the last section was that interesting subshifts of finite-type could be constructed which are metrically conjugate to given substitution subshifts. And in fact Mozes has given a rather general prescription for this. Before we discuss this further, it is appropriate to make the transition to tilings.

First some notation. We will consider (finite) alphabets $\mathcal{A}$ whose letters are polyhedra in $\mathbf{R}^d$. A "tiling (from $\mathcal{A}$)" will then be a countable collection of "tiles" (sets which are congruent to letters of $\mathcal{A}$), which is simultaneously a covering and a packing of $\mathbf{R}^d$.

Now although in the last section I presented the work of Berger *et al* as results about subshifts, in fact this early work was formulated in terms of tilings of the plane. The letters in $\mathcal{A}$ are there all basically unit squares, but each has a different pattern of bumps and dents on the edges so that, like jigsaw puzzle pieces, they only fit together in certain ways. This is easily seen to determine a set $\phi \subset \mathcal{A}^K$, where K is a 2×2 subset of $\mathbf{Z}^2$, and thus a subshift X_ϕ of finite-type. We will refer to such tilings as "square-ish".

In order to study symmetry properties it is necessary to generalize the above mathematics to include polyhedra which are not basically squares or cubes. We will introduce the ideas through two examples. In both cases there is a substitution system and an associated finite-type system. As with subshifts, a substitution tiling system will be a special set of tilings associated with a substitution function, and a finite-type tiling system will consist of **all** tilings by the letters in the given alphabet, the finite-type condition consisting of the geometric restriction that the tiles fit together and cover space. We will begin by describing the two substitution systems.

For the Penrose tilings the alphabet $\mathcal{A}$ for the substitution system consists of two letters, called the "kite" and the "dart"; see Figure 3.

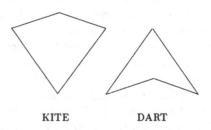

KITE   DART

FIGURE 3.   PENROSE LETTERS FOR SUBSTITUTION

The family B of words is the set of all finite collections of tiles (that is, sets congruent to letters of A), whose union is connected and simply connected, and in which tiles have pairwise disjoint interiors. The substitution function $F : A \longrightarrow B$ should be thought of as obtained by first associating with each letter (shown by dotted lines in Figure 4) a collection of "shrunken tiles" (shown by solid lines in Figure 4), small by a factor $(1 + \sqrt{5})/2$, and then expanding the collection about the origin by $(1 + \sqrt{5})/2$. With this convention we define the family B_F of words of the form $F^k(a)$, $a \in A$, $k \geq 1$. Finally we define the substitution system X_F associated with F to be the set of all tilings x of the plane such that every subword of x is congruent to a subword of some word in B_F.

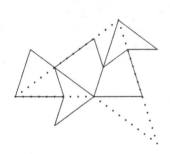 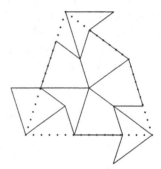

SUBSTITUTION FOR DART   SUBSTITUTION FOR KITE

FIGURE 4.   PENROSE SUBSTITUTIONS

For the pinwheel tilings the alphabet for the substitution system consists of a $1, 2, \sqrt{5}$ right triangle and its reflection, as in Figure 5. The substitution function, indicated in Figure 6, should again be considered a composite of two maps. The substitution system X_F is defined as above.

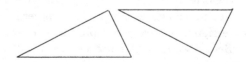

FIGURE 5.   PINWHEEL LETTERS FOR SUBSTITUTION

We have thus specified two substitution systems. The main question in this section is how to associate finite-type systems with such substitution systems, in analogy with the special case of square-ish tilings (*i.e. subshifts*)

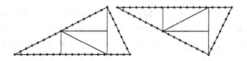

FIGURE 6.   PINWHEEL LETTERS OF LEVEL ONE FOR SUBSTITUTION

discussed in the last section.   The vague idea is that one wants a new
alphabet similar to the given one, but altered so that the tiles can *only* fit
together in the special way the original letters do in the given substitution
tilings.

For the Penrose system the answer is easy. By examining small words
in any tiling one sees that the kite and dart can be prevented from abutting
incorrectly (for example from forming a rhombus, copies of which could then
tile periodically) by adding bumps and dents to them as in Figure 7. And
with these letters one can only tile the plane as in the substitution.

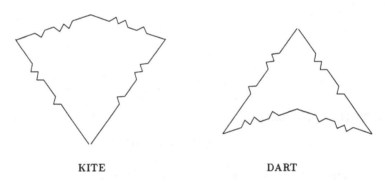

KITE   DART

FIGURE 7.   PENROSE LETTERS FOR FINITE-TYPE

So for the Penrose substitution system it is easy, by examining small
words in the tilings, to create a new alphabet which defines a finite-type
system conjugate to the original. It is then natural to ask whether a similar
analysis can be made for the pinwheel tilings. The answer is no, and we will
present a simplified version of an argument proving this, an argument due
to Ludwig Danzer. Before we do this however, we consider the following
generalization of the above analysis.

Given a tiling x let W be the set of all words in x with upper bound D
on the diameter. We use W to create the dynamical system X_W, defined
to contain all tilings y by the letters in x such that all words of diameter at
most D in y are congruent to words in W. It is then not hard to construct a
new alphabet of polyhedra such that the dynamical system consisting of all

tilings by these letters is conjugate to X_W. One aspect of this generalized method should be emphasized. For the Penrose system we examined words of two letters in the substitution system, and ended up with an alphabet for the finite-type system of the same cardinality as the original, just distortions of the original letters. But in this generalized method, where we examine larger words in the substitution system, we may end up with a much larger alphabet than for the substitution system.

We now return to the pinwheel substitution system. Consider the periodic tiling P of which the unit cell is either of the two words called B and C in Figure 8.

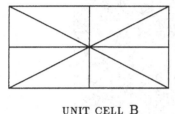

UNIT CELL B UNIT CELL C

FIGURE 8

$F^5(a)$ is shown in Figure 9. In this word one sees several copies of B and C. Now given words of diameter D in a pinwheel tiling, apply the substitution at least $2D\sqrt{5}$ times to P, creating a new periodic tiling P', with B and C leading to two unit cells of P' which we call B' and C'. It is easy to see that all words of diameter at most D in P' are interior to one of these two unit cells, and *so must appear in the pinwheel*. This proves the above assertion since any finite-type rule, based on a set of words of a certain size appearing in the pinwheel, will be compatible with the periodic tiling P' – that is, P' would belong to the space of tilings, in contradiction to Lemma 3.

Now as we noted there is a finite-type pinwheel system conjugate to the pinwheel substitution, although it cannot be obtained directly from words of fixed size by the above method. Furthermore, the technique used by Mozes for square-ish tiles has also proved to be too simple. Next we will sketch a method which works for the pinwheel [Rad4].

First an overview. The letters in the new alphabet $\mathcal{A}'$ come in two families, one for each of the two triangles in the original $\mathcal{A}$. The letters in each family can be thought of as perturbed versions of the original triangle, that is, the original triangle with some pattern of bumps and dents on each edge. But this is not the best way to think of them; it is preferable to view each letter as a triangle with certain information associated with each vertex. (It is proven in [Rad4] that the information can transformed to

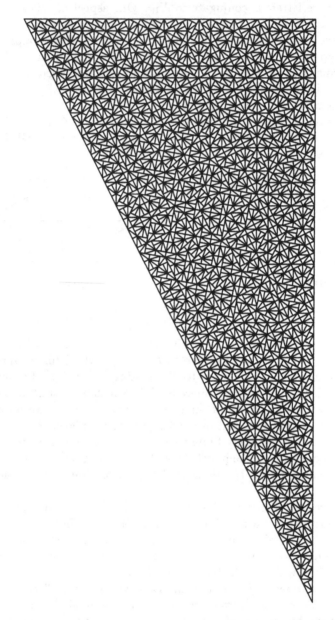

FIGURE 9.   PINWHEEL LETTER OF LEVEL 5 FOR SUBSTITUTION

bumps and dents.) The finite-type system is then defined by requiring that
the triangles may abut if and only if the information associated with vertices

of abutting triangles is "consistent" in a certain precise sense. Roughly speaking, the information can be thought of as a picture of what "should be" the environment of that vertex, referring implicitly to any triangles abutting the given one at that vertex. (By "consistent" we mean that their pictures agree.) We write "should be" because the information can only be thought of as a wish list, and it is necessary to prove that in any tiling with such marked triangles all the wish lists are automatically satisfied. Next we want to indicate what information is included at vertices.

Before this can be done we will pick out one tiling from the substitution system and define a certain hierarchical structure in it. The substitution function (Figure 6) associates five triangles with each letter of $\mathcal{A}$. We give them labels, A through E, as in Figure 10. Next we construct a certain tiling of the plane, called "the tesselation", as follows. Start with a letter sitting in the plane and use the substitution. Reorient the word obtained so the "C" triangle is sitting over the original letter. Apply the substitution to this word, reorienting again so the "C" associated with the first "C" again lies over the original letter. Repeating this process indefinitely produces the tesselation.

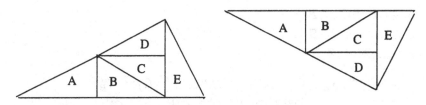

FIGURE 10.   TILES WITH LABELS A THROUGH E IN TESSELATION

We introduce the following notation with reference to the above process. We say the original letter is a letter of level 0. The first application of the substitution produces a word of five letters, a letter of level 1, containing the five letters of level 0. The five letters of level 0 are given labels A, B, C, D or E. When the substitution is applied again, the letter of level 1 is (by choice) seen to be a letter with label C inside a letter of level 2, *etc.* So the tesselation can be understood to be a tiling of the plane by letters of all levels, each labelled by A through E. Finally we introduce the notion of "complete edges" for these letters. Referring to the three edges of the letters as "small", "medium" and "large" (or S, M and L respectively) in the obvious way, we define the complete edges to be the small edges of letters (of any level) with label A, B, C and D, the medium edges of letters with label C, D and E, and the large edges of letters with label B and C. These happen to be the "internal" edges in the letters of level n as these sit in a letter of level $n + 1$; see Figure 10.

It is easy to see from the construction that in the tesselation each edge of each letter of level 0 is geometrically part of precisely one complete edge. Next we show how this fact is used in the information associated with the letters of the alphabet $\mathcal{A}'$ of the finite-type system.

The information associated with each edge of the letters of $\mathcal{A}'$ is incorporated in a tree-like structure. The highest level in the tree is the variable μ. With each of the three vertices of the letter, labelled S, M and L referring to the size of the angle of the vertex, there is the variable μ_S or μ_M or μ_L. These variables each contain other variables, some of which contain other variables *etc.* A complete list of the variables involved is:

$$\mu = (\mu_S, \mu_M, \mu_L)$$
$$\mu_j = (A_1, E_1^1, E_1^2; A_2, E_2^1, E_2^2; \cdots, A_e, E_e^1, E_e^2)$$
$$3 \le e \le 8$$

$$A_j \in (S, M, L, \pi)$$

$$E_k^j = (J, N, P, F, G, H)$$
$$J \in (S, M, L)$$
$$N \in (A, B, C, D, E)$$
$$P \in (A, B, C, D, E)$$
$$F \in (S, M, L, Z, R)$$

$$G = (G_A^{\alpha 1 M}, G_E^{\alpha 1 S}, G_B^{\alpha 1 S}, G_D^{\alpha 1 S}, G_D^{\alpha 1 M}, G_A^{\alpha 2 M}, G_E^{\alpha 2 S}, G_B^{\alpha 2 S},$$
$$G_D^{\alpha 2 S}, G_D^{\alpha 2 M}, G_A^{\epsilon 1 M}, G_E^{\epsilon 1 S}, G_B^{\epsilon 1 S}, G_D^{\epsilon 1 S},$$
$$G_D^{\epsilon 1 M}, G_A^{\epsilon 2 M}, G_E^{\epsilon 2 S}, G_B^{\epsilon 2 S}, G_D^{\epsilon 2 S}, G_D^{\epsilon 2 M})$$
$$G_m^{jkn} \in \{+1, -1\}$$

$$H = (H_A^{1 M}, H_E^{1 S}, H_B^{1 S}, H_D^{1 S}, H_D^{1 M}, H_A^{2 M}, H_E^{2 S}, H_B^{2 S}, H_D^{2 S}, H_D^{2 M})$$
$$H_m^{kn} \in \{+1, -1\}$$

There do not exist letters with all possible values of these variables; there are restrictions, forty two of them, which spell out explicitly which combinations of values may occur. It would be hard to keep track of the variables if they did not have simple geometric meaning, so we next indicate this interpretation of the variables.

The basic structure of a variable such as μ_S is a sequence of triples: an A followed by two E's. This refers to the letters of $\mathcal{A}'$ which "should" abut the given letter, and gives their geometric relationships. The A's refer to the sizes of the angles of each of the abutting letters at the vertex in question. Each edge which "should" emanate from the vertex is split into two to enable reference to each of the two letters sharing that edge, and this is what the pair of E's refer to in each triple.

Now each E_k^j is a complicated quantity, but essentially all the information in it relates to certain complete edges. For example the variable J gives the size (small, medium or large) of the complete edge that the letter edge is part of; N gives the label of the letter, (A through E), of whatever level, that that complete edge is part of; and P gives the label of the letter (A through E) that that letter, of whatever level, is part of. The variable F describes how the complete edge which contains the given level 0 letter behaves as it meets the vertex in question; it might end at an angle of size S, M or L, or it might not end in which case F has the value z, or it might not end and not even have a vertex there (value R.) This takes care of all the variables except for G and H. The "gun" variables G "shoot" information out of vertices. The information is oscillatory – that is, the value of $+1$ or -1 does not carry intrinsic meaning, but is like a phase, and what is significant is if two variables meet at a vertex in phase or out of phase. The variables travel along channels prepared, along certain complete edges, by the H variables. The gun variables are needed in the induction proof which shows that in any tiling with the letters from $\mathcal{A}'$ each of these letters is contained in a unique letter of level k for every $k \geq 1$. This means that letters of level 0 must group together uniquely in groups of 5, then letters of level 1 must also, *etc.* The gun variables are needed to force this grouping. Roughly, the label which a given letter of level k is supposed to have is kept in a certain geometric place, and that information has to travel around in a reliable way to ensure that the neighboring letters of that level are correct.

So much for our sketch of the proof that the pinwheel is finite-type; for the full story one must read [Rad4]. We end this section with the hope that just as the simpler technique of Mozes works quite generally for square-ish substitutions, some version of the pinwheel technique will work for a large class of substitutions of general shapes, thereby giving an understanding for the way hierarchical ("fractal") structures could originate from local, finite-type, rules.

III. Statistical mechanics and tilings

This section is concerned with certain ideas from statistical physics which give an unconventional and useful viewpoint to the general area of subshifts of finite-type with $\mathbf{Z}^d$ action, or finite-type tiling systems with $\mathbf{R}^d$ action. We begin with a short, elementary introduction to (classical) statistical mechanics.

Heuristically, assume we wish to analyze a piece of matter composed of many atoms (for simplicity we assume they are all of the same element), at various temperatures $\tau > 0$ and chemical potentials c. Assume there are N of these atoms constrained in a cube L of volume $|L|$, which means their positions constitute N variable points in L, or a point in L^N.

We somehow assign a "(potential) energy" $E^L(x;c)$ to all possible configurations x of L^N. (This will be discussed further below. We are ignoring the kinetic energy as its contribution to the following is relatively minor and easy to insert if desired.) Then for each $\tau > 0$ we consider

$$f_L(x) \equiv \frac{\exp[-E^L(x;c)/\tau]}{\int_{L^N} \exp[-E^L(x;c)/\tau]\,d^N x} \qquad 6)$$

as a probability density for x. For fixed c we take the (weak-$*$) limit of this probability measure as $L \to \mathbf{R}^d$, getting a Gibbs measure $\mu_{\tau,c}$ on configurations in $\mathbf{R}^d$. We are mainly interested in the weak-$*$ limit $\tau \downarrow 0$, giving $\mu_{0,c}$.

Let X be the collection of all countable subsets of $\mathbf{R}^d$ such that no two points are less than unit distance apart. Let H be the subcollection of finite subsets in X which contain the origin.

Define an "interaction" as a function $e : H \longrightarrow \mathbf{R}$ such that $e(h) = 0$ if $|q| \geq r$ for some $q \in h$. (Here r is some fixed positive number, the "range of the many-body interaction".) For each cube L define E^L on X:

$$E^L(x;c) \equiv \sum_{q \in x \cap L} \{e[T^{-q}x \cap D_{r,0}] - c\} \qquad 7)$$

where $D_{r,0}$ is the open ball centered at the origin 0 of radius r, and as usual T denotes translation. Finally, define $X_{GS} \subset X$ by: $x \in X_{GS}$ if for every L, and every $x' \in X$ such that $x \cap (\mathbf{R}^d \backslash L) = x' \cap (\mathbf{R}^d \backslash L)$, it follows that $E^L(x;c) \leq E^L(x';c)$. GS stands for "ground state", and the notation ignores the dependence on c. Note: it can be proven [Rad2] under rather general conditions on E^L that $\mu_{0,c}(X_{GS}) = 1$.

It is not hard to check that X_{GS} is always nonempty, and closed under translation. We will postpone until the next section the definition of the natural topology for spaces like X_{GS}; suffice it to say that two configurations are "close" if they differ in some large neighborhood of the origin, and then only by a small rigid motion. We note that X_{GS} is compact and metrizable in that topology [R-W].

Consider the following example. Let $d = 2$, $c = 0$ and

$$e(h) \equiv \begin{cases} -25 + 24|h_1 - h_2|, & \text{if } h = \{h_1,\ h_2\} \text{ and } 1 \leq |h_1 - h_2| \leq 25/24; \\ 0, & \text{otherwise} \end{cases}$$
$$8)$$

Then if x is the triangular lattice as in Figure 11 a) then $x \in X_{GS}$. Also, if x' is the configuration in Figure 11 b) then still $x' \in X_{GS}$. Notice the

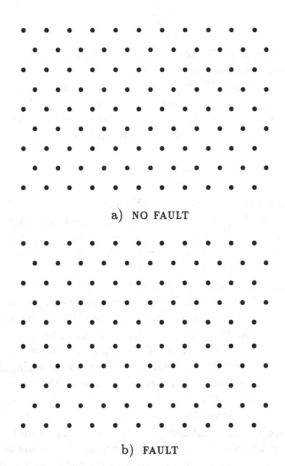

a) NO FAULT

b) FAULT

FIGURE 11. TWO GROUND STATE CONFIGURATIONS

horizontal "fault line" in x'. The ergodic measures on X_{GS} each have as support the translations of x rotated by some fixed angle.

We now discuss one way in which configurations enter into physical theory – how they respond to beams of waves directed at them. We will assume the points $\{t_j\}$ in a configuration $x \in X_{GS}$ represent the centers of atoms, all with the same electron density. Let the electron density at $t \in \mathbf{R}^3$ of an atom centered at t_j be

$$\rho_j(t) = k(t - t_j) = \begin{cases} 1/10 - |t - t_j|, & |t - t_j| \le 1/10 \\ 0, & |t - t_j| > 1/10 \end{cases} \qquad 9)$$

Define $\psi : X_{GS} \longrightarrow \mathbf{R}$ by:

$$\psi(x) = \sum_{t_j \in x} k(-2t_j) \qquad\qquad 10)$$

Assume X_{GS} is uniquely ergodic, with measure μ_{GS}; this is a nondegeneracy assumption, essentially equivalent to assuming that the value of c is not that of a phase transition [Rad2]. Fix $x = \{t_j\} \in X_{GS}$ and an increasing sequence of cubes L_i. Steven Dworkin has shown [Dwo1] that if a plane wave of wavelength s in the direction W_0 is scattered off the atoms in L_i the intensity I_i scattered in the direction W satisfies:

$$\frac{I_i}{|L_i|} \xrightarrow[|L_i| \to \infty]{} d<E_\lambda\psi, \psi> \qquad\qquad 11)$$

where $\lambda = (W - W_0)/s$ and $\{E_\lambda : \lambda \in \mathbf{R}^3\}$ is the spectral family associated with translations on X_{GS}. (That is, $\{E_\lambda\}$ is the spectral family, guaranteed by Stone's theorem [R-N], of the unitary representation of the translations on the complex Hilbert space of complex valued functions on X_{GS} square integrable with respect to the unique invariant measure on X_{GS}.)

This is intimately connected with the history of research on tilings as follows. Assume we have some finite-type tiling system X_F such as the Penrose or pinwheel. With each of the letters in the alphabet we associate one or more "atoms", enough to avoid accidental symmetries (in a sense to be clarified below). Then for each tiling of the tiling system we can associate a configuration of points in $\mathbf{R}^d$, the centers of the atoms associated with each tile. If the atoms are placed generically in the letters, the system of configurations $\tilde{X}_F$ thus produced is naturally isomorphic to the tiling system X_F, including of course the action $\mathbf{R}^d$ on it. Now $\tilde{X}_F$ is intuitively like a ground state system X_{GS} in that configurations belong to $\tilde{X}_F$ if and only if they satisfy the local rules of the tiles, which is very much like the local rules of a ground state system. This idea occurred to physicists and is the basis of their interest in Penrose tilings to model quasicrystals. They use 3 dimensional versions of the Penrose system to which they associate "atoms" as above, creating a space $\tilde{X}_F$ of configurations of "atoms". And they do this precisely because the dynamical spectrum of the 2 or 3 dimensional Penrose systems has unusual rotational symmetries, namely symmetry under rotations by $2\pi/10$ about certain axes. Such symmetries are impossible for the scattering intensities of ordinary crystals, but are in fact the hallmark of certain real materials called quasicrystals. To summarize, tilings such as those of Penrose are of interest in the modeling of quasicrystals because of the symmetries of their dynamical spectra. (There is also interest in the *smoothness* of their spectra; see [Rue], [B-R] and [Rad2]. And another related direction suggested by physics is that of "deceptions" of tilings; see

[D-S] and [Dwo2].)

Before ending this section we make one more observation. Consider the dynamical system of configurations under space translations together with a Gibbs measure $\mu_{\tau,c}$. We have been concerned with the case $\tau = 0$ and have been led to interesting uniquely ergodic finite-type systems, with zero entropy. (The zero entropy result of the first section has been generalized to tiling systems by Jiunn-I Shieh [Shi].) It can be proven that entropy increases with τ, and that at high temperature or entropy the positions become statistically independent (the particles become "free"). The high entropy regime is the one best understood in ergodic theory and physics, with results on Bernoulli systems *etc.* But the low or zero entropy regime seems to contain a great deal of new and exciting possibilities, such as these tiling systems; just as the physics is more interesting in that regime, so too seems to be the mathematics.

IV. A new form of symmetry

We will now draw upon elements from the last two sections to elucidate a new form of geometrical symmetry and its relevance for ergodic theory [Rad6]. We will illustrate the idea by tilings of d dimensional Euclidean space by polyhedral tiles.

One usually says that a tiling (or other pattern) in space has a certain symmetry if it is invariant under the corresponding rigid motion of the space. The new form of symmetry we will describe is different, and is of a statistical nature. We will use the Penrose and pinwheel tilings as motivation, but first we need some notation.

Let X be some tiling dynamical system made from some finite alphabet $\mathcal{A}$ of polyhedra in $\mathbf{R}^d$. (We do not require at this stage that X be either a finite-type or substitution system, but only require that it be locally finite in the following sense. Given any "radius" $r > 0$, look at all the clusters of tiles that are a distance less than r from the center of mass of one of the tiles in any tiling in X. We require that there be only finitely many possible such clusters up to congruence. This is automatic for substitution systems.) Since we distinguish tilings which are rigid motions of one another there is a natural representation on X of G, the connected part of the Euclidean group, which we denote $T^g : X \longrightarrow X$ for $g \in G$. We put a topology on X using a neighborhood basis with three parameters: $x \in X$, a neighborhood Θ of the identity of G, and a neighborhood R of the origin of $\mathbf{R}^d$. The corresponding open set of tilings is:

$$C_{x,\Theta,R} \equiv \{x' \in X : T^g x'|_R = x|_R, \text{ for some } g \in \Theta\} \qquad 12)$$

where "$|_R$" refers to restriction to R. It is not hard to show that X is compact and metrizable in this topology [R-W].

We noted that the Euclidean group is naturally represented on X. Now we assume another feature of the Penrose and pinwheel tilings, the representation of a similarity. That is, we assume that for each $x \in X$ there is a *unique* $y(x) \in X$ associated, made from tiles which are larger by some factor $\gamma > 1$. Let σ be the similarity on $\mathbf{R}^d$ which takes t to t/γ, and let $T^\sigma : X \longrightarrow X$ be defined by $T^\sigma x = (1/\gamma)y(x)$. This homeomorphism is thought of as a representation of the similarity σ since it has the correct relations with the representation of G, for example:

$$(T^\sigma)T^t(T^\sigma)^{-1} = T^{\sigma(t)} = T^{t/\gamma} \tag{13}$$

for all translations t. One use of this is the following. Assume for some tiling x_0 that $T^{t_0}x_0 = x_0$. Then for all n, $T^{\sigma^n(t_0)}[T^{\sigma^n}x_0] = T^{\sigma^n}x_0$. But $T^{\sigma^n(t_0)} = T^{t_0/\gamma^n}$. And this implies that $t_0 = 0$ since otherwise $T^{\sigma^n}x_0$ would be invariant under a translation by an amount smaller than the size of any tile.

We need one more set of ideas before we can consider the new symmetry. We must examine the content of Birkhoff's theorem in this tiling context. Assume μ is an ergodic Borel probability measure on X. Then for each basic open set $C = C_{x,\Theta,R}$ we have:

$$\mu(C) = \int_X \chi_C(x')\, d\mu = \lim_{L \to \mathbf{R}^d} \frac{1}{|L|} \int_L \chi_C(T^t x')\, dt \tag{14}$$

where χ_C is the characteristic function for C. Consider the fraction

$$\frac{1}{|L|} \int_L \chi_C(T^t x')\, dt \tag{15}$$

It is roughly

$$\frac{\epsilon^d \times (\text{the number of times the "picture of } C \text{ appears" in } x')}{|L|} \tag{16}$$

where the "picture of C" refers to the part of the tiling in the definition of C, and the ϵ^d refers to the amount of translation and rotation the picture is allowed to move, in the definition of C.

In summary, this fraction is, up to a factor, just the *frequency* with which the picture of C appears in x'. Now although in general Birkhoff's theorem only holds for μ almost all x', if X is uniquely ergodic under translations then the convergence is not only valid for all x' but is in fact uniform in x' [Fur]. This is relevant because of the following, in which X_{pin} refers to the pinwheel system and X_{Pen} refers to the Penrose system.

Theorem 3. X_{pin} is uniquely ergodic. $X_{Pen} = \bigcup_{\alpha \in [0, 2\pi/10)} X_\alpha$, where the X_α are uniquely ergodic and rotations of one another.

We will give a general technique for proving such results but first we want to discuss its use. Note that any rotation on X can be lifted to a rotation on the set of translation invariant probability measures on X. So if there is only one such measure, as the theorem claims is the case for the pinwheel, then that measure must be invariant under all rotations of X. A similar argument shows for the Penrose system that all translation invariant measures are invariant under rotations by multiples of $2\pi/10$. And from the above analysis of the geometric meaning of such measures we obtain the following statistical interpretation of these symmetries. For the Penrose tilings, any finite cluster of tiles appears in each tiling with the *same frequency* in each of 10 equally separated orientations. For the pinwheel, given any two equal-size intervals I_1 and I_2 of orientations, and any finite cluster of tiles in a tiling, the cluster will appear in that (or any other) tiling having an orientation in I_1 with the *same frequency* as it does with an orientation in I_2. In other words, the rotational symmetries are not leaving invariant tilings, but translation invariant measures on a space of tilings; and the invariance of all such measures is equivalent to the invariance of the above "frequencies" of finite portions of the tilings. Notice that this is a fundamentally new meaning of symmetry for a tiling or pattern, justified at least by its application in the structural study of materials [Rad6].

We now sketch the proof of a general argument [Rad5] for proving unique ergodicity of 2 dimensional tilings, which implies Theorem 3. We assume a general tiling system X as above, and will make a few assumptions as needed.

To prove unique ergodicity of X, given 15), we need to prove that for each finite cluster of tiles and interval of orientation of that cluster (with respect to some arbitrary standard), there is an associated frequency which is independent of the tiling and independent of rotation of the interval. We begin by ignoring the orientations, or what is the same thing, only considering the case where the interval of orientations is the full circle. One key idea is a refinement of the requirement concerning the similarity σ. We further require that this representation of σ be effected by a substitution in the usual way. Namely, the "small size" tiling x is obtained from the "large size" tiling $y(x)$ by simultaneously replacing each of the tiles in $y(x)$ by an appropriate word of small size letters. When such a word is expanded about the origin by the factor γ one gets a tile or letter "of level 1", *etc.* (We note that in simple examples like the pinwheel the letters of all levels are geometrically similar; this is not the case for the Penrose tilings, and is not necessary for the method.) If the finite alphabet of X is $\mathcal{A} = \{a_1, a_2, \ldots, a_K\}$ we introduce the notation that the letters of all levels, obtained by starting with a_k at level 0, are tiles or letters of "type k". The above hierarchical assumption on X is useful in that all we need prove now is that each given finite cluster of tiles has the same frequency of occurance *within each tile of level* $N \gg 1$, and the frequency is independent of the type of the level N tile. (The reasoning is that we are just neglecting the occurances of the cluster on the edges between the tiles of level N, and for large N this is negligible.)

We now introduce the $K \times K$ matrix M for which M_{jk} is the number of type j tiles (level 0) in a type k tile of level 1. It follows that M_{jk}^N is the number of type j tiles in a type k tile of level N. For the pinwheel,

$$M = \begin{pmatrix} 2 & 3 \\ 3 & 2 \end{pmatrix} \qquad \text{17)}$$

Next we assume there is some p such that $M_{jk}^p > 0$ for all j, k. It then follows from the Perron-Frobenius theorem that there is some $\delta > 0$, and functions g, h, such that for all j, k:

$$\frac{M_{jk}^N}{\delta^N} \xrightarrow[N \to \infty]{} g(j)h(k) > 0 \qquad \text{18)}$$

Therefore

$$\frac{M_{jk}^N}{M_{j'k}^N} \xrightarrow[N \to \infty]{} \frac{g(j)}{g(j')} > 0 \qquad \text{19)}$$

The important points are that the limits in 19) exist, and that they are independent of k, which are precisely what was needed to prove the result for clusters consisting of single tiles. (The generalization to larger clusters is routine and we refer to [Rad5] for the details.) This is a well-known argument for substitution subshifts [Que], and suffices to prove the result on Penrose systems in Theorem 3; one just enlarges the Penrose alphabet to include the kite and dart rotated by multiples of $2\pi/10$ and only considers tiles which are *translates* of letters, not *congruent* to letters. But to deal with the pinwheel, or other tilings in which rotations play an essential role, we must now deal with the orientations. To do this we generalize the matrix M above to a family of matrices parametrized by $m \in \mathbf{Z}$:

$$M(m)_{jk} \equiv \sum_{\ell} \exp(im\alpha_{\ell k}) \qquad \text{20)}$$

where $\alpha_{\ell k}$ is the angle of the ℓth copy of the tile of type j in a tile of type k of level 1. Note that $M(0)$ is just the previous matrix M and that $M(1)$ keeps track of the angles of the constituents of the level 1 tiles. We now make the further assumption that there is some tile of level q which contains two tiles of level 0 with irrational relative orientation. (For the pinwheel $q = 2$ suffices.) With this assumption we can use the Weyl criterion on equidistribution of points on a circle, and an old matrix result of Wielandt, to prove that the orientations of tiles (and then finite clusters of tiles) are equally distributed in every tiling in X; see [Rad5] for details. ∎

At this point we would like to mention three open problems that are suggested by the above. First, although the method of this section makes sense in any dimension $d \geq 2$, we do not have a tool as effective as Weyl's criterion in higher dimensions. And yet tiling in higher dimensions could well bring new phenomena since their rotation groups are more interesting. This suggests the study of some 3 dimensional version of the pinwheel, in which the added richness of rotations in 3 dimensions comes into play. And finally there is the obvious need to generalize to a wide class of substitutions the method used to solve the pinwheel.

Summary. The main theme of this article has been the generalization of subshifts of finite-type (with $\mathbf{Z}^d$ actions) to tiling systems (with $\mathbf{R}^d$ actions), and in particular the new role played by rotations. This new marriage of groups of rotations with ergodic theory should prove beneficial in many ways. It clearly enriches ergodic theory, with at least new "symmetry" invariants, but also provides new lines of research. And in the other direction ergodic theory seems to provide a new way to analyze the geometric symmetries of patterns in space, through "statistical" symmetries, in which the invariance of a pattern is replaced by the invariance of the frequencies of all its finite subpatterns [Rad6].

Bibliography

[B-R] D. Berend and C. Radin, Are there chaotic tilings? *Comm. Math. Phys.* 152 (1993), 215-219.

[Ber] R. Berger, The undecidability of the domino problem, *Mem. Amer. Math. Soc. no. 66*, (1966).

[Dwo1] S. Dworkin, Spectral theory and x-ray diffraction, *J. Math. Phys.* 34 (1993), 2965-2967.

[Dwo2] S. Dworkin, Deceptions and choice in tiling systems of finite type, University of Texas preprint, 1994.

[D-S] S. Dworkin and J.-I. Shieh, Deceptions in quasicrystal growth, *Comm. Math. Phys.* to appear.

[Fur] H. Furstenberg, *Recurrence in Ergodic Theory and Combinatorial Number Theory*, Princeton University Press, Princeton, 1981.

[Moz] S. Mozes, Tilings, substitution systems and dynamical systems generated by them, *J. d'Analyse Math.* 53 (1989), 139-186.

[Que] M. Queffélec, *Substitution Dynamical Systems - Spectral Analysis*, Lecture Notes in Mathematics, Vol. 1294, Springer-Verlag, Berlin, 1987.

[Rad1] C. Radin, Disordered Ground States of Classical Lattice Models, *Revs. Math. Phys.* 3 (1991), 125-135.

[Rad2] C. Radin, Global Order from Local Sources, *Bull. Amer. Math. Soc.* 25 (1991), 335-364.

[Rad3] C. Radin, Z^n versus Z actions for systems of finite type, *Contemporary Mathematics* 135 (1992), 339-342.

[Rad4] C. Radin, The pinwheel tilings of the plane, *Annals of Math.* 139 (1994), 661-702.

[Rad5] C. Radin, Space tilings and substitutions, *Geometriae Dedicata* to appear.

[Rad6] C. Radin, Symmetry and tilings, *Notices Amer. Math. Soc.* 42 (1995), 26-31.

[R-W] C. Radin and M. Wolff, Space tilings and local isomorphism, *Geometriae Dedicata* 42 (1992), 355-360.

[R-N] F. Riesz and B. Sz-Nagy, *Functional Analysis*, tr. by L.F. Boron, Frederick Ungar, New York, 1955, 392-393.

[Rob] R.M. Robinson, Undecidability and nonperiodicity for tilings of the plane, *Invent. Math.* 12 (1971), 177-209.

[Rue] D. Ruelle, Do turbulent crystals exist?, *Physica* 113A (1982), 619-623.

[Shi] J.-I. Shieh, The entropy of uniquely ergodic tiling systems, University of Texas preprint, 1994.

Overlapping cylinders: the size of a dynamically defined Cantor-set

Károly Simon

Abstract

This is a survey article on the results about the Hausdorff dimension or Lebesque measure of the attractors of some non-invertible hyperbolic maps and other fractals of overlapping construction.

1 Introduction

It is well known that under some regularity conditions we can use the pressure formula [MM, PeWe] to compute the Hausdorff dimension of dynamically defined Cantor-sets of the real line.

In Section 2 we examine whether or not the same is true if the cylinders of the Cantor-set under consideration intersect each other.

In Section 3 we consider attractors of some hyperbolic non-invertible maps of the plane (whose cylinders intersect each other). When we compute their Hausdorff dimension we face a similar problem to that considered in section 2. Furthermore, we see how we can trace back the computation of the Hausdorff dimension of attractors of some axiom-A diffeomorphisms of the space (generalized solenoids) to the problem of computation of the Hausdorff dimension of the attractors of non-invertible hyperbolic endomorphisms of the plane.

We denote the Hausdorff and the box dimension of a set F by $dim_H(F)$. $dim_B(F)$ respectively. (For the definition of Hausdorff and box dimension see [Falcb1].

1.1 The non-overlapping case

Here we give a brief review of the most important results when the cylinders of the dynamically defined Cantor-set are well separated; that is they are disjoint or in the self-similar case the *open set condition* holds. (For further reading see [Bedf1, Bedf2, Falcb1, Falcb2, Hutch, MaWi, Mor,P., PaTeb, PeWe].)

First we consider the self-similar case. Suppose that $S_1, \ldots, S_k : \mathbf{R}^n \to \mathbf{R}^n$ are similarities with ratios $0 < c_1, \ldots, c_k < 1$

$$|S_i(x) - S_i(x)| = c_i|x - y|.$$

We say that the *Open Set Condition* holds for $\{S_i\}_{i=1}^k$ if there exists a non-empty bounded open set V such that

$$\bigcup_{i=1}^{k} S_i(V) \subset V, S_i(V) \cap S_j(V) \neq \emptyset \tag{1}$$

holds for $i \neq j$. The following theorem is due to P.Moran [Mor,P.] and J.E.Hutchinson [Hutch] and it shows, that if the open set condition holds, the possible best covering in the definition of the Hausdorff dimension is the trivial one, that is the covering by cylinders.

Theorem 1.1 *Suppose that the Open Set Condition (1) holds for the similarities* $\{S_i\}_{i=1}^k$ *on* $\mathbf{R}^n$ *with ratios* $0 < c_1, \ldots, c_k < 1$. *If F is the invariant set, that is the unique non-empty compact set satisfying*

$$F = \bigcup_{i=1}^{k} S_i(F) \tag{2}$$

then $dim_H(F) = dim_B(F) = s$, *where s is given by*

$$\sum_{i=1}^{k} c_i^s = 1. \tag{3}$$

If we consider contractions instead of similarities we do not have such an explicit formula for the Hausdorff dimension. To express the fact that the best covering, in the definition of the Hausdorff dimension, is the trivial one (i.e. the covering by cylinders) we use the notion of topological pressure. See [Bedf1, Theorem 1] or [PeWe].

The pressure formula was used first by R.Bowen [Bow, page 21]. A few years later H.McCluskey and A.Manning proved that for a C^1 axiom-A diffeomorphism of a surface, we can compute the Hausdorff dimension of the intersection of a basic set with an unstable manifold by the pressure formula. More precisely, (see [MM, page 252]) let f be a C^1 axiom-A diffeomorphism of a surface and let Λ be one of its basic sets. For a continuous real valued function ψ on Λ we define the pressure of ψ by

$$P(\psi) = \limsup n^{-1} \log \sup_E \sum_{x \in E} \exp\left(\sum_{i=0}^{n-1} \psi\left(f^i x\right)\right) \tag{4}$$

where the supremum is taken over (n, δ) separated sets E and δ is an expansive constant for $f|\Lambda$.

Theorem 1.2 (H.McCluskey,A.Manning) *Let Λ be a basic set for a C^1 axiom-A diffeomorphism $f : M^2 \to M^2$ with $(1,1)$ splitting*

$$T_\Lambda M = E^s \bigoplus E^u. \tag{5}$$

Then $dim_H(\Lambda \cap W^u) = s$ where s is the unique solution of the equation

$$P\left(-s \log \|Df_x \mid E_x^u\|\right) = 0. \tag{6}$$

2 The self similar case in dimension one

In this section we are working in dimension one.

2.1 Falconer's Theorem

Let $\{S_i\}_{i=1}^k$ be similatitudes on $\mathbf{R}$ with ratios $0 < c_1, \ldots, c_k < 1$ and let F be the invariant set for $\{S_i\}_{i=1}^k$. That is, F is the only non-empty compact set satisfying (2). One can easily see that if there is a $k \in \mathbf{N}$ and two distinct $(i_1, \ldots, i_m)$, $(j_1, \ldots, j_m) \in \{1, 2, \ldots, k\}^m$ such that $S_{i_1} \circ, \ldots, \circ S_{i_m}(F) = S_{j_1} \circ, \ldots, \circ S_{j_m}(F)$ (i.e. two cylinders coincide) then $dim_H(F) < s$, where s is the solution of (3). However it is still unsolved whether or not this is the only reason for the drop of the Hausdorff dimension of F (in dimension one)? More precisely:

Open Problem 1 *Is it true that if $dim_H(F) < s$ then $S_{i_1} \circ, \ldots, S_{i_k}(F) = S_{j_1} \circ, \ldots, S_{j_k}(F)$ holds for some $k \in \mathbf{N}$ and two distinct $(i_1, \ldots, i_k)$, $(j_1, \ldots, j_k)$?*

The best result we know about this question is the Falconer's Theorem (see [Falc1])

Theorem 2.1 (Falconer) *Let $T_i : \mathbf{R} \to \mathbf{R}$ $(1 \le i \le k)$ be linear contractions $T_i(x) = \lambda_i x$, with $0 < |\lambda_i| < 1$ and $\sum_{i=1}^k |\lambda_i| < 1$. Then for almost all $(c_1, \ldots, c_k) \in \mathbf{R}^k$ in the sense of k-dimensional Lebesgue measure the non-empty compact set $E \subset \mathbf{R}$ satisfying*

$$E = \bigcup_{i=1}^k (T_i + c_i) E$$

has Hausdorff dimension s, where $\sum_{i=1}^k |\lambda_i|^s = 1$.

For the higher dimensional analog of this theorem see [Falc2].

2.2   The (0,1,3) problem

M.Keane,M.Smorodinsky and B.Solomyak considered [KeSmSo] the following one parameter family of Cantor-sets with overlapping cylinders:

$$\Lambda(\lambda) = \left\{ \sum_{k=1}^{\infty} i_k \lambda^k \mid i_k = 0,1,3 \right\}$$

Obviously for $\lambda < \frac{1}{4}$ we may apply the Open Set Condition. Thus in this case $dim_H \Lambda(\lambda) = \frac{\log 3}{-\log \lambda}$. On the other hand, if $\lambda \geq \frac{2}{5}$ then $\Lambda(\lambda)$ is an interval. However, for the parameters between $\frac{1}{4}$ and $\frac{2}{5}$ the structure of $\Lambda(\lambda)$ is much more interesting. As the authors in [KeSmSo] have pointed out, for $\lambda < \frac{1}{3}$, $\Lambda(\lambda)$ is a set of zero Lebesgue measure, furthermore they constructed a sequence $\{\gamma_n\}_{n=1}^{\infty} \subset [\frac{1}{3}, \frac{2}{5}]$ such that $\gamma_n \downarrow \frac{1}{3}$ and $m(\Lambda(\lambda)) = 0$, where m is the Lebesgue measure. M. Keane posed the following question: is the function $\lambda \to dim_H \Lambda(\lambda)$ continuous on the interval $[\frac{1}{4}, \frac{1}{3}]$? Answering to this problem M.Pollicott and the author have pointed out (see [PoSi]) that for almost all $\lambda \in [\frac{1}{4}, \frac{1}{3}]$, $dim_H \Lambda(\lambda) = \frac{\log 3}{-\log \lambda}$. However, there exists a dense subset of the interval $[\frac{1}{4}, \frac{1}{3}]$ on which $dim_H \Lambda(\lambda) < \frac{\log 3}{-\log \lambda}$. Thus we obtained a negative answer to Keane's question. Then B.Solomyak pointed out that for almost all $\lambda \in [\frac{1}{4}, \frac{1}{3}]$ $m(\Lambda(\lambda)) > 0$ holds. We can summarize what we know about the set $\Lambda(\lambda)$ for different parameters, in the following theorem;

Theorem 2.2   *1. If $\lambda \leq \frac{1}{4}$ then $dim_H \Lambda(\lambda) = \frac{\log 3}{-\log \lambda}$*

2. If $\lambda \in [\frac{1}{4}, \frac{1}{3}]$ then the set of those λ's for which $dim_H \Lambda(\lambda) \neq \frac{\log 3}{-\log \lambda}$ is a dense subset of $[\frac{1}{4}, \frac{1}{3}]$ of zero Lebesgue measure. (See [PoSi].)

3. If $\lambda \in [\frac{1}{3}, \frac{2}{5}]$ then for almost all λ's $m(\Lambda(\lambda)) > 0$ (see [So]), but $\exists \gamma_n \downarrow \frac{1}{3}$ such that $m(\Lambda(\gamma_n)) = 0$. (See [KeSmSo].)

4. If $\lambda \geq \frac{2}{5}$ then $\Lambda(\lambda) = [0, \frac{3\lambda}{1-\lambda}]$.

The only thing we know about the exceptional set in case 2, it is (see [PoSi]), that for every $t < \frac{1}{3}$ the following holds:

$$dim_H \left\{ \lambda \in [\frac{1}{4}, t] \mid dim_H(\Lambda(\lambda)) < \frac{\log 3}{-\log \lambda} \right\} < 1.$$

Finally we mention an open problem posed by M.Keane.

Open Problem 2 (Keane) *Let us call a set "large" if it contains an interval and call it "small" if it is a set of zero Lebesgue measure. Is there any λ for which the set $\Lambda(\lambda)$ is neither large nor small?*

It was B. Solomyak who found some connections between this (0,1,3)-problem and a very old Erdős problem what is the object of the following section.

2.3 On a problem of Erdős $(\sum \pm \lambda^n)$

In the 1930's people started to examine the distribution ν_λ of the following random variable:

$$Y_\lambda = \sum_{n=0}^{\infty} \pm \lambda^n, \qquad (7)$$

where the $+, -$ signs are chosen independently with probability $\frac{1}{2}$. It is obvious that for $\lambda < \frac{1}{2}$ the distribution ν_λ is singular. Furthermore, B.Jessen and A.Wintner have proved (see [JeWi]) that ν_λ is either purely absolutely continuous or purely singular (with respect to the Lebesgue measure). Then it is naturel to ask: for which λ's is the distribution ν_λ absolutely continuous? P.Erdős has proved the following theorems about this problem:

Theorem 2.3 (Erdős) ν_λ *is singular if* $\frac{1}{\lambda}$ *is a PV number.*

Theorem 2.4 (Erdős) *There exists a* $t < 1$ *such that for almost every* $\lambda \in (t, 1)$ *the distribution* ν_λ *is absolutely continuous.*

The very interesting problem, whetrer we may write $t = \frac{1}{2}$ in Erdős theorem above, had been open for more than fifty years, when B.Solomyak (see [So]) solved it, giving a positive answer.

Theorem 2.5 (Solomyak) *The measure* ν_λ *has a density in* $L^2(\mathbf{R})$ *for Lebesgue a.e.* $\lambda \in (0.5, 1)$.

We mention just two open problems here.

Open Problem 3 *Is it true that for all but countable many* $\lambda \in (0.5, 1)$ *the distribution* ν_λ *is absolutely continuous?*

Open Problem 4 (R.D.Mauldin) *If* ν_λ *is absolutely continuous, is it equivalent to the Lebesgue measure?*

2.4 On a conjecture of Furstenberg

Furstenberg has conjectured [unpublished], that if S_u is the invariant set of the following similarities: $x \to \frac{x}{3}$, $x \to \frac{x+1}{3}$, $x \to \frac{x+u}{3}$, then

$$dim_H S_u = 1 \qquad (8)$$

holds for every irrational u. Although this is still an open problem, R.Kenyon (see [Ken]) has obtained some remarkable results. At first glance this problem seems to be similar to the (0,1,3) problem. However, there are some very significant differences; in this problem the parameter is the translation of the third cylinder, but the contraction rate is constant $\frac{1}{3}$, while in the $(0,1,3)$ problem the contraction rate was the parameter.

As there are three contractions with ratios $\frac{1}{3}$ one could think that (8) should hold for each u. However, because of the overlapping between the cylinders, sometimes the Hausdorff dimension is less than one. More precisely, one can see that

$$S_u = \left\{ \sum_{k=1}^{\infty} \frac{a_k}{3^k} \mid a_k = 0, 1, u \right\}. \tag{9}$$

Let $S_u^v = \{ x \in S_u \mid a_1 = v \}$ where $v \in \{0, 1, u\}$. Then for $u > 2$, $S_u^0 \cap S_u^1 \neq \emptyset$, and for $1 < u \leq 2$, $S_u^1 \cap S_u^u \neq \emptyset$ hold. (One can verify this in a similar way to the proof of [PoSi, Proposition 2].)

So the question is: how much does this overlap count?

Since S_u is the projection of the set

$$S = \left\{ \sum_{i=1}^{\infty} \alpha_i 3^{-i} \mid \alpha_i \in \{(0,0), (1,0), (0,1)\} \right\} \tag{10}$$

(see [Ken]) with the projection $\Pi_u = \begin{pmatrix} 1 & u \\ 0 & 0 \end{pmatrix}$ i.e. $S_u = \Pi_u(S)$. Notice that the well known Projection Theorem (see [Falcb2, page 83]) implies that $dim_H S_u = 1$ for a.e. u. (Since $dim_H S = 1$ follows from Theorem 1.1.) Furthermore, R.Kenyon (see [Ken]) managed to prove a topological analog of this statement:

Theorem 2.6 *The set of u's for which $dim_H S_u = 1$ is a residual subset of* **R**.

3 Horseshoes with overlapping cylinders

It follows from the McCluskey-Manning Theorem (Theorem 1.2) that we can compute the Hausdorff dimension of the attractor of an axiom-A diffeomorphism with $(1,1)$-splitting of a surface with the formula

$$dim_H \Lambda = 1 + s \tag{11}$$

where $P\left(s \log \|Df_x \mid E_x^s\|\right) = 0$ (see Theorem 1.2). In this section we will see that we may apply (11) to some horseshoes with overlapping cylinders.

3.1 The non-uniformly hyperbolic case

M.Yakobson examined (see [Ya]) the following non-invertible and non-uniformly hyperbolic maps (Yakobson's twisted horseshoe maps) of the unit square $S = [0,1] \times [0,1]$:

$$F(x,y) = \left(\varphi(y) + \lambda(x - \frac{1}{2}), g(y)\right)$$

where $g : [0,1] \to [0,1]$ is defined by

1. $\|g(y) - 4y(1-y)\|_{C^2} < \varepsilon$ with a small ε
2. $\max g(y) = 1$
3. $g(0) = g(1) = 0$

further,

$$\varphi(y) = \begin{cases} \frac{1}{4} & for \ y \leq \frac{1}{4} - \delta \\ y & for \ \frac{1}{4} \leq y \leq \frac{3}{4} \\ \frac{3}{4} & for \ y \geq \frac{3}{4} + \delta \end{cases}$$

where δ is a small positive number and λ is a small positive parameter (see [Ya]).

In fact the function F is a two dimensional analog of the map $x \to 4x(1 - x)$. Yakobson described the geometry of the attractor $\Lambda = \cap_{n=0}^{\infty} F^n S$ and proved that there is an invariant measure similar to the Bowen-Ruelle-Sinai measure. Then Yakobson asked whether or not we can apply the formula (11) in this case? The author gave the positive answer to this question in [Si1]. In fact in [Si1, Corrolary 2] it is proved that $dim_H \Lambda = 1 + s$ holds even if we consider a small perturbation of F in its first coordinate, making it non-linear. More precisely:

Theorem 3.1 (Simon) *Let $F(x,y) = (h(x,y), g(y))$ where g is the same as in Yakobson's twisted horseshoe map and h is an arbitrary function satisfying:*

1. *h is a C^2 map;*
2. *$0 < h'_x < \frac{1}{5}$;*
3. *$0 \leq h'_y \leq 1$ and $h'_y \geq \frac{1}{2}$ if $|y - c| < \frac{1}{4}$;*
4. *If $F(x_1, y_1) = F(x_2, y_2)$ then $|y_i - c| < \frac{1}{8}$ $(i = 1, 2)$.*

Then we can compute the Hausdorff dimension of the attractor Λ of F with the formula (11).

Even in this more general version F is "near-hyperbolic". We do not know the answer if the hyperbolicity is dropped completely, that is:

Open Problem 5 *Does the formula (11) give the Hausdorff dimension for the attractor* Λ *of the following neither hyperbolic nor invertible self map of the square?*

$$F(x,y) = \left(\varphi(y) + \lambda(x - \frac{1}{2}), g_a(y) \right)$$

where φ *is the same as in the definition of the Yakobson's twisted horseshoe maps, and* $g_a(y) = ay(1-y)$ $a < 4$.

In what follows we shall consider hyperbolic, non-invertible maps.

3.2   The hyperbolic case

First we state a general theorem and then we will see some of its applications in this section and in the following one. This theorem says (roughly speaking) that for a hyperbolic horseshoe with overlapping cylinders we can compute the Hausdorff dimension of the attractor Λ with the formula (11), if there is no tangential intersection between the unstable manifolds, and the contraction in the stable direction is uniformly stronger than the expansion in the unstable direction, i.e. the area is decreased in each step.

To state this Theorem precisely, we need some definitions.

Let $I, J \subset \mathbf{R}$ be intervals. We consider functions $F : I \times J \to I \times J$ of the form

$$F(x,y) = (h(x,y), g(y)),\qquad (12)$$

where $h : I \times J \to I$ is a C^2 map, and $0 < h'_x < 1$. Further we suppose that there exists a partition $\{J_i\}_{i=1}^m$ of the interval J such that $\min |g'(y)| > c > 1$, and $g(J_i) = J$ for $1 \le i \le m$. Then the attractor Λ of F consists of curves from the bottom to the top of the rectangle $I \times J$. These curves are the unstable manifolds of Λ.

Theorem 3.2 (Simon) *Using the notations above suppose that:*

1. $\max h'_x < \min \frac{1}{|g'(y)|}$
2. *there is no tangential intersection between two unstable fibers.*

Then $dim_H(\Lambda) = 1 + s$ *where* s *is the solution of the pressure formula* $P(s\psi) = 0$ *and* $\psi = \log h'_x$.

Bothe proved (see [Bot, Theorem B]) that if we assume that the contractions in the stable directions are "strong" enough, then assumption 2 in the previous theorem holds for an open and dense subset, that is it holds "typically".

Now we will see some applications of this theorem.

3.2.1 Slanting baker transformations

K Falconer has introduced the family of slanting baker transformations. The difference between baker and slanting baker transformations is that in latter case the cylinders are slanting. More precisely: fix $\lambda_1, \lambda_2, \mu_1, \mu_2 \in \mathbf{R}$, with $|\lambda_1| + |\lambda_2| < 1$. For each $c_1, c_2 \in \mathbf{R}$ Falconer has defined the following map:

$$T(x,y) = \begin{cases} (\lambda_1 x + \mu_1 y + c_1, 2y - 1) & if \ 0 \leq y \leq 1 \\ (\lambda_2 x + \mu_2 y + c_2, 2y + 1) & if \ -1 \leq y < 0 \end{cases} \tag{13}$$

Then for a sufficiently large interval K

$$T : K \times [-1,1] \to K \times [-1,1].$$

Falconer proved in [Falc1] that for almost all (c_1, c_2) we can compute the Hausdorff dimension of the attractor $\Lambda = \cap_{k=o}^{\infty} T^k(K \times [-1,1])$ as if the cylinders did not intersect each other.

Theorem 3.3 (Falconer) *For almost all $(c_1, c_2) \in \mathbf{R}$ $dim_H \Lambda = 1 + s$ where s is the solution of the equation $|\lambda_1|^s + |\lambda_2|^s = 1$, or what is the same, s is the solution of the pressure formula:*

$$P\left(s \log \|DT_x \mid E_x^s\|\right) = 0 \tag{14}$$

One can immediately see that we may apply Theorem 3.1 to the slanting baker transformation, if the contraction rates are less than $\frac{1}{2}$ (since the expansion in the unstable direction is 2).

Theorem 3.4 (Simon) *Suppose that $0 < |\lambda_1|, |\lambda_2| < \min\{\frac{1}{2}, |\mu_1 - \mu_2|\}$. Then*

$$dim_H \Lambda = 1 + s$$

where Λ is the attractor of the slanting baker transformation, and $|\lambda_1|^s + |\lambda_2|^s = 1$.

3.2.2 Cylinder maps

P.Carter and R.D.Mauldin investigated [CaMa] the following family of non-invertible endomorphisms of the cylinder: $T : S^1 \times \mathbf{R} \to S^1 \times \mathbf{R}$

$$T(x,y) = \left(e^{2\pi i x}, b(y - \Phi(e^{2\pi i x}))\right)$$

where a is an integer and Φ is a continuous map from S^1 into $\mathbf{R}$. Equivalently, one can think of T as

$$T(x,y) = (ax(mod1), b(y - \Phi(x)))$$

where Φ is a continuous function of $\mathbf{R}$ into $\mathbf{R}$ with period one. P.Carter and R.D.Mauldin have proved (see [CaMa, Theorem 1.1]) the following theorem:

Theorem 3.5 (Carter, Mauldin) *If $b > 1$, then the expansive map T has an invariant repelling set Γ, which is the graph of the function*

$$f(x) = \sum_{p=0}^{\infty} \frac{1}{b^p} \Phi(a^p x).$$

If $f(x) < y$, then $T^n(x, y)$ diverges to $+\infty$, and if $y < f(x)$ then $T^n(x, y)$ diverges to $-\infty$. Moreover, T is chaotic and expansive on Γ.

Furthermore, Carter and Mauldin have proved, that for $0 < b < 1$ the map T is uniformly hyperbolic and has an attractor M. Namely, let $K = S^1 \times [-\frac{b}{1-b}, \frac{b}{1-b}]$. Then $T(K) \subset K$ and $M = \cap_{n=0}^{\infty} T^n(K)$.

The authors in [CaMa] gave a more detailed description of M when Φ was the tent map, Φ had period one and:

$$\Phi(x) = \begin{cases} 2x & \text{if } 0 \leq x \leq \frac{1}{2} \\ 2 - 2x & \text{if } \frac{1}{2} \leq x \leq 1 \end{cases}$$

Theorem 3.6 (Carter,Mauldin) *If $b \geq \frac{2}{3}$ then M is an annulus.*

For $b \in (\frac{1}{2}, \frac{2}{3})$ almost nothing is known about M. One can easily see that if $0 < b < \frac{1}{2}$ then T satisfies the assumptions of Theorem3.1. Namely, the contraction rate b is less than the expansion rate. Further, one can easily check that the unstable manifolds do not intersect each other tangentially. (Certainly they do intersect each other but all of these intersections are transversal.) Thus as a consequence of Theorem 3.1 we have that:

Theorem 3.7 *For $0 < b < \frac{1}{2}$ we can compute the Hausdorff dimension of M with the following formula:*

$$dim_H M = 1 + \frac{\log 2}{-\log b}$$

3.3   Solenoids

One can easily compute the Hausdorff dimension of the standard or Smale-Williams solenoid, i.e. the attractor Λ of the following map $f : P \to P$ (P is the solid torus)

$$f(\Theta, r, s) = (2\Theta(mod 1), \lambda r + \varepsilon \cos(2\pi\Theta), \lambda s + \varepsilon \sin(2\pi\Theta)) \qquad (15)$$

Angular sections are covered by 2^n disjoint circles of radius λ^n. After some elementary observations, one can see that these circles provide the "best covering" in the definition of the Hausdorff dimension, i.e.

$$dim_H \Lambda = 1 + \frac{\log 2}{\log \lambda}. \qquad (16)$$

However, if we do not assume in the definition of f (15) that the contractions are the same constant, then it is much more difficult to obtain the Hausdorff dimension of the attractor. Bothe proved (see [Bot]) a rather general theorem about this problem. To introduce it, we need some notations:

Let P be the solid torus and define the function $f : P \to P$ by

$$f(t, x_1, x_2) = (\varphi(t), \lambda_1(t)x_1(t) + z_1(t), \lambda_2(t)x_2(t) + z_2(t)) \qquad (17)$$

with $\varphi'(t) > 1, 0 < \lambda_i(t) < 1$. The intersection of the attractor $\Lambda = \cap_{i=0}^{\infty} f^i(P)$ with a disk $D(t) = \{\{t\} \times (x_1, x_2) \mid x_1^2 + x_2^2 \leq 1\}$ $(t \in S^1)$ is denoted by $\Lambda(t)$. So $\Lambda(t) = \Lambda \cap D(t)$. Further, let $\Pi : P \to S^1, \rho_1, \rho_2 : P \to S^1 \times [-1, 1]$ are defined by $\Pi(t, x, y) = t, \rho_1(t, x, y) = (t, x), \rho_2(t, x, y) = (t, y)$.

Now we can state Bothe's Theorem:

Theorem 3.8 (Bothe) *Suppose that*

1. *For any arc B in S^1 and any two components B_1, B_2 of $\Lambda \cap \Pi^{-1}(B)$ the arcs $\rho_i(B_1), \rho_i(B_2)$ are transverse in $S^1 \times [-1, 1]$ at each point of $\rho_i(B_1) \cap \rho_i(B_2)$ (f is intrinsically transverse),*

2. $\sup \lambda_i < \inf \dot{\varphi} \sup \dot{\varphi}^{\frac{-4 \log \inf \lambda_i}{\log \sup \lambda_i}}$.

Then $dim_H \Lambda(t) = \max(p_1, p_2)$, where $P(p_i \log \lambda_i) = 0$ (and P is the topological pressure). Thus $dim_H \Lambda = 1 + \max(p_1, p_2)$.

At the same time Y.Pesin and H.Weiss investigated the so called General Smale-Williams solenoids (see [PeWe]). They are similar to the standard solenoid but in this general case, the contraction rates in the two stable directions are **two distinct** constants. They gave a lower and an upper bound on the Hausdorff dimension of the attractor of the General Smale-Williams solenoid (see [PeWe]). The author proved in [Si2] that the upper bound given by Weiss and Pesin is just the Hausdorff dimension of this attractor. More precisely: Let $f : P \to P$ be defined by

$$f(\Theta, r, s) = (2\Theta(mod 1), \lambda r + \varepsilon \cos(2\pi\Theta), \mu s + \varepsilon \sin(2\pi\Theta)) \qquad (18)$$

and denote $\Lambda = \cap_{n=0}^{\infty} f^n(P)$. We can see that the restriction of f to the plane (Θ, r) is a function $F(\Theta, r) = (2\Theta(mod 1), \lambda r + \varepsilon \cos(2\pi\Theta))$ which semiconjugates to f. Namely, if $\Pi(\Theta, r, s) = (\Theta, r)$ then

$$F \circ \Pi = \Pi \circ f.$$

We denote $\Lambda' = \cap_{n=0}^{\infty} F^n (S^1 \times [-1, 1])$. Then $\Pi(\Lambda) = \Lambda'$. Since F satisfies the assumptions of the Theorem 3.1 (see [Si2]), we have that $dim_H \Lambda' = 1 + p_1$.

Therefore $dim_H\Lambda \geq dim_H\Lambda' = 1 + p_1$. In the same way we can see that $dim_H\Lambda \geq 1 + p_2$. On the other hand, Y.Pesin and H.Weiss have proved in [PeWe] that $dim_H\Lambda \leq 1 + \max(p_1,p_2)$. Thus we have seen that

$$dim_H\Lambda = 1 + \max(p_1,p_2)$$

The idea, to trace back the computation of the Hausdorff dimension of an attractor of a hyperbolic invertible map in $\mathbf{R}^3$ to the attractor of a non-invertible map in $\mathbf{R}^2$, appeared first in [PoWe] and [Bot]. It is well known that McCluskey-Manning Theorem (Theorem 1.2) does not work in $\mathbf{R}^3$. However we saw in this section a method to compute the Hausdorff dimension of some axiom-A attractor in $\mathbf{R}^3$.

References

[Bedf1] T.Bedford. **Hausdorff dimension and box dimension in self-similar sets.** *Proc. Conf. topology and Measure V. (Binz,GDB,1987)*

[Bedf2] T. Bedford. **Applications of dynamical sysytems theory of fractals a study of cukie-cutter Cantor-sets.** *Delft University of technology Report 98-76*

[Bot] H.G.Bothe. **The Hausdorff dimension of certain attractors.** *Preprint 1993.*

[Bow] R.Bowen. **Hausdorff dimension of quasy-circles.** *IHES* **50** (1979), 11-25.

[CaMa] P.Carter & R.D.Mauldin. **Cylinder maps.** *Preprint 1990.*

[Erd1] P.Erdős. **On a family of symmetric Bernoulli convolutions,** *Amer. J. Math.* **61** (1939),974-975.

[Erd2] P.Erdős. **On the smoothneesproperties of Bernoulli convolutions,** *Trans. Amer.Math. Soc.* **62** (1940), 180-186.

[Falcb1] K.Falconer. **The geometry of fractal sets** *Cambridge University Press, Cambridge 1985*

[Falcb2] K.Falconer. **Fractal Geometry** *John Willey & sons 1990*

[Falc1] K.Falconer. **Hausdorff dimension of fractals of overlapping constraction.** *J. Stat. Phys.* **47** (1987),123-132.

[Falc2] K.Falconer. The Hausdorff dimension of self-affine fractals. *Math. Proc. Camb.Soc.* **103** (1989), 339-350.

[Hutch] J.E.Hutchinsion. Fractals and self-similarity. *Indiana Univ. Math. J.* 30,1985,713-747.

[JeWi] B.Jessen, A.Wintner. Distribution functins and the Riemann zeta function. *Trans. Amer. Math. Soc.* 38 (1935),48-88.

[KeSmSo] M.Keane,M.Smorodinsky & B.Solomyak. On the morphology of γ-expansions with deleted digits. *Preprint 1993.*

[Ken] R.Kenyon. Projecting the one-dimensional Sierpienski gasket. *Preprint 1993.*

[MaWi] R.Mauldin & S.Williams. Hausdorff dimension in graph directed constructions. *Transactions of the AMS* **309**:2 (1988), 811-829.

[MM] H.McCluskey , M.Manning. Hausdorff dimension for horseshoes. *Ergod. Th & Dyn. Sys.* 3,(1983),251—260.

[Mor,P.] P.Moran. Additive functions of intervals and Hausdorff dimension. *Proceedings of the Cambridge Philosophical Society* **42** (1946),15-23.

[Mor,M.] M. Moran. Fractal series and infinite products. *Proceedings of the European Conference of Iteration Theory, 1989* **246-256.**

[PaTeb] J.Palis & F.Takens. Hyperbolic & sensitive chaotic dynamics at homoclinic bifurcations. *Cambridge studies in advanced mathematics 35.*

[PeWe] Y.Pesin, H.Weiss. On the dimension of deterministic and random Cantor-like sets, symbolic dynamics, and the Eckmann–Ruelle Conjecture. *Preprint 1994*

[PoSi] M.Pollicott & K.Simon. The Hausdorff dimension of λ-expansions with deleted digits. *accepted for publication in the Transactions of the AMS*

[PoWe] M.Pollicott & H.Weiss. The dimension of some self affine limit sets in the plane and hyperbolic sets. *Journal of Statistical Physics* **77**,(1994),841-866.

[Si1] K.Simon. Hausdorff dimension for non-invertible maps. *Ergod. Th. and Dyn. Systems* 13,(1993),199-212.

[Si2]   K.Simon. The Hausdorff dimension of the General Smale-
          Williams solenoid. *Preprint 1994.*

[So]   B.Solomyak. On the random series $\sum \pm \lambda^n$ (an Edrős prob-
          lem). *Preprint 1994.*

[Ya]   M.V.Yakobson. Invariant measures for some one-dimen-
          sional attractors. *Ergod. Th. and Dyn. Systems* **3** (1983), 155-
          186.

[Yo]   L.S.Young. Dimension, entropy, Lyapunov exponents. *Er-
          dod. Th. and Dynam. Systems* **2** (1982), 109-124.

Uniformity in the Polynomial Szemerédi Theorem

VITALY BERGELSON
The Ohio State University, Columbus, OH 43210 U.S.A.

RANDALL McCUTCHEON
The Ohio State University, Columbus, OH 43210 U.S.A.

0. Introduction

The purpose of this paper is to give a concise proof and some combinatorial consequences of the following theorem.

Theorem 0.1 Assume that $(X, \mathcal{A}, \mu, T)$ is an invertible probability measure preserving system, $k \in \mathbf{N}$, $A \in \mathcal{A}$ with $\mu(A) > 0$, and $p_i(x) \in \mathbf{Q}[x]$ are polynomials satisfying $p_i(\mathbf{Z}) \subset \mathbf{Z}$ and $p_i(0) = 0$, $1 \le i \le k$. Then

$$\liminf_{N-M\to\infty} \frac{1}{N-M} \sum_{n=M}^{N-1} \mu(A \cap T^{p_1(n)}A \cap \cdots \cap T^{p_k(n)}A) > 0.$$

Being less general than the polynomial Szemerédi theorem obtained in [BL1] (see Theorem 1.19 in [B2], these Proceedings) in that it deals with one rather than a commuting family of invertible measure preserving transformations of $(X, \mathcal{A}, \mu)$ and thereby has combinatorial applications in $\mathbf{Z}$ and $\mathbf{R}$ rather than in $\mathbf{Z}^t$ and $\mathbf{R}^t$, Theorem 0.1 has nevertheless a novel stronger although subtle feature: the limit appearing there is a uniform limit, whereas the main theorem of [BL1] would merely give, in the case of a single measure preserving transformation,

$$\liminf_{N\to\infty} \frac{1}{N} \sum_{n=0}^{N-1} \mu(A \cap T^{p_1(n)}A \cap \cdots \cap T^{p_k(n)}A) > 0.$$

The reasons for undertaking the task of presenting a proof of Theorem 0.1 are twofold. First of all, the argument that we are going to give has some new and in our opinion promising features developed for our proof of an IP polynomial Szemerédi theorem (see section 4 in [B2]). These features, which allow for the attainment of uniformity of the limit, are more general than the methods of [BL1]. What one would like to have, of course, is an extension of Theorem 0.1 to a multi-operator situation, namely, one would like to show (for example) that for commuting invertible measure preserving transformations $T_1, \cdots, T_k$ of $(X, \mathcal{A}, \mu)$ one has

$$\liminf_{N-M\to\infty} \frac{1}{N-M} \sum_{n=M}^{N-1} \mu(A \cap T_1^{p_1(n)} A \cap \cdots \cap T_k^{p_k(n)} A) > 0.$$

Indeed, with extra effort we could, combining the techniques we present here with those of [BL1], give a full generalization of [BL1], namely we could show that for polynomials $p_{ij}(x) \in \mathbf{Q}[x]$ with $p_{ij}(\mathbf{Z}) \subset \mathbf{Z}$ and $p_{ij}(0) = 0$, $1 \leq i \leq s$, $1 \leq j \leq k$, and $A \in \mathcal{A}$ with $\mu(A) > 0$ one has

$$\liminf_{N-M\to\infty} \frac{1}{N-M} \sum_{n=M}^{N-1} \mu\left(\bigcap_{i=1}^{s} \left(\prod_{j=1}^{k} T_j^{p_{ij}(n)} \right) A \right) > 0.$$

However, the proof in this general case would be much more cumbersome than that of [BL1], to the point of obscuring the new features. Therefore, we choose to confine ourselves to the single operator case.

The other reason for presenting a proof of the uniformity of the limit is that this seems to be the right or most desirable thing to have in any ergodic theorem dealing with weak or norm Cesàro convergence. For example, Furstenberg's ergodic Szemerédi theorem [F1] established uniformity of the limit appearing in Theorem 0.1 for first degree polynomials $p_i(n)$. We would like to say that

$$\lim_{N-M\to\infty} \frac{1}{N-M} \sum_{n=M}^{N-1} \mu(A \cap T^{p_1(n)} A \cap \cdots \cap T^{p_k(n)} A)$$

exists, but this seems to be presently out of reach even in the linear case when $k \geq 4$ (for the non-uniform limit as well).

Two major tools which we use are:

(i) The structure theorem for measure preserving systems established by Furstenberg in [F1], and which will be used in the form appearing in [FKO], which, roughly speaking, tells us that $(X, \mathcal{A}, \mu, T)$ can be "exhausted" by a chain of factors so that at every link in the chain there is either *relative compactness* or *relative weak mixing*.

(ii) An elaboration of a special case of a *Polynomial Hales-Jewett Theorem*, recently obtained in [BL2] which plays in our treatment the role analogous to that of the polynomial van der Waerden theorem in the proof of the polynomial Szemerédi theorem in [BL1], and which allows us to push uniformity of the limit through compact extensions. See Theorem 3.1, Corollary 3.2, and the appendix (Section 4).

We wish to conclude this introduction by giving some of the combinatorial consequences of Theorem 0.1. Each is proved using a *correspondence principle* due to Furstenberg (see Theorem 0.2 below). In order to formulate this correspondence principle, as well as our applications, we will remind the reader of a few definitions.

Suppose that $E \subset \mathbf{Z}$ is a set. The *upper density*, $\overline{d}(E)$, and *lower density* $\underline{d}(E)$ of E are defined by

$$\overline{d}(E) = \limsup_{N \to \infty} \frac{|E \cap [-N, -N+1, , \cdots, N]|}{2N+1},$$

$$\underline{d}(E) = \liminf_{N \to \infty} \frac{|E \cap [-N, -N+1, \cdots, N]|}{2N+1}.$$

The *upper Banach density* of E is given by

$$d^*(E) = \limsup_{N-M \to \infty} \frac{|E \cap [M, \cdots, N-1]|}{N-M}.$$

E is said to be *syndetic* if it has bounded gaps, or, more formally, if for some finite set $F \subset \mathbf{Z}$ one has

$$E + F = \{x + y : x \in E, y \in F\} = \mathbf{Z}.$$

Clearly, any syndetic set has positive lower density, any set of positive lower density has postive upper density, and any set of positive upper density has positive upper Banach density. It is also completely clear that these are all different notions.

Here now is Furstenberg's correspondence principle.

Theorem 0.2 Given a set $E \subset \mathbf{Z}$ with $d^*(E) > 0$ there exists a probability measure preserving system $(X, \mathcal{A}, \mu, T)$ and a set $A \in \mathcal{A}$, $\mu(A) = d^*(E)$, such that for any $k \in \mathbf{N}$ and any $n_1, \cdots, n_k \in \mathbf{Z}$ one has:

$$d^*\big(E \cap (E - n_1) \cap \cdots \cap (E - n_k)\big) \geq \mu(A \cap T^{n_1} A \cap \cdots \cap T^{n_k} A).$$

The first of our applications follows easily from Theorems 0.1 and 0.2.

Theorem 0.3 Let $E \subset \mathbf{Z}$ with $d^*(E) > 0$. Then for any polynomials $p_i(x) \in \mathbf{Q}[x]$ with $p_i(\mathbf{Z}) \subset \mathbf{Z}$ and $p_i(0) = 0$, $1 \leq i \leq k$, the set

$$\{n \in \mathbf{Z} : \text{ for some } x \in \mathbf{Z}, \{x, x + p_1(n), \cdots, x + p_k(n)\} \subset E\}$$

is syndetic.

Theorem 0.4 Let $p_i(x) \in \mathbf{Q}[x]$ with $p_i(\mathbf{Z}) \subset \mathbf{Z}$ and $p_i(0) = 0$, $1 \le i \le k$. Suppose that $r \in \mathbf{N}$ and that $\mathbf{N} = \bigcup_{i=1}^r C_i$ is any partition of $\mathbf{N}$ into r cells. Then there exists some $L \in \mathbf{N}$ and some $\alpha > 0$ with the property that for any interval $I = [M, N] \subset \mathbf{Z}$ with $N - M \ge L$ there exists i, $1 \le i \le r$, and $n \in C_i \cap I$ such that

$$d^* \Big(C_i \cap (C_i - p_1(n)) \cap \cdots \cap (C_i - p_k(n)) \Big) \ge \alpha.$$

In particular, the system of polynomial equations

$$x_0 = n,$$
$$x_2 - x_1 = p_1(n),$$
$$x_3 - x_1 = p_2(n),$$
$$\vdots$$
$$x_{k+1} - x_1 = p_k(n)$$

has monochromatic solutions $\{x_0, \cdots, x_{k+1}\}$ with $n = x_0$ choosable from any long enough interval.

Proof. Renumbering the sets C_i if needed, let $(C_i)_{i=1}^s$, where $s \le r$, be those C_i for which $d^*(C_i) > 0$. For all i, $1 \le i \le s$, let $(X_i, \mathcal{A}_i, \mu_i, T_i)$ and $A_i \in \mathcal{A}_i$, $\mu(A_i) = d^*(C_i)$ be measure preserving systems and sets having the property that for any $k \in \mathbf{N}$ and any $n_1, \cdots, n_k \in \mathbf{Z}$ one has:

$$d^* \big(C_i \cap (C_i - n_1) \cap \cdots \cap (C_i - n_k) \big) \ge \mu_i(A_i \cap T_i^{n_1} A_i \cap \cdots \cap T_i^{n_k} A_i).$$

This is possible by Theorem 0.2. Applying Theorem 0.1 to $X = X_1 \times \cdots \times X_s$, $\mathcal{A} = \mathcal{A}_1 \otimes \cdots \otimes \mathcal{A}_s$, $\mu = \mu_1 \times \cdots \times \mu_s$, $A = A_1 \times \cdots \times A_s$, and $T = T_1 \times \cdots \times T_s$ we have:

$$\liminf_{N-M \to \infty} \frac{1}{N-M} \sum_{n=M}^{N-1} \mu(A \cap T^{p_1(n)} A \cap \cdots \cap T^{p_s(n)} A)$$

$$= \liminf_{N-M \to \infty} \frac{1}{N-M} \sum_{n=M}^{N-1} \prod_{i=1}^s \mu_i(A_i \cap T_i^{p_1(n)} A_i \cap \cdots \cap T_i^{p_s(n)} A_i) = 3\alpha > 0. \tag{1}$$

Let

$$S = \{n : \mu(A \cap T^{p_1(n)} A \cap \cdots \cap T^{p_s(n)} A) \ge \alpha\}.$$

Let L now be large enough that if $I = [M, N] \subset \mathbf{Z}$ is any interval with $N - M > L$ then

$$|S \cap I| > \alpha |I| \text{ and } \Big| \Big(\bigcup_{i=1}^s C_i \Big) \cap I \Big| > (1 - \alpha)|I|.$$

(The former is possible by (1) which, as one may check, cannot hold unless $|S \cap I| > \alpha|I|$ for large intervals I. The latter is possible since C_i, $1 \leq i \leq s$ consist of *all* C_i for which $d^*(C_i) > 0$.) Of course, it follows that for any interval I with $|I| > L$, there exists i, $1 \leq i \leq s$, and $n \in C_i \cap I \cap S$. For this n we have

$$d^* \Big(C_i \cap (C_i - p_1(n)) \cap \cdots \cap (C_i - p_k(n)) \Big)$$
$$\geq \mu_i(A_i \cap T_i^{p_1(n)} A_i \cap \cdots \cap T_i^{p_k(n)} A_i)$$
$$\geq \mu(A \cap T^{p_1(n)} A \cap \cdots \cap T^{p_k(n)} A) \geq \alpha.$$

$\square$

1. Measure theoretic preliminaries.

In this section we collect the facts concerning measure spaces and their factors which we will be using. For more details, the reader may wish to consult [FKO]. First of all, we remark that for the proof of Theorem 0.1 it suffices to assume that the measure space $(X, \mathcal{A}, \mu)$ is a Lebesgue space. This is a result of the fact that we may always pass to the σ-algebra generated by $(T^n A)_{n \in A}$, which is separable, the fact that μ may clearly be assumed non-atomic, and the fact that Theorem 0.1 is a result about the measure algebra induced by $(X, \mathcal{A}, \mu)$ (with no reference to the points of X). Therefore, as any separable non-atomic probability measure algebra is isomorphic to that induced by Lebesgue measure on the unit interval, one may freely choose $(X, \mathcal{A}, \mu)$ to be any measure space having a separable, non-atomic measure algebra; in particular, one may assume that $(X, \mathcal{A}, \mu)$ is Lebesgue. Furthermore, we may, in view of ergodic decomposition, assume that T is ergodic.

Suppose that $(X, \mathcal{A}, \mu, T)$ is an ergodic measure preserving system, where $(X, \mathcal{A}, \mu)$ is a Lebesgue space, and that $\mathcal{B} \subset \mathcal{A}$ is a complete, T-invariant σ-algebra. Then $\mathcal{B}$ determines a factor $(Y, \mathcal{B}_1, \nu, S)$ of $(X, \mathcal{A}, \mu, T)$, the construction of which we now indicate. Let $(B_i)_{i=1}^{\infty} \subset \mathcal{B}$ be a T-invariant sequence of sets which is dense in $\mathcal{B}$ (in the sense that for any $B \in \mathcal{B}$ and $\epsilon > 0$ there exists $i \in \mathbf{N}$ such that $\mu(B \triangle B_i) < \epsilon$), and denote by Y the set of equivalence classes under the equivalence relation which identifies x_1 with x_2, $x_1 \approx x_2$, when for all $i \in \mathbf{N}$, $x_1 \in B_i$ if and only if $x_2 \in B_i$. Let $\pi : X \to Y$ be the natural projection and let $\mathcal{B}_1 = \{B \subset Y : \pi^{-1}(B) \in \mathcal{B}\}$. For $B \in \mathcal{B}_1$, let $\nu(B) = \mu(\pi^{-1}B)$. Finally, write $S\pi(x) = \pi(Tx)$. Then $(Y, \mathcal{B}_1, \nu, S)$ is a factor of $(X, \mathcal{A}, \mu, T)$.

Since any complete, T-invariant σ-algebra $\mathcal{B} \subset \mathcal{A}$ determines such a factor, we will simply say that $\mathcal{B}$ is a factor of $\mathcal{A}$, or that $\mathcal{A}$ is an extension of $\mathcal{B}$, and will identify $\mathcal{B}_1$ with $\mathcal{B}$ when referring to the induced system, which we now write as $(Y, \mathcal{B}, \nu, S)$. If $x \in X$ and $y \in Y$, with $y = \pi(x)$, we will say that "x is in the fiber over y."

If $(Y, \mathcal{B}, \nu)$ is a factor of $(X, \mathcal{A}, \mu)$, then there is a uniquely (up to null sets in Y) determined family of probability measures $\{\mu_y : y \in Y\}$ on X with the property that μ_y is supported on $\pi^{-1}(y)$ for a.e. $y \in Y$ and such that for every $f \in L^1(X, \mathcal{A}, \mu)$ we have

$$\int_X f(x) \, d\mu(x) = \int_Y \left(\int_X f(z) \, d\mu_y(z) \right) d\nu(y).$$

Sometimes we write μ_x for μ_y when x is in the fiber over y. The decomposition gives, for any $\mathcal{A}$-measurable function f, the *conditional expectation* $E(f|\mathcal{B})$:

$$E(f|\mathcal{B})(y) = \int_X f(x) d\mu_y(x) \quad a.e.$$

Equivalently, the conditional expectation $E(\cdot|\mathcal{B}) : L^2(X, \mathcal{A}, \mu) \to L^2(X, \mathcal{B}, \mu)$ is the orthogonal projection onto $L^2(X, \mathcal{B}, \mu)$. In particular, $E\big(E(f|\mathcal{B})|\mathcal{B}\big) = E(f|\mathcal{B})$.

Let $\mathcal{A} \otimes \mathcal{A}$ be the completion of the σ-algebra of subsets of $X \times X$ generated by all rectangles $C \times D$, $C, D \in \mathcal{A}$. Now define a $T \times T$-invariant measure $\tilde{\mu}$ on $(X \times X, \mathcal{A} \otimes \mathcal{A})$ by letting, for $f_1, f_2 \in L^\infty(X, \mathcal{A}, \mu)$

$$\int_{X \times X} f_1(x_1) f_2(x_2) \, d\tilde{\mu} = \int_Y \int_X \int_X f_1(x_1) f_2(x_2) \, d\mu_y(x_1) d\mu_y(x_2) d\nu(y).$$

(It is clear that there is one and only one measure which satisfies this condition.) We write $X \times_Y X$ for the set of pairs $(x_1, x_2) \in X \times X$ with $x_1 \approx x_2$. One checks that $X \times_Y X$ is the support of $\tilde{\mu}$, and we speak of the measure preserving system $(X \times_Y X, \mathcal{A} \otimes_\mathcal{B} \mathcal{A}, \tilde{\mu}, \tilde{T})$, where $\mathcal{A} \otimes_\mathcal{B} \mathcal{A}$ is the $\tilde{\mu}$-completion of the σ-algebra $\{(X \times_Y X) \cap C : C \in \mathcal{A} \otimes \mathcal{A}\}$ and $\tilde{T}$ is the restriction of $T \times T$ to $X \times_Y X$.

We now procede to introduce the basic elements of the Furstenberg structure theory. The specific format we adopt is from [FKO].

Definition 1.1 Suppose that $(Y, \mathcal{B}, \nu, S)$ is a factor of an ergodic system $(X, \mathcal{A}, \mu, T)$ arising from a complete, T-invariant σ-algebra $\mathcal{B} \subset \mathcal{A}$, and $f \in L^2(X, \mathcal{A}, \mu)$. We will say that f is *almost periodic over* Y, and write $f \in AP$, if for every $\delta > 0$ there exist functions $g_1, \cdots, g_k \in L^2(X, \mathcal{A}, \mu)$ such that for every $n \in \mathbf{Z}$ and a.e. $y \in Y$ there exists some $s = s(n, y)$, $1 \le s \le k$, such that $\|T^n f - g_s\|_{L^2(X, \mathcal{B}, \mu_y)} < \delta$. If the set AP of almost periodic over Y functions is dense in $L^2(X, \mathcal{A}, \mu)$, we say that $(X, \mathcal{A}, \mu, T)$ is a *compact extension* of $(Y, \mathcal{B}, \nu, S)$, or simply that $\mathcal{A}$ is a compact extension of $\mathcal{B}$. If $(X \times_Y X, \mathcal{A} \otimes_\mathcal{B} \mathcal{A}, \tilde{\mu}, \tilde{T})$ is ergodic, then $(X, \mathcal{A}, \mu, T)$ is said to be a *weakly mixing extension* of $(Y, \mathcal{B}, \nu, S)$, or, $\mathcal{A}$ is said to be a weakly mixing extension of $\mathcal{B}$.

For proofs of the following two propositions, see [FKO].

Proposition 1.2 Suppose that an ergodic system $(X, \mathcal{A}, \mu, T)$ is a weakly mixing extension of $(Y, \mathcal{B}, \nu, S)$. Then $(X \times_Y X, \mathcal{A} \otimes_\mathcal{B} \mathcal{A}, \tilde{\mu}, \tilde{T})$ is also a weakly mixing extension of $(Y, \mathcal{B}, \nu, S)$.

Proposition 1.3 Suppose that $(X, \mathcal{A}, \mu, T)$ is a compact extension of $(Y, \mathcal{B}, \nu, S)$. Then for every $A \in \mathcal{A}$ with $\mu(A) > 0$ there exists some $A' \subset A$ with $\mu(A') > 0$ and $1_{A'} \in AP$.

The notions of relative weak mixing and relative compactness are mutually exclusive. Moreover, one may show that $(X, \mathcal{A}, \mu, T)$ is a weakly mixing extension of $(Y, \mathcal{B}, \nu, S)$ if and only if there is no intermediate factor $(Z, \mathcal{C}, \gamma, U)$ between $(X, \mathcal{A}, \mu, T)$ and $(Y, \mathcal{B}, \nu, S)$ which is a proper compact extension of $(Y, \mathcal{B}, \nu, S)$. (This is the relativized version of the fact that a system is weakly mixing if and only if it has no non-trivial compact factor.) The structure theorem (see Theorem 6.17 in [F2] and remarks following) we need may now be formulated as follows:

Theorem 1.4 Suppose that $(X, \mathcal{A}, \mu, T)$ is a separable measure preserving system. There is an ordinal η and a system of T-invariant sub-σ algebras $\{\mathcal{A}_\xi \subset \mathcal{A} : \xi \leq \eta\}$ such that:

(i) $\mathcal{A}_0 = \{\emptyset, X\}$

(ii) For every $\xi < \eta$, $\mathcal{A}_{\xi+1}$ is a compact extension of $\mathcal{A}_\xi$.

(iii) If $\xi \leq \eta$ is a limit ordinal then $\mathcal{A}_\xi$ is the completion of the σ-algebra generated by $\bigcup_{\xi' < \xi} \mathcal{A}_{\xi'}$.

(iv) Either $\mathcal{A}_\eta = \mathcal{A}$ or else $\mathcal{A}$ is a weakly mixing extension of $\mathcal{A}_\eta$.

The factor $\mathcal{A}_\eta$ appearing in the structure theorem is called the *maximal distal factor* of $\mathcal{A}$. In the next section, we show that in order to prove Theorem 0.1 for the system $(X, \mathcal{A}, \mu, T)$, it suffices to establish that the conclusion holds when A is taken from its maximal distal factor.

2. Weakly mixing extensions.

In this section, we will prove the following relativized version of Theorem 3.1 from [B1].

Theorem 2.1 Suppose an ergodic system $(X, \mathcal{A}, \mu, T)$ is a weakly mixing extension of $(Y, \mathcal{B}, \nu, S)$, and that $p_1(x), \cdots, p_k(x) \in \mathbf{Q}[x]$ are non-zero, pairwise distinct polynomials with $p_i(\mathbf{Z}) \subset \mathbf{Z}$ and $p_i(0) = 0$, $1 \leq i \leq k$. Then for any $f_1, \cdots, f_k \in L^\infty(X, \mathcal{A}, \mu)$,

$$\lim_{N-M \to \infty} \left\| \frac{1}{N-M} \sum_{n=M}^{N-1} \left(\prod_{i=1}^{k} T^{p_i(n)} f_i - \prod_{i=1}^{k} S^{p_i(n)} E(f_i | \mathcal{B}) \right) \right\| = 0.$$

Using Theorem 2.1, we will then show (Corollary 2.5) that if the conclusion of Theorem 0.1 holds for the maximal distal factor of a system, then it also holds for the full system. In other words, the validity of Theorem 0.1 passes through weakly mixing extensions.

We will be using the following concept of *convergence in density*.

Definition 2.2 Suppose that $(x_h)_{h \in \mathbf{N}} \subset \mathbf{R}$. If for every $\epsilon > 0$ the set $\{h \in \mathbf{Z} : |x_h - x| < \epsilon\}$ has (lower) density 1, we write

$$D-\lim_{h \to \infty} x_h = x.$$

Equivalently, $x_h \to x$ as $h \to \infty$, $h \notin E$ for some $E \subset \mathbf{Z}$ with $\overline{d}(E) = 0$.

We call the following lemma a "van der Corput type trick", because it is motivated by van der Corput's fundamental inequality.

Lemma 2.3 Suppose that $\{x_n : n \in \mathbf{Z}\}$ is a bounded sequence of vectors in a Hilbert space $\mathcal{H}$. If

$$D-\lim_{h \to \infty} \limsup_{N-M \to \infty} \frac{1}{N-M} \sum_{n=M}^{N-1} \langle x_n, x_{n+h} \rangle = 0,$$

then

$$\lim_{N-M \to \infty} \left\| \frac{1}{N-M} \sum_{n=M}^{N-1} x_n \right\| = 0.$$

Proof. Let $\epsilon > 0$. Fix H large enough that

$$\sum_{r=-H}^{H} \frac{H - |r|}{H^2} \limsup_{N-M \to \infty} \frac{1}{N-M} \sum_{u=M}^{N-1} \langle x_u, x_{u+r} \rangle < \epsilon.$$

We have

$$\frac{1}{N-M} \sum_{n=M}^{N-1} x_n = \frac{1}{N-M} \sum_{n=M}^{N-1} \left(\frac{1}{H} \sum_{h=1}^{H} x_{n+h} \right) + \Psi'_{M,N} = \Psi_{M,N} + \Psi'_{M,N},$$

where $\limsup_{N-M \to \infty} \|\Psi'_{M,N}\| = 0$. We will show that

$$\limsup_{N-M \to \infty} \|\Psi_{M,N}\| < \epsilon.$$

We have

$$||\Psi_{M,N}||^2 \leq \frac{1}{N-M} \sum_{n=M}^{N-1} ||\frac{1}{H} \sum_{h=1}^{H} x_{n+h}||^2$$

$$= \frac{1}{N-M} \sum_{n=M}^{N-1} \frac{1}{H^2} \sum_{h,k=1}^{H} \langle x_{n+h}, x_{n+k} \rangle$$

$$= \sum_{r=-H}^{H} \frac{H-|r|}{H^2(N-M)} \sum_{u=M}^{N-1} \langle x_u, x_{u+r} \rangle + \Psi_{M,N}'',$$

where $\Psi_{M,N}'' \to 0$ as $N-M \to \infty$. By choice of H the last expression is less than ϵ when $N-M$ is sufficiently large.

$\square$

The following lemma will serve as a starting point in our proof of Theorem 2.1. Here and throughout (for the sake of convenience and without loss of generality) we take $L^\infty(X,\mathcal{A},\mu)$, $L^2(X,\mathcal{A},\mu)$, etc. to consist of real-valued functions only. Also, if $f,g \in L^2(X,\mathcal{A},\mu)$, we will write $f \otimes g(x_1,x_2) = f(x_1)g(x_2)$.

Lemma 2.4 Suppose that an ergodic system $(X,\mathcal{A},\mu,T)$ is a weakly mixing extension of $(Y,\mathcal{B},\nu,S)$ and that $f,g \in L^\infty(X,\mathcal{A},\mu)$. If either $E(f|\mathcal{B}) = 0$ or $E(g|\mathcal{B}) = 0$ then

$$D-\lim_{h\to\infty} ||E(fT^h g|\mathcal{B})|| = 0.$$

Proof. Note that it suffices to show that

$$\lim_{N\to\infty} \frac{1}{N} \sum_{n=1}^{N} ||E(fT^n g|\mathcal{B})||^2 = 0.$$

We have

$$\lim_{N\to\infty} \frac{1}{N} \sum_{n=1}^{N} ||E(fT^n g|\mathcal{B})||^2$$

$$= \lim_{N\to\infty} \frac{1}{N} \sum_{n=1}^{N} \int (f \otimes f)\tilde{T}^n(g \otimes g)d\tilde{\mu}$$

$$= \left(\int (f \otimes f)d\tilde{\mu} \right)\left(\int (g \otimes g)d\tilde{\mu} \right)$$

$$= \left(\int E(f|\mathcal{B})^2 \, d\nu \right)\left(\int E(g|\mathcal{B})^2 \, d\nu \right) = 0.$$

□

We now describe the inductive technique whereby we will prove Theorem 2.1. This technique was also utilized in [B1]. First of all we may and shall assume without loss of generality that all polynomials are in $\mathbf{Z}[x]$. The induction will be on $P = \{p_1(x), \cdots, p_k(x)\} \subset \mathbf{Z}[x]$, the set of polynomials appearing in the theorem, using a partial ordering on the family of all such sets which we now describe.

Suppose that $P = \{p_1(x), \cdots, p_k(x)\} \subset \mathbf{Z}[x]$ is a finite set of pairwise distinct polynomials having zero constant term. We associate with P the (infinite) *weight vector* $(a_1, \cdots, a_m, \cdots)$ where, for each $i \in \mathbf{N}$, a_i is the number of distinct integers which occur as the leading coefficient of some polynomial from P which is of degree i. For example,

$$P = \{4x, 9x, 3x^2 - 5x, 3x^2 + 12x, -7x^2, 2x^4, 2x^4 + 3x, 2x^4 - 10x^3, 17x^5\}$$

has weight vector $(2, 2, 0, 1, 1, 0, 0, \cdots)$. (Notice that the weight vector ends in zeros.) If P has weight vector $(a_1, \cdots, a_m, \cdots)$ and Q has weight vector $(b_1, \cdots, b_m, \cdots)$, we write $P < Q$, and say that "P precedes Q", if for some n, $a_n < b_n$, and $a_m = b_m$ for all $m > n$. This partial order comes from a well-ordering on the set of weight vectors

$$\{(a_1, \cdots, a_m, 0, 0, \cdots) : m \in \mathbf{N}, a_i \in \mathbf{N} \cup \{0\}, 1 \le i \le m, a_m > 0\},$$

and therefore gives rise to the inductive technique we are after. This technique, which we call *PET-induction*, works as follows: in order to prove any assertion $\mathcal{W}(P)$ for all finite sets of pairwise distinct polynomials P having zero constant term, it suffices to show that

(i) $\mathcal{W}(P)$ holds for all P having minimal weight vector $(1, 0, 0, \cdots)$, and

(ii) If $\mathcal{W}(P)$ holds for all $P < Q$ then $\mathcal{W}(Q)$ holds as well.

We now proceed to prove Theorem 2.1 via PET-induction.

Proof of Theorem 2.1. First we show that the conclusion holds if the weight vector of $P = \{p_1(x), \cdots, p_k(x)\}$ is $(1, 0, 0, \cdots)$. In this case $k = 1$ and $p_1(x) = jx$ for some non-zero integer j. We may write f_1 as the sum of two functions, one of which has zero conditional expectation over $\mathcal{B}$ and the other of which is $\mathcal{B}$-measurable, namely $f_1 = (f_1 - E(f_1|\mathcal{B})) + E(f_1|\mathcal{B})$. Since the conclusion obviously holds when f_1 is replaced by $E(f_1|\mathcal{B})$ (recall that E is idempotent), we need only show that the conclusion holds when f_1 is replaced by $(f_1 - E(f_1|\mathcal{B}))$, i.e. we may assume without loss of generality that $E(f_1|\mathcal{B}) = 0$. What we must show, then, is that

$$\lim_{N-M \to \infty} \left\| \frac{1}{N-M} \sum_{n=M}^{N-1} T^{jn} f_1 \right\| = 0.$$

However, by the uniform mean ergodic theorem,

$$\lim_{N-M\to\infty} \frac{1}{N-M} \sum_{n=M}^{N-1} T^{jn} f_1 = P f_1,$$

in norm, where P is the projection onto the set of T^j-invariant functions. Since $(X, \mathcal{A}, \mu, T)$ is a weakly mixing extension of $(Y, \mathcal{B}, \nu, S)$, and $E(f_1|\mathcal{B}) = 0$, we have $P f_1 = 0$. This completes the minimal weight vector case.

Suppose now that $Q = \{p_1(x), \cdots, p_k(x)\}$ is a family of non-zero, pairwise distinct polynomials having zero constant term, and that the conclusion holds for all P with $P < Q$. Reindexing if necessary, we may assume that $1 \leq \deg p_1 \leq \deg p_2 \leq \cdots \leq \deg p_k$. Let $f_1, \cdots, f_k \in L^\infty(X, \mathcal{A}, \mu)$. Suppose that $E(f_a|\mathcal{B}) = 0$ for some a, $1 \leq a \leq k$. We then must show that

$$\lim_{N-M\to\infty} \left\| \frac{1}{N-M} \sum_{n=M}^{N-1} \left(\prod_{i=1}^{k} T^{p_i(n)} f_i \right) \right\| = 0.$$

To see that the supposition is made without loss of generality, consider the identity

$$\prod_{i=1}^{k} a_i - \prod_{i=1}^{k} b_i = (a_1 - b_1) b_2 b_3 \cdots b_k + a_1 (a_2 - b_2) b_3 b_4 \cdots b_k + \cdots$$
$$+ a_1 a_2 \cdots a_{k-1} (a_k - b_k)$$

with $a_i = T^{p_i(n)} f_i$ and $b_i = S^{p_i(n)} E(f_i|\mathcal{B})$, noting that on the right hand side we have a sum of terms each of which has at least one factor with zero expectation relative to $\mathcal{B}$.

We use Lemma 2.3. Let $x_n = \prod_{i=1}^{k} T^{p_i(n)} f_i$. Then

$$\limsup_{N-M\to\infty} \frac{1}{N-M} \sum_{n=M}^{N-1} \langle x_n, x_{n+h} \rangle$$

$$= \limsup_{N-M\to\infty} \frac{1}{N-M} \sum_{n=M}^{N-1} \int \left(\prod_{i=1}^{k} T^{p_i(n)} f_i \right) \left(\prod_{i=1}^{k} T^{p_i(n+h)} f_i \right) d\mu$$

$$= \limsup_{N-M\to\infty} \frac{1}{N-M} \sum_{n=M}^{N-1} \int f_1 \left(\prod_{i=2}^{k} T^{p_i(n)-p_1(n)} f_i \right)$$
$$\left(\prod_{i=1}^{k} T^{p_i(n+h)-p_1(n)-p_i(h)} (T^{p_i(h)} f_i) \right) d\mu. \tag{2}$$

For any $h \in \mathbf{Z}$ let

$$P_h = \{p_i(n) - p_1(n) : 2 \leq i \leq k\}$$
$$\cup \{p_i(n + h) - p_1(n) - p_i(h) : \deg p_i \geq 2, 1 \leq i \leq k\}.$$

P_h consist of polynomials with zero constant term. Furthermore, the equivalence class of polynomials in Q with degree and leading coefficient the same as $p_1(n)$ has been annihilated in P_h. All other equivalence classes consisting of polynomials in Q of the same degree as $p_1(n)$ have been preserved (although the leading coefficients of these classes have changed). Equivalence classes of higher degree are completely intact. New equivalence classes may exist, but if so they will be of lesser degree than $p_1(n)$. It follows that $P_h < Q$. We now consider two cases:

Case 1. $\deg p_1 \geq 2$. Then $\deg p_i \geq 2$, $1 \leq i \leq k$, and one may check that for all h outside of some finite set, P_h consists of $2k - 1$ distinct polynomials. For these h, we use our induction hypothesis for the validity of the theorem conclusion for the family P_h (utilizing weak convergence only) and continue from (2):

$$= \limsup_{N-M\to\infty} \frac{1}{N - M} \sum_{n=M}^{N-1} \int E(f_1|\mathcal{B}) \left(\prod_{i=2}^{k} S^{p_i(n)-p_1(n)} E(f_i|\mathcal{B}) \right)$$

$$\left(\prod_{i=1}^{k} S^{p_i(n+h)-p_1(n)-p_i(h)} E(T^{p_i(h)} f_1|\mathcal{B}) \right) d\nu = 0.$$

This since $E(f_a|\mathcal{B}) = 0$.

Case 2. $\deg p_1 = \deg p_2 = \cdots = \deg p_t = 1 < \deg p_{t+1}$. (Of course, if all the p_i are of degree 1 then $t = k$ and there is no p_{t+1}.) In this case $p_1(n + h) - p_1(n) - p_1(h) = 0$, and $p_i(n + h) - p_1(n) - p_i(h) = p_i(n) - p_1(n)$, $2 \leq i \leq t$, so that P_h will consist of $2k - t - 1$ elements (again, excepting a finite set of h's for which other relations might hold). In this case we write $p_i(n) = c_i n$, $1 \leq i \leq t$, and proceed from (2):

$$= \limsup_{N-M\to\infty} \frac{1}{N - M} \sum_{n=M}^{N-1} \int E(f_1 T^{c_1 h} f_1|\mathcal{B}) \left(\prod_{i=2}^{t} S^{p_i(n)-p_1(n)} E(f_i T^{c_i h} f_i|\mathcal{B}) \right)$$

$$\left(\prod_{i=t+1}^{k} S^{p_i(n)-p_1(n)} E(f_i|\mathcal{B}) S^{p_i(n+h)-p_1(n)-p_i(h)} E(T^{p_i(h)} f_i|\mathcal{B}) \right) d\nu.$$

If $t + 1 \leq a \leq k$, this is zero. If $1 \leq a \leq t$, however, it will still be at most

$$\left(\left\| E(f_a T^{c_a h} f_a|\mathcal{B}) \right\|_{L^2(Y,\mathcal{B},\nu)} \right) \prod_{l \neq a} \|f_l\|_\infty^2,$$

so that, by Lemma 2.4,

$$D\text{--}\lim_{h\to\infty} \limsup_{N-M\to\infty} \frac{1}{N-M} \sum_{n=M}^{N-1} \langle x_n, x_{n+h} \rangle$$

$$\leq D\text{--}\lim_{h\to\infty} \|E(f_a T^{c_a h} f_a | \mathcal{B})\|_{L^2(Y,\mathcal{B},\nu)} \cdot \prod_{l\neq a} \|f_l\|_\infty^2 = 0.$$

In either of these two cases, Lemma 2.3 gives

$$\lim_{N-M\to\infty} \left\| \frac{1}{N-M} \sum_{n=M}^{N-1} \left(\prod_{i=1}^{k} T^{p_i(n)} f_i \right) \right\| = 0.$$

$\square$

The following corollary is what we have been aiming for in this section.

Corollary 2.5 Suppose that $(X, \mathcal{A}, \mu, T)$ is an ergodic measure preserving system and denote by $(Y, \mathcal{A}_\eta, \nu, S)$ its maximal distal factor. If for all $A \in \mathcal{A}_\eta$ with $\nu(A) > 0$, we have

$$\limsup_{N-M\to\infty} \frac{1}{N-M} \sum_{n=M}^{N-1} \nu(A \cap S^{p_1(n)} A \cap \cdots \cap S^{p_k(n)} A)$$

$$= \limsup_{N-M\to\infty} \frac{1}{N-M} \sum_{n=M}^{N-1} \int 1_A S^{-p_1(n)} 1_A \cdots S^{-p_k(n)} 1_A \, d\nu > 0,$$

then for all $A \in \mathcal{A}$ with $\mu(A) > 0$,

$$\limsup_{N-M\to\infty} \frac{1}{N-M} \sum_{n=M}^{N-1} \mu(A \cap T^{p_1(n)} A \cap \cdots \cap T^{p_k(n)} A)$$

$$= \limsup_{N-M\to\infty} \frac{1}{N-M} \sum_{n=M}^{N-1} \int 1_A T^{-p_1(n)} 1_A \cdots T^{-p_k(n)} 1_A \, d\mu > 0.$$

Proof. By Theorem 1.4, $(X, \mathcal{A}, \mu, T)$ is either isomorphic to, or is a non-trivial weakly mixing extension of, $(Y, \mathcal{A}_\eta, \nu, S)$. In the former case there is nothing to prove, so we assume the latter. If $A \in \mathcal{A}$, then for some $\delta > 0$ we have

$$\nu(A_\delta) = \nu(\{y \in Y : \mu_y(A) \geq \delta\}) > 0.$$

We have $E(1_A|\mathcal{A}_\eta) > \delta 1_{A_\delta}$, so that by Theorem 2.1 (utilizing only weak convergence),

$$\limsup_{N-M\to\infty} \frac{1}{N-M} \sum_{n=M}^{N-1} \int 1_A T^{-p_1(n)} 1_A \cdots T^{-p_k(n)} 1_A \, d\mu$$

$$= \limsup_{N-M\to\infty} \frac{1}{N-M} \sum_{n=M}^{N-1} \int 1_A S^{-p_1(n)} E(1_A|\mathcal{A}_\eta) \cdots S^{-p_k(n)} E(1_A|\mathcal{A}_\eta) \, d\nu$$

$$= \limsup_{N-M\to\infty} \frac{1}{N-M} \sum_{n=M}^{N-1} \int E(1_A|\mathcal{A}_\eta) S^{-p_1(n)} E(1_A|\mathcal{A}_\eta) \cdots$$

$$\cdots S^{-p_k(n)} E(1_A|\mathcal{A}_\eta) \, d\nu$$

$$\geq \delta^{k+1} \limsup_{N-M\to\infty} \frac{1}{N-M} \sum_{n=M}^{N-1} \int 1_{A_\delta} S^{-p_1(n)} 1_{A_\delta} \cdots S^{-p_k(n)} 1_{A_\delta} \, d\nu > 0.$$

$\square$

3. Uniform polynomial Szemerédi theorem for distal systems.

According to Corollary 2.5, in order to establish Theorem 0.1 for an arbitrary system $(X, \mathcal{A}, \mu, T)$, it suffices to establish that the conclusion holds for its maximal distal factor $(X, \mathcal{A}_\eta, \mu, T)$. That is what we shall do in this section, using transfinite induction on the set of ordinals $\{\xi : \xi \leq \eta\}$ appearing in Theorem 1.4. As Theorem 0.1 trivially holds for the one point system, there are two cases to check: that the validity of the theorem is not lost in the passage to successor ordinals, namely, through compact extensions, and that the validity of the theorem is not lost in the passage to limit ordinals. Again, in this section we assume without loss of generality that all polynomials are in $\mathbf{Z}[x]$.

In order to show that the validity of Theorem 0.1 passes through compact extensions, we will use a combinatorial result, the polynomial Hales-Jewett theorem obtained in [BL2]. A special case of it is given as Theorem 3.1. If A is a set, we denote by $F(A)$ the set of all finite subsets of A. (Of course if A is finite this is just $\mathcal{P}(A)$.)

Theorem 3.1 Suppose numbers $k, d, r \in \mathbf{N}$ are given. Then there exists a number $N = N(k, d, r) \in \mathbf{N}$ having the property that whenever we have an r-cell partition

$$F\left(\{1, \cdots, k\} \times \{1, \cdots, N\}^d\right) = \bigcup_{i=1}^{r} C_i,$$

one of the sets C_i, $1 \leq i \leq r$ contains a configuration of the form

$$\left\{ A \cup (B \times S^d) : B \text{ ranges over subsets of } \{1, \cdots, k\} \right\}$$

for some $A \subset \left(\{1, \cdots, k\} \times \{1, \cdots, N\}^d \right)$ and some non-empty set $S \subset \{1, \cdots, N\}$ satisfying

$$A \cap \left(\{1, \cdots, k\} \times S^d \right) = \emptyset.$$

Theorem 3.1 is a set-theoretic version and generalization of the polynomial van der Waerden theorem proved in [BL1]. To give some of the flavor of how one uses this theorem to help with polynomial dealings, we show first, as an example, how Theorem 3.1 guarantees that for any finite partition of $\mathbf{Z}$, $\mathbf{Z} = \bigcup_{i=1}^{r} D_i$, we may find i, $1 \leq i \leq r$ and $x, n \in \mathbf{Z}$, $n \neq 0$, with $\{x, x + n^2\} \subset D_i$. Namely, let $N = N(1, 2, r)$ as in Theorem 3.1 and create a partition

$$F\left(\{1\} \times \{1, \cdots, N\}^2 \right) = \bigcup_{i=1}^{r} C_i$$

according to the rule:

$$A \in C_i \text{ if and only if } \sum_{(1,t,s) \in A} ts \in D_i, 1 \leq i \leq r.$$

According to Theorem 3.1, there exists i, $1 \leq i \leq r$, $A \subset \left(\{1\} \times \{1, \cdots, N\}^d \right)$, and a non-empty set $S \subset \{1, \cdots, N\}$, satisfying $A \cap \left(\{1\} \times S^2 \right) = \emptyset$, such that

$$\left\{ A, A \cup \left(\{1\} \times S^2 \right) \right\} \subset C_i.$$

Letting $x = \sum_{(1,t,s) \in A} ts$ and $n = \sum_{t \in S} t$, we then have $\{x, x + n^2\} \subset D_i$.

This example uses very little of the strength of Theorem 3.1, in particular it only needs the case $k = 1$ there. By considering general k, one may prove in a completely analogous fashion that for any finite set of polynomials $p_1(x), \cdots, p_k(x) \in \mathbf{Z}[x]$ with $p_i(0) = 0$, $1 \leq k$, and any finite partition of $\mathbf{Z}$, $\mathbf{Z} = \bigcup_{i=1}^{r} D_i$, one may find i, $1 \leq i \leq r$, and $x, n \in \mathbf{N}$, $n \neq 0$, with $\{x, x + p_1(n), \cdots, x + p_k(n)\} \subset D_i$. The following consequence of Theorem 3.1 is all we shall need from it and is a still further elaboration of the method introduced in the previous paragraph. In an appendix we will show how it can be derived from Theorem 3.1.

Corollary 3.2 Suppose that $r, k, t \in \mathbf{N}$ and that

$$p_1(x_1, \cdots, x_t), \cdots, p_k(x_1, \cdots, x_t) \in \mathbf{Z}[x_1, \cdots, x_t]$$

with $p_i(0, \cdots, 0) = 0$, $1 \leq i \leq k$. Then there exist numbers $w, N \in \mathbf{N}$, and a set of polynomials

$$Q = \{q_1(y_1, \cdots, y_N), \cdots, q_w(y_1, \cdots, y_N)\} \subset \mathbf{Z}[y_1, \cdots, y_N]$$

with $q_1(0, \cdots, 0) = 0$, $1 \leq i \leq w$, such that for any r-cell partition $Q = \bigcup_{i=1}^r C_i$, there exists some i, $1 \leq i \leq r$, $q \in Q$, and non-empty, pairwise disjoint subsets $S_1, \cdots, S_t \subset \{1, \cdots, N\}$, such that, under the symbolic substitution $x_m = \sum_{n \in S_m} y_n$, $1 \leq m \leq t$, we have

$$\Big\{q(y_1, \cdots, y_N), q(y_1, \cdots, y_N) + p_1(x_1, \cdots, x_t), \cdots,$$

$$q(y_1, \cdots, y_N) + p_k(x_1, \cdots, x_t)\Big\} \subset C_i.$$

We now make some definitions.

Definition 3.3 A subset $E \subset \mathbf{Z}$ will be called *thick* if for every $M \in \mathbf{N}$, there exists $a \in \mathbf{Z}$ such that $\{a, a+1, a+2, \cdots, a+M\} \subset E$.

Note that a set is thick if and only if it contains arbitrarily large intervals, and a set is syndetic if and only if it intersects every thick set non-trivially.

The following definition is tailored to fit into the framework of the usage of Corollary 3.2 in passing to compact extensions. Here for any $n_1, \cdots, n_t \in \mathbf{N}$ (or any additive group) we write

$$FS(n_1, \cdots, n_t) = \{n_{i_1} + \cdots + n_{i_m} : 1 \leq m \leq t, 1 \leq i_1 < \cdots < i_m \leq t\}.$$

Definition 3.4 Suppose $(X, \mathcal{A}, \mu, T)$ is an invertible measure preserving system and that $\mathcal{B} \subset \mathcal{A}$ is a complete T-invariant sub-σ-algebra. $\mathcal{B}$ is said to have the *PSZ property* if for every $A \in \mathcal{B}$ with $\mu(A) > 0$, $t \in \mathbf{Z}$, and polynomials $p_1(x_1, \cdots x_t), \cdots, p_k(x_1, \cdots, x_t) \in \mathbf{Z}[x_1, \cdots, x_t]$ having zero constant term, there exists $\delta > 0$ such that in every thick set $E \subset \mathbf{Z}$, there exist $n_1, \cdots, n_t \in \mathbf{Z}$ such that $FS(n_1, \cdots, n_t) \subset E$ and

$$\mu(A \cap T^{p_1(n_1, \cdots, n_t)} A \cap \cdots \cap T^{p_k(n_1, \cdots, n_t)} A) > \delta.$$

The case $t = 1$, in particular, gives some $\delta > 0$ for which the set

$$\{n \in \mathbf{Z} : \mu(A \cap T^{p_1(n)} A \cap \cdots \cap T^{p_k(n)} A) > \delta\}$$

is syndetic, which insures that

$$\liminf_{N-M \to \infty} \frac{1}{N-M} \sum_{n=M}^{N-1} \mu(A \cap T^{p_1(n)} A \cap \cdots \cap T^{p_k(n)} A) > 0.$$

Therefore, in light of Corollary 2.5, we will have established Theorem 0.1 if we are able to prove that the maximal distal factor of any system has the PSZ property. This is exactly what we shall do in the remainder of this section. The reader may wonder why this definition is stronger than appears necessary, that is, why we choose to deal with polynomials of many variables. The reason is that our method of proof requires the PSZ property in this strength in order to preserve itself under compact extensions. Corollary 3.2 is the key to this method, as we shall now see.

Theorem 3.5 Suppose that $(X, \mathcal{A}, \mu, T)$ is an ergodic measure preserving system and that $\mathcal{B} \subset \mathcal{A}$ is a complete, T-invariant sub-σ-algebra having the PSZ property. If $(X, \mathcal{A}, \mu, T)$ is a compact extension of the factor $(Y, \mathcal{B}, \nu, S)$ determined by $\mathcal{B}$, then $\mathcal{A}$ has the PSZ property as well.

Proof. Suppose that $A \in \mathcal{A}$, $\mu(A) > 0$. By Proposition 1.3 there exists a subset $A' \subset A$, $\mu(A') > 0$, such that $1_{A'} \in AP$. Therefore we may assume without loss of generality that $f = 1_A \in AP$. Suppose that $t, k \in \mathbf{N}$ and that

$$p_1(x_1, \cdots, x_t), \cdots, p_k(x_1, \cdots, x_t) \in \mathbf{Z}[x_1, \cdots, x_t]$$

have zero constant term. There exists some $c > 0$ and a set $B \in \mathcal{B}$, $\nu(B) > 0$, such that for all $y \in B$, $\mu_y(A) > c$. Let $\epsilon = \sqrt{\frac{c}{8k}}$. Since $1_A \in AP$, there exist functions $g_1, \cdots, g_r \in L^2(X, \mathcal{A}, \mu)$ having the property that for any $n \in \mathbf{N}$, and a.e. $y \in Y$, there exists $s = s(n, y)$, $1 \leq s \leq r$, such that $\|T^n f - g_s\|_y < \epsilon$. For these numbers r, k, t and polynomials p_i, let $w, N \in \mathbf{N}$ and

$$Q = \{q_1(y_1, \cdots, y_N), \cdots, q_w(y_1, \cdots, y_N)\} \subset \mathbf{Z}[y_1, \cdots, y_N],$$
$$q_i(0, \cdots, 0) = 0, \ 1 \leq i \leq w$$

have the property that for any r-cell partition $Q = \bigcup_{i=1}^r C_i$, there exists i, $1 \leq i \leq r$, $q \in Q$, and pairwise disjoint subsets $S_1, \cdots, S_t \subset \{1, \cdots, N\}$ such that substituting $x_m = \sum_{n \in S_m} y_n$, $1 \leq m \leq t$, we have

$$\big\{ q(y_1, \cdots, y_N), q(y_1, \cdots, y_N) - p_1(x_1, \cdots, x_t), \cdots,$$
$$q(y_1, \cdots, y_N) - p_k(x_1, \cdots, x_t) \big\} \subset C_i.$$

(This is possible by Corollary 3.2.)

Since $\mathcal{B}$ has the PSZ property, there exists $\eta > 0$ such that for every thick set $E \subset \mathbf{Z}$, there exists $u_1, \cdots, u_N \in \mathbf{Z}$ such that $FS(u_1, \cdots, u_N) \subset E$ and

$$\nu\big(B \cap S^{q_1(u_1, \cdots, u_N)} B \cap \cdots \cap S^{q_w(u_1, \cdots, u_N)} B \big) > \eta.$$

Let D be the number of ways of choosing t non-empty, pairwise disjoint sets $S_1, \cdots, S_t \subset \{1, \cdots, N\}$, and set $\delta = \frac{c\eta}{2D}$. We want to show that in any thick set $E \subset \mathbf{Z}$ there exist $n_1, \cdots, n_t \in \mathbf{Z}$ such that $FS(n_1, \cdots, n_t) \subset E$ and

$$\mu\big(A \cap T^{p_1(n_1, \cdots, n_t)} A \cap \cdots \cap T^{p_k(n_1, \cdots, n_t)} A\big) > \delta.$$

Let E be any thick set. There exist $u_1, \cdots, u_N \in \mathbf{Z}$ such that $FS(u_1, \cdots, u_N) \subset E$ and

$$\nu\big(B \cap S^{q_1(u_1, \cdots, u_N)} B \cap \cdots \cap S^{q_w(u_1, \cdots, u_N)} B\big) > \eta.$$

Pick any $y \in \big(B \cap S^{q_1(u_1, \cdots, u_N)} B \cap \cdots \cap S^{q_w(u_1, \cdots, u_N)} B\big)$. Form an r-cell partition of Q, $Q = \bigcup_{i=1}^{r} C_i$, by the rule $q_a(y_1, \cdots, y_N) \in C_i$ if and only if $s\big(q_a(u_1, \cdots, u_N), y\big) = i$, $1 \leq a \leq w$. In particular, if $q_a \in C_i$ then $\|T^{q_a(u_1, \cdots, u_N)} f - g_i\|_y < \epsilon$. For this partition, there exists some i, $1 \leq i \leq r$, some $q \in Q$, and pairwise disjoint subsets $S_1, \cdots, S_t \subset \{1, \cdots, N\}$ such that, under the substitution $x_m = \sum_{n \in S_m} y_n$, $1 \leq m \leq t$, we have

$$\{q(y_1, \cdots, y_N), q(y_1, \cdots, y_N) - p_1(x_1, \cdots, x_t), \cdots,$$
$$q(y_1, \cdots, y_N) - p_k(x_1, \cdots, x_t)\} \subset C_i.$$

In particular, making the analogous substitutions $n_m = \sum_{n \in S_m} u_n$, $1 \leq m \leq t$, we have $FS(n_1, \cdots, n_t) \subset E$, and furthermore, we have, setting $p_0(x_1, \cdots, x_t) = 0$,

$$\left\|T^{q(u_1, \cdots, u_N) - p_b(n_1, \cdots, n_t)} f - g_i\right\|_y < \epsilon; \quad 0 \leq b \leq k.$$

Setting $\tilde{y} = S^{-q(u_1, \cdots, u_N)} y$, we have

$$\left\|T^{-p_b(n_1, \cdots, n_t)} f - T^{-q(u_1, \cdots, u_N)} g_i\right\|_{\tilde{y}} < \epsilon; \quad 0 \leq b \leq k.$$

In particular, since this holds for $b = 0$, we have by the triangle inequality

$$\left\|T^{-p_b(n_1, \cdots, n_t)} f - f\right\|_{\tilde{y}} < 2\epsilon, \quad 1 \leq b \leq k.$$

It follows that

$$\mu_{\tilde{y}}\big(A \setminus T^{p_b(n_1, \cdots, n_t)} A\big) \leq \left\|T^{-p_b(n_1, \cdots, n_t)} f - f\right\|_{\tilde{y}}^2 \leq 4\epsilon^2; \quad 1 \leq b \leq k.$$

Moreover, $\tilde{y} \in B$, so that $\mu_{\tilde{y}}(A) \geq c$, therefore, since $\epsilon = \sqrt{\frac{c}{8k}}$,

$$\mu_{\tilde{y}}\big(A \cap T^{p_1(n_1, \cdots, n_t)} A \cap \cdots \cap T^{p_k(n_1, \cdots, n_t)} A\big) \geq c - 4k\epsilon^2 = \frac{c}{2}.$$

The sets $S_1, \cdots, S_t$ depend measurably on y, and therefore the numbers $n_1, \cdots, n_t$ are measurable functions of y defined on the set

$$\left(B \cap S^{q_1(u_1, \cdots, u_N)} B \cap \cdots \cap S^{q_w(u_1, \cdots, u_N)} B \right),$$

which, recall, is of measure greater than η. Hence, as there are only D choices possible for $S_1, \cdots, S_t$, we may assume that for all $y \in H$, where $H \in \mathcal{B}$ satisfies $\nu(H) > \frac{\eta}{D}$, $n_1, \cdots, n_t$ are constant. For this choice of $n_1, \cdots, n_t$ we have

$$\mu\left(A \cap T^{p_1(n_1, \cdots, n_t)} A \cap \cdots \cap T^{p_t(n_1, \cdots, n_t)} A \right) \geq \frac{c}{2}\nu(H) > \frac{c\eta}{2D} = \delta.$$

$\square$

Looking again at Theorem 1.4, we see that, having proved that the PSZ property is preserved under compact extensions, we have left only to prove that, given a totally ordered chain of T-invariant sub-σ-algebras with the PSZ property, the completion of the σ-algebra generated by the chain again possesses the PSZ property.

Proposition 3.6 Suppose that $(X, \mathcal{A}, \mu, T)$ is a measure preserving system and that $\mathcal{A}_\xi$ is a totally ordered chain of sub-σ-algebras of $\mathcal{A}$ having the PSZ property. If $\bigcup_\xi \mathcal{A}_\xi$ is dense in $\mathcal{A}$, that is, if $\mathcal{A}$ is the completion of the σ-algebra generated by $\bigcup_\xi \mathcal{A}_\xi$, then $\mathcal{A}$ has the PSZ property.

Proof. Suppose $A \in \mathcal{A}$, $\mu(A) > 0$, $t, k \in \mathbf{N}$, and that

$$p_1(x_1, \cdots, x_t), \cdots, p_k(x_1, \cdots, x_t) \in \mathbf{Z}[x_1, \cdots, x_t]$$

are polynomials with zero constant term. There exists ξ and $B \in \mathcal{A}_\xi$ such that

$$\mu\left((A \setminus B) \cup (B \setminus A) \right) \leq \frac{\mu(A)}{4(k+1)}.$$

Let $\int d\mu = \int_Y \int_X d\mu_y d\nu(y)$ be the decomposition of the measure μ over the factor $\mathcal{A}_\xi$. Let $C = \{y \in B : \mu_y(A) \geq 1 - \frac{1}{2(k+1)}\}$. It is easy to see that $\nu(C) > 0$. Since $\mathcal{A}_\xi$ has the PSZ property, there exists some $\alpha > 0$ having the property that in any thick set E we may find $n_1, \cdots, n_t \in \mathbf{Z}$ such that $FS(n_1, \cdots, n_t) \subset E$ and

$$\nu\left(C \cap S^{p_1(n_1, \cdots, n_t)} C \cap \cdots \cap S^{p_k(n_1, \cdots, n_t)} C \right) > \alpha. \qquad (3)$$

Set $\delta = \frac{\alpha}{2}$ and let E be any thick set. Find $n_1, \cdots, n_t \in \mathbf{Z}$ satisfying (3) and with $FS(n_1, \cdots, n_t) \subset E$. For any $y \in \left(C \cap S^{p_1(n_1, \cdots, n_t)} C \cap \cdots \cap \right.$

$S^{p_k(n_1,\cdots,n_t)}C)$ we have $\mu_y(A), \mu_y(T^{p_1(n_1,\cdots,n_t)}A), \cdots, \mu_y(T^{p_k(n_1,\cdots,n_t)}A)$ all not less than $1 - \frac{1}{2(k+1)}$, from which it follows that

$$\mu_y\big(A \cap T^{p_1(n_1,\cdots,n_t)}A\big) \cap \cdots \cap T^{p_k(n_1,\cdots,n_t)}A\big) \geq \frac{1}{2}.$$

Therefore,

$$\mu\big(A \cap T^{p_1(n_1,\cdots,n_t)}A\big) \cap \cdots \cap T^{p_k(n_1,\cdots,n_t)}A\big) > \frac{\alpha}{2} = \delta.$$

$\square$

4. Appendix: Proof of Theorem 3.2.

We will now derive Corollary 3.2 from Theorem 3.1. First, we derive from Theorem 3.1 its natural "multi-parameter" version.

Proposition 4.1 Suppose $k, d, r, t \in \mathbf{N}$ are given. Then there exists a number $N = N(k, d, r) \in \mathbf{N}$ having the property that for any r-cell partition

$$F\big(\{1, 2, \cdots, k\} \times \{1, 2, \cdots, tN\}^d\big) = \bigcup_{i=1}^{r} C_i,$$

one of the cells C_i, $1 \leq i \leq r$ contains a configuration of the form

$$\Big\{A \cup \big(B \times (S_1 \cup \cdots \cup S_t)^d\big) : B \subset \{1, \cdots, k\}\Big\},$$

where $A \subset \big(\{1, \cdots, k\} \times \{1, \cdots, tN\}^d\big)$ and $\emptyset \neq S_j \subset \{(j-1)N+1, \cdots, jN\}$, $1 \leq j \leq t$ satisfy

$$A \cap \Big(\{1, \cdots, k\} \times (S_1 \cup \cdots \cup S_t)^d\Big) = \emptyset.$$

Proof. Let $N = N(k, d, r)$ be as in Theorem 3.1. Given any partition

$$F\big(\{1, 2, \cdots, k\} \times \{1, 2, \cdots, tN\}^d\big) = \bigcup_{i=1}^{r} C_i,$$

we will construct a partition

$$F\big(\{1, 2, \cdots, k\} \times \{1, 2, \cdots, N\}^d\big) = \bigcup_{i=1}^{r} D_i.$$

in the following way: given any set $U \subset \left(\{1, 2, \cdots, k\} \times \{1, 2, \cdots, N\}^d\right)$, let

$$U_j = \left\{v \in \{1, \cdots, N\}^d : (j, v) \in U\right\}, 1 \le j \le k.$$

Also, for any set $E \subset \{1, \cdots, N\}^d$, let $\gamma(E) \subset \{1, \cdots, tN\}^d$ be the set which is obtained by taking the union of t^d shifts of E,

$$\gamma(E) = \bigcup_{0 \le i_1, \cdots, i_d < t} \left(E + (i_1 N, \cdots, i_d N)\right).$$

(Notice, in particular, that $\gamma(\{1, \cdots, N\}^d) = \{1, \cdots, tN\}^d$.) Now let

$$\gamma(U) = \bigcup_{j=1}^{k} \{j\} \times \gamma(U_j),$$

and put $U \in D_i$ if and only if $\gamma(U) \in C_i$. According to the property whereby N was chosen, one of the sets D_i, $1 \le i \le r$, say D_j, contains a configuration of the form

$$\left\{H \cup (B \times S^d) : B \subset \{1, \cdots, k\}\right\},$$

for some $H \subset \left(\{1, \cdots, k\} \times \{1, \cdots, N\}^d\right)$ and some non-empty $S \subset \{1, \cdots, N\}$ satisfying

$$H \cap \left(\{1, \cdots, k\} \times S^d\right) = \emptyset.$$

Let $S_i = S + (i - 1)N$, $1 \le i \le t$, and put $A = \gamma(H)$. Then $\emptyset \ne S_i \subset \{(i - 1)N + 1, \cdots, iN\}$, $1 \le i \le t$, and one may check that for every $B \subset \{1, \cdots, k\}$,

$$\gamma\big(H \cup (B \times S^d)\big) = A \cup \big(B \times (S_1 \cup \cdots \cup S_t)^d\big).$$

It follows that

$$\left\{A \cup \big(B \times (S_1 \cup \cdots \cup S_t)^d\big) : B \subset \{1, \cdots, k\}\right\} \subset C_j.$$

Furthermore, $A \cap \left(\{1, \cdots, k\} \times (S_1 \cup \cdots \cup S_t)^d\right) = \emptyset$, so we are done.

$\square$

We are now in position to prove Corollary 3.2.

Proof of Corollary 3.2. Let d be the maximum degree of the polynomials p_j, $1 \le j \le k$, and let $N = N(k, d, r)$ be as in Proposition 4.1 above. We claim there exists a map

$$\varphi : F\big(\{1, \cdots, k\} \times \{1, \cdots, tN\}^d\big) \to \mathbf{Z}[y_1, \cdots, y_{tN}]$$

satisfying $\varphi(A \cup B) = \varphi(A) + \varphi(B)$ whenever $A \cap B = \emptyset$, and such that

$$\varphi(\{j\} \times \{m_1^{(1)}, \cdots, m_{s_1}^{(1)}, \cdots, m_1^{(t)}, \cdots, m_{s_t}^{(t)}\}^d)$$
$$= p_j\big(y_{m_1^{(1)}} + \cdots + y_{m_{s_1}^{(1)}}, \cdots, y_{m_1^{(t)}} + \cdots + y_{m_{s_t}^{(t)}}\big)$$

whenever $1 \le j \le k$ and $\{m_1^{(l)}, \cdots, m_{s_l}^{(l)}\} \subset \{(l-1)N+1, \cdots, lN\}$, $1 \le l \le t$. We now proceed to define the function φ on singletons.

Fix j, $1 \le j \le k$. There exist polynomials

$$\alpha_{i_1, \cdots, i_t}(u_1, u_2, \cdots, u_{i_1 + \cdots + i_t}) \in \mathbf{Z}[u_1, \cdots, u_{i_1 + \cdots + i_t}], \quad 1 \le i_1 + \cdots + i_t \le d,$$

such that whenever $\{m_1^{(l)}, \cdots, m_{s_l}^{(l)}\} \subset \{(l-1)N+1, \cdots, lN\}$, $1 \le l \le t$, we have

$$p_j\big(u_{m_1^{(1)}} + \cdots + u_{m_{s_1}^{(1)}}, \cdots, u_{m_1^{(t)}} + \cdots + u_{m_{s_t}^{(t)}}\big)$$
$$= \sum_{1 \le i_1 + \cdots + i_t \le d} \;\; \sum_{\{l_1^{(l)}, \cdots, l_{i_l}^{(l)}\} \subset \{m_1^{(l)}, \cdots, m_{s_l}^{(l)}\}} \alpha_{i_1, \cdots, i_t}(u_{l_1^{(1)}}, \cdots$$
$$\cdots, u_{l_{i_1}^{(1)}}, \cdots, u_{l_1^{(t)}}, \cdots, u_{l_{i_t}^{(t)}}).$$

For example, say that $p_j(x_1, x_2) = x_1^2 x_2 + x_1^2 + x_2$. Then $\alpha_{1,0}(u_1) = u_1^2$, $\alpha_{0,1}(u_1) = u_1$, $\alpha_{2,0}(u_1, u_2) = 2u_1 u_2$, $\alpha_{1,1}(u_1, u_2) = u_1^2 u_2$, and $\alpha_{2,1}(u_1, u_2, u_3) = 2u_1 u_2 u_3$. Now for each t-tuple of subsets

$$\{l_1^{(l)}, \cdots, l_{i_l}^{(l)}\} \subset \{(l-1)N+1, \cdots, lN\}, \quad 1 \le l \le t, \quad 1 \le i_1 + \cdots + i_t \le d,$$

we pick exactly one representative point $(j, a_1, \cdots, a_d) \in (\{j\} \times \{1, \cdots, tN\}^d)$ which satisfies

$$\{a_1, \cdots, a_d\} = \bigcup_{l=1}^{t} \{l_1^{(l)}, \cdots, l_{i_l}^{(l)}\},$$

and define

$$\varphi\big(\{(j, a_1, \cdots, a_d)\}\big) = \alpha_{i_1, \cdots, i_t}(y_{l_1^{(1)}}, \cdots, y_{l_{i_1}^{(1)}}, \cdots, y_{l_1^{(t)}}, \cdots, y_{l_{i_t}^{(t)}})$$
$$\in \mathbf{Z}[y_1, \cdots, y_{tN}].$$

φ is defined to be zero on all singletons in $(\{j\} \times \{1, \cdots, tN\}^d)$ not so chosen.

Repeat these steps for $1 \le j \le k$. Now extend φ to all of $F(\{1, \cdots, k\} \times \{1, \cdots, tN\}^d)$ according to the additivity condition we require of φ.

Now, whenever $1 \leq j \leq k$, and $\{m_1^{(l)}, \cdots, m_{s_l}^{(l)}\} \subset \{(l-1)N+1, \cdots, lN\}$, $1 \leq l \leq t$, we have

$$p_j\big(y_{m_1^{(1)}} + \cdots + y_{m_{s_1}^{(1)}}, \cdots, y_{m_1^{(t)}} + \cdots + y_{m_{s_t}^{(t)}}\big)$$

$$= \sum_{1 \leq i_1 + \cdots + i_t \leq d} \sum_{\{l_1^{(l)}, \cdots, l_{i_l}^{(l)}\} \subset \{m_1^{(l)}, \cdots, m_{s_l}^{(l)}\}} \alpha_{i_1, \cdots, i_t}\big(y_{l_1^{(1)}}, \cdots$$

$$\cdots, y_{l_{i_1}^{(1)}}, \cdots, y_{l_1^{(t)}}, \cdots, y_{l_{i_t}^{(t)}}\big)$$

$$= \sum_{(a_1, \cdots, a_d) \in \{m_1^{(1)}, \cdots, m_{s_1}^{(1)}, \cdots, m_1^{(t)}, \cdots, m_{s_t}^{(t)}\}^d} \varphi\big(\{(j, a_1, \cdots, a_d)\}\big)$$

$$= \varphi\big(\{j\} \times \{m_1^{(1)}, \cdots, m_{s_1}^{(1)}, \cdots, m_1^{(t)}, \cdots, m_{s_t}^{(t)}\}^d\big).$$

Let $w = k(tN)^d$ and let $\big(q_i(y_1, \cdots, y_{tN})\big)_{i=1}^w$ be the images of the singletons under φ, so that

$$\mathcal{F} = \varphi\Big(F\big(\{1, \cdots, k\} \times \{1, \cdots, tN\}^d\big)\Big) = FS\big(q_i(y_1, \cdots, y_{tN})\big)_{i=1}^w.$$

If now $\mathcal{F} = \bigcup_{i=1}^r C_i$, then

$$F\big(\{1, \cdots, k\} \times \{1, \cdots, tN\}^d\big) = \bigcup_{i=1}^r \varphi^{-1}(C_i),$$

and by Proposition 4.1, there exists i, $1 \leq i \leq r$, $A \in \varphi^{-1}(C_i)$, and sets

$$S_l \subset \{(l-1)N+1, \cdots, lN\}, \quad 1 \leq l \leq t,$$

having the property that for all j, $1 \leq j \leq k$,

$$A \cap \Big(\{j\} \times \big(S_1 \cup \cdots \cup S_t\big)^d\Big) = \emptyset, \quad A \cup \Big(\{j\} \times \big(S_1 \cup \cdots \cup S_t\big)^d\Big) \in \varphi^{-1}(C_i).$$

Let $q(y_1, \cdots, y_{tN}) = \varphi(A)$. Put $p_0(x_1, \cdots, x_t) = 0$. Then for all j, $0 \leq j \leq k$, we have, substituting $x_l = \sum_{n \in S_l} y_n$, $1 \leq l \leq t$,

$$q(y_1, \cdots, y_{tN}) + p_j(x_1, \cdots, x_t) = \varphi\Big(A \cup \big(\{j\} \times (S_1 \cup \cdots \cup S_t)^d\big)\Big) \in C_i.$$

$\square$

REFERENCES

[B1] V. Bergelson, Weakly mixing PET, *Ergodic Theory and Dynamical Systems* **7** (1987), 337-349.

[B2] V. Bergelson, Ergodic Ramsey theory–an update.

[BL1] V. Bergelson and A. Leibman, Polynomial extensions of van der Waerden's and Szemerédi's theorem, *Journal of AMS* (to appear).

[BL2] V. Bergelson and A. Leibman, Polynomial Hales-Jewett theorem, preprint.

[F1] H. Furstenberg, Ergodic behavior of diagonal measures and a theorem of Szemerédi on arithmetic progressions, *J. d'Analyse Math.* **31** (1977), 204-256.

[F2] H. Furstenberg, *Recurrence in Ergodic Theory and Combinatorial Number Theory*, Princeton University Press, 1981.

[FKO] H. Furstenberg, Y. Katznelson and D. Ornstein, The ergodic theoretical proof of Szemerédi's theorem, *Bull. Amer. Math. Soc.* **7** (1982), 527-552.

Some 2-d Symbolic Dynamical Systems: Entropy and Mixing

Robert Burton*   Jeffrey E. Steif[†]

Abstract

We give a survey of a number of recent results concerning measures of maximal entropy for certain symbolic dynamical systems and their respective mixing properties.

1 Introduction

In this introduction, we discuss four models (or subshifts) that have been recently investigated and discuss their measures of maximal entropy and the corresponding mixing properties. We will stick for concreteness to two dimensions but mention when results also hold in higher dimensions (which will typically be the case). These examples demonstate that when one moves from one to two dimensions, many new phenomena can occur.

Before discussing these examples, we first describe (very briefly) the general set up (the reader should see [4] for all missing details). We will consider symbolic dynamical systems given by a closed $\mathbf{Z}^2$ invariant subset X of $F^{\mathbf{Z}^2}$ (F is a finite set) where $\mathbf{Z}^2$ acts in the obvious way. As is usual, these are called **subshifts**.

Definition 1.1: *The* **topological entropy** *of X is*

$$H(X) = \lim_{n \to \infty} \frac{\log N_n}{(2n+1)^2}$$

*Department of Mathematics, Kidder Hall, Oregon State University, Corvallis, OR 97331, burton@math.orst.edu

†Department of Mathematics, Chalmers University of Technology, S-412 96 Gothenburg, Sweden, steif@math.chalmers.se

The research of the first author was supported in part by AFOSR grant # 91-0215 and NSF grant # DMS-9103738. The research of the second author was supported by a grant from the Swedish Natural Science Research Council.

where N_n is the number of elements of $F^{[-n,n]^2}$ which extend to an element of X. If μ is a translation invariant probability measure on X, then the **measure theoretic entropy** of μ is

$$H(\mu) = \lim_{n \to \infty} \frac{H_n(\mu)}{(2n+1)^2}$$

where $H_n(\mu)$ is the entropy of μ restricted to $F^{[-n,n]^2}$.

Both of these limits exist by subadditivity and clearly $H(\mu) \leq H(X)$ for any such μ. In fact we have the following variational principle. See [14] for an elementary proof.

Theorem 1.2: *Let X be a subshift. Let $\mathcal{M}_X$ be the set of translation invariant measures on X. Then $H(X) = \sup_{\mu \in \mathcal{M}_X} H(\mu)$ and moreover the supremum is achieved at some measure.*

We mention here that measures of maximal entropy for symbolic dynamical systems which are given by local rules (so called **subshifts of finite type**) possess a very important property which allows one to analyze them. This property is called having **uniform conditional probabilities** and means that the conditional distribution on a finite box given the outside is uniform distribution among those configurations which extend to the boundary condition. See [4] and [5].

There are a number of mixing properties that one is interested in. ¿From a purely ergodic theoretic point of view, it is interesting to know if a given translation invariant measure is **Bernoulli**. This means that it is isomorphic to a process of independent and identically distributed (i.i.d.) random variables (i.e., there is a measure preserving invertible coding to an i.i.d. process which commutes with spatial translations). In order to show that certain concrete processes are Bernoulli, a number of properties have been defined which were proven to imply the Bernoulli property. For example, the property of **Very Weak Bernoulli** characterizes Bernoulli. Another important property which implies (and is strictly stronger than) Bernoulli is **Weak Bernoulli**. There is some problem in extending this concept to more than one dimension but it is possible to do so. See [7] for a description of this problem and for a way of extending this definition to higher dimensions in a reasonable way. One of the key ideas in obtaining this extension is the notion of a "coupling surface". We will not give here the definition of Weak Bernoulli in higher dimensions but rather refer to [7].

In this paper, we will consider four concrete examples of such systems which have been investigated to some extent and discuss their measures of maximal entropy and their corresponding mixing properties.

Example 1 (The Iceberg Model): Let M be a positive integer and $F = \{-M, -M+1, \ldots, -2, -1, 0, 1, 2, \ldots, M-1, M\}$. We take X to be the subset of $F^{\mathbf{Z}^2}$ where no positive is allowed to sit next to a negative.

Example 2 (The Generalized Hard Core Model): Let m be a positive integer and $F = \{0, 1, \ldots, m\}$. Let X be the subset of $F^{\mathbf{Z}^2}$ where no two positive integers are allowed to sit next to each other. (When $m = 1$, we recover the usual Hard Core Model.)

Example 3 (Dominoes): This symbolic dynamical system is the set of domino tilings of the plane, i.e. the set of tilings by rectangles of side lengths 1 and 2. This is equivalent to the set of perfect matchings of the lattice $\mathbf{Z}^2$. We consider the lattice graph in which the vertex set is $\mathbf{Z}^2$ and two vertices are joined by an edge if they are one unit apart. We define Ω_D to be the set of subgraphs of the lattice with the property that every vertex is connected to exactly one other vertex. Notice that Ω_D may be recoded as a subshift as follows.

Let $F = \{1, 2, 3, 4\}$ and let

$$G_1 = \{(3, 1), (2, 3), (4, 1), (2, 4), (1, 1), (2, 2)\}$$

and

$$G_2 = \{(3, 1), (3, 2), (1, 4), (2, 4), (3, 3), (4, 4)\}.$$

Let $X = \{\eta \in F^{\mathbf{Z}^2} : x - y = (1, 0)$ implies $(\eta(x), \eta(y)) \notin G_1$ and $x - y = (0, 1)$ implies $(\eta(x), \eta(y)) \notin G_2\}$.

Example 4 (Essential Spanning Forests): This symbolic dynamical system is not given by a local rule as all of our above systems but has the advantage of being susceptible to harmonic analysis. Ω_F is the set of subgraphs of $\mathbf{Z}^2$ that are essential spanning forests. A **forest** is a subgraph without circuits, which means that either two vertices have no path in the subgraph connecting them or else there is a unique non-self-intersecting such path. A **spanning** forest is a forest such that every vertex touches at least one edge. For example, the set Ω_D is a collection of spanning forests. An **essential** spanning forest is a spanning forest in which each component is infinite. As in the domino process, Ω_F may be recoded as a symbolic dynamical system; in particular it is closed.

In §2 and §3, we give some results concerning the measures of maximal entropy and their mixing properties for the Iceberg Model and Generalized Hard Core Model respectively. In §4, we give results concerning measures of maximal entropy for the Essential Spanning Forest process and their mixing properties. We then describe a correspondence between Dominoes and Spanning Trees that allows us to transfer the results for the Essential Spanning Forests to corresponding properties for Dominoes.

2   The Iceberg Model: Measures of Maximal Entropy and Mixing

The following result can be found in [5].

Theorem 2.1: *In two dimensions, the Iceberg Model with $M = 1$ has a unique measure of maximal entropy while for M sufficiently large, there is more than one measure of maximal entropy.*

It is also the case (see [5]) that in any dimension, there is more than one measure of maximal entropy if M is sufficiently large. As far as the proof is concerned, the larger d is, the larger M must be for this to occur. However, this is certainly a consequence of only the proof and in some sense it should be easier in higher dimensions for the existence of more than one measure of maximal entropy. An interesting question is how many measures of maximal entropy there are when $M = 1$ in higher dimensions. Actually, as indicated in [5], the proof of Theorem 2.1 is not given there but rather it is explained that the proofs for an almost identical model work in this case as well. As far as this almost identical model, a result is obtained in [11] whose analogue for the Iceberg Model is the statement that if there is more than one measure of maximal entropy for M, then there is more than one measure of maximal entropy for $M + 1$. One would think that the proof in [11] could be extended to this analogous statement for the Iceberg Model.

Using the method of proof in [16], one can prove the following. Whenever we refer to [16] in this paper, if one wants to see published work, one can refer to [1] which extends the work in [16] to an amenable group setting.

Theorem 2.2: *If the Iceberg Model has a unique measure of maximal entropy, then it is Bernoulli.*

We did not specify two dimensions here since this result holds for all dimensions. For large M, as Theorem 2.1 tells us, there is more than one measure of maximal entropy. One of these is obtained by placing boundary conditions M outside $[-n, n]^2$, taking uniform distribution on the configurations on $[-n, n]^2$ consistent with this boundary condition and taking a limit as $n \to \infty$. Similarly, we can do the same thing with boundary conditions $-M$. It is easy to see using monotonicity (see [4] and [5]) that these two limits exist and that they are the same if and only if there is a unique measure of maximal entropy. We call these two (possibly equal) measures the plus and minus state respectively. Using the methods in [16], one can also prove the following which is again valid in all dimensions.

Theorem 2.3: *In the Iceberg Model, the plus and minus states are Bernoulli.*

Clearly Theorem 2.3 contains Theorem 2.2. In fact, one can prove, using the methods in [4], [6] and [7], the following stronger result which is valid in all dimensions but only for sufficiently large M (at least as far as the proof is concerned). Note that Theorem 2.3 is valid for all M.

Theorem 2.4: *In the Iceberg Model, if M is sufficiently large, then the plus and minus states are the only ergodic measures of maximal entropy, and they are Weak Bernoulli.*

3   The Generalized Hard Core Model: Measures of Maximal Entropy and Mixing

We first discuss the usual Hard Core Model (where $m = 1$).

Theorem 3.1: *The usual Hard Core Model (where $m = 1$) has a unique measure of maximal entropy in two dimensions.*

This was proved in [9] with a computer assisted proof. A more modern proof (not computer assisted) using ideas from percolation is given in [2]. Both of these proofs use in a very strong way the fact that the dimension of the lattice is 2. For $m = 1$, the methods given in [16] (together with monotonicity properties described in [2]) allow one to the show that the measure of maximal entropy given in Theorem 3.1 is Bernoulli. It would be interesting to know in which dimensions the usual Hard Core Model has a unique measure of maximal entropy.

As far as the generalized Hard Core Model (and sticking to two dimensions), an interesting phenomenon occurs (which we noticed after some comments that J. van den Berg made to us). When one takes the Generalized Hard Core Model with m large, one obtains an ergodic measure of maximal entropy which is not weakly mixing. We briefly describe the reason that this eigenvalue appears for large m.

We can take $[-n, n]^2$, place boundary conditions outside this box by placing 1's on the locations outside this box which are contained in the sublattice $\{(n, m) \in \mathbf{Z}^2 : n + m$ is even and 0's on the locations outside this box which are contained in the sublattice $\{(n, m) \in \mathbf{Z}^2 : n + m$ is odd and consider uniform distribution among the configurations on $[-n, n]^2$ which respect this boundary condition. We can then let $n \to \infty$ and obtain a limiting measure which is different than that which we would obtain by reversing 0's and 1's in the boundary condition. The fact that these limiting measures exist for all m follows from the methods in [2]. The fact that the limiting measures that one obtains are different for large m can be proved along the same lines as the proof in [8] (see also [13]) of a phase transition in a hard core model with an external field. If T and S denote translations in the horizontal and vertical directions respectively and if these limiting measures μ_o and μ_e are different, then one can show (using the methods in [2]) that μ_o and μ_e are not invariant under $\{T^i S^j\}_{i,j\in\mathbf{Z}}$ but are invariant under the index 2 subgroup $\{T^i S^j\}_{i,j\in\mathbf{Z}, i+j\equiv 0 \pmod 2}$ and moreover that $\mu_o = T(\mu_e) = S(\mu_e)$. Next, if one takes a convex combination of these two measures (giving each weight 1/2), one obtains a measure of maximal entropy which is ergodic but not weakly mixing. The moral of this story is that the existence of certain nontranslation invariant measures can yield information about the mixing properties of a given translation invariant measure.

We would expect that the methods in [4], [5], [6], and [7] could be extended to show that for large m, there is a unique measure of maximal entropy and that two measures μ_o and μ_e described above are Weak Bernoulli with respect to $\{T^i S^j\}_{i,j\in\mathbf{Z}, i+j\equiv 0 \pmod 2}$.

4   Spanning Trees and Dominoes: Measures of Maximal Entropy and Mixing

We first describe the measures of maximal entropy for the essential spanning tree system. A component of an essential spanning forest is **one-ended** if each vertex has a unique non-self-intersecting infinite path within the subgraph.

The following results may be found in [15] and [3].

Theorem 4.1: *There is a unique measure of maximal entropy on the essential spanning tree system. This measure is concentrated on one-ended trees.*

The measure of maximal entropy is also Very Weak Bernoulli using an argument of Pemantle related to that [16]. This relies on an inequality in [10] which we describe here.

Suppose that we have a finite graph G and we put the probability measure P on subgraphs of G that gives uniform probability to each spanning tree of G. We say that an event A is an **increasing event** if for each subgraph $\omega \in A$ and each ω' that includes ω as a subgraph we have $\omega' \in A$. The next result is contained in [10].

Theorem 4.2: *If E is the event that the edge e is included in the random subgraph and if A is an increasing event defined on coordinates that do not depend on e then E and A are negatively correlated, i.e. $P(E\cap A) \leq P(E)P(A)$.*

This is intuitively plausible because spanning trees are maximal subgraphs without loops and conditioning on the presence of an edge at one location should make it harder to have edges at other locations.

To show that a measure μ is Very Weak Bernoulli it suffices to prove the following condition: for each ϵ there is an N_0 so that for all $N \geq N_0$ if we take a square of side length N, then for each pair of configurations outside the square, there is a coupling ν of the conditional measures on the subgraphs of the square, μ_1, μ_2 so that ν−almost everywhere the coordinates have at most ϵN^2 disagreements.

In the case of the uniform spanning tree measure the conditional distribution inside the square only depends upon which boundary vertices are joined outside the square. These boundary vertices divide into equivalence classes of outside connected points. If there are k classes, $0 < k < 4N$ then the conditional measure inside the square is the measure that gives uniform probability to all spanning trees that connect all points inside to each of these k classes. Applying Theorem 4.2 a finite number of times one can show that the conditional measure when there is one class on the boundary is stochastically dominated by any conditional measure that arises when there are k classes. A domination result in [12] shows that there is a coupling of these two conditional measures so that the first subgraph is a subset of the second. Since any spanning tree with k classes has $k - 1$ more edges than a spanning tree with 1 class the number of disagreements is less than $4N$ which is less than ϵN^2 for large N.

Thus the uniform spanning tree measure is Very Weak Bernoulli. We wish to transfer this result to the domino tiling system. This uses a trick of Temperley [17] and is carried out in [3] but we present it here anyway.

Recall that Ω_D is the set of graphs which represent domino tilings and Ω_F is the set of all subgraphs of $\mathbf{Z}^2$ that are essential spanning forests. We describe a way to assign an essential spanning forest to each domino tiling, giving us a map $\Phi : \Omega_D \to \Omega_F$. Given $\omega \in \Omega_D$ we let $\Phi(\omega)$ be the subgraph of $\mathbf{Z}^2$ with an edge between neighbors (n, m) and (n', m') if one of the two edges connecting $(2n, 2m)$ and $(2n', 2m')$ is an edge in the subgraph ω.

Let T and S denote translations in the horizontal and vertical directions respectively. Then by definition Φ satisfies $\Phi(T^2(\omega)) = T(\Phi(\omega))$ and $\Phi(S^2(\omega)) = S(\Phi(\omega))$. The construction may be described as follows: Begin with any point $(2n, 2m)$ on the even sublattice $\{(2n, 2m)\}_{(n,m)\in\mathbf{Z}^2}$. There is a domino that emanates from this point. Paint a line along this domino continuing the line until reaching the next point on this even sublattice. At this point we are on a new domino and we repeat the process. This process is unending. To see that an essential spanning forest results from all the painted lines, we need only be sure that no loops are created. A little thought and diagramming shows that any such loop will contain an odd number of vertices in the lattice. This is an impossible situation because any finite set of dominoes covers an even number of vertices. This gives us an essential spanning forest which lives on the even sublattice from which we obtain an essential spanning forest on $\mathbf{Z}^2$ by simply scaling it down.

The map Φ is not invertible but it does take a shift invariant measure to a shift invariant measure whose entropy is 4 times larger.

Let μ be the measure of maximal entropy for the essential spanning tree process. Recall that μ is supported on the set of one-ended trees. We show that the map Φ is invertible on this set. This implies that $\Phi^{-1}(\mu)$ is the unique measure of maximal entropy on Ω_D. Otherwise there would be another measure ν on Ω_D whose entropy was at least as large as one fourth the entropy of μ. Then $\Phi(\nu)$ would have entropy at least as large as the entropy of μ giving a contradiction.

To see the invertibility of Φ on the support of μ take any one-ended spanning tree on $\mathbf{Z}^2$. Transfer this tree to the even sublattice of $\mathbf{Z}^2$ by putting two edges joining neighbors $(2n, 2m)$ and $(2n', 2m')$ if there was an edge joining (n, m) and (n', m') in the spanning tree. Now for any neighbors (n, m) and (n', m'), place two edges joining $(2n + 1, 2m + 1)$ and $(2n' + 1, 2m' + 1)$ if and only if such edges would not traverse an edge pair of the tree already placed down. This gives a one-ended essential spanning tree on the odd sublattice $\{(2n + 1, 2m + 1)\}_{(n,m)\in\mathbf{Z}^2}$ which together with the first tree creates an essential spanning forest in $\mathbf{Z}^2$ consisting of two one-ended components which are dual to each other. A *leaf* will be any pair of edges that are connected to only one other pair of edges. Place dominoes at the ends of each leaf and erase the remaining edge of the leaf. This will create new leaves which we turn into dominoes as before. After applying this operation countably many times we have a domino tiling ω so that $\Phi(\omega)$ is the original spanning tree.

References

[1] Adams, S. (1992), Fölner Independence and the amenable Ising model, *Ergodic Theory and Dynamical Systems*, **12**, 633-657.

[2] Berg, J. D. van den and Steif, J. E., (1994). On Hard-Core Lattice Gas Models, Percolation and Certain Loss Networks, *Stochastic Processes and Their Applications*, **49**, No. 2, 179–197.

[3] Burton R. M. and Pemantle, R., (1993). Local Characteristics, Entropy and Limit Theorems for Spanning Trees and Domino Tilings via Transfer–Impedances, *Annals of Probability*, **21**, 1329–1371.

[4] Burton, R. M. and Steif, J. E., (1994). Nonuniqueness of Measures of Maximal Entropy for Subshifts of Finite Type, *Ergodic Theory and Dynamical Systems*, **14**, 213–235.

[5] Burton, R. M. and Steif, J. E., New Results on Measures of Maximal Entropy, *Israel Journal of Mathematics*, **89**, (1995), 275–300.

[6] Burton, R. M. and Steif, J. E., Quite Weak Bernoulli with Exponential Rate and Percolation for Random Fields, *Stochastic Processes and Their Applications*, to appear.

[7] Burton, R. M. and Steif, J. E., Coupling Surfaces and Weak Bernoulli in One and Higher Dimensions, submitted.

[8] Dobrushin, R. L., (1968). The problem of uniqueness of a Gibbs random field and the problem of phase transition, *Functional Analysis and Applications* **2**, 302–312.

[9] Dobrushin, R. L., Kolafa, J. and Shlosman, S. B., (1985). Phase diagram of the two–dimensional Ising antiferromagnet (computer–assisted proof), *Communications in Mathematical Physics*, **102**, 89–103.

[10] Feder, T. and Mihail M., (1992). Balanced Matroids, *STOC* 26–38.

[11] Häggström, O., On phase transitions for Subshifts of finite type, *Israel Journal of Mathematics*, to appear.

[12] Kamae, T., Krengel, U. and O'Brien, G. L., (1977). Stochastic inequalities on partially ordered spaces, *Annals of Probability*, **5**, 899-912.

[13] Louth, G. M., (1990). *Stochastic networks: complexity, dependence and routing*, Ph.D. thesis, Cambridge University.

[14] Misiurewicz, M., (1975). A short proof of the variational principle for a $\mathbf{Z}_+^N$–action on a compact space, *Asterisque* **40**, 147–157.

[15] Pemantle, R., (1991). Choosing a spanning tree for the integer lattice uniformly, *Annals of Probability*, **19**, 1559–1574.

[16] Ornstein, D. S. and Weiss, B., (1973). $\mathbf{Z}^d$-actions and the Ising model, unpublished manuscript.

[17] Temperley, H., (1974). In *Combinatorics: Proceedings of the British Combinatorial Conference, 1973* 202-204. Cambridge University Press.

A note on certain rigid subshifts

Kari Eloranta*
Institute of Mathematics
Helsinki University of Technology
02150 Espoo, Finland

eloranta@janus.hut.fi

Abstract

We first study the subshifts of finite type on $\{0,1\}^{\mathbf{Z}^2}$ having the property that in every 3×3 square there is a fixed number k of ones. Due to certain weak periodicity properties these configurations can be analyzed. In particular we investigate the connectedness of the set of configurations under slide-deformations and show that it depends critically on k. We also prove the topological non-transitivity of the shift-action. Some of these properties are shown to hold for two-dimensional subshifts of the same generalized type.

Keywords: subshift of finite type, $\mathbf{Z}^2$-action, zero entropy, topological transitivity, ground state, tiling
AMS Classification: 52C20, 54H20
Running head: Rigid subshifts

* Research partially supported by the Finnish Cultural Foundation and University of Warwick

Introduction

The space of **configurations** $X = \{0,1\}^{\mathbf{Z}^2}$ together with the product topology is a compact metric space. The following non-empty subsets of X are of interest to us.

Definition: *For $k = 1,2,3$ or 4 **the rule** k is satisfied if in a 3×3 square there are exactly k ones. The set of configurations which satisfy rule k everywhere is denoted by X_k.*

Remark: Properties of rules 5,6,7 and 8 obviously follow from the properties of the defined rules by considering zeros instead of ones.

The 3×3 square in which we impose the consistency requirement according to one of the rules is called a **window** and the piece of configuration in it a **scenery**.

It is natural to equip the configuration space with the horizontal and vertical coordinate shifts

$$\sigma_h(x)_{(i,j)} = x_{(i+1,j)} \quad \text{and} \quad \sigma_v(x)_{(i,j)} = x_{(i,j+1)}, \qquad x \in X.$$

These actions obviously commute. Since X_k are closed under the action of the shifts we obtain **subshifts of finite type** $(X_k, \sigma_h, \sigma_v)$.

Surprisingly little is known about the structure of these spaces and the action on them. Indeed this seems to be the case for two-dimensional subshifts in general. In this note we alleviate this situation a little bit by establishing a few basic results for the stated rules and their generalizations. The original motivation was provided by K. Schmidt in [**S**].

1. Spaces

All the defined spaces are non-empty. We first investigate how the allowed or legal configurations can be characterized and what is their interrelation in each of the spaces X_k.

There are two different types of periodic configurations: (horizontally or vertically) periodic and doubly-periodic. The first is only invariant under $\sigma_h{}^p\sigma_v{}^q$ for some minimal $(p, q) \neq (0, 0)$ (and its multiples), the second under $\sigma_h{}^{p_1}\sigma_v{}^{q_1}$ and $\sigma_h{}^{p_2}\sigma_v{}^{q_2}$ where $(p_i, q_i) \neq (0, 0)$ and (p_i, q_i) are not rationally related.

Given a configuration in X_k suppose that we keep all the columns/rows fixed except one which we slide (as in a slide rule) an arbitrary integral amount. The resulting object, a **slide deformation** of the originriginal configuration, is of course in X. But what is important is that X_k is closed under some slide deformations:

Lemma 1.1.: *Given a configuration in X_k any vertical or horizontal slide of a bi-infinite column or a row of period three results in a configuration in X_k.*

The proof is immediate since a slide of a period three column/row leaves the number of ones seen in any 3×3 window unchanged. Call these slides legal.

Proposition 1.2.: *The configurations in X_1 are either periodic or doubly-periodic. In X_k, $k = 2, 3, 4$ there are also aperiodic configurations.*

Proof: Let $k = 1$. Consider in a legal configuration an arbitrary one say at (x, y) and its nearest neighboring one to the right at $(x + 3, y')$ (this is uniquely determined). If they have the same ordinates test $(x + 3, y')$ and the nearest one to its right for the same property and so on until a pair with different ordinates is found. If none is found check the left neighbors the same way. If no such pair is found we have a bi-infinite periodic row $\underline{001}$. But this implies that the rows at distance one and two are $\underline{0}$ and the rows at distance three are $\underline{001}$. By induction the global configuration is horizontally periodic or doubly-periodic. If a vertically unaligned pair is found it forces a bi-infinite vertical column of width eight and of period three. Extending this we see that the configuration is vertically periodic or doubly-periodic.

In the case of rules $2, 3$ and 4 a doubly-periodic configuration is generated for example from the prototiles shown in Figure 1 a,b and c respectively. All non-trivial rows are period three so a slide by one to the right of any such row gives a horizontally periodic configuration which is not vertically periodic. In these configurations there are still period three columns (through the middle of a deformed tile). Hence a vertical slide by one results in an aperiodic legal configuration.   ∎

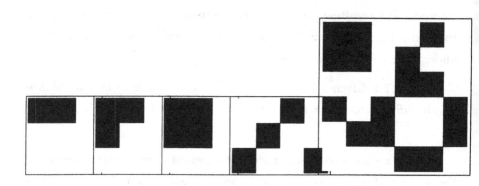

Figure 1 a,b,c,d,e. Prototiles. Black is one, white is zero.

As indicated in the proof it is natural to think of the configurations in terms of tilings. These are made of copies from a finite set of **prototiles**. We do not allow rotations or reflections of these as is sometimes done ([GS]) and we only consider rectangular shapes. Doubly-periodic configurations of course have the key property of being immediately globally constructible from a single tile. Periodic but non-doubly periodic configurations cannot be generated from a single tile but they may be generated from a tile and its periodic shift as in the proof above. As the proof indicates some of the aperiodic configurations are also generated from a small number of prototiles.

The following result resolves the possible sizes of prototiles.

Theorem 1.3.: *For X_1 the prototile generating a doubly-periodic configuration is of size 3×3. For X_k, $k = 2$ or 4 it can be of any size $3w \times 3h$ where w, h are natural numbers. For X_3 all prototiles have at least one of the sidelengths a multiple of three.*

Proof: By the structure of configurations in X_1 the first assertion is clear. For $k = 2$ or 4 consider a legal period segment of height 3 and width (period) l. Let it contain n ones. Then necessarily $n/(3l) = k/9$ which is the density of ones in any strip of height/width three. But this implies that l is divisible by 3. For a $l_1 \times l_2$ prototile apply this argument along any column or row of width 3 to obtain the second part of the statement. By using legal column and row slides as in the proof of Proposition 1.2. it is easy to show that prototiles with any w, h can be constructed. If $k = 3$ the density argument applied to a $l_1 \times l_2$ prototile implies that one of the sidelengths is necessarily divisible by three. ∎

Remark: In X_3 there are indeed configurations generated from prototiles with one of the sidelengths not a multiple of three, e.g. the one shown in Figure 1d.

The doubly-periodic configurations would seem to be exceptional configurations in the spaces X_k but is this really the case? Note that by the proof of Proposition 1.2. X_1 is **slide-connected** i.e. every configuration can be deformed to a unique doubly-periodic configuration with a sequence of legal column and row slides. Moreover the question about slide-connectivity is of some subtlety since seemingly the only way of constructing a configuration for any X_k is by legal moves from a doubly-periodic configuration.

The argument for X_1 can be refined to establish the following. Due to the combinatorial nature of the proof we only give the outline.

Proposition 1.4.: X_2 *is slide-connected.*

Sketch of the Proof: Given an arbitrary configuration $x \in X_2$ there are two cases. 1. Suppose that we can find the prototile of Figure 1a. (call it P) somewhere in x or in its rotation. There are five essentially different alternatives for the 3×3-block immediately to its right. Suppose that the block is not PP. Every one of the other four 3×6-blocks either (i) generates a unique period 3 column of width 6 or (ii) generates a column of width 6 where there are period 3 columns. In both cases we can slide the columns in such a way that the 3×6-block becomes PP. By continuing inductively to the right and to the left we obtain the doubly-periodic configuration generated by P.

2. Call by P' the block where the right one in P has been slid to the center. If P cannot be found in x there must be P' (or its reflection) somewhere. Consider the 3×6-block where P' is the left half. There are six essentially different alternatives for this block. It turns out that three of them generate a period three column of width 3. By column-slides we can then make the tile P appear on the right half of the 3×6-block and the argument reduces to the first case.

In each of the remaining three 3×6-blocks there is at least one nontrivial period 3 row. Now extend this block to the right to a 3×9-block. If the periodic row(s) do not continue to the extension then one can again check that the column of width 9 is vertically periodic and we reduce to case 1. If the periodic row(s) continue we carry on the extension to the right and to the left by three columns at a time. If we do not reduce to the case 1. in this process our configuration must have at least every third row periodic (and these are shifts of the same row) and the rest vertically of period 3. Then align the periodic rows. The configuration is now of vertical period three and by column slides reduces to the doubly periodic configuration generated by P.   ∎

Remark: There is good reason to believe that X_3 is slide-connected as well. The same argument should work but the combinatorics seem to become quite unwieldy.

However the more interesting discovery is that X_4 **is not slide-connected!** After solving (with computer) all the legal 6×6 prototiles a few "exotic" ones like the one shown in Figure 1e. were found. In the global configuration it generates there are only rows and columns of period six hence it cannot be slide-deformed to a period three configuration. In view of this it seems that X_4 is a more complex space than the others. The finding is compatible with the general physical idea of such phenomena can appear as the density $(k/9)$

reaches a certain neighborhood of its equilibrium level (here the density of ones in the symmetric Bernoulli field, $1/2$). It would be quite interesting to know if there is a hierarchy of more complex structures here; for example is there also an aperiodic configuration that is not slide-connected to a doubly periodic one?

In solving the prototiles it was also found that in every 6×4, 6×5, 6×6 and 6×7 prototile for X_3 there are period three rows or columns, an observation supporting the claim above for X_3.

Finally we note that the slide-argument works for rectangular windows and results 1.1-1.3 generalize immediately. However it is not clear how to resolve the slide-connectivity in the general case. No non-trivial finite window admits the slide-argument to the three coordinate directions in the triangular lattice case.

2. Shift action

We now consider the basic properties of the (σ_h, σ_v)-action on the configuration spaces.

If in a window we see n_1 ones in the leftmost, n_2 in the center and n_3 in the rightmost column we call it **type** (n_1, n_2, n_3).

The following observation is of some consequence.

Lemma 2.1.: *Let x be a configuration in X_k, any k. Given a scenery in x with column distribution (n_1, n_2, n_3) consider the bi-infinite horizontal strip of height three containing it. Every scenery in this strip has column distribution (n_1, n_2, n_3) or it's cyclic permutation. Analogously for the scenery's row-distribution.*

Proof: By shifting the window to the right by one the scenery loses n_1 ones hence the same number has to enter from the right. ∎

It is easy to see that each X_k, $k > 1$, contains configurations with k adjacent (horizontal or vertical) ones somewhere.

Proposition 2.2.: *For rule 2 no configuration with two adjacent horizontal ones can contain two adjacent vertical ones. For rules 3 and 4 no configuration with three adjacent horizontal ones can contain three adjacent vertical ones.*

Proof: The argument is identical for all three cases so let us consider $k = 2$. Let H be the horizontal strip of height three containing a scenery with two vertical adjacent ones and let L be the vertical strip of width three containing a scenery with two horizontal adjacent ones. By applying Lemma 2.1. we see that in $H \cap V$ there are both two adjacent horizontal and vertical ones. Hence the rule 2 is violated. ∎

Remark: Before going into our main result we note the phenomenon (and thereby answer a question by Mike Boyle) that in the systems $(X_k, \sigma_h, \sigma_v)$ there are one-dimensional shifts embedded that are not subshifts of finite type. Consider e.g. a strip of width five around the diagonal $y = x$. On this we define the natural one-dimensional system: legal sceneries together with the shift $\sigma_h \sigma_v$. But by the Proposition symbols from arbitrary far in the past (southwest) can forbid certain symbols at the origin. Hence no finite collection of legal words suffices to check the one-dimensional sequences.

For each X_k, $k > 1$, let $X_k^{(h)}$ and $X_k^{(v)}$ be the sets of configurations containing somewhere at least $k \wedge 3$ horizontally or vertically adjacent ones. Let $X_k^{(c)}$ be the complement of their union. In view of Proposition 2.2. these three sets partition the configuration space. Subsequently we drop the index k in $X_k^{(\cdot)}$ once it has been declared which rule is considered.

The **orbit** of an element $x \in X_k$ is the set $O(x) = \{\sigma_h^i \sigma_v^j(x) | \ (i,j) \in \mathbf{Z}^2\}$. Denote its closure by $\overline{O(x)}$. A continuous transformation on a metric space is called **topologically transitive** if there is a element in the space whose orbit under the transformation is dense ([**W**]). Note that the shift-action on any X_k is continuous in the inherited topology so it make sense to ask whether this minimal mixing prevails in our set-up.

Recall that X can be metrized by many equivalent ways e.g. by

$$d(x,x) = 0 \quad \text{and} \quad d(x,y) = 2^{-\min\{\max\{|i|,|j|\}| \ x_{(i,j)} \neq y_{(i,j)}\}} \quad \text{for} \quad x \neq y.$$

Hence for density it suffices to consider matching on arbitrarily large squares centered at origin.

Theorem 2.3.: *None of the systems* $(X_k, \sigma_h, \sigma_v)$, $k = 1, 2, 3$ *or* 4 *is topologically transitive.* X_1 *has two maximal transitive components; the vertically and horizontally periodic configurations. For* X_k, $k > 1$, *the following exclusions hold:*

(i) $X^{(h)} \cap \overline{O(x)} = \emptyset$ $\forall x \in X^{(v)}$ *and* $X^{(v)} \cap \overline{O(x)} = \emptyset$ $\forall x \in X^{(h)}$,

(ii) $X^{(h)} \cap \overline{O(x)} = \emptyset$ *and* $X^{(v)} \cap \overline{O(x)} = \emptyset$ $\forall x \in X^{(c)}$.

For $k > 2$ *we also have the weaker exclusion:* $X^{(c)} \not\subset \overline{O(x)}$ $\forall x \in X^{(h)} \cup X^{(v)}$.

Proof: In view of Proposition 1.2. we know that every vertically (or doubly) periodic configuration is uniquely determined by specifying by how much (± 1 or 0) every third column is shifted with respect to its neighbor three to the left. This set of vertical off-sets $\{-1, 0, 1\}^{\mathbf{Z}}$ is in a one-to-one correspondence with the set of vertically (and doubly-) periodic configurations up to a horizontal shift to match the non-zero columns. Call a legal column of width three with all ones in the left column a slab. The desired x is built simply by placing first slabs with different off-sets next to each other, then next to them pairs of slabs with all different pairs of off-sets and so on. Hereby we exhaust all finite blocks of off-sets and consequently x must have a dense orbit in vertically periodic configurations.

For $k > 1$ it is an immediate consequence of the definition of $X^{(c)}$ that the orbits of its elements are bounded away from elements in both $X^{(h)}$ and $X^{(v)}$. Moreover the exclusion property of Proposition 2.2. clearly implies that no element in $X^{(h)}$ $(X^{(v)})$ can have its orbit arbitrarily close to any element in $X^{(v)}$ $(X^{(h)}$ respectively). For $k > 2$ this can be further refined as follows. Note that for each element x in $X^{(h)}$ every scenery must by Lemma 2.1. be of the type $(1, 1, 1)$ for rule 3 and $(2, 1, 1)$ or its cyclic permutation for rule 4. But then $O(x)$ is bounded away from those configurations in $X^{(c)}$ in which every scenery is of type $(2, 1, 0)$ or $(2, 2, 0)$ respectively (for example the doubly periodic configurations generated by the prototiles in Figure 1b and 1c). ∎

Remarks: 1. Also in the $k > 1$ case at least two of the transitive components are simply related: the one in the complement of $X^{(h)}$ is of course just a 90 degree rotation of the one outside $X^{(v)}$. Whether these or the other components further split by a more subtle exclusion/conservation law remains open.

2. The results indicate that configurations are indeed quite rigid. Another sign of this is that the systems (any k) are of zero topological entropy. This follows from the observation that a diagonal strip of horizontal width four determines the entire configuration. But the number of different such diagonals of length N is proportional to 2^{cN}. Hence the number of different configurations is a $N \times N$ square is bounded by $2^{c'N}$ and the entropy must be zero.

By interpreting entropy as temperature it is perhaps justified to view our systems as the "frozen" groundstates of more general systems where defects i.e. tiling errors are allowed. Here they are not present hence the minimal level of periodicity as formulated in lemmas 1.1. and 2.1. which in turn give the rest. The positive-entropy case is elaborated in the upcoming work.

The methods used for Theorem 2.3. suitably refined actually enable one to prove the basic non-transitivity result in greater generality. We present it here and give a streamlined proof indicating the modifications.

Suppose that we have $s < \infty$ symbols $S = \{0, 1, 2, \ldots, s - 1\}$ and consider the subsets of S^{Z^2} defined by the **rule** $(k_0, k_1, \ldots, k_{s-1})$ which requires that in every $p \times q$ rectangle $(p, q \geq 2)$ there are exactly k_i copies of the symbol i. Obviously $k_i \geq 1$ and $\sum k_i = pq$. Call the subspace $X_{(k_0, k_1, \ldots, k_{s-1})}$.

Theorem 2.4.: *The shift-action on any of the spaces* $X_{(k_0, k_1, \ldots, k_{s-1})}$ *is topologically non-transitive.*

Proof: If $k_i = 1$ for some symbol i the argument is as in Theorem 2.3. Suppose that $k_i \geq 2 \ \forall i$. Given a symbol i define $X^{(h)}(i)$ for it essentially as before: it contains all configurations in which there is somewhere at least $k_i \wedge p$ i's in a horizontal p-block (and in $x \in X^{(v)}(i)$ there is somewhere at least $k_i \wedge q$ i's in a vertical q-block). The exclusion argument of Proposition 2.2. works for this symbol if $k_i \wedge p + k_i \wedge q - 1 > k_i$. This is equivalent to $1 < k_i < p + q - 1$ holding. Hence if k_i satisfies these inequalities the sets $X^{(h)}(i)$ and $X^{(v)}(i)$ are disjoint (and contain the orbits starting from them). As before elements in the complement of their union cannot have dense orbits in either set and the non-transitivity follows.

The remaining (and novel) case is the one where $k_i < p + q - 1$ fails for all symbols. So we consider the case $s \leq pq/(p+q-1)$. Given a symbol i the configurations in $X^{(h)}(i)$ have in every scenery the symbol i-distributions of the form $(n_1, n_2, \ldots, n_p)$, $\forall n_l \geq 1$ (or its cyclic permutation). On the other hand $X^{(c)}(i)$ clearly contains e.g. doubly periodic configurations where the i-distribution is of the form $(n_1', n_2', \ldots, 0)$ (or its cyclic permutation) in every scenery. This is just because $k_i \geq p + q - 1$ $\forall i$ so in particular any symbol can form an L -shaped area like the symbols 0 does in Figure 1c. Hence elements of $X^{(h)}(i)$ cannot have dense orbits in $X^{(c)}(i)$ and the converse holds again by definition. ∎

The type-preservation in shifting the window seems to produce a strong enough argument only in two dimensions. The natural generalization of the definition for $X^{(h)}$ for $\mathbf{Z}^d$-actions, $d \geq 3$ is not useful for this reason and it is unclear how to argue the higher dimensional case.

Acknowledgement

The author would like to thank Mats Nordahl for stimulating discussions on some of these problems and a referee for comments.

References

[GS] Grünbaum, B., Shephard, G. C.: Tilings and patterns, Freeman, 1987.

[S] Schmidt, K.: Algebraic ideas in ergodic theory, CBMS Reg. Conf. Ser. in Math. **76**, American Mathematical Society, 1990.

[W] Walters, P.: An introduction to ergodic theory, Springer-Verlag, 1982.

ENTROPY OF GRAPHS, SEMIGROUPS AND GROUPS

SHMUEL FRIEDLAND

Institut des Hautes Études Scientifiques
and University of Illinois at Chicago

§0. Introduction

Let X be a compact metric space and $T : X \to X$ is continuous transformation. Then the dynamics of T is a widely studied subject. In particular, $h(T)$ - the entropy of T is a well understood object. Let $\Gamma \subset X \times X$ be a closed set. Then Γ induces certain dynamics and entropy $h(\Gamma)$. If X is a finite set then Γ can be naturally viewed as a directed graph. That is, if $X = \{1, ..., n\}$ then Γ consists of all directed arcs $i \to j$ so that $(i, j) \in \Gamma$. Then Γ induces a subshift of finite type which is a widely studied subject. However, in the case that X is infinite, the subject of dynamic of Γ and its entropy are relatively new. The first paper treating the entropy of a graph is due to [**Gro**]. In that context X is a compact Riemannian manifold and Γ can be viewed as a Riemannian submanifold. (Actually, Γ can have singularities.) We treated this subject in [**Fri1-3**]. See Bullet [**Bul1-2**] for the dynamics of quadratic correspondences and [**M-R**] for iterated algebraic functions.

The object of this paper is to study the entropy of a corresponding map induced by Γ. We now describe briefly the main results of the paper. Let X be a compact metric space and assume that $\Gamma \subset X \times X$ is a closed set. Set

$$\Gamma_+^\infty = \{(x_i)_1^\infty : (x_i, x_{i+1}) \in \Gamma, i = 1, ..., \}.$$

Let $\sigma : \Gamma_+^\infty \to \Gamma_+^\infty$ be the shift map. Denote by $h(\Gamma)$ be the topological entropy of $\sigma | \Gamma_+^\infty$. It then follows that σ unifies in a natural way the notion of a (continuous) map $T : X \to X$ and a (finitely generated) semigroup or group of (continuous) transformations $S : X \to X$. Indeed, let $T_i : X \to X, i = 1, ..., m$, be m continuous transformations. Denote by $\Gamma(T_i)$ the graphs corresponding to $T_i, i = 1, ..., m$. Set $\Gamma = \cup_1^m \Gamma(T_i)$.

I would like to thank to M. Boyle, M. Gromov and F. Przytycki for useful remarks.

Typeset by $\mathcal{A}\mathcal{M}\mathcal{S}$-TEX

319

Then the dynamics of σ is the dynamics of the semigroup generated by $\mathcal{T} = \{T_1, ..., T_m\}$. If $\mathcal{T}$ is a set of homeomorphisms and $\mathcal{T}^{-1} = \mathcal{T}$ then the dynamics of σ is the dynamics of the group $\mathcal{G}(\mathcal{T})$ generated by $\mathcal{T}$. In particular, we let $h(\mathcal{G}(\mathcal{T})) = h(\Gamma)$ be the entropy of $\mathcal{G}(\mathcal{T})$ using the particular set of generators $\mathcal{T}$. For a finitely generated group $\mathcal{G}$ of homeomorphisms of X we define

$$h(\mathcal{G}) = \inf_{\mathcal{T}, \mathcal{G} = \mathcal{G}(\mathcal{T})} h(\mathcal{G}(\mathcal{T})).$$

In the second section we study the entropy of graphs, semigroups and groups acting on the finite space X. The results of this section give a good motivation for the general case. In particular we have the following simple inequality

$$h(\cup_{i=1}^m \Gamma_i) \leq h(\cup_{i=1}^m (\Gamma_i \cup \Gamma_i^T)) \leq \log \sum_{i=1}^m e^{h(\Gamma_i \cup \Gamma_i^T)}. \tag{0.1}$$

Here $\Gamma^T = \{(y, x) : (x, y) \in \Gamma\}$. Let $Card(X) = n$. Then any group of homeomorphisms $\mathcal{G}$ of X is a subgroup of the symmetric group S_n acting on X as a group of permutations. We then show that if $\mathcal{G}$ is commutative then $h(\mathcal{G}) = \log k$ for some integer k. If $\mathcal{G}$ acts transitively on X then k is the minimal number of generators for $\mathcal{G}$. Moreover, $h(\mathcal{G}) = 0$ iff $\mathcal{G}$ is a cyclic group. For each $n \geq 3$ we produce a group $\mathcal{G}$ generated by two elements so that $0 < h(\mathcal{G}) < \log 2$.

In §3 we discuss the entropy of graphs on compact metric spaces. We show that if $T_i : X \to X, i = 1, ..., m$, is a set of Lipschitzian transformations of a compact Riemannian manifold X of dimension n then

$$h(\cup_1^m \Gamma(T_i)) \leq \log \sum_1^m L_+(T_i)^n. \tag{0.2}$$

Here, $L_+(T_i)$ is the maximum of the Lipschitz constant of T_i and 1. Thus, $L_+(T_i)^n$ is analogous to the norm of a graph on a finite space X. The above inequality generalizes to semi-Riemannian manifolds which have a Hausdorff dimension $n \in \mathbf{R}_+$ and a finite volume with respect to a given metric d on X. Thus, if X is a compact smooth Riemannian manifold and $\mathcal{G}$ is a finitely generated group of diffeomorphisms (0.2) yields that $h(\mathcal{G}) < \infty$. Let X be a compact metric space and $T : X \to X$ a noninvolutive homeomorphism ($T^2 \neq Id$). We then show that $h(\Gamma(T) \cup \Gamma(T^{-1})) \geq \log 2$. The following example due to M. Boyle shows that (0.1) does not apply in general. Let X be a compact metric space for which there exists a homeomorphism $T : Y \to Y$ with $h(T) = h(T^2) = \infty$. (See for example [**Wal**, p. 192].) Set

$$X = X_1 \cup X_2, X_1 = Y, X_2 = Y, T_i(X_1) = X_2, T_i(X_2) = X_1,$$

$$T_1(x_1) = Tx_1, T_1(x_2) = T^{-1}x_2,$$

$$T_2(x_1) = T^{-1}x_1, T_2(x_2) = Tx_2, x_1 \in X_1, x_2 \in X_2.$$

As $T_1^2 = T_2^2 = Id$ it follows that $2h(T_1) = 2h(T_2) = h(Id) = 0$. Clearly, $T_2 T_1 | X_1 = T^2$ and $h(\Gamma) = \infty$. The last section discusses mainly the entropy of semigroups and groups of Möbius transformations on the Riemann sphere. Let $\mathcal{T} = \{T_1, ..., T_k\}$ is a set of Möbius transformations. Inequality (0.2) yield that $h(\mathcal{G}(\mathcal{T})) \leq \log k$. Let $T_i(z) = z + a_i, i = 1, 2$, be two translations of $\mathbf{C}$. Assume that $\frac{a_1}{a_2}$ is a negative rational number. We then show that

$$h(\Gamma(T_1) \cup \Gamma(T_2)) = -\frac{|a|}{|a| + |b|} \log \frac{|a|}{|a| + |b|} - \frac{|b|}{|a| + |b|} \log \frac{|b|}{|a| + |b|}.$$

Assume now that a_1 and a_2 are linearly independent over $\mathbf{R}$. We then show that

$$h(\cup_1^2 (\Gamma(T_i) \cup \Gamma(T_i^{-1}))) = \log 4.$$

It is of great interest to see if $h(\mathcal{G})$ has any geometric meaning for a finitely generated Kleinian group $\mathcal{G}$. Consult with [**G-L-W**], [**L-W**], [**N-P**], [**L-P**] and [**Hur**] for other definitions of the entropy of relations and foliations.

§1. Basic definitions

Let X be a compact metric space and assume that $\Gamma \subset X \times X$ is a closed set. Set

$$X^k = \prod_1^k X_i, X_+^\infty = \prod_1^\infty X_i, X^\infty = \prod_{i \in \mathbf{Z}} X_i, X_i = X, i \in \mathbf{Z},$$

$$\Gamma^k = \{(x_i)_1^k : (x_i, x_{i+1}) \in \Gamma, i = 1, ..., k - 1, \}, k = 2, ...,$$

$$\Gamma_+^\infty = \{(x_i)_1^\infty : (x_i, x_{i+1}) \in \Gamma, i = 1, ..., \},$$

$$\Gamma^\infty = \{(x_i)_{i \in \mathbf{Z}} : (x_i, x_{i+1}) \in \Gamma, i \in \mathbf{Z}\}.$$

We shall assume that $\Gamma^k \neq \emptyset, k = 2, ...,$ unless stated otherwise. (In any case, if this assumption does not hold we set $h(\Gamma) = 0$.) This in particular implies that $\Gamma_+^\infty \neq \emptyset, \Gamma^\infty \neq \emptyset$. Let

$$\pi_{p,q}^l : X^l \to X^{q-p+1}, \{x_i\}_1^l \mapsto \{x_i\}_p^q, 1 \leq p \leq q \leq l.$$

If no ambiguity arise we shall denote $\pi_{p,q}^l$ by $\pi_{p,q}$. The maps $\pi_{p,q}$ are well defined for X_+^∞, X^∞. For $p \leq 0, p \leq q$ we let $\pi_{p,q} : X^\infty \to X^{q-p+1}$. Similarly, for a finite p we have the obvious maps $\pi_{-\infty,p}, \pi_{p,\infty}$ whose range is Γ_+^∞. Let $d : X \times X \to \mathbf{R}_+$ be a metric on X. As X is compact we have that X is a bounded diameter $0 < D < \infty$. That is, $d(x,y) \leq D, \forall x, y \in X$. On $X^k, X^\infty_+, X^\infty$ one has the induced metric

$$d(\{x_i\}_1^k, \{y_i\}_1^k) = \max_{1 \leq i \leq k} \frac{d(x_i, y_i)}{\rho^{i-1}},$$

$$d(\{x_i\}_1^\infty, \{y_i\}_1^\infty) = \sup_{1 \leq i} \frac{d(x_i, y_i)}{\rho^{i-1}},$$

$$d(\{x_i\}_{i \in \mathbf{Z}}, \{y_i\}_{i \in \mathbf{Z}}) = \sup_{i \in \mathbf{Z}} \frac{d(x_i, y_i)}{\rho^{|i-1|}}.$$

Here $\rho > 1$ to be fixed later. Since X is compact it follows that $X^k, X^\infty_+, X^\infty$ are compact metric spaces where the infinite products have the Tychonoff topology. Let

$$\sigma : X^\infty_+ \to X^\infty_+, \sigma((x_i)_1^\infty) = (x_{i+1})_1^\infty,$$

$$\sigma : X^\infty \to X^\infty, \sigma((x_i)_{i \in \mathbf{Z}}) = (x_{i+1})_{i \in \mathbf{Z}}$$

be the one sided shift and two sided shift respectively. We refer to Walters [**Wal**] for the definitions and properties of dynamical systems used here. Note that $\Gamma^\infty_+, \Gamma^\infty$ are invariant subsets of one sided and two sided shifts, i.e.

$$\sigma : \Gamma^\infty_+ \to \Gamma^\infty_+, \sigma : \Gamma^\infty \to \Gamma^\infty.$$

We call the above restrictons of σ as the dynamics (maps) induced by Γ. As Γ was assumed to be closed it follows that $\Gamma^\infty_+, \Gamma^\infty$ are closed too. Hence, we can define the topological entropies $h(\sigma|\Gamma^\infty_+), h(\sigma|\Gamma^\infty)$ of the corresponding restrictions. We shall show that these two entropies are equal. The above entropy is $h(\Gamma)$.

Denote by $C(X)$ the Banach space of all continuous functions $f : X \to \mathbf{R}$. For $f \in C(X)$ it is possible to define the topological pressure $P(\Gamma, f)$ as follows. First observe that f induces the following continuous functions

$$f_1 : \Gamma^\infty_+ \to \mathbf{R}, f_1((x_i)_1^\infty) = f(x_1),$$

$$f_2 : \Gamma^\infty \to \mathbf{R}, f_2((x_i)_{i \in \mathbf{Z}}) = f(x_1).$$

Let $P(\sigma, f_1), P(\sigma, f_2)$ be the topological pressures of f_1, f_2 with respect to the map σ acting on $\Gamma^\infty_+, \Gamma^\infty$ respectively. We shall show that the above topological pressures coincide. We then let $P(\Gamma, f) = P(\sigma, f_1) = P(\sigma, f_2)$.

Let $T : X \to X$ be a continuous map. Set $\Gamma = \Gamma(T) = \{(x, y) : x \in X, y = T(x)\}$ be the graph of T. Denote by $h(T)$ the topological entropy of T. It then follows that $h(T) = h(\Gamma)$. Indeed, observe that $x \mapsto orb_T(x) = (T^{i-1}(x))_1^\infty$ induces a homeomorphism $\phi : X \to \Gamma(T)^\infty_+$ such that $T = \phi^{-1} \circ \sigma \circ \phi$ and the equality $h(T) = h(\sigma|\Gamma_+^\infty)$ follows. Similarly, for $f \in C(X)$ we have the equality $P(T, f) = P(\sigma, f_1) = P(\Gamma(T), f)$.

Let $\Gamma_\alpha, \alpha \in \mathcal{A}$ be a family of closed graphs in $X \times X$. Set

$$\vee_{\alpha \in \mathcal{A}} \Gamma_\alpha = Closure(\cup_{\alpha \in \mathcal{A}} \Gamma_\alpha).$$

Note that if $\mathcal{A}$ is finite then $\vee \Gamma_\alpha = \cup \Gamma_\alpha$. The dynamics of $\Gamma = \vee \Gamma_\alpha$ is called the product dynamics induced by $\Gamma_\alpha, \alpha \in \mathcal{A}$. Let $T_\alpha : X \to X, \alpha \in \mathcal{A}$ be a set of continuous maps. Set

$$\mathcal{T} = \cup_{\alpha \in \mathcal{A}} T_\alpha, \Gamma(\mathcal{T}) = Closure(\cup_{\alpha \in \mathcal{A}} \Gamma(T_\alpha)).$$

Then the dynamics of $\Gamma(\mathcal{T})$ is the dynamics of a semigroup $\mathcal{S}(\mathcal{T})$ generated by $\mathcal{T}$. If each $T_\alpha, \alpha \in \mathcal{A}$ is a homeomorphism and $\mathcal{T}^{-1} = \mathcal{T}$ then the dynamics of $\Gamma(\mathcal{T})$ is the dynamics of a group $\mathcal{G}(\mathcal{T})$ generated by $\mathcal{T}$. Note that for a fixed $x \in X$ the orbit of x is given by the formula

$$orb_\mathcal{T}(x) = \{(x_i)_1^\infty, x_1 = x,$$
$$x_i \in Closure(T_{\alpha_{i-1}} \circ \cdots \circ T_{\alpha_1}(x)), \alpha_1, ..., \alpha_{i-1} \in \mathcal{A}, i = 2, ..., \}.$$

If $\mathcal{A}$ is finite then we can drop the closure in the above definition.

Let $\mathcal{T}$ be a set of continuous transformations of X as above. We then define

$$h(\mathcal{S}(\mathcal{T})) = h(\Gamma(\mathcal{T})), \quad P(\mathcal{S}(\mathcal{T}), f) = P(\Gamma(\mathcal{T}), f), f \in C(X)$$

to be the entropy of $\mathcal{S}(\mathcal{T})$ and the topological pressure of f with respect to the set of generators $\mathcal{T}$. In order to ensure that the above quantities are finite we shall assume that $\mathcal{T}$ is a finite set. Given a finitely generated semigroup $\mathcal{S}$ of $T : X \to X$ let

$$h(\mathcal{S}) = \inf_{\mathcal{T}, \mathcal{S} = \mathcal{S}(\mathcal{T})} h(\mathcal{S}(\mathcal{T})), \quad P(\mathcal{S}, f) = \inf_{\mathcal{T}, \mathcal{S} = \mathcal{S}(\mathcal{T})} P(\mathcal{S}(\mathcal{T}), f), f \in C(X).$$

Here, the infimum is taken over all finite generators of $\mathcal{S}$.

§2. Entropy of graphs on finite spaces

Let X be a finite space. We assume that $X = \{1, ..., n\}$. Then each $\Gamma \subset X \times X$ is in one to one correspondence with a $n \times n$ $0 - 1$ matrix $A = (a_{ij})_1^n$. That is $(i, j) \in \Gamma \iff a_{ij} = 1$. As usual we let $M_n(\{0-1\})$ be the set of $0 - 1$ $n \times n$ matrices. For $\Gamma \subset X \times X$ we let $A(\Gamma) \in M_n(\{0-1\})$ to be the matrix induced by Γ and for $A \in M_n(\{0 - 1\})$ we let $\Gamma(A)$ to be the graph induced by A. The assumption that $\Gamma^k \neq \emptyset, k = 1, 2, ...,$ is equivalent to $\rho(A(\Gamma)) > 0 \iff \rho(A(\Gamma)) \geq 1$. Here, for any A in the set of $n \times n$ complex valued matrices $M_n(\mathbf{C})$ we let $\rho(A)$ to be the spectral radius of A. For $\Gamma \subset X \times X$ consider the sets $X_l = \pi_{l,l}(\Gamma^l), l = 2, ..., $. It easily follows that $X_2 \supset X_3 \supset \cdots X_n = X_{n+1} = \cdots = X'$. Then $\Gamma^l \neq \emptyset, l = 2, ...,$ iff $X' \neq \emptyset$. Set $\Gamma' = \Gamma \cap X' \times X'$. It then follows that $\Gamma^\infty = \Gamma'^\infty$. Moreover,

$$\pi_{1,\infty}(\Gamma^\infty) = \pi_{1,\infty}(\Gamma'^\infty) = \Gamma'^\infty_+ \subset \Gamma^\infty_+.$$

Here the containment is strict iff $X' \neq X$. It is well known fact in symbolic dynamics that if $X' \neq \emptyset$ then

$$h(\sigma|\Gamma^\infty_+) = h(\sigma|\Gamma^\infty) = \log \rho(A(\Gamma)) = \log \rho(A(\Gamma')) = h(\sigma|\Gamma'^\infty) = h(\sigma|\Gamma'^\infty_+).$$

See for example [**Wal**]. We thus let $h(\Gamma)$ - the entropy of the graph Γ to be any of the above numbers. In fact, X' can be viewed as a limit set of the "transformation" induced by Γ on X'. If $\rho(A(\Gamma)) = 0$, i.e. $X' = \emptyset$ we then let $h(\Gamma) = \log^+ \rho(A(\Gamma))$. Here, $\log^+ x = \log \max(x, 1)$.

Let $\Gamma_\alpha \subset X \times X, \alpha \in \mathcal{A}$ be a family of graphs. Set $A_\alpha = (a_{ij}^{(\alpha)})_1^n = A(\Gamma_\alpha), \alpha \in \mathcal{A}$. It then follows that

$$\vee_{\alpha \in \mathcal{A}} A_\alpha \overset{\text{def}}{=} (\max_{\alpha \in \mathcal{A}} a_{ij}^{(\alpha)})_1^n = A(\vee_{\alpha \in \mathcal{A}}\Gamma_\alpha).$$

The Perron-Frobenius theory of nonnegative matrices yields straightforward that $\rho(A_\alpha) \leq \rho(\vee A_\beta)$. This is equivalent to the obvious inequality $h(\Gamma_\alpha) \leq h(\vee \Gamma_\beta)$. We now point out that we can not obtain an upper bound on $h(\vee \Gamma_\alpha)$ as a function of $h(\Gamma_\alpha), \alpha \in \mathcal{A}$. It suffices to pass to the corresponding matrices and their spectral radii. Let $A = (a_{ij})_1^n \in M_n(\{0-1\})$ matrix such that $a_{ij} = 1 \iff i \leq j$. Assume that $B = A^T$. Then $\rho(A) = \rho(B) = 1, \rho(A \vee B) = n$.

Let $\| \cdot \| : \mathbf{C}^n \to \mathbf{R}_+$ be a norm on $\mathbf{C}^n$. Denote by $\| \cdot \| : M_n(\mathbf{C}) \to \mathbf{R}_+$ the induced operator norm. Clearly, $\rho(A) \leq \|A\|$. Hence

$$\rho(\vee_{\alpha \in \mathcal{A}} A_\alpha) \leq \rho(\sum_{\alpha \in \mathcal{A}} A_\alpha) \leq \sum_{\alpha \in \mathcal{A}} \|A_\alpha\|.$$

Thus

$$h(\vee_{\alpha \in \mathcal{A}}\Gamma_\alpha) \leq \log^+ \sum_{\alpha \in \mathcal{A}} \|A_\alpha\|. \tag{2.1}$$

In the next section we shall consider analogs of $\|A(\Gamma)\|$ for which we have the inequality (2.1) for any set $\mathcal{A}$. For a graph $\Gamma \subset X \times X$ let $\Gamma^T = \{(x,y) : (y,x) \in \Gamma\}$. That is, $A(\Gamma^T) = A^T(\Gamma)$. A graph Γ is symmetric if $\Gamma^T = \Gamma$. Assume that Γ is symmetric. It then follows that $\rho(A(\Gamma)) = \|A(\Gamma)\|$ where

$\|\cdot\|$ is the spectral norm on $M_n(\mathbf{C})$, i.e. $\|A\| = \rho(AA^*)^{\frac{1}{2}}$. Thus, for a family $\Gamma_\alpha, \alpha \in \mathcal{A}$ of symmetric graphs we have the inequalites

$$h(\vee_{\alpha \in \mathcal{A}} \Gamma_\alpha) \leq \log \sum_{\alpha \in \mathcal{A}} e^{h(\Gamma_\alpha)}. \tag{2.2}$$

More generally, for any family of graphs we have the inequalites

$$h(\vee_{\alpha \in \mathcal{A}} \Gamma_\alpha) \leq h(\vee_{\alpha \in \mathcal{A}}(\Gamma_\alpha \vee \Gamma_\alpha^T)) \leq \log \sum_{\alpha \in \mathcal{A}} e^{h(\Gamma_\alpha \vee \Gamma_\alpha^T)}. \tag{2.3}$$

Let $T : X \to X$ be a transformation. Then $A(T) = A(\Gamma(T))$ is a $0 - 1$ stochastic matrix, i.e. each row of $A(T)$ contains exactly one 1. Vice versa, if $A \in M_n(\{0 - 1\})$ is a stochastic matrix then $A = A(T)$ for some transformation $T : X \to X$. Furthermore, $T : X \to X$ is a homeomorphism iff $A(T)$ is a permutation matrix. For $\mathcal{T} = \{T_1, ..., T_k\}$ $\mathcal{S}(\mathcal{T})$ is a group iff each T_i is a homeomorphism, i.e. $A(T_i)$ is a permutation matrix for $i = 1, ..., k$. Clearly, any group of homeomorphisms $\mathcal{S}$ of X is a subgroup of the symmetric group $S_n, n = Card(X)$.

(2.4) Theorem. *Let X be a finite space and assume that $T_i : X \to X, i = 1, ..., k$, be a set of transformation. Set*

$$\mathcal{T} = \{T_1, ..., T_k\}, \Gamma = \Gamma(\mathcal{T}) = \cup_1^k \Gamma(T_i), A = A(\Gamma).$$

Then $h(\mathcal{S}(\mathcal{T})) \leq \log k$. Furthermore, $h(\mathcal{S}(\mathcal{T})) = 0$ iff $A(\Gamma')$ is a permutation matrix. Assume that $k \geq 2$. Then $h(\mathcal{S}(\mathcal{T})) = \log k$ iff there exists an irreducible component $\hat{X} \subset X'$ on which $\mathcal{S}(\mathcal{T})$ acts transitively such that $A(\Gamma \cap \hat{X} \times \hat{X})$ is $0 - 1$ matrix with k ones in each row. In particular, $h(\mathcal{S}(\{T, T^{-1}\})) = \log 2$ for $T^2 \neq Id$. Assume finally that $\mathcal{S}(\mathcal{T})$ is a commutative group. Then $h(\mathcal{S}(\mathcal{T})) = \log k'$ for some integer $1 \leq k' \leq k$.

Proof. Recall that $h(\mathcal{S}(\mathcal{T})) = \log \rho(A)$. As $A(T_i)$ is a stochasic matrix it follows that $\rho(A(T_i)) = 1, i = 1, ..., k$. Since $A \geq A(T_i)$ we deduce that $\rho(A) \geq 1$. Thus, $X' \neq \emptyset$. Then $X' = \cup_1^m X_i, X_i \cap X_j = \emptyset, 1 \leq i < j \leq m$. Here, A acts transitively on each X_i. Set $\Gamma_i = \Gamma \cap X_i \times X_i, A_i = A(\Gamma_i), i = 1, ..., m$. Note that each A_i is an irreducible matrix. It then follow that $h(\Gamma) = \max \log \rho(A_i)$. Set $u_i : X_i \to \{1\}$. Then $A_i u_i \leq k u_i$. The *minmax* characterization of Wielandt for an irreducible A_i yields that $\rho(A_i) \leq k$. The equality holds iff each row of A_i has exactly k ones. Thus, $h(\Gamma) = \log k, k > 1$ iff each row of some A_i has k ones.

Assume next that T is a homeomorphism such that $T^2 \neq Id$. Set $\Gamma = \Gamma(T) \cup \Gamma(T^{-1})$. Then $X' = X = \cup_1^m X_i$ and least one X_i contains more then one point. Clearly, this A_i has two ones in each row and column. Hence, $h(\Gamma) = \log 2$.

Assume now that $\mathcal{G} = \mathcal{S}(\mathcal{T})$ is a commutative group. Then $X = X' = \cup_1^m X_l$. We claim that the following dichotomy holds for each pair $T_i, T_j, i \neq j$. Either $T_i(x) \neq T_j(x) \forall x \in X_l$ or $T_i(x) = T_j(x) \forall x \in X_l$. Indeed, assume that $T_i(x) = T_j(x)$ for some $x \in X_l$. As $\mathcal{G}$ acts transitively on X_l and is commutative we deduce that $T_i(x) = T_j(x) \forall x \in T_l$. Thus $\Gamma(T_i) \cap X_l \times X_l, i = 1, ..., k$, consists of k_l distinct permutation matrices which do not have any 1 in common. That is $\Gamma_l = \Gamma \cap X_l \times X_l$ is a matrix with k_l ones in each row and column. Hence,

$$h(\Gamma_l) = \log k_l, l = 1, ..., m, h(\Gamma) = \log \max_{1 \leq l \leq m} k_l.$$

(2.5) Theorem. *Let X be a finite space of n points. If $\mathcal{G}$ is commutative then $h(\mathcal{G}) = \log k$ for some integer k which is not greater then the number of the minimal generators of $\mathcal{G}$. If $\mathcal{G}$ acts transitively on X or the restriciton of $\mathcal{G}$ to one of the irreducible (transitive) components is faithful then k is the minimal number of generators of $\mathcal{G}$. In particular, for any $\mathcal{G}$ $h(\mathcal{G}) = 0$ iff $\mathcal{G}$ is cyclic. For each $n \geq 3$ there exists a group $\mathcal{G}$ which acts transitively on X so that $0 < h(\mathcal{G}) < \log 2$.*

Proof. Assume first that $\mathcal{G}$ is commutative. Let $\mathcal{T} = \{T_1, ..., T_p\}$ be a set of generators. Theorem 2.4 yields that $h(\mathcal{G}(\mathcal{T})) = \log k(\mathcal{T}), k(\mathcal{T}) \leq p$. Choose a minimal subset of generators $\mathcal{T}' \subset \mathcal{T}$. Clearly, $h(\mathcal{G}(\mathcal{T}')) \leq h(\mathcal{G}(\mathcal{T}))$. Thus, to compute $h(\mathcal{G})$ it is enough to assume that $\mathcal{T}$ consists of a minimal set of generators of $\mathcal{G}$. Hence, $h(\mathcal{G}) = \log k$ and k is at most the number of the minimal generators of $\mathcal{G}$.

Assume now that $\mathcal{G}$ acts transitively on X. The arguments of the proof of Theorem 2.4 yield that $x \in X, T_i(x) \neq T_j(x)$ for $i \neq j$. Therefore, $h(\mathcal{G}(\mathcal{T})) = \log p$. In particular, $h(\mathcal{G}) = \log k$ where k is the minimal number of generators for $\mathcal{G}$. Suppose now that X is reducible under the action of $\mathcal{G}$ and the restriction of $\mathcal{G}$ to one of its irreducible components is faithful. Then the above results yield that $h(\mathcal{G}) = \log k$ where k is the minimal number of generators of $\mathcal{G}$.

Assume now that $h(\mathcal{G}) = 0$. Let $h(\mathcal{G}) = h(\mathcal{G}(\mathcal{T}))$. Assume first that $\mathcal{G}$ acts irreducibly on X. If $\mathcal{T}$ consists of one element T we are done. Assume to the contrary that $\mathcal{T} = \{T_1, ..., T_q\}, q > 1$. Then $A(\Gamma) \geq A(T_1)$. Since $A(\Gamma)$ is irreducible as $\mathcal{G}$ acts transitively, and $A(\Gamma) \neq A(T_1)$ we deduce that $\rho(A(\Gamma)) > 1$. See for example [**Gan**]. This contradicts our assumption that $h(\mathcal{G}) = 0$. Hence, $\mathcal{G}$ is generated by one element, i.e. $\mathcal{G}$ is cyclic. Assume now that $X = \cup_1^m X_i$ is the decompostion of X to its irreducible components.

According to the above arguments $\Gamma(T) \cap X_i \times X_i$ is a permutation matrix. Hence $\Gamma(T)$ is a permutation matrix corresponding to the homeomorphism T. Thus $\mathcal{G}$ is generated by T.

Assume that $Card(X) = n \geq 3$. Let $T : X \to X$ be a homeomorphism that acts transitively on X, i.e. $T^n = Id, T^{n-1} \neq Id$. Let $Q : X \to X, Q \neq T$ be another homeomorphism so that $Q(x) = T(x)$ for some $x \in X$. Set $\mathcal{G} = \mathcal{G}(\{T, Q\})$. According to Theorem 2.4 $h(\mathcal{G}(\{T, Q\})) < \log 2$. Hence, $h(\mathcal{G}) < \log 2$. As $\mathcal{G}$ is not cyclic it follows that $h(\mathcal{G}) > 0$. $\diamond$

It is an interesting problem to determine the entropy of a commutative group in the general case.

§3. Entropy of graphs on compact spaces

Let X be a compact metric space and $\Gamma \subset X \times X$ be a closed graph. As in the previous section set $X_l = \pi_{l,l}(\Gamma^l), l = 2, ...,$. Then $\{X_l\}_2^\infty$ is a sequence of decreasing closed spaces. Let $X' = \cap_2^\infty X_l, \Gamma' = \Gamma \cap X' \times X'$. Clearly,

$$\Gamma^\infty = \Gamma'^\infty, \pi_{1,\infty}(\Gamma^\infty) = \pi_{1,\infty}(\Gamma'^\infty) = \Gamma'^\infty_+ \subset \Gamma^\infty_+.$$

(3.1) Theorem. *Let X be a compact metric space and $\Gamma \subset X \times X$ be a closed set. Then*

$$h(\sigma|\Gamma^\infty_+) = h(\sigma|\Gamma'^\infty_+) = h(\sigma|\Gamma^\infty),$$
$$P(\Gamma^\infty_+, f) = P(\Gamma'^\infty_+, f) = P(\Gamma^\infty, f), f \in C(X).$$

Proof. The equality $h(\sigma|\Gamma^\infty_+) = h(\sigma|\Gamma'^\infty_+)$ follows from the observation that $\Gamma'^\infty_+ = \cap_0^\infty \sigma^l(\Gamma^\infty_+)$. See [**Wal**, Cor. 8.6.1.]. We now prove the equality $h(\sigma|\Gamma'^\infty) = h(\sigma|\Gamma^\infty)$ It is enough to assume that $X' = X$. Set $X_1 = \Gamma^\infty_+, X_2 = \Gamma^\infty$. Let $\pi : X_2 \to X_1$ be the projection $\pi_{1,\infty}$. It then follows that $\pi(X_2) = X_1, \pi \circ \sigma_2 = \sigma_1 \circ \pi$. Denote by σ_i the restriction of σ to X_i and let $h_i = h(\sigma_i)$ be the topological entropy of σ_i. As σ_1 is a factor of σ_2 one deduces $h_1 \leq h_2$.

We now prove the reversed inequality $h_1 \geq h_2$. Let Y be a compact metric space and assume that $T : Y \to Y$ is a continuous transformation. Denote by $\Pi(Y)$ the set of all probability measures on the Borel σ-algebra generated by all open sets of Y. Let $\mathcal{M}(T) \subset \Pi(Y)$ be the set of all T-invariant probability measures. Assume that $\mu \in \mathcal{M}(T)$. Then one defines the Kolmogorov-Sinai entropy $h_\mu(T)$. The variational principle states that

$$h(T) = \sup_{\mu \in \mathcal{M}(T)} h_\mu(T), \ P(T, f) = \sup_{\mu \in \mathcal{M}(T)} \left(h_\mu(T) + \int f d\mu \right), f \in C(X).$$

Let $\mathcal{B}_2$ be the σ-algebra generated by open sets in X_2. An open set $A \subset X_2$ is called cylindrical if there exist $p \leq q$ with the following property. Let $y \in \pi_{i,i}(A)$. Then for $i \leq p$
we have the property $\pi_{1,1}^2((\pi_{2,2}^2)^{-1}(y)) \subset \pi_{i-1,i-1}(A)$. For $i \geq q$ we have the property $\pi_{2,2}^2((\pi_{1,1}^2)^{-1}(y)) \subset \pi_{i+1,i+1}(A)$. Let $\mathcal{C} \subset \mathcal{B}_2$ be the finite Borel subalgebra generated by open cylindrical sets. Note that each set in $\mathcal{C}$ is cylindrical. Since σ_2 is a homeomorphism it follows that for any $\mu \in \mathcal{M}(\sigma_2)$ $\mathcal{B}(\mathcal{C}) \overset{o}{=} \mathcal{B}_2$. That is up a set of zero μ-measure every set in $\mathcal{B}_2$ can be presented as a set in σ-Borel algebra generated by $\mathcal{C}$. Let $\alpha \subset \mathcal{C}$ be a finite partition of X_2. One then can define the entropy $h(\sigma_2, \alpha)$ with respect to the measure μ [**Wal**, Ch.4]. Since σ_2 is a homeomorphism and μ is σ_2 invariant it follows that $h(\sigma_2, \alpha) = h(\sigma_2, \sigma_2^m(\alpha))$ for any $m \in \mathbf{Z}$. The assumption that $\mathcal{B}(\mathcal{C}) \overset{o}{=} \mathcal{B}_2$ implies that $\sup_{\alpha \in \mathcal{C}} h(\sigma_2, \alpha) = h_\mu(\sigma_2)$. Taking m big enough in the previous equality we deduce that it is enough to consider all finite partitions $\alpha \subset \mathcal{C}$ with the following property. For each $A \in \alpha$ and each $i \leq 1, y \in \pi_{i,i}(A)$ we have the condition $\pi_{1,1}^2((\pi_{2,2}^2)^{-1}(y)) \subset \pi_{i-1,i-1}(A)$. It then follows that μ projects on $\mu' \in \mathcal{M}(\sigma_1)$ and $h_\mu(\sigma_2) = h_{\mu'}(\sigma_1)$. The variational principle yields $h_2 \leq h_1$ and the equalities of all three entropies are established.

To prove the three equalities on the topological pressure we use the analogous arguments for the topological pressure. $\diamond$

Let $h(\Gamma)$ to be one of the entropies in Theorem 3.1. We call $h(\Gamma)$ the entropy of Γ. For $f \in C(X)$ we denote by $P(\Gamma, f)$ to be one of the topological in Theorem 3.1. Let X be a complete metric space with a metric d. Denote by $B(x,r)$ the open ball of radius r centered in x. Let $\bar{B}(x,r) = Closure(B(x,r))$. We say that X is semi-Riemannian of Hausdorff dimension $n \geq 0$ if for every open ball $B(x,r), 0 < r < \delta$ the Hausdorff dimension of $\bar{B}(x,r)$ is n and its Hausdorf volume $vol(\bar{B}(x,r))$ satisfies the inequality

$$\alpha r^n \leq vol(\bar{B}(x,r))$$

for some $0 < \alpha$. Recall that if the Hausdorff dimension of a compact set $Y \subset X$ is m then its Hausdorff volume is defined as follows.

$$vol(Y) = \lim_{\epsilon \to 0} \liminf_{x_i, 0 < \epsilon_i \leq \epsilon, i=1,\ldots,k, \cup B(x_i,\epsilon_i) \supset Y} \sum_1^k \epsilon_i^m.$$

The following lemma is a straightforward generalization of Bowen's inequality [**Bow**], [**Wal**, Thm. 7.15].

(3.2) Lemma. *Let X be a semi-Riemannian compact metric space of Hausdorff dimension n. Assume that $T : X \to X$ is Lipschitzian -* $d(T(x), T(y))$

$\leq \lambda d(x, y)$ *for all $x, y \in X$ and some $\lambda \geq 1$. Suppose furthermore that X has a finite n dimensional Hausdorff volume. Then $h(T) \leq \log \lambda^n$.*

Proof. As X is compact and semi-Riemannian it follows that X has the Hausdorff dimension n. Let $N(k, \epsilon)$ be the cardinality of the maximal (k, ϵ) separated set. Assume that $\{x_1, ..., x_{N(k,\epsilon)}\}$ is a maximal (k, ϵ) separated set. That is for $i \neq j$

$$\max_{0 \leq l \leq k-1} d(T^l(x_i), T^l(x_j)) > \epsilon.$$

We claim that

$$\bar{B}(x_i, \epsilon_k) \cap \bar{B}(x_j, \epsilon_k) = \emptyset, i \neq j, \epsilon_k = \frac{\epsilon}{3\lambda^{k-1}}.$$

This is immediate from the inequality $d(T^l(x), T^l(y)) \leq \lambda^l d(x, y)$ and the (k, ϵ) separability of $\{x_1, ..., x_{N(k,\epsilon)}\}$. We thus deduce the obvious inequality

$$\sum_{l=1}^{N(k,\epsilon)} vol(\bar{B}(x_l, \epsilon_k)) \leq vol(X).$$

In the above inequality assume that $\epsilon \leq \delta$. Then the lower bound on $vol(\bar{B}(x_l, \epsilon_k))$ yields

$$N(k, \epsilon) \leq \frac{vol(X) 3^n \lambda^{n(k-1)}}{\alpha \epsilon^n}.$$

Thus

$$h(T) = \lim_{\epsilon \to 0} \limsup_{k \to \infty} \frac{\log N(k, \epsilon)}{k} \leq n \log \lambda$$

and the proof of the lemma is completed. $\diamond$

The above estimate can be improved as follows. Let X be a compact metric space and $T : X \to X$. Set

$$L(T) = \sup_{x \neq y \in X} \frac{d(T(x), T(y))}{d(x, y)}, \quad L_+(T) = \max(L(T), 1).$$

Thus T is Lipschitzian iff $L(T) < \infty$. Let

$$l(T) = \liminf_{k \to \infty} L_+^{\frac{1}{k}}(T^k).$$

Note that T^k is Lipschtzian for some $k \geq 1$ iff $l(T) < \infty$. $l(T)$ can be considered as a generalization of the maximal Lyapunov exponent for the mapping T. As $h(T^k) = kh(T), k \geq 0$ from Lemma 3.2 we obtain.

(3.3) Theorem. *Let X be a semi-Riemannian compact metric space of Hausdorff dimension n. Assume that $T : X \to X$ is a continuous map. Suppose furthermore that X has a finite n dimensional Hausdorff measure. Then $h(T) \leq n \log l(T)$.*

We have in mind the following application. Let $T : \mathbf{C}P^1 \to \mathbf{C}P^1$ be a rational map of the Riemann sphere $\mathbf{C}P^1$. Let $X = J(T)$ be its Julia set. It is plausible to assume that $\log l(T)$ on X is the Lyapunov exponent corresponding to T and the maximal T-invariant measure on X. Suppose that the Hausdorff dimension of X is n and X has a finite Hausdorff volume. Assume furthermore that X is semi-Riemannian of Hausdorff dimension n. We then can apply Theorem 3.3. As $h(T) = \log deg(T)$ we have the inequality $deg(f) \leq l(f)^n$.

(3.4) Theorem. *Let X be a semi-Riemannian compact metric space of Hausdorff dimension n. Assume that $T_i : X \to X, i = 1, ..., m$, are continuous maps. Let $\Gamma(T_i)$ be the graph of $T_i = 1, ..., m$. Set $\Gamma = \cup_1^m \Gamma(T_i)$. Suppose furthermore that X has a finite n dimensional Hausdorff volume. Then*

$$h(\Gamma) \leq \log \sum_1^m L_+(T_i)^n.$$

Proof. It is enough to consider the nontrivial case where each T_i is Lipschitzian. In the definitions of the metrics on $\Gamma^k, \Gamma_+^\infty$ set

$$\rho > \max_{1 \leq i \leq m} L_+(T_i).$$

Let $M = \{1, ..., m\}$. Then for $\omega = (\omega_1, ..., \omega_{k-1}) \in M^{k-1}$ we let

$$\Gamma(\omega) = \{(x_i)_1^k : x_1 \in X, x_i = T_{\omega_{i-1}} \circ \cdots \circ T_{\omega_1}(x_1), i = 2, ..., k\} \subset \Gamma^k,$$
$$\omega \in M^{k-1}.$$

Clearly, each $\Gamma(\omega)$ is isometric to X. Hence, the Hausdorff dimension of $\Gamma(\omega)$ is n and $vol(\Gamma(\omega)) = vol(X)$. Furthermore, $\cup_{\omega \in M^{k-1}}\Gamma(\omega) = \Gamma^k$. It then follows that each Γ^k has Hausdorff dimension n, has finite Hausdorff volume not exceeding $m^{k-1}vol(X)$ and is semi-Riemannian compact metric space of Hausdorff dimension n. Moreover, the volume of any closed ball

$\bar{B}(y,r) \subset \Gamma^k$ is at least αr^n where α is the constant for X. Let $Y = \Gamma_+^\infty$ and consider a maximal (k, ϵ) separated set in Y of cardinality $N(k, \epsilon)$ - $y^j \in Y, j = 1, ..., N(k, \epsilon)$. That is

$$y^j = (x_i^j)_{i=1}^\infty, (x_i^j, x_{i+1}^j) \in \Gamma, i = 1, ..., j = 1, ..., N(k, \epsilon),$$

$$\max_{1 \le i} \frac{d(x_i^j, x_i^l)}{\rho^{(i-k)^+}} > \epsilon, 1 \le j \ne l \le N(k, \epsilon).$$

Here, $a^+ = \max(a, 0), a \in \mathbf{R}$. Fix $\epsilon, 0 < \epsilon < \delta$. Assume that D is the diameter of X and let $K(\epsilon) = \lceil log_\rho D - log_\rho \epsilon \rceil$. It then follows that

$$\max_{1 \le i \le k+K(\epsilon)} d(x_i^j, x_i^l) > \epsilon, 1 \le j \ne l \le N(k, \epsilon). \tag{3.5}$$

Set $z^j = (x_i^j)_{i=1}^{k+K(\epsilon)} \subset \Gamma^{k+K(\epsilon)}, j = 1, ..., N(k, \epsilon)$. Clearly,

$$\{z^j\}_1^{N(k+K(\epsilon))} = \cup_{\omega \in M^{k+K(\epsilon)-1}}(\{z^j\}_1^{N(k,\epsilon)} \cap \Gamma(\omega)) \Rightarrow$$

$$N(k, \epsilon) \le \sum_{\omega \in M^{k+K(\epsilon)-1}} Card(\{z^j\}_1^{N(k,\epsilon)} \cap \Gamma(\omega)).$$

We now estimate $Card(\{z^j\}_1^{N(k,\epsilon)} \cap \Gamma(\omega))$ for a fixed $\omega = (\omega_1, ..., \omega_{k+K(\epsilon)-1})$ $\in M^{k+K(\epsilon)-1}$. For each $z^j = (x_i^j)_{i=1}^{k+K(\epsilon)} \in \Gamma(\omega)$ consider the closed set ball

$$\bar{B}(z^j, \epsilon(\omega)) \subset \Gamma(\omega), \epsilon(\omega) = \frac{\epsilon}{3 \prod_1^{k+K(\epsilon)-1} L_+(T_{\omega_i})}.$$

(We restrict here our discussion to the compact metric space $\Gamma(\omega)$ with the metric induced from $\Gamma^{k+K(\epsilon)}$.) Let $z^j \ne z^l \in \Gamma(\omega)$. The condition (3.5) yields that $\bar{B}(z^j, \epsilon(\omega)) \cap \bar{B}(z^l, \epsilon(\omega)) = \emptyset$. As $\Gamma(\omega)$ is isometric to X we deduce that

$$Card(\{z^j\}_1^{N(k,\epsilon)} \cap \Gamma(\omega)) \le \frac{vol(X)3^n \prod_{i=1}^{k+K(\epsilon)-1} L_+(T_{\omega_i})^n}{\alpha \epsilon^n}.$$

Hence,

$$N(k, \epsilon) \le \sum_{\omega \in M^{k+K(\epsilon)-1}} \frac{vol(X)3^n \prod_{i=1}^{k+K(\epsilon)-1} L_+(T_{\omega_i})^n}{\alpha \epsilon^n} =$$

$$\frac{vol(X)3^n \left(\sum_{i=1}^m L_+(T_i)^n\right)^{k+K(\epsilon)-1}}{\alpha \epsilon^n}.$$

Thus

$$h(\Gamma) = \lim_{\epsilon \to 0} \limsup_{k \to \infty} \frac{log N(k, \epsilon)}{k} \le \log \sum_{i=1}^{n} L_+(T_i)^n$$

and the theorem is proved. ◇

We remark that the inequality of Theorem 3.4 holds if we replace the assumption that X has a finite n-Hausdorff volume by the following one: the number of points of every $r - separated$ set in X does not exceed Cr^{-n} for some positive constant C.

Let X satisfies the assumptions of Theorem 3.4. It then follows that for the Lipschitzian maps $f : X \to X$ the quantity $L_+(T)^n$ is the "norm" of the graph $\Gamma(f)$ discussed in §2.

(3.6) Lemma. *Let X be a compact metric space and $T : X \to X$ be a noninvolutive homeomorphism ($T^2 \neq Id$). Then $\log 2 \le h(\Gamma(T) \cup \Gamma(T^{-1}))$. If $T, T^{-1} : X \to X$ are noninvolutive isometries then $h(\Gamma(T) \cup \Gamma(T^{-1})) = \log 2$.*

Proof. Assume first that T has a periodic orbit $Y = \{y_1, ..., y_p\}$ of period $p > 2$. Restrict T, T^{-1} to this orbit. Theorem 2.4 yields the desired inequality. Assume now that we have an infinite orbit $y_i = T^i(y), i = 1, 2, ...,$. Fix $n \ge 3$. Let $Y_n = \{y_1, ..., y_n\}$. Denote by $\Gamma_n \subset Y_n \times Y_n$ the graph corresponding to the undirected linear graph on the vertices $y_1, ..., y_n$. That $(i, j) \in \Gamma_n \iff |i - j| = 1$. Clearly

$$\Gamma_n^\infty \subset \Gamma^\infty, \quad \Gamma = \Gamma(T) \cup \Gamma(T^{-1}).$$

Hence $h(\Gamma_n) \le h(\Gamma)$. Obviuosly, $h(\Gamma_n) = log \rho(A(\Gamma_n))$. It is well known that $\rho(A(\Gamma_n)) = 2cos\frac{\pi}{n+1}$. (The eigenvalues of $A(\Gamma_n)$ are the roots of the Chebycheff polynomial.) Let $n \to \infty$ and deduce $h(\Gamma) \ge \log 2$. Assume now that T and T^{-1} are noninvolutive isometries. Then Theorem 3.4 and the above inequality implies that $h(\Gamma(T) \cup \Gamma(T^{-1})) = \log 2$. ◇

Thus, Theorem 3.4 is sharp for $m = 2$. Similar examples using isometries and Theorem 2.4 show that Theorem 3.4 is sharp in general.

Let X be a compact metric space and $T_i : X \to X, i = 1, ..., m$, be a set of continuous transformations. Let $\mathcal{T} = \{T_1, ..., T_m\}$. Then $h(\mathcal{S}(\mathcal{T}))$ was defined to be the entropy of the graph $\Gamma = \cup_1^m \Gamma(T_i)$. As in the case of $m = 1$ this entropy can be defined in terms of $"(k, \epsilon)"$ separated (spanning) sets as follows. Set

$$d_{k+1}(x, y) = \max(\max_{1 \le i_1, j_1, ..., i_k, j_k \le m,} d(T_{i_1} ... T_{i_k}(x), T_{j_1} ... T_{j_k}(y)), d(x, y)),$$

$$k = 1, 2, ...,.$$

Let $M(k, \epsilon)$ be the maximal cardinality the ϵ separated set in the metric d_k.

(3.7) Lemma. *Let X be a compact metric space and assume that $T_i : X \to X, i = 1, ..., m$, are continuous transformations. Then*

$$h(\mathcal{S}(\{T_1, ..., T_m\})) = \lim_{\epsilon \to 0} \limsup_{k \to \infty} \frac{\log M(k, \epsilon)}{k}.$$

Proof. From the definiton of the (k, ϵ) separated set for Γ_+^∞ it immediately follows that

$$M(k, \epsilon) \le N(k, \epsilon).$$

The arguments in the proof of Theorem 3.4 yield that

$$N(k, \epsilon) \le M(k + K(\epsilon))$$

and the lemma follows. $\diamond$

§4. Approximating entropy of graphs by entropy of subshifts of finite type

Let X be a set. $\mathcal{U} = \{U_1, ..., U_m\} \subset 2^X$ is called a finite cover of X if $X = \cup_1^m U_i$. The cover $\mathcal{U}$ is called minimal if any strict subset of $\mathcal{U}$ is not a cover of X. Let $\Gamma \subset X \times X$ be any subset. Introduce the following graph and its corresponding matrix on the space $< m >= \{1, ..., m\}$:

$$\mathcal{U} = \{U_1, ..., U_m\}, \ \Gamma(\mathcal{U}) = \{(i, j) : \Gamma \cap U_i \times U_j \ne \emptyset\} \subset < m > \times < m >,$$

$$A(\Gamma(\mathcal{U})) = (a_{ij})_1^m \in M_m(\{0 - 1\}), a_{ij} = 1 \iff (i, j) \in \Gamma(\mathcal{U})\}.$$

Note that $\Gamma(\mathcal{U})$ induces a subshift of a finite type on $< m >$. Thus, $\log^+ \rho(\Gamma(\mathcal{U}))$ is the entropy of Γ induced by the cover $\mathcal{U}$. Let $\mathcal{V}$ be also a finite cover of X. Then $\mathcal{V}$ is called a refinement of $\mathcal{U}$, written $\mathcal{U} < \mathcal{V}$, if every member of $\mathcal{V}$ is a subset of a member of $\mathcal{U}$. Assume that $\mathcal{V} = \{V_1, ..., V_m\}$ is a refinement of $\mathcal{U}$ such that $V_i \subset U_i, i = 1, ..., m$. It then follows that $A(\Gamma(\mathcal{U})) \ge A(\Gamma(\mathcal{V}))$ for any $\Gamma \subset X \times X$. Hence, $\rho(A(\Gamma(\mathcal{U}))) \ge \rho(A(\Gamma(\mathcal{V})))$. If $U_i \cap U_j = \emptyset, 1 \le i < j \le m$, then $\mathcal{U}$ is called a finite partition of X. Given a finite minimal cover $\mathcal{U} = \{U_1, ..., U_m\}$ there always exist a partition $\mathcal{V} = \{V_1, ..., V_m\}$ such that $V_i \subset U_i, i = 1, ..., m$. Indeed, consider a partition $\mathcal{U}'$ corresponding to the subalgebra generated by $\mathcal{U}$. This partition is a refinement of $\mathcal{U}$. Then each U_i is union of some sets in $\mathcal{U}'$. Set $V_1 = U_1$. Let $V_2 \subset U_2$ be the union of sets of $\mathcal{U}'$ which are subsets of $U_2 \backslash U_1$. Continue this process to construct $\mathcal{V}$. In particular, $\rho(A(T, \mathcal{U})) \ge \rho(A(T, \mathcal{V}))$.

Let $\mathcal{U} < \mathcal{V}$ be finite partitions of X. Assume that $\Gamma \subset X \times X$. In general, there is no relation between $\rho(A(\Gamma(\mathcal{U})))$ and $\rho(A(\Gamma(\mathcal{V})))$. Indeed, if $A(\Gamma(\mathcal{V}))$ is a matrix whose all entries are equal to 1 then $A(\Gamma(\mathcal{U}))$ is also a matrix whose all entries are equal to 1. Hence

$$\rho(A(\Gamma(\mathcal{V}))) = Card(\mathcal{V}) > \rho(A(\Gamma(\mathcal{U}))) = Card(\mathcal{U}) \iff \mathcal{U} \neq \mathcal{V}.$$

Assume now that $Card(\mathcal{V}) = n, A(\Gamma(\mathcal{V})) = (\delta_{(i+1)j})_1^n, n+1 \equiv 1$ be the matrix corresponding to a cyclic graph on $< n >$. Suppose furthermore that $n \geq 3$ and let $U_1 = V_1 \cup V_2, U_i = V_{i+1}, i = 2, ..., n-1$. It then follows that $\rho(A(\Gamma(\mathcal{U}))) > \rho(A(\Gamma(\mathcal{V}))) = 1$.

Let $\mathcal{F}_\epsilon, 0 < \epsilon < 1$ be a family of finite covers of X increasing in ϵ. That is, $\mathcal{F}_\delta \subset \mathcal{F}_\epsilon, 0 < \delta \leq \epsilon < 1$. Assume that $\Gamma \subset X \times X$ be any set. We then set

$$e(\Gamma, \mathcal{F}) = \lim_{\epsilon \to 0^+} \inf_{\mathcal{U} \in \mathcal{F}_\epsilon} \log^+ \rho(A(\Gamma(\mathcal{U}))).$$

Thus, $e(\Gamma, \mathcal{F})$ can be considered as the entropy of Γ induced by the family $\mathcal{F}_\epsilon$. Its definition is reminiscent of the definition of the Hausdorff dimension of a metric space X. Let $\mathcal{U}$ be a finite cover of X. Clearly, $A(\Gamma^T(\mathcal{U})) = A^T(\Gamma(\mathcal{U}))$. Hence, $\rho(A(\Gamma(\mathcal{U}))) = \rho(A(\Gamma^T(\mathcal{U})))$ and $e(\Gamma, \mathcal{F}) = e(\Gamma^T, \mathcal{F})$.

(4.1) Lemma. *Let $\mathcal{U} = \{U_1, ..., U_m\}$ be a finite cover of compact metric space X. Assume that $\operatorname{diam}(\mathcal{U}) \overset{\text{def}}{=} \max \operatorname{diam}(U_i) \leq \frac{\delta}{2}$. Let $\Gamma \subset X \times X$ be a closed set. Assume that $N_k(\delta)$ is the maximal cardinality of (k, δ) separated set for $\sigma : \Gamma_+^\infty \to \Gamma_+^\infty$. Then*

$$\limsup_{k \to \infty} \frac{\log N_k(\delta)}{k} \leq \log \rho(A(\Gamma(\mathcal{U}))).$$

Proof. Set

$$A^k(\Gamma(\mathcal{U})) = (a_{ij}^{(k)})_1^m, \nu_k(\mathcal{U}) = \sum_1^m a_{ij}^{(k-1)}.$$

Then $\nu_k(\mathcal{U})$ is counting the number of distinct point

$$(y_i)_1^k \in < m >^k, (y_i, y_{i+1}) \in \Gamma(\mathcal{U}), i = 1, ..., k-1.$$

Let $K(\delta)$ be defined as in the proof of Theorem 3.4. We claim that $N(k, \delta) \leq \nu_{k+K(\delta)}(\mathcal{U})$. Indeed, assume that $x^i = (x_j^i)_{j=1}^\infty, i = 1, ..., N(k, \delta)$, is a (k, δ) separated set. Then each x^i generates at least one point $y^i = (y_1^i, ..., y_p^i) \in < m >^p$ as follows: $x_j^i \in U_{y_j^i}, j = 1, ..., p$. From (3.5) and the assumption that $\operatorname{diam}(\mathcal{U}) < \frac{\delta}{2}$ we deduce that for $p = k + K(\delta)$ $i \neq l \Rightarrow y^i \neq y^l$. Hence

$N(k, \delta) \leq \nu_{k+K(\delta)}(\mathcal{U})$. As a point x^i may generate more then one point y^i in general we have strict inequality. Since $A(T, \mathcal{U})$ is a nonnegative matrix it is well known that

$$K_1 \rho(A)^k \leq \nu_k \leq K_2 k^{m-1} \rho(A(T, \mathcal{U}))^k, k = 1, ..., .$$

See for example [**F-S**]. The above inequalities yield the lemma. ◇

Let $\{\mathcal{U}_i\}_1^\infty$ be sequence of finite open covers such $\text{diam}(\mathcal{U}_i) \to 0$. Assume that $\Gamma \subset X \times X$ is closed. Then $\{\mathcal{U}_i\}_1^\infty$ is called an approximation cover sequence for Γ if

$$lim_{i \to \infty} \log^+ \rho(A(\Gamma(\mathcal{U}_i))) = h(\Gamma).$$

Note as $\rho(A^T) = \rho(A), \forall A \in M_n(\mathbf{C})$ and $h(\Gamma) = h(\Gamma^T)$ we deduce that $\{\mathcal{U}_i\}_1^\infty$ is also an approximation cover for Γ^T. Use Lemma 4.1 and (2.2) for finite graphs to obtain sufficient conditions for the validity of the inequality (2.2) for infinite graphs.

(4.2) Corollary. *Let X be a compact metric space and $\Gamma_j^T = \Gamma_j \subset X \times X, j = 1, ..., m$ be closed sets. Assume that there exist a sequence of open finite covers*

$$\{\mathcal{U}_i\}_1^\infty, \lim_{i \to \infty} \text{diam}(\mathcal{U}_i) = 0,$$

which is an approximation cover for $\Gamma_1, ..., \Gamma_m$. Then

$$h(\cup_1^m \Gamma_j) \leq \log \sum_1^m e^{h(\Gamma_j)}.$$

Let Z be a compact metric space and $T : Z \to Z$ is a homeomorphism. Then T is called expansive if there exists $\delta > 0$ such that

$$\sup_{n \in \mathbf{Z}} d(T^n(x), T^n(y)) > \delta, \forall x, y \in Z, x \neq y.$$

A finite open cover $\mathcal{U}$ of Z is called a generator for homeomorphism T if for every bisequence $\{U_n\}_{-\infty}^\infty$ of members of $\mathcal{U}$ the set $\cap_{n=-\infty}^\infty T^{-n} \bar{U}_n$ contains at most one point of X. If this condition is replaced by $\cap_{n=-\infty}^\infty U_n$ then $\mathcal{U}$ is called a weak generator. A basic result due to Keynes and Robertson [**K-R**] and Reddy [**Red**] claims that T is expansive iff T has a generator iff T has a weak generator. See [**Wal**, §5.6]. Moreover, T is a factor of the restriction of a shift S on a finite number of symbols to a closed S−invariant set Δ [**Wal**, Thm 5.24]. If Δ is a subshift of a finite type then T is called

FP. See [**Fr**] for the theory of FP maps. In particular, for any expansive T, $h(T) < \infty$.

Let $\Gamma \subset X \times X$ be a closed set such that $\Gamma^\infty \neq \emptyset$. Then Γ is called expansive if

$$\sup_{n \in \mathbf{Z}} d(\sigma^n(x), \sigma^n(y)) > \delta, \forall x, y \in \Gamma^\infty, x \neq y$$

for some $\delta > 0$. A finite open cover $\mathcal{U}$ of X is called a generator for Γ if for every bisequence $\{U_n\}_{-\infty}^\infty$ of members of $\mathcal{U}$ the set

$$x = (x_n)_{-\infty}^\infty \in \Gamma^\infty, x_n \in \bar{U}_n, n \in \mathbf{Z}$$

contains at most one point of Γ^∞. If this condition is replaced by $x_n \in U_n$ then $\mathcal{U}$ is called a weak generator. We claim that Γ is expansive iff Γ has a generator iff Γ has a weak generator. Indeed, observe first that the condition that Γ is expansive is equivalent to the assumption that σ is expansive on Γ^∞. Let $V_i = \pi_{1,1}^{-1}(U_i) \subset X^\infty, i = 1, ..., m$. That is, V_i is an open cylindrical set in X^∞ whose projection on the first coordinate is U_i while on all other coordinates is X. Set $W_i = V_i \cap \Gamma^\infty, i = 1, ..., m$. It now follows that $W_1, ..., W_m$ is a standard set of generators for the map $\sigma : \Gamma^\infty \to \Gamma^\infty$.

Assume that $T : X \to X$ is expansive with the expansive constant δ. It is known [**Wal**, Thm. 7.11] that

$$h(T) = \limsup_{k \to \infty} \frac{\log N(k, \delta_0)}{k}, \delta_0 < \frac{\delta}{4}.$$

Thus, according to Lemma 4.1 $h(\Gamma) \leq \log \rho(A(\Gamma(\mathcal{U})))$ if Γ is expansive with an expansive constant δ and $\text{diam}(\mathcal{U}) < \frac{\delta}{8}$. Assume that $T_i : X \to X, i = 1, ..., m$, are expansive maps. We claim that for $m > 1$ it can happen that $h(\cup_1^m \Gamma(T_i))$ is infinite. Let T_1 be Anosov map on the 2-torus X in the standard coordinates. Now change the coordinates in X by a homeomorphism and let T_2 be Anosov with respect to the new coordinates. It is possible to choose a homeomorphism (which is not diffeo!) so that that $T_2 \circ T_1$ contains horseshoes of arbitrary many folds. Hence $h(\Gamma(T_1) \cup \Gamma(T_2)) \geq h(T_2 \circ T_1) = \infty$.

§5. Entropy of semigroups of Möbius transformations

Let $X \subset \mathbf{C}P^n$ be an irreducible smooth projective variety of complex dimension n. Assume that $\Gamma \subset X \times X$ be a projective variety such that the projections $\pi_{i,i} : \Gamma \to X, i = 1, 2$ are onto and finite to one. Then Γ can be viewed as a graph of an algebraic function. In algebraic geometry such a graph is called a correspondence. Furthermore, Γ induces a linear operator

$$\Gamma^* : H_{*,a}(X) \to H_{*,a}(X), \quad H_{*,a}(X) = \sum_{j=0}^n H_{2j,a}(X),$$

$$\Gamma^* : H_{2j,a}(X) \to H_{2j,a}(X), j = 0, ..., n.$$

Here, $H_{2j,a}(X)$ is the homology generated by the algebraic cycles of X of complex dimension j over the rationals $\mathbf{Q}$. Indeed, if $Y \subset X$ is an irreducible projective variety then $\Gamma^*([Y]) = [\pi_{2,2}^2((\pi_{1,1}^2)^{-1}(Y))]$. Let $\rho(\Gamma^*)$ be the spectral radius of Γ^*. Assume that first that Γ is irreducible. In [**Fri3**] we showed that $h(\Gamma) \leq \log \rho(\Gamma^*)$. However our arguments apply also to the case Γ is reducible. We also conjectured in [**Fri3**] that in the case that Γ is irreducible we have the equality $h(\Gamma) = \log \rho(\Gamma^*)$. We now doubt the validity of this conjecture. We will show that in the reducible case we can have a strict inequality $h(\Gamma) < \log \rho(\Gamma^*)$. Let $\Gamma_i \subset X \times X, i = 1, ..., m$, be algebraic correspondences as above. Set $\Gamma = \cup_1^m \Gamma_i$. Then

$$\Gamma^* = \sum_1^m \Gamma_i^*, \quad h(\Gamma) \leq log\rho(\sum_1^m \Gamma_i^*).$$

Thus, there is a close analogy between the entropy of algebraic (finite to one) correspondences and entropy of shifts of finite types. Consider the simplest case of the above situation. Let $X = \mathbf{C}P^1$ be the Riemann sphere and Γ be an algebraic curve given by a polynomial $p(x, y) = 0$ on some chart $\mathbf{C}^2 \subset \mathbf{C}P^1 \times \mathbf{C}P^1$. Let $d_1 = deg_y(p), d_2 = deg_x(p), d_1 \geq 1, d_2 \geq 1$. It then follows that $\rho(\Gamma^*) = \max(d_1, d_2)$. Note that $\rho(\Gamma^*) = 1$ iff Γ is the graph of a Möbius transformation. Observe next that if $f_i : \mathbf{C}P^1 \to \mathbf{C}P^1, i = 1, ..., m$, are nonconstant rational maps then the correspondance given by $p(x, y) = \prod_1^m (y - f_i(x))$ is induced by $\Gamma = \cup_1^m \Gamma(f_i)$. In particular,

$$h(\Gamma) \leq \log \sum_1^m deg(f_i). \tag{5.1}$$

Here, by $deg(f_i)$ we denote the topological degree of the map f_i. Combine the above inequality with Lemma 3.6 to deduce that for any noninvolutive Möbius transformation f we have the equality $h(\Gamma(f) \cup \Gamma(f^{-1})) = \log 2$.

(5.2) Lemma. *Let $f, g : \mathbf{C}P^1 \to \mathbf{C}P^1$ be two Möbius transformations such that x as a common fixed attracting point of f and g and y is a common repelling point of f and g, Then $h(\Gamma(f) \cup \Gamma(g)) = 0$.*

Proof. We may assume that

$$f = az, g = bz, 0 < |a|, |b| < 1.$$

Set $\Gamma = \Gamma(f) \cup \Gamma(g)$. It the follows that for any point $\zeta = (z_i)_1^\infty \neq \eta = (\infty)_1^\infty$ $\sigma^l(z)$ converges to the fixed point $\xi = (0)_1^\infty$. That is, the nonwondering set of σ is the set $\{\xi, \eta\}$ on which σ acts trivially. Hence $h(\Gamma) = 0$. $\diamond$

(5.3) Lemma. *Let* $f, g : \mathbf{C}P^1 \to \mathbf{C}P^1$ *be two parabolic Möbius transformation with the same fixed point* $-\infty$, *i.e.* $f = z + a, g = z + b$. *If either* a, b *are linearly independent over* $\mathbf{R}$ *or* $b = \alpha a, \alpha \geq 0$ *then* $h(\Gamma(f) \cup \Gamma(g)) = 0$.

Proof. Let $\Gamma = \Gamma(f) \cup \Gamma(g), \eta = (\infty)_1^\infty$. If ether a, b are linearly independent over $\mathbf{R}$ or $b = \alpha a, \alpha > 0, a \neq 0$ then for any point $\zeta \in \Gamma_+^\infty$ $\sigma^l(\zeta)$ converges to the fixed point η. Hence $h(\Gamma) = 0$. Suppose next that $a = b = 0$. Then σ is the identity map on Γ_+^∞ and $h(\Gamma) = 0$. Assume finally that $b = 0, a \neq 0$. Then Ω limit set of σ consists of all points $\zeta = (z_i)_1^\infty, z_i = z_1, i = 2, ..., $. So $\sigma | \Omega$ is identity and $h(\Gamma) = 0$. $\diamond$

(5.4) Theorem. *Let* $T = z + a, Q = z + b, ab \neq 0$ *be two Möbius transformations of* $\mathbf{C}P^1$. *Assume that there* $\frac{b}{a}$ *is a negative rational number. Then*

$$h(\Gamma) = -\frac{|a|}{|a| + |b|} \log \frac{|a|}{|a| + |b|} - \frac{|b|}{|a| + |b|} \log \frac{|b|}{|a| + |b|}.$$

We first state an approximation lemma which will be used later.

(5.5) Lemma. *Let* X *be compact metric space and* $T : X \to X$ *be a continuous transformation. Assume that we have a sequence of closed subsets* $X_i \subset X, i = 1, ...,$ *which are* $T-$ *invariant, i.e.* $T(X_i) \subset X_i, i = 1, 2, ...,$. *Suppose furthermore that* $\forall \delta > 0 \exists M(\delta)$ *with the following property.* $\forall x \in X \backslash X_i \exists y = y(x, i) \in X_i, \sup_{n \geq 0} d(T^n(x), T^n(y)) \leq \delta$ *for each* $i > M(\delta)$. *Then* $\lim_{i \to \infty} h(T | X_i) = h(T)$.

Proof. Observe first that $h(T) \geq h(T | X_i)$. Thus it is left to show

$$\liminf_{i \to \infty} h(T | X_i) \geq h(T).$$

Let $N(k, \epsilon), N_i(k, \epsilon)$ be the cardinality of maximal (k, ϵ) separating set of X and X_i respectively. Clearly, $N_i(k, \epsilon) \leq N(k, \epsilon)$. Let $x_1, ..., x_{N(k,\epsilon)}$ be a (k, ϵ) separating set of X. Then

$$\forall i > M(\frac{\epsilon}{4}), \ \forall x_j \exists y_{j,i} \in X_i, \sup_{n \geq 0} d(T^n(x_j), T^n(y_{j,i})) \leq \frac{\epsilon}{4}.$$

Hence, $y_{j,i}, j = 1, ..., N(k, \epsilon)$, is $\frac{\epsilon}{2}$ separated set in X_i. In particular, $N(k, \epsilon) \leq N_i(k, \frac{\epsilon}{2}), i > M(\frac{\epsilon}{4})$. Thus

$$\limsup_{k \to \infty} \frac{\log N(k, \epsilon)}{k} \leq \limsup_{k \to \infty} \frac{\log N_i(k, \frac{\epsilon}{2})}{k} \leq h(T | X_i), i > M(\frac{\epsilon}{4}).$$

The characterization of $h(T)$ yields the lemma. $\diamond$

Proof of Theorem 5.4. W.l.o.g. (without loss of generality) we may assume that $a = p, b = -q$ where p, q are two positive coprime integers. First note that $\mathbf{C}P^1$ is foliated by the invariant lines $\Im z = Const$. Hence, the maximal characterization of $h(\sigma)$ as the supremum over all measure entropy $h_\mu(\sigma)$ for all extremal σ invariant measures yields that it enough to restrict ourselves to the action of T, Q on (closure of) the real line. Using the same argument again it is enough to consider the action on the lattice $\mathbf{Z} \subset \mathbf{R}$ plus the point at ∞. We may view $Y = \mathbf{Z} \cup \{\infty\}$ as a compact subspace of $S^1 = \{z : |z| = 1\}$.

$$0 \mapsto 1, \infty \mapsto -1, j \mapsto e^{\frac{\pi\sqrt{-1}(1+2j)}{2j}}, 0 \neq j \in \mathbf{Z}.$$

For a positive integer i let $Y_i = \{-ipq, -ipq + 1, ..., ipq - 1, ipq\}$. Set

$$\Gamma = \Gamma(T) \cup \Gamma(Q) \subset Y \times Y, X = \Gamma_+^\infty, \Gamma_i = \Gamma \cap Y_i \times Y_i, X_i = (\Gamma_i)_+^\infty, i = 1, ..., .$$

We will view a point $x = (x_j)_1^\infty \in X$ a path of a particle who starts at time 1 at x_1 and jumps from the place x_i at time i to the place x_{i+1} at time $i + 1$. At each point of the lattice $\mathbf{Z}$ a particle is allowed to jump p steps forward and q backwards. The point $\xi = (\infty)_1^\infty$ is the fixed point of our random walk. Observe next that Γ_i is a subshift of a finite type on $2ipq + 1$ points corresponding to the random walk in which a particle stays in the space Y_i. Note that $A_i = A(\Gamma_i)$ is a matrix whose almost each row (column) sums to two, except the first and the last $\max(p, q) - 1$ rows (columns). Moreover, $h(\sigma|X_i) = \log \rho(A_i)$. We claim that $X, X_i = 1, ...,$ satisfy the assumption of Lemma 5.5. That is any point $x = (x_j)_1^\infty \in X$ can be approximated up to an arbitrary $\epsilon > 0$ by $y_i = (y_{j,i})_{j=1}^\infty \in X_i$ for $i > M(\epsilon)$. We assume that $i > L$ some fixed big L. Suppose first that $x_j > ipq, j = 1, ..., .$ That is the path described by the vector x never enters X_i. Then consider the following path $y_i = (y_{j,i})_{j=1}^\infty \in X_i$. It starts at the point ipq, i.e. $y_{1,i} = ipq$. Then it jumps p times to the left to the point $(i-1)pq$. Then it the particle jumps q time to the right back to the the point ipq and so on. Clearly, $\sup_{n \geq 0} d(\sigma^n(x), \sigma^n(y_i)) \leq d((i - 1)pq, \infty)$. Hence for i big enough the above distance is less than ϵ. Same arguments apply to the case $x_j < -ipq, j = 1, ..., .$ Consider next a path $x = (x_j)_1^\infty$ which starts outside X_i and then enters X_i at some time. If the particle enters to X_i and then stays for a short time, e.g. $\leq pq$, every time it enters X_i then we can approximate this path by a path looping around the vertex ipq or $-ipq$ in X_i as above. Now suppose that we have a path which enters to X_i at least one time for a longer period of time. We then approximate this path by a path $(y_{i,j})_{j=1}^\infty \in X_i$ such that this path coincide with x for all time when x

is in X_i except the short period when x leaves X_i. One can show that such path exists. (Start with the simple example $p = 1, q = 2$.) It then follows that $\sup_{n \geq 0} d(\sigma^n(x), \sigma^n(y_i)) \leq d((i - K)pq, \infty)$ for some $K = K(p, q)$. If i is big enough then we have the desired approximation. Lemma 5.5 yields

$$h(\Gamma) = \lim_{i \to \infty} \log \rho(A_i).$$

We now estimate $\log \rho(A_i)$ from above and from below. Recall the well known formula for the spectral radius of a nonnegative $n \times n$ matrix A:

$$\rho = \limsup_{m \to \infty} \left(trace(A^m)\right)^{\frac{1}{m}} = \limsup_{m \to \infty} \left(\max_{1 \leq j \leq n} a_{jj}^{(m)}\right)^{\frac{1}{m}}, A^m = (a_{ij}^{(m)})_1^n.$$

Let $A = A_i$. We now estimate $a_{jj}^{(m)}$. Obviously, $a_{jj}^{(m)}$ is positive if $m = (p + q)k$ as we have to move kq times to the right and kp times to the left. Assume that $m = (p + q)k$. To estimate $a_{jj}^{(m)}$ we assume that we have an uncostrained motion on $\mathbf{Z}$. Then the number of all possible moves on $\mathbf{Z}$ bringing us back to the original point is equal to

$$\frac{((p + q)k)!}{(qk)!(pk)!} \leq K\sqrt{p + q}\frac{(p + q)^{(p+q)k}}{q^{qk}p^{pk}}.$$

The last part of inequality follows from the Stirling formula for some suitable K. The characterization of $\rho(A)$ gives the inequality

$$\log \rho(A_i) \leq \log \alpha = \log(p + q) - \frac{p}{p + q}\log p - \frac{q}{p + q}\log q.$$

We thus deduce the upper bound on $h(\Gamma) \leq \log \alpha$. Let $0 < \delta < \alpha$. The Stirling formula yields that for $k > M(\delta)$

$$\frac{((p + q)k)!}{(qk)!(pk)!} \geq (\alpha - \delta)^{(p+q)k}.$$

Fix $k > M(\delta)$ and let $i > k$. Then for $m = (p + q)k$

$$a_{00}^{(m)} = \frac{((p + q)k)!}{(qk)!(pk)!}.$$

Clearly,

$$\rho(A)^m = \rho(A^m) \geq a_{00}^{(m)}.$$

Thus, $h(\Gamma) \geq \log \rho(A_i) \geq \log(\alpha - \delta)$. Let $\delta \to 0$ and deduce the theorem. ◇

Note that $h(\Gamma)$ is the entropy of the Bernoulli shift on two symbols with the distribution $(\frac{p}{p+q}, \frac{q}{p+q})$. This can be explained by the fact that to have a closed orbit of length $k(p + q)$ we need move to the right kq times and to the left kp. That is, the frequency of the right motion is $\frac{q}{p+q}$ and the left motion is $\frac{p}{p+q}$. It seems that Theorem 5.4 remains valid as long as $\frac{a}{b}$ is a real negative number.

(5.6) Theorem. *Let $f, g : \mathbb{C}P^1 \to \mathbb{C}P^1$ be two parabolic Möbius transformations with the same fixed point - ∞, i.e. $f = z + a, g = z + b$ where a, b are linearly independent over* **R**. *Let $\Gamma = \Gamma(f) \cup \Gamma(f^{-1}) \cup \Gamma(g) \cup \Gamma(g^{-1})$. Then $h(\Gamma) = \log 4$.*

Proof. The orbit of any fixed point $z \in \mathbb{C}$ under the action of the group generated by f, g is a lattice in $\mathbb{C}$ which has one accumulation point $\infty \in \mathbb{C}P^1$. Let Y is defined in the proof of Theorem 5.4. Consider the dynamics of $\sigma \times \sigma$ on $Y_j \times Y_j$, for $j = 1, ...,$ as in the proof Theorem 5.4. It then follows that $h(\Gamma) = 2h(\sigma | X) = 2 \log 2$. $\diamond$

Let $\mathcal{T} = \{f_1, ..., f_k\}$ be a set of k - Möbius transformations. Set $\Gamma = \cup_1^k \Gamma(f_i)$. Then (5.1) yields $h(\Gamma) \leq \log k$. Our examples show that we may have a strict inequality even for the case $k = 2$. Let Γ be the correspondence of the Gauss arithmetic-geometric mean $y^2 = \frac{(x+1)^2}{4x}$ [**Bul2**]. Our inequality in [**Fri3**] yield that $h(\Gamma) \leq \log 2$. According to Bullet [**Bul2**] it is possible to view the dynamics of Γ as a factor of the dynamics of $\tilde{\Gamma} = \Gamma(f_1) \cup \Gamma(f_2)$ for some two Möbius transformations f_1, f_2. Hence, $h(\Gamma) \leq h(\tilde{\Gamma})$. If $h(\tilde{\Gamma}) < \log 2$ we will have a counterexample to our conjecture that $h(\Gamma) = \log 2$. Even if $h(\tilde{\Gamma}) = \log 2$ we can still have the inequality $h(\Gamma) < \log 2$ as the dynamics of Γ is a subfactor of the dynamics of $h(\tilde{\Gamma})$. Thus, it would be very interesting to compute $h(\Gamma)$.

Assume that $\mathcal{T}$ generates nonelementary Kleinian group. Theorem 2.5 suggests that $e^{h(\Gamma)}$ may have a noninteger value. It would be very interesting to find such a Kleinian group.

We now state an open problem which is inspired by Furstenberg's conjecture [**Fur**]. Assume that $1 < p < q$ are two co-prime integers. (More generally $p^m = q^n \Rightarrow m = n = 0$.) Let

$$f, g : \mathbb{C}P^1 \to \mathbb{C}P^1, T_1(z) = z^p, T_2(z) = z^q, z \in \mathbb{C}^1, f(\infty) = g(\infty). = \infty$$

Note that for f and g $0, \infty$ are two attractive points with the interior and the exterior of the unit disk as basins of attraction respectively. Thus, the nontrivial dynamics takes place on the unit circle S^1. Note that $f \circ g = g \circ f$. Hence f and g have common invariant probability measures. Let $\mathcal{M}$ be the convex set of all probability measures invariant under f, g. Denote by $\mathcal{E} \subset \mathcal{M}$ the set of the extreme points of $\mathcal{M}$ in the standard w^* topology. Then $\mathcal{E}$ is the set of ergodic measures with respect to f, g. (For a recent discussion on the common invariant measure of a semigroup of commuting transformation see [Fri4]). Furstenberg's conjecture (for $p = 2, q = 3$) is that any ergodic measure $\mu \in \mathcal{E}$ is either supported on a finite number of points or is the Lebesgue (Haar) measure on S^1. See [Rud] and [K-S] for the recent results on this conjecture. Let $\mathcal{G}$ be the semigroup generated by $\mathcal{T} = \{f, g\}$. Then (0.2) for $X = S^1$ or the results of [Fri3] yield the inequality $h(\mathcal{G}(\mathcal{T})) \leq \log(p + q)$. What is the value of $h(\mathcal{G}(\mathcal{T}))$? It is plausible to conjecture equality in this inequality.

References

[Bow] R. Bowen, Entropy for group endomorphisms and homogeneous spaces, *Trans. Amer. Math. Soc.* 153 (1971), 404-414.

[Bul1] S. Bullett, Dynamics of quadratic correspondences, *Nonlinearity* 1 (1988), 27-50.

[Bul2] S. Bullett, Dynamics of the arithmetic-geometric mean, *Topology* 30 (1991), 171-190.

[Fr] D. Fried, Finitely presented dynamical systems, *Ergod. Th. & Dynam. Sys.* 7 (1987), 489-507.

[Fri1] S. Friedland, Entropy of polynomial and rational maps, *Annals Math.* 133 (1991), 359-368.

[Fri2] S. Friedland, Entropy of rational selfmaps of projective varieties, *Advanced Series in Dynamical Systems* Vol. 9, pp. 128-140, World Scientific, Singapore 1991.

[Fri3] S. Friedland, Entropy of algebraic maps, *J. Fourier Anal. Appl.* 1995, to appear.

[Fri4] Invariant measures of groups of homeomorphisms and Auslander's conjecture, *J. Ergod. Th. & Dynam. Sys.* 1995, to appear.

[F-S] S. Friedland and H. Schneider, The growth of powers of nonnegative matrix, *SIAM J. Algebraic Discrete Methods* 1 (1980), 185-200.

[Fur] H. Furstenberg, Disjointness in ergodic theory, minimal sets, and a problem on diophantine approximation, *Math. Sys. Theory* 1 (1967), 1-49.

[Gan] F.R. Gantmacher, *Theory of Matrices*, II, Chelsea Pub. Co., New York, 1960.

[G-L-W] E. Ghys, R. Langenvin and P. Walczak, Entropie géométrique des feuilletages, *Acta Math.* 160 (1988), 105-142.

[Gro] M. Gromov, On the entropy of holomorphic maps, preprint, 1977.

[**Hur**] M. Hurley, On topological entropy of maps, *Ergod. Th. & Dynam. Sys.* 15 (1995), 557-568.

[**K-S**] A. Katok and R.J. Spatzier, Invariant measures for higher rank hyperbolic abelian actions, *preprint*.

[**K-R**] B. Keynes and J.B. Robertson, Generators for topological entropy and expansiveness, *Math. Systems Theory* 3 (1969), 51-59.

[**L-P**] R. Langevin and F. Przytycki, Entropie de l'image inverse d'une application, *Bull Soc. Math. France* 120 (1992), 237-250.

[**L-W**] R. Langevin and Walczak, Entropie d'une dynamique, *C.R. Acad. Sci. Paris*, 1991.

[**M-R**] H.F. Münzner and H.M. Rasch, Iterated algebraic functions and functional equations, *Internat. J. Bifur. Chaos Appl. Sci. Engrg.* 1 (1991), 803-822.

[**N-P**] Z. Nitecki and F. Przytycki, The entropy of the relation inverse to a map II, *Preprint*, 1990.

[**Red**] W.L. Reddy, Lifting homeomorphisms to symbolic flows, *Math. Systems Theory* 2 (1968), 91-92.

[**Rud**] D.J. Rudolph, ×2 and ×3 invariant measures and entropy, *Ergod. Th. & Dynam. Sys.* 10 (1990), 823-827.

[**Wal**] P. Walters, *An Introduction to Ergodic Theory*, Springer 1982.

On Representation of Integers in Linear Numeration Systems

Christiane Frougny * and Boris Solomyak †

Abstract

Linear numeration systems defined by a linear recurrence relation with integer coefficients are considered. The normalization function maps any representation of a positive integer with respect to a linear numeration system onto the normal one, obtained by the greedy algorithm. Addition is a particular case of normalization. We show that if the characteristic polynomial of the linear recurrence is the minimal polynomial of a Pisot number, then normalization is a function computable by a finite 2-tape automaton on any finite alphabet of integers. Conversely, if the characteristic polynomial is the minimal polynomial of a Perron number which is not a Pisot number, then there exist alphabets on which normalization is not computable by a finite 2-tape automaton.

1 Introduction

In this paper we study numeration systems defined by a linear recurrence relation with integer coefficients. These numeration systems have also been considered in [Fra] and [PT]. The best known example is the Fibonacci numeration system defined from the sequence of Fibonacci numbers. In the Fibonacci numeration system, every integer can be represented using digits 0 and 1. The representation is not unique, but one of them is distinguished : the one which does not contain two consecutive 1's.

Let U be an integer sequence satisfying a linear recurrence. By a greedy algorithm, every positive integer has a representation in the system U that

*Université Paris 8 and Laboratoire Informatique Théorique et Programmation, Institut Blaise Pascal, 4 place Jussieu, 75252 Paris Cedex 05, France. Supported in part by the PRC Mathématiques et Informatique of the Ministère de la Recherche et de l'Espace.

†Department of Mathematics GN-50, University of Washington, Seattle, Washington 98195, USA. Supported in part by USNSF Grant 9201369.

we call the *normal* representation. The *normalization* is the function which transforms any representation on any finite alphabet of integers into the normal one. From now on, by alphabet we mean finite alphabet of integers.

Addition of two integers represented in the system U can be performed as follows : add the two representations digit by digit, without carry, and then normalize this word to obtain the normal representation of the sum. Thus addition can be viewed as a particular case of normalization.

In this paper we continue with the study of the process started in [F1], [F2] of normalization in linear numeration systems from the point of view of computability by finite automata. Finite automata are a "simple" model of computation, since only a finite memory is used. It is known that in the standard k-ary numeration system, where k is an integer ≥ 2, addition and more generally normalization on any alphabet are computable by a finite 2-tape automaton (see [E]).

The results presented here are strongly connected with symbolic dynamics and representation of real numbers. Let us recall the following. Let θ be a real number > 1. Every real number has a normal representation in base θ called its θ-*expansion* ([R], [P]). A *symbolic dynamical system* is a closed shift-invariant subset of $A^{\mathbf{N}}$, the set of infinite sequences on an alphabet A. The θ-*shift* is a symbolic dynamical system which is the closure of the set of infinite sequences which are θ-expansions of numbers of $[0, 1[$. A symbolic dynamical system is said to be of *finite type* if the set of its finite factors is defined by the interdiction of a finite set of words. It is said to be *sofic* if the set of its finite factors is recognized by a finite automaton.

The nature of the θ-shift is related to the arithmetical properties of θ. A *Pisot* number is an algebraic integer > 1 such that every algebraic conjugate has modulus < 1. A *Salem* number is an algebraic integer > 1 such that every algebraic conjugate has modulus ≤ 1, with at least one conjugate of modulus 1. A *Perron* number is an algebraic integer which is bigger in modulus than all other algebraic conjugates. If θ is a Pisot number then the θ-shift S_θ is sofic [Bel, Sc]. If S_θ is sofic then θ is a Perron number [L].

For normalization the following result is known : normalization of real numbers in base θ is a function computable by a finite 2-tape automaton on any alphabet if and only if θ is a Pisot number [BF].

We give here results for the normalization of integers which are similar to the above mentioned result for the normalization of real numbers. Let U be an integer sequence satisfying a linear recurrence. If the characteristic polynomial P of U is the minimal polynomial of a Pisot number $\theta > 1$, then the set of normal U-representations of nonnegative integers is recognizable by a finite automaton (Theorem 4), and normalization in the system U is a function computable by a finite 2-tape automaton on any alphabet (Theorem 2). If the characteristic polynomial P of U is the minimal polynomial of a number $\theta > 1$ which is a Perron number, but is not a Pisot number, then there exist alphabets on which normalization is not computable by a finite 2-tape automaton (Theorem 3).

Thus the question when normalization is computable by a finite 2-tape automaton is completely settled, given that the polynomial of the recurrence is irreducible and has a positive dominant root. Although some of our arguments work when P is reducible, we don't have a complete answer for that case.

Connections with dynamics are also present in the methods we use. Results from [F1, F2] enable us to reduce the problem to the study of polynomial division on a sequence of polynomials associated to a word in the alphabet. This in turn becomes a problem of dynamics on the integral lattice in $\mathbf{R}^m$ (m is the degree of P), and the Pisot condition translates into having a one-dimensional unstable subspace.

2 Definitions

2.1 Representation of integers and linear numeration systems

Let $U = (u_n)_{n \geq 0}$ be a strictly increasing sequence of integers with $u_0 = 1$. A *representation in the system* U — or a *U-representation* — of a nonnegative integer N is a finite sequence of integers $(d_n)_{0 \leq n \leq k}$ such that

$$N = \sum_{n=0}^{k} d_n u_{k-n}.$$

Such a representation will be written $d_0 \cdots d_k$, most significant digit first.

We say that a word $d = d_0 \cdots d_k$ is *lexicographically greater* than a word $f = f_0 \cdots f_k$, and this will be denoted by $d >_{lex} f$, if there exists an index $0 \leq i \leq k$ such that $d_0 = f_0, \ldots, d_{i-1} = f_{i-1}$ and $d_i > f_i$. Among all possible U-representations $d_0 \cdots d_k$ of a given nonnegative integer N one is distinguished and called the *normal U-representation* of N : the greatest in the lexicographical ordering. It is sometimes (see [Sh]) called the *greedy* representation, since it can be obtained by the following greedy algorithm (see [Fra]) :

Given integers m and p let us denote by $q(m, p)$ and $r(m, p)$ the quotient and the remainder of the Euclidean division of m by p.

Let $n \geq 0$ such that $u_n \leq N < u_{n+1}$ and let $d_0 = q(N, u_n)$ and $r_0 = r(N, u_n)$, $d_i = q(r_{i-1}, u_{n-i})$ and $r_i = r(r_{i-1}, u_{n-i})$ for $i = 1, \cdots, n$. Then $N = d_0 u_n + \cdots + d_n u_0$.

By convention the normal representation of 0 is the empty word ε. Under the hypothesis that the ratio u_{n+1}/u_n is bounded by a constant as n tends to infinity, the integers of the normal U-representation of any integer N are bounded and contained in a *canonical* finite alphabet A_U associated with U. The set of normal U-representations of all the nonnegative integers is denoted by $L(U)$.

Conversely, let C be a finite alphabet of integers (possibly containing negative integers); any sequence of integers, or *word* in C^* (the free monoid generated by C), is given a *numerical value* by the function $\pi_U : C^* \longrightarrow \mathbf{Z}$ which is defined by

$$\pi_U(w) = \sum_{n=0}^{k} d_n u_{k-n} \qquad \text{where} \qquad w = d_0 \cdots d_k.$$

For any alphabet C one can define a partial function

$$\nu_{U,C} : C^* \longrightarrow A_U^*$$

that maps a word w of C^* such that $\pi_U(w)$ is nonnegative onto the normal U-representation of $\pi_U(w)$. This function is called the *normalization*. A word is said to be in *normal form* if it is the normal U-representation of some nonnegative integer.

A *linear numeration system* is defined by a strictly increasing sequence of integers $U = (u_n)_{n \geq 0}$ with $u_0 = 1$ satisfying the linear recurrence relation

$$u_n = a_1 u_{n-1} + a_2 u_{n-2} + \cdots + a_m u_{n-m}, \quad a_i \in \mathbf{Z}, \quad a_m \neq 0, \quad n \geq m. \qquad (1)$$

In that case, the canonical alphabet A_U associated with U is $A_U = \{0, \cdots, K\}$ where $K < \max(u_{i+1}/u_i)$ is bounded. The polynomial $P(X) = X^m - a_1 X^{m-1} - \cdots - a_m$ will be called the *characteristic polynomial* of the recurrence relation (1).

2.2 θ-expansions

Now let θ be a real number > 1. A *representation in base* θ (or a θ-*representation*) of a real number $0 \leq x \leq 1$ is an infinite sequence $(x_n)_{n \geq 1}$ of integers such that

$$x = \sum_{n \geq 1} x_n \theta^{-n}.$$

The *numerical value* of an infinite sequence $(x_n)_{n \geq 1} \in C^{\mathbf{N}}$ where C is an alphabet of integers, is given by the function $\pi_\theta : C^{\mathbf{N}} \longrightarrow \mathbf{R}$ which is defined by

$$\pi_\theta((x_n)_{n \geq 1}) = \sum_{n \geq 1} x_n \theta^{-n}.$$

We say that $(x_n)_{n\geq 1}$ is *greater in the lexicographical ordering* than $(s_n)_{n\geq 1}$ if there exists an index $i \geq 1$ such that $x_1 = s_1, \ldots, x_{i-1} = s_{i-1}$ and $x_i > s_i$. As above, the greatest of all θ-representations of a given $x \in [0,1]$ in the lexicographical ordering is distinguished as the *normal θ-representation* of x, usually called the *θ-expansion* of x. It is obtained by means of the *θ-transformation* of the unit interval

$$T_\theta(x) = \theta x \pmod 1, \ x \in [0,1].$$

For $x \in [0,1]$, we have $x_n = \lfloor \theta T_\theta^{n-1}(x) \rfloor$. The digits x_n of a θ-expansion are elements of the set $A_\theta = \{0, \cdots, \lfloor \theta \rfloor\}$ when θ is not an integer; when θ is an integer, $A_\theta = \{0, \cdots, \theta - 1\}$ (see [R]).

We denote by $d(1, \theta) = (t_n)_{n\geq 1}$ the θ-expansion of 1. If $d(1, \theta)$ ends with infinitely many zeroes, it is said to be *finite*, and the ending zeroes are omitted. Let D_θ be the set of θ-expansions of numbers of $[0, 1[$, and let S_θ be the θ-*shift*, which is the closure of D_θ. Recall that the θ-shift S_θ is a system of finite type if and only if $d(1, \theta)$ is finite [P], and is a sofic system if and only if $d(1, \theta)$ is eventually periodic [Be2].

Now, for any alphabet of integers C the *normalization function*

$$\nu_{\theta,C} : C^{\mathbf{N}} \longrightarrow A_\theta^{\mathbf{N}}$$

is the partial function that maps a sequence s of $C^{\mathbf{N}}$ such that $\pi_\theta(s) \geq 0$ onto the θ-expansion of $\pi_\theta(s)$.

2.3 Finite automata

We recall some definitions. More details can be found in [E] or in [HU]. An *automaton over a finite alphabet* A, $\mathcal{A} = (Q, A, E, I, T)$ is a directed graph labelled by elements of A; Q is the set of *states*, $I \subset Q$ is the set of *initial states*, $T \subset Q$ is the set of *terminal* states and $E \subset Q \times A \times Q$ is the set of labelled *edges*. The automaton is *finite* if Q is finite, and this will always be the case in this paper. A subset L of A^* is said to be *recognizable by a finite automaton* if there exists a finite automaton $\mathcal{A}$ such that L is equal to the set of labels of paths starting in an initial state and ending in a terminal state.

Let L be a subset of A^*. The *right congruence* modulo L is defined by

$$f \sim_L g \Leftrightarrow [\forall h \in A^*, \ fh \in L \text{ iff } gh \in L].$$

It is known that the set L is recognizable by a finite automaton if and only if the right congruence modulo L has finite index (Myhill-Nerode Theorem, see [E] or [HU]).

A *2-tape automaton* is an automaton over the non-free monoid $A^* \times B^*$: $\mathcal{A} = (Q, A^* \times B^*, E, I, T)$ is a directed graph the edges of which are labelled by elements of $A^* \times B^*$. The automaton is finite if the set of edges E is finite (and thus Q is finite). These 2-tape automata are also known as *transducers*. Such an automaton is said to be *letter-to-letter* if the edges are labelled by couples of letters, that is, by elements of $A \times B$. Letter-to-letter 2-tape automata are called *synchronous* automata in [ECHLPT]. A relation R of $A^* \times B^*$ is said to be *computable by a finite 2-tape automaton* if there exists a finite 2-tape automaton $\mathcal{A}$ such that R is equal to the set of labels of paths starting in an initial state and ending in a terminal state. A function is *computable by a finite 2-tape automaton* if its graph is a relation computable by a finite 2-tape automaton.

Similar notions for infinite words can be defined. A subset of $A^{\mathbf{N}}$ (resp. of $A^{\mathbf{N}} \times B^{\mathbf{N}}$) is *recognizable by a finite automaton* (resp. *is computable by a finite 2-tape automaton*) if it is the set of labels of infinite paths starting in an initial state and going infinitely often through a terminal state in such an automaton (see [E]).

A *factor* of a word w is a word f such that there exist words w' and w'' with $w = w'fw''$. When $w' = \varepsilon$, f is said to be a *left factor* of w. If L is a subset of A^* or of $A^{\mathbf{N}}$, we denote by $F(L)$ (resp. $LF(L)$) the set of finite factors (resp. finite left factors) of words of L. The *length* of a word $w = w_1 \cdots w_n$ of A^*, where $w_i \in A$ for $1 \leq i \leq n$, is denoted by $|w|$ and is equal to n.

2.4   Previous results on normalization

Let us first recall what is already known on normalization in a linear numeration system U. If $\nu_{U,C}$ is computable by a finite (letter-to-letter) 2-tape automaton then for every subalphabet $D \subseteq C$, $\nu_{U,D}$ has the same property. If the normalization is computable by a finite 2-tape automaton then the set $L(U)$ of normal U-representations of nonnegative integers is recognizable by a finite automaton because $L(U) = \nu_{U,A_U}(A_U^*)$ and the image by a function computable by a finite 2-tape automaton of a set recognizable by a finite automaton is also recognizable by a finite automaton (see [E]).

Let c be an integer ≥ 1. Set

$$Z(U,c) = \{d_0 \cdots d_k \mid d_i \in \mathbf{Z}, \ |d_i| \leq c, \ \sum_{n=0}^{k} d_n u_{k-n} = 0\}.$$

Then the following holds.

PROPOSITION 1 . — [F1, F2] *Let U be a linear recurrent sequence of integers satisfying (1).*
A. *If $L(U)$ and $Z(U,c)$ are recognizable by a finite automaton then normal-*

ization on $C = \{0, \cdots, c\}$

$$\nu_{U,C} : C^* \longrightarrow A_U^*$$

is a function computable by a finite 2-tape automaton.

B. *Let us assume that the characteristic polynomial* P *has a dominant root* > 1. *If* $\nu_{U,C}$ *is computable by a finite 2-tape automaton then this automaton can be chosen letter-to-letter, and* $Z(U, c)$ *is recognizable by a finite automaton.*

The fact that normalization can be realized by a *letter-to-letter* 2-tape automaton is important because it means that computations are done without unbounded delay (see also [ECHLPT] for the use of automata of that kind in group theory).

A recent result shows that there exists an alphabet playing a special role. Recall that $A_U = \{0, \ldots, K\}$ is the canonical alphabet.

PROPOSITION 2 . — [FSa] *Under the same hypotheses as in Part B of Proposition 1 above, let* $A'_U = \{0, \cdots, K, K+1\}$. *Then normalization* $\nu_{U,C}$: $C^* \longrightarrow A_U^*$ *is a function computable by a finite letter-to-letter 2-tape automaton for every alphabet* C *of (possibly negative) integers if and only if* $\nu_{U,A'_U} : A'^*_U \longrightarrow A_U^*$ *is a function computable by a finite letter-to-letter 2-tape automaton.*

We now turn to the case of θ-expansions. Recall that $A_\theta = \{0, \cdots, \lfloor \theta \rfloor\}$ and let $A'_\theta = \{0, \cdots, \lfloor \theta \rfloor, \lfloor \theta \rfloor + 1\}$.

THEOREM 1 . — [BF] *Normalization* $\nu_{\theta,C} : C^{\mathbf{N}} \longrightarrow A_\theta^{\mathbf{N}}$ *is a function computable by a finite letter-to-letter 2-tape automaton on any alphabet* C *of integers if and only if* $\nu_{\theta,A'_\theta} : A'^{\mathbf{N}}_\theta \longrightarrow A_\theta^{\mathbf{N}}$ *is a function computable by a finite letter-to-letter 2-tape automaton if and only if* θ *is a Pisot number.*

3 Main results

In this paper we give results on normalization in linear numerations systems which answer a question of [F1, F2] and are the counterparts of Theorem 1.

THEOREM 2 . — *Let* U *be a linear recurrent sequence of integers satisfying (1). If the characteristic polynomial* P *of* U *is the minimal polynomial of a Pisot number* $\theta > 1$, *then for any alphabet* C, *normalization* $\nu_{U,C}$ *is a function computable by a finite letter-to-letter 2-tape automaton.*

This is the generalization of the old result about normalization in the Fibonacci numeration system (see [F2]).

THEOREM 3 . — *If the characteristic polynomial P of U is the minimal polynomial of a Perron number $\theta > 1$ which is not a Pisot number, then for every integer $c \geq K + 1$, normalization $\nu_{U,C}$ on $C = \{0, \cdots, c\}$ is not computable by a finite 2-tape automaton.*

EXAMPLE 1 . — Let $u_{n+4} = 3u_{n+3} + 2u_{n+2} + 3u_n$, with (for instance) $u_0 = 1$, $u_1 = 4$, $u_2 = 15$, $u_3 = 54$. In that case, the canonical alphabet is $A_U = \{0, \cdots, 3\}$. The characteristic polynomial has a dominant root θ which is a Perron number, but is not a Pisot number. We have $d(1, \theta) = 3203$, so the θ-shift S_θ is a system of finite type.

From Theorem 3 we know that for any alphabet $C \supseteq \{0, \cdots, 4\}$, normalization on C is not computable by a finite 2-tape automaton. Nevertheless, it can be shown with an *adhoc* proof that normalization on A_U is computable by a finite letter-to-letter 2-tape automaton. ☐

3.1　Proofs

Let us recall the following well-known facts about linear recurrent sequences.

In both theorems the characteristic polynomial P of the linear recurrent sequence is the minimal polynomial of a Perron number $\theta > 1$. Then P is irreducible over $\mathbf{Q}$ and so it has m distinct (possibly complex) zeros $\theta_1 = \theta, \theta_2, \cdots, \theta_m$. Then

$$u_n = b_1\theta^n + b_2\theta_2^n + \cdots + b_m\theta_m^n, \tag{2}$$

where b_i are (in general, complex) constants. As θ is the dominant zero of P,

$$\lim_{n\to\infty} \frac{u_n}{\theta^n} = b_1 > 0,$$

and thus $u_n \sim b_1\theta^n$ when $n \to \infty$. It follows from (2) that

$$|u_j - b_1\theta^j| \leq d\rho^j, \ j \geq 0, \tag{3}$$

where $d = (m - 1)\max\{|b_i|, i \leq m\}$ and $\rho = \max\{|\theta_j|, j \geq 2\}$. Notice that $\rho < 1$ if θ is a Pisot number.

In view of Proposition 1, we first prove the following result (which answers in a particular case a question raised in [Sh]).

THEOREM 4 . — *Let U be a linear recurrent sequence of integers satisfying (1). If the characteristic polynomial P is the minimal polynomial of a Pisot number θ, then the set $L(U)$ of normal representations is recognizable by a finite automaton.*

REMARK. Recently Hollander [Ho] gave a complete characterization of linear recurrent systems with the set $L(U)$ recognizable by a finite automaton, under the sole condition of dominant root. His proof is different from ours.

Proof of Theorem 4. The proof relies on the theory of θ-expansions (discussed in Subsection 2.2 above). Recall that D_θ is the set of all θ-expansions of numbers of $[0,1[$. The following description of the set D_θ is due to Parry [P]. Let $d(1,\theta) = (t_n)_{n\geq 1}$ be the expansion of 1. If $d(1,\theta)$ is finite of length m, let $d^*(1,\theta) = (t_1 \cdots t_{m-1}(t_m - 1))^\omega$.

LEMMA 1 . — [P] *A sequence $(x_n)_{n\geq 1}$ belongs to the set D_θ if and only if*

$$\forall k \geq 1, \ (x_n)_{n\geq k} <_{lex} \begin{cases} d^*(1,\theta), & \text{if } d(1,\theta) \text{ is finite;} \\ d(1,\theta), & \text{otherwise} \end{cases}$$

Since θ is a Pisot number, the θ-expansion of 1 is eventually periodic ([Be1], [Sc]) : $d(1,\theta) = (t_n)_{n\geq 1} = t_1 \cdots t_N(t_{N+1} \cdots t_{N+p})^\omega$ with $t_{N+jp+k} = t_{N+k}$ for $1 \leq k \leq p$, $j \geq 0$. Then the set $F(D_\theta)$ of finite factors of D_θ is recognizable by a finite automaton, because S_θ is sofic [Be2]. In fact, we will need the precise form of this automaton which we denote $\mathcal{A}_1$.

The automaton $\mathcal{A}_1$ has $N+p$ states $q_1, \ldots, q_{N+p}$. For each i, $1 \leq i \leq N+p$, there are edges labelled by $0, 1, \ldots, t_i - 1$ from q_i to q_1, and an edge labelled t_i from q_i to q_{i+1} if $i < N+p$. Finally, there is an edge labelled t_{N+p} from q_{N+p} to q_{N+1}. Let q_1 be the only initial state, and all states be terminal. When the θ-expansion of 1 happens to be finite, say $d(1,\theta) = (t_n)_{n\geq 1} = t_1 \cdots t_m$, with $t_n = 0$ for $n > m$, the same construction applies with $N = m, p = 0$ and all edges from q_m (labelled $0, 1, \ldots, t_m - 1$) leading to q_1. Notice that $\mathcal{A}_1$ is deterministic (*i.e.* there are no two edges with the same label from the same state). That $F(D_\theta)$ is precisely the set recognized by this automaton follows from Lemma 1.

The automaton $\mathcal{A}_2$ recognizing $L(U)$ will be built upon $\mathcal{A}_1$. Fix $b \in \mathbf{N}$, and let $B_b^{(j)}$, $1 \leq j \leq N+p$, be the set of U-normal representations $g_1 \ldots g_b$ with the following property :

$$g_1 u_{b-1} + \cdots + g_b u_0 < \begin{cases} u_b, & \text{if } j = 1; \\ u_{b+j-1} - t_1 u_{b+j-2} - \cdots - t_{j-1} u_b, & \text{if } j > 1. \end{cases} \quad (4)$$

Thus, $B_b^{(1)} = B_b$, the set of all U-normal representations of length b, and $B_b^{(j)}, j > 1$, are some of its subsets. We also denote by $B_{<b}$ the set of all U-normal representations of length less than b.

Next, let $F^{(j)}(D_\theta)$ be the set of finite factors of D_θ for which the path in the automaton $\mathcal{A}_1$ starting at q_1 ends at the vertex q_j (this is well-defined

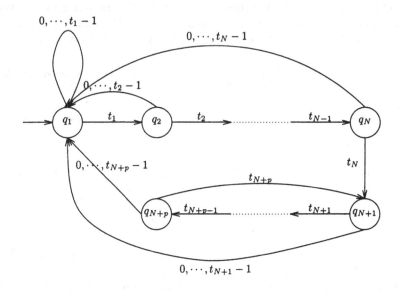

Figure 1: Automaton $\mathcal{A}_1$

since the automaton is deterministic). Notice that a word in $F^{(j)}(D_\theta)$ ends with $t_1 \ldots t_{j-1}$ for $j > 1$.

CLAIM. *There exists $b > 0$ such that*

$$L(U) = \bigcup_{j=1}^{N+p} F^{(j)}(D_\theta) B_b^{(j)} \cup B_{<b} \tag{5}$$

where $F^{(j)}(D_\theta) B_b^{(j)}$ means concatenation of $F^{(j)}(D_\theta)$ with words of $B_b^{(j)}$.

This claim will imply our theorem since $F^{(j)}(D_\theta)$ clearly are all recognizable, and finite unions and concatenations preserve recognizability.

Proof of the claim. 1) First we show that for all $b \in \mathbf{N}$ sufficiently large

$$L(U) \subseteq F(D_\theta) B_b \cup B_{<b}.$$

Let f be in $L(U)$. We can assume $|f| > b$ and proceed by induction on $|f|$, using $|f| = b$ as the base case. Let $f = f_0 \cdots f_n$, $n \geq b$ with $f_0 \neq 0$. Let $q = \pi_U(f)$. Since f is in normal form, $u_n \leq q < u_{n+1}$ and $q - f_0 u_n$ is normally represented by $f_1 \cdots f_n$. By induction $f_1 \cdots f_n \in F(D_\theta) B_b$.

Suppose that f is not in $F(D_\theta) B_b$. Then, by Lemma 1, either the θ-expansion of 1 is finite : $d(1, \theta) = t_1 \cdots t_m$ and $f_0 \cdots f_{m-1} = t_1 \cdots t_m$, or for some $h \geq 1$ and some i, $0 \leq i \leq n-b+1$, we have $f_0 \cdots f_i = t_1 \cdots t_i(t_{i+1}+h)$. Let us analyze these possibilities separately.

If $d(1,\theta) = t_1 \ldots t_m$, then $Q(X) = X^m - t_1 x^{m-1} - \cdots - t_m$ has θ as a zero, and so $P(X)$, the minimal polynomial for θ, divides $Q(X)$. Therefore, the sequence $\{u_i\}$ satisfies the linear recurrence relation

$$u_{l+m} = t_1 u_{l+m-1} + \cdots + t_m u_l, \quad l \geq 0.$$

Then $f_0 \cdots f_{m-1} = t_1 \cdots t_m$ leads to a contradiction with the assumption that f is normal since $f_0 u_n + \cdots + f_{m-1} u_{n-m+1} = u_{n+1}$.

The remaining possibility to have $f \notin F(D_\theta)B_b$, is, as mentioned above, that for some $h \geq 1$, $0 \leq i \leq n-b+1$,

$$f_0 \cdots f_i = t_1 \cdots t_i(t_{i+1} + h).$$

Then we get

$$q = \pi_U(f) \geq t_1 u_n + \cdots + t_i u_{n-i+1} + (t_{i+1}+1)u_{n-i}.$$

By (3) $|u_j - b_1 \theta^j| \leq d\rho^j$, $\rho < 1$, hence

$$q \geq b_1(t_1 \theta^n + \cdots + t_{i+1}\theta^{n-i}) + b_1 \theta^{n-i} - (\lfloor\theta\rfloor+1)d(\rho^n + \cdots + \rho^{n-i}).$$

It is easy to see from the definition of the θ-transformation T_θ that

$$T_\theta^j(1) = \sum_{k=1}^{\infty} t_{j+k}\theta^{-k}, \quad j \geq 0. \tag{6}$$

It follows that

$$t_1 \theta^n + \cdots + t_{i+1}\theta^{n-i} = \theta^{n+1} - \theta^{n-i}(T_\theta^{i+1}(1))$$

Applying a crude estimate $\rho^n + \cdots + \rho^{n-i} \leq (1-\rho)^{-1}$ and setting $C_1 = (\lfloor\theta\rfloor+1)d(1-\rho)^{-1}$, we obtain

$$q \geq b_1 \theta^{n+1} + b_1 \theta^{n-i}(1 - T_\theta^{i+1}(1)) - C_1.$$

Since the θ-expansion of 1 is eventually periodic, the orbit $\{T_\theta^j(1)\}$ is finite, so

$$\min\{1 - T_\theta^{i+1}(1), \ i \geq 0\} = \delta > 0.$$

Therefore

$$b_1 \theta^{n+1} + d \geq b_1 \theta^{n+1} + d\rho^{n+1} \geq u_{n+1} > q \geq b_1 \theta^{n+1} + \delta b_1 \theta^{n-i} - C_1.$$

It follows that

$$d + C_1 > \delta b_1 \theta^{n-i} \geq \delta b_1 \theta^{b-1}.$$

Thus, if $b > 0$ is such that $\theta^{b-1} > (d + C_1)/(\delta b_1)$, we get a contradiction.

2) Now we are going to prove the inclusion $\subseteq$ in (5) for the same b as in 1).

Let $f \in L(U)$. The case $|f| < b$ is obvious. If $|f| = b$ we have $f \in F^{(1)}(D_\theta)B_b^{(1)}$ since $B_b^{(1)} = B_b$, and an empty word is assumed to belong to $F^{(1)}(D_\theta)$. If $|f| > b$, then using 1) we can write $f = f'g$ where $|g| = b$, $f' \in F^{(j)}(D_\theta)$ for some j. Check that $g \in B_b^{(j)}$. If $j = 1$ this is immediate since $B_b^{(1)} = B_b$, so suppose that $j > 1$. Let $g = g_1 \ldots g_b$. If $g \notin B_b^{(j)}$ then

$$g_1 u_{b-1} + \cdots + g_b u_0 \geq u_{b+j-1} - t_1 u_{b+j-2} - \cdots - t_{j-1} u_b,$$

and by definition of $F^{(j)}(D_\theta)$, $f' = f'' t_1 \ldots t_{j-1}$. Then

$$\pi_U(t_1 \cdots t_{j-1} g) = t_1 u_{b+j-2} + \cdots + t_{j-1} u_b + g_1 u_{b-1} + \cdots + g_b u_0 \geq u_{b+j-1}$$

which contradicts the assumption that f is a U-normal representation.

3) It remains to prove the inclusion $\supseteq$ in (5) for b sufficiently large. Suppose that f is a word which belongs to the right-hand side of (5). The case $|f| < b$ is obvious. The case $|f| = b$ is also clear, and it will serve as induction base. Let $|f| > b$, $f = f_0 \cdots f_n$ with $n \geq b$ and $f_0 \neq 0$. Suppose that f is not in normal form; let g be the normal form of f. If $|f| = |g|$ then $g >_{lex} f$. Let $g = g_0 \cdots g_n$. There exists i, $0 \leq i \leq n$ such that $g_0 \cdots g_{i-1} = f_0 \cdots f_{i-1}$ and $g_i > f_i$. Set $q = \pi_U(f_i \cdots f_n) = \pi_U(g_i \cdots g_n)$. As g is in normal form, $q = g_i u_{n-i} + r$, with $0 \leq r < u_{n-i}$. By induction $f_{i+1} \cdots f_n \in L(U)$, and so $q' = \pi_U(f_{i+1} \cdots f_n) < u_{n-i}$. We get $q = f_i u_{n-i} + q'$, which is in contradiction with $q = g_i u_{n-i} + r \geq (f_i + 1)u_{n-i}$. Now let $|g| > |f|$. Then $\pi_U(g) = \pi_U(f) \geq u_{n+1}$.

Case A. Suppose that there exists i such that

$$f = t_1 \cdots t_i(t_{i+1} - h)f', \quad h \geq 1, \ 1 \leq i \leq n - b.$$

Then $f' \in F^{(j)}(D_\theta)B_b^{(j)}$ for some j. By induction $f' \in L(U)$ so $\pi_U(f') \leq u_{n-i} - 1$. Thus

$$\pi_U(f) \leq t_1 u_n + \cdots + t_i u_{n-i+1} + t_{i+1} u_{n-i} - 1$$

$$\leq b_1(t_1 \theta^n + \cdots + t_{i+1} \theta^{n-i}) - 1 + d\lfloor \theta \rfloor (\rho^n + \cdots + \rho^{n-i}).$$

Since $d(1, \theta) = (t_n)_{n \geq 1}$, we have $\theta^{n+1} \geq t_1 \theta^n + \cdots + t_{i+1} \theta^{n-i}$. Thus we can write

$$\pi_U(f) \leq b_1 \theta^{n+1} - 1 + C_2 \rho^{n-i},$$

where $C_2 = d\lfloor \theta \rfloor (1 - \rho)^{-1}$. On the other hand,

$$b_1 \theta^{n+1} - d\rho^{n-i} \leq b_1 \theta^{n+1} - d\rho^{n+1} \leq u_{n+1} \leq \pi_U(f),$$

which implies

$$1 \le (d + C_2)\rho^{n-i} \le (d + C_2)\rho^b.$$

If $b > 0$ is chosen so that $\rho^b < (d + C_2)^{-1}$ (and this is possible since $\rho < 1$), we get a contradiction.

Case B. Suppose that $f \in F^{(j)}(D_\theta)B_b^{(j)}$ but cannot be written as in Case A. Recalling the construction of the automaton $\mathcal{A}_1$, we see that $j \ge 2$ and

$$f = t_1 \ldots t_{n-b+1}g, \quad g = g_1 \ldots g_b \in B_b^{(j)}.$$

The path $t_1 \cdots t_{n-b+1}$ in the automaton $\mathcal{A}_1$ must end at the state q_j, so

$$j = \begin{cases} n - b + 2, & \text{if } n - b + 2 \le N + p; \\ N + l, & \text{if } n - b + 2 = N + kp + l, \ k \ge 1, \ l = 1, \ldots, p. \end{cases} \quad (7)$$

Let us show that if $b > 0$ is sufficiently large, then

$$\pi_U(g) = g_1 u_{b-1} + \cdots + g_b u_0 < u_{n+1} - t_1 u_n - \cdots - t_{n-b+1} u_b. \quad (8)$$

This will finish the proof since we would get

$$\pi_U(f) = t_1 u_n + \cdots + t_{n-b+1} u_b + \pi_U(g) < u_{n+1},$$

a contradiction. By definition of $B_b^{(j)}$ (4), we have

$$\pi_U(g) \le u_{b+j-1} - t_1 u_{b+j-2} - \cdots - t_{j-1} u_b - 1. \quad (9)$$

If $j = n - b + 2$, then (8) and (9) are identical, and we are done. Otherwise, notice that by (3)

$$u_{b+j-1} - t_1 u_{b-j+2} - \cdots - t_{j-1} u_b = b_1(\theta^{b+j-1} - t_1 \theta^{b+j-2} - \cdots - t_{j-1}\theta^b) + R,$$

where $|R| \le d\lfloor\theta\rfloor\rho^b(1-\rho)^{-1}$. Using (6) this can be rewritten as

$$u_{b+j-1} - t_1 u_{b-j+2} - \cdots - t_{j-1} u_b = b_1 \theta^b(T_\theta^{j-1}(1)) + R. \quad (10)$$

Similarly,

$$u_{n+1} - t_1 u_n - \cdots - t_{n-b+1} u_b = b_1 \theta^b(T_\theta^{n-b+1}(1)) + R', \quad (11)$$

where $|R'| \le d\lfloor\theta\rfloor\rho^b(1-\rho)^{-1}$. Using (7) and the periodicity of $d(1,\theta)$, we have $T_\theta^{n-b+1}(1) = T_\theta^{j-1}(1)$, so the expressions in (10) and (11) differ by no more than $|R| + |R'| \le 2d\lfloor\theta\rfloor\rho^b(1-\rho)^{-1}$. For b large enough, this quantity will be less than 1, because $\rho < 1$, so (9) implies (8). The proof of Theorem 4 is complete. ∎

The idea of using division of polynomials in recognizability problems was used in [F1, F2]. Here we develop it further.

Let $f = f_0 \cdots f_k$ be a word. The *polynomial associated with* f is $F(X) = f_0 X^k + \cdots + f_k$. The *value* of F with respect to U is $F_{|U} = \pi_U(f)$. The following simple lemma will be useful in the proof of both Theorem 2 and Theorem 3.

LEMMA 2 . —   *Let* $f_0 \ldots f_n$ *be a word in an arbitrary integral alphabet. Denote by* $F^{(k)}$ *the polynomial associated with the left factor* $f_0 \ldots f_k$, $0 \le k \le n$, *and consider the division of* $F^{(k)}$ *by* $P(X) = X^m - a_1 X^{m-1} - \cdots - a_m$. *Let* $R^{(k)}$ *be the remainder in this division and* $\mathbf{r}^{(k)}$ *be the vector of its coefficients :*

$$R^{(k)}(X) = r_0^{(k)} X^{m-1} + \cdots + r_{m-1}^{(k)}, \quad \mathbf{r}^{(k)} = [r_0^{(k)} r_1^{(k)} \ldots r_{m-1}^{(k)}]^t.$$

Then

$$\mathbf{r}^{(k+1)} = M\mathbf{r}^{(k)} + \mathbf{f}_{k+1}, \quad 0 \le k \le n-1, \tag{12}$$

where M *is the companion matrix of* P :

$$M = \begin{bmatrix} a_1 & 1 & & \\ a_2 & 0 & \cdot & \\ \cdot & & \cdot & \cdot \\ \cdot & & & \cdot & 1 \\ a_m & & & & 0 \end{bmatrix} \tag{13}$$

and $\mathbf{f}_{k+1} = [0, \ldots, 0, f_{k+1}]^t$.

Proof. We have

$$F^{(k)}(X) = P(X) Q^{(k)}(X) + R^{(k)}(X). \tag{14}$$

It is clear that $F^{(k+1)}(X) = X F^{(k)}(X) + f_{k+1}$, so

$$X R^{(k)}(X) + f_{k+1} = a P(X) + R^{(k+1)}(X).$$

Evidently, $a = r_0^{(k)}$, so

$$R^{(k+1)}(X) = -r_0^{(k)} P(X) + X R^{(k)} + f_{k+1}.$$

Rewriting this in vector form yields (12).   ∎

Proof of Theorem 2.

In view of Proposition 1 and Theorem 4 it is enough to prove that, for any c, the set $Z(U, c)$ is recognizable by a finite automaton, and this is equivalent to proving that right congruence mod $Z(U, c)$ has finite index [see Subsection 2.3 on finite automata].

LEMMA 3 . — *Let f and g be two left factors of $Z(U,c)$. Let $F(X)$ and $G(X)$ be the polynomials associated with f and g. Suppose that the remainders of F and G in division by P coincide. Then f and g are right congruent modulo $Z(U,c)$.*

Proof. Let $f = f_0 \cdots f_k$, $g = g_0 \cdots g_l$, $F(X) = f_0 X^k + \cdots + f_k$, $G(X) = g_0 X^l + \cdots + g_l$. We are given that there exists a polynomial Q such that

$$F(X) = G(X) + P(X)Q(X). \tag{15}$$

We want to check that $fh \in Z(U,c)$ if and only if $gh \in Z(U,c)$. Let $h = h_1 \ldots h_n$, then

$$
\begin{aligned}
\pi_U(fh) &= f_0 u_{k+n} + f_1 u_{k+n-1} + \cdots + f_k u_n + h_1 u_{n-1} + \cdots + h_n u_0 \\
&= (F(X)X^n + H(X))_{|U}; \\
\pi_U(gh) &= g_0 u_{l+n} + g_1 u_{l+n-1} + \cdots + g_l u_n + h_1 u_{n-1} + \cdots + h_n u_0 \\
&= (G(X)X^n + H(X))_{|U}.
\end{aligned}
$$

From (15) we get

$$F(X)X^n + H(X) = G(X)X^n + P(X)Q(X)X^n + H(X).$$

Since P is the characteristic polynomial of U, we have $P(X)X^i{}_{|U} = 0$ for all $i \geq 0$, hence $P(X)Q(X)X^n{}_{|U} = 0$. Thus $\pi_U(fh) = 0$ if and only if $\pi_U(gh) = 0$ and the lemma is proved. ∎

Before proceeding with the proof we need to introduce some notations and recall standard material from linear algebra. Let M be the companion matrix of the polynomial $P(X)$ (13). Its eigenvalues $\theta_1 = \theta, \theta_2, \ldots, \theta_m$ are all distinct since P is a minimal polynomial of θ and so irreducible over $\mathbf{Q}$. The eigenvalues are all real or they form pairs of complex conjugates. Enumerate them in such a way that if $\operatorname{Im} \theta_j > 0$ then $\theta_{j+1} = \overline{\theta}_j$. Let $\mathbf{f}_j$ be the corresponding eigenvectors. We need a basis $\{\mathbf{e}_j\}_1^m$ for $\mathbf{R}^m$ defined as follows : If θ_j is real choose $\mathbf{f}_j$ to be real and set $\mathbf{e}_j = \mathbf{f}_j/\|\mathbf{f}_j\|$ (here $\|\cdot\|$ is the Euclidean norm). If $\operatorname{Im}\theta_j > 0$, set

$$\mathbf{e}_j = \operatorname{Re}\mathbf{f}_j/\|\operatorname{Re}\mathbf{f}_j\|, \quad \mathbf{e}_{j+1} = \operatorname{Im}\mathbf{f}_j/\|\operatorname{Im}\mathbf{f}_j\|$$

(notice that $\operatorname{Re}\mathbf{f}_j \neq 0$, $\operatorname{Im}\mathbf{f}_j \neq 0$ for any nonreal eigenvalue θ_j of the real matrix M). Introduce the norm in $\mathbf{R}^m$:

$$|\|\mathbf{x}\|| = \left(\sum_{j=1}^m |x_j|^2\right)^{1/2}, \quad \text{where } \mathbf{x} = \sum_{j=1}^m x_j \mathbf{e}_j.$$

Let λ be an eigenvalue of M. We will use the seminorms $||\mathbf{x}||_\lambda$ for $\lambda = \theta_j$, $j \leq m$, defined as follows :

$$||\mathbf{x}||_\lambda = \begin{cases} |x_j|, & \text{if } \lambda = \theta_j \in \mathbf{R}; \\ (|x_j|^2/2 + |x_{j+1}|^2/2)^{1/2}, & \text{if } \lambda \in \{\theta_j, \theta_{j+1}\} \text{ with } \theta_j = \overline{\theta}_{j+1} \notin \mathbf{R}. \end{cases}$$

Clearly $|||\mathbf{x}|||^2 = \sum_\lambda ||\mathbf{x}||_\lambda^2$, where the summation is over all eigenvalues $\lambda = \theta_j$, $j = 1, \ldots, m$. The main property of these seminorms is

$$||M\mathbf{x}||_\lambda = |\lambda| \, ||\mathbf{x}||_\lambda, \quad \mathbf{x} \in \mathbf{R}^m, \tag{16}$$

which is checked by a direct computation.

We will need an explicit formula for $||\mathbf{x}||_\theta$. Let $\mathbf{e}_j^*$ be a left eigenvector (row-vector) of the matrix M corresponding to θ_j. We have $\mathbf{e}_j^* \mathbf{e}_k = 0$ for $k \neq j$, so $\mathbf{x} = \sum x_j \mathbf{e}_j$ implies

$$x_j = \frac{\mathbf{e}_j^* \mathbf{x}}{\mathbf{e}_j^* \mathbf{e}_j}.$$

Since M is the companion matrix (13), one can readily see that $\mathbf{e}_j^* = [\theta_j^{m-1}, \theta_j^{m-2}, \ldots, 1]$ is a left eigenvector. Therefore, for the largest eigenvalue $\theta = \theta_1 > 0$ we have

$$||\mathbf{x}||_\theta = |x_1| = |\mathbf{e}_1^* \mathbf{x}/\mathbf{e}_1^* \mathbf{e}_1| = \gamma_1 |x_1 \theta^{m-1} + x_2 \theta^{m-2} + \cdots + x_m|, \tag{17}$$

where $\gamma_1 = |\mathbf{e}_1^* \mathbf{e}_1|^{-1}$ is a constant depending on $P(X)$.

The core of the proof is the following.

LEMMA 4 . — *Fix c and consider all left factors of $Z(U, c)$. If P is the minimal polynomial of a Pisot number θ, then there are finitely many possible remainders in the division of associated polynomials by P.*

Proof. Let $f = f_0 \ldots f_n \in Z(U, c)$. Let $\mathbf{r}^{(k)}$ be the vectors corresponding to the remainders in the division of polynomials associated with left factors $f_0 \ldots f_k$ by P, as in Lemma 2. Since they are all integral vectors, it is enough to show that norms of $\mathbf{r}^{(k)}$ are bounded by a constant, depending on P and c but not on n. Let

$$||\mathbf{x}||_s = (\sum_{|\lambda| < 1} ||\mathbf{x}||_\lambda^2)^{1/2}$$

($||\mathbf{x}||_s$ measures the stable component of $\mathbf{x}$ with respect to the action by M). Since θ is the only eigenvalue with modulus ≥ 1, we have

$$|||\mathbf{x}|||^2 = ||\mathbf{x}||_\theta^2 + ||\mathbf{x}||_s^2.$$

It is immediate from (16) that

$$||Mx||_s \leq \rho||x||_s, \quad x \in \mathbf{R}^m, \tag{18}$$

where $\rho = \max\{|\theta_j|, j \geq 2\} < 1$. Since all norms on $\mathbf{R}^m$ are equivalent, it is enough to estimate $|||r^{(k)}|||$, hence $||r^{(k)}||_\theta$ and $||r^{(k)}||_s$. The idea of looking at "stable" and "unstable" components in a similar context was used by Thurston [T] in his proof of the theorem on eventual periodicity of θ-expansions of rational numbers in the Pisot base (proved previously by Bertrand and Schmidt in [Be1, Sc]).

First let us estimate $||r^{(k)}||_s$. Using Lemma 2 and (18), and setting $\gamma_2 = |||[0,\ldots,0,1]^t||_s$, we get, keeping in mind that $|f_i| \leq c$,

$$||r^{(k+1)}||_s \leq ||Mr^{(k)}||_s + ||f_{k+1}||_s \leq \rho||r^{(k)}||_s + c\gamma_2.$$

It follows that for $\mu \geq c\gamma_2(1-\rho)^{-1}$,

$$||r^{(k)}||_s \leq \mu \Rightarrow ||r^{(k+1)}||_s \leq \mu.$$

Let $\gamma_3 = \max\{||x||_s : x = [x_1,\ldots,x_m]^t, |x_i| \leq c\}$. Since $R^{(m-1)} = F^{(m-1)}$, we have $||r^{(m-1)}||_s \leq \gamma_3$. Therefore, if $\mu = \max\{c\gamma_2(1-\rho)^{-1}, \gamma_3\}$, then $||r^{(k)}||_s \leq \mu$ for all k.

It remains to estimate $||r^{(k)}||_\theta$. We need the following

CLAIM. *There exists a constant $D = D(P,c)$ such that for any $f = f_0\ldots f_n \in Z(U,c)$ and any $k \leq n$,*

$$|F^{(k)}(\theta)| = |f_0\theta^k + f_1\theta^{k-1} + \cdots + f_k| < D.$$

Proof of the claim. By the definition of $Z(U,c)$ we have

$$f_0u_n + \cdots + f_nu_0 = 0, \quad |f_i| \leq c.$$

Using (3) we get

$$|f_0\theta^n + \cdots + f_n| \leq \frac{d}{b_1}(|f_0|\rho^n + \cdots + |f_n|) \leq \frac{cd}{b_1(1-\rho)}.$$

Next,

$$f_0\theta^k + \cdots + f_k = \theta^{k-n}(f_0\theta^n + \cdots + f_n) - (f_{k+1}\theta^{-1} + \cdots + f_n\theta^{k-n}),$$

hence

$$|f_0\theta^k + \cdots + f_k| \leq \theta^{k-n}\frac{cd}{b_1(1-\rho)} + \frac{c}{\theta-1} \leq \frac{cd}{b_1(1-\rho)} + \frac{c}{\theta-1}.$$

Since the constants b_1, d, ρ depend on $P(X)$ only, the claim is verified. ∎

Now we can estimate $||\mathbf{r}^{(k)}||_\theta$. Combining (17), (14) and the claim above we obtain

$$||\mathbf{r}^{(k)}||_\theta = \gamma_1 |R^{(k)}(\theta)| = \gamma_1 |F^{(k)}(\theta)| < \gamma_1 D.$$

This completes the proof of the lemma. ∎

It follows from Lemmas 3 and 4 that right congruence modulo $Z(U, c)$ has finite index in $\{-c, \cdots, c\}^*$ so Theorem 2 is proved. ∎

Proof of Theorem 3.
Let $I(P, c)$ be the set of words $f = f_0 \cdots f_k$ such that $|f_i| \le c$ and the associated polynomial F is divisible by P. Clearly $I(P, c) \subseteq Z(U, c)$. We recall results from [F1, F2] (here we do not need the irreducibility of P).

LEMMA 5 . — [F1, F2] *Right classes mod $I(P, c)$ of elements of $LF(I(P, c))$ are exactly the remainders in the division of associated polynomials by P.*

PROPOSITION 3 . — [F1, F2] $I(P, c)$ *is recognizable by a finite automaton for every c if and only if P has no root of modulus 1.*

Now we give a result which was incorrectly asserted in [F1, F2] to hold when P is the minimal polynomial of the recurrence.

LEMMA 6 . — *Let us assume that P is the minimal polynomial of a Perron number θ. A right class mod $Z(U, c)$ can contain only a finite number of right classes mod $I(P, c)$.*

Proof. First, let us show that, if f is a word of $LF(I(P, c))$ then $[f]_{I(P,c)} \subset [f]_{Z(U,c)}$. Clearly $LF(I(P, c))$ is included in $LF(Z(U, c))$. Let f and g be two words of $LF(I(P, c))$ equivalent modulo $I(P, c)$, and let F and G be the associated polynomials. Then there exists a polynomial Q such that $F = PQ + G$, and then, for every y of $\{-c, \cdots, c\}^*$, $FX^{|y|} + Y = GX^{|y|} + PQX^{|y|} + Y$. Thus $(FX^{|y|} + Y)|_U = 0$ if and only if $(GX^{|y|} + Y)|_U = 0$ so $f \sim_{Z(U,c)} g$.
Now, suppose that there exist infinitely many different classes modulo $I(P, c)$ with representatives $f_1, f_2, \cdots$, in the same class modulo $Z(U, c)$. Thus there exist infinitely many words $w_1, w_2, \cdots$ such that $f_i w_i \in I(P, c)$ and $f_j w_i \notin I(P, c)$ for $i \ne j$, else we would get $f_i w_i$ and $f_j w_i \in I(P, c)$, which would imply that $F_i - F_j$ is divisible by P, and then $f_i \sim_{I(P,c)} f_j$, contrary to the hypothesis. For every i and j we have $f_i \sim_{Z(U,c)} f_j$, so $\pi(f_i w_i) = 0$ implies $\pi(f_k w_i) = 0$ and $\pi(f_j w_i) = 0$. Thus $\pi((f_j - f_k)0^{|w_i|}) = 0$, for every i, j, and

k. Let $h = h_1 \cdots h_n$ be the word $f_j - f_k$ for fixed j and k, and let H be the associated polynomial. Then, for $N = |w_i|$ we have

$$h_1 u_{N+n-1} + \cdots + h_n u_N = 0.$$

Since P is the minimal polynomial of θ, we have from (2) that $u_j = \sum_{i=1}^m b_i \theta_i^j$, where $\theta_1 = \theta$, and thus

$$b_1 H(\theta)\theta^N + \cdots + b_m H(\theta_m)\theta_m^N = 0.$$

Since $|\theta_i| < \theta$ for $i > 1$ and N can be chosen arbitrarily large, we get $H(\theta) = 0$. Since P is the minimal polynomial for θ, $H = F_j - F_k$ is divisible by P, a contradiction. ∎

COROLLARY 1 . — *Let P be the minimal polynomial of a Perron number θ. If $Z(U,c)$ is recognizable by a finite automaton, so is $I(P,c)$.*

Proof. If $Z(U,c)$ is recognizable by a finite automaton, then the number of classes modulo $Z(U,c)$ is finite. From Lemma 6, the number of classes modulo $I(P,c)$ is finite and $I(P,c)$ is recognizable by a finite automaton. ∎

So, in order to prove Theorem 3, by Lemmas 5 and 6 it is enough to show that in $LF(Z(U,c))$ there are infinitely many remainders in the division by P.

If P has a zero which is a Salem number, then Theorem 3 is a consequence of Proposition 3 and Corollary 1. So we assume that the second largest zero $\xi = \theta_2$ has modulus bigger than one. We are going to use the basis $\{e_j\}_1^m$, the norm $||| \cdot |||$ and the seminorms $|| \cdot ||_\lambda$ introduced above, in the proof of Theorem 2.

LEMMA 7 . — *For any $\mu > 0$, $\varepsilon > 0$, there exists a vector $\mathbf{x} \in \mathbf{Z}^m$ such that $||\mathbf{x}||_\xi \geq \mu$, $||\mathbf{x}||_\theta < \varepsilon$.*

Proof. Let H_ξ be the subspace of $\mathbf{R}^n$ spanned by e_2 if $\xi = \theta_2$ is real and by e_2, e_3 otherwise. (Recall that in the latter case e_2, e_3 arise from the complex eigenvectors corresponding to $\xi, \bar{\xi}$.) Fix an arbitrary nonzero vector $\mathbf{y} \in H_\xi$ (so $||\mathbf{y}||_\xi > 0$). The norm $||| \cdot |||$ is equivalent to the Euclidean norm, so for some $\tau > 0$ we have

$$\max_{\mathbf{a} \in \mathbf{R}^m} \min_{\mathbf{z} \in \mathbf{Z}^m} |||\mathbf{a} - \mathbf{z}||| \leq \tau.$$

Let $\mathbf{z}_k \in \mathbf{Z}^m$, $k \geq 1$, be such that $|||k\mathbf{y} - \mathbf{z}_k||| = \min\{|||k\mathbf{y} - \mathbf{z}||| : \mathbf{z} \in \mathbf{Z}^m\}$. Since $k\mathbf{y} \in H_\xi$, we have

$$||\mathbf{z}_k||_\theta = ||\mathbf{z}_k - k\mathbf{y}||_\theta \leq |||\mathbf{z}_k - k\mathbf{y}||| \leq \tau.$$

Let $\mathbf{z}_k = \sum_{j=1}^m c_j^{(k)} \mathbf{e}_j$. The eigemvalue $\theta_1 = \theta$ is real, so $|c_1^{(k)}| = \|\mathbf{z}_k\|_\theta \leq \tau$. Hence the sequence $\{c_1^{(k)}\}_{k\geq 1}$ is bounded and has a limit point. This implies that for any $\mu_1 > 0$, $\varepsilon > 0$, there exist k, l such that

$$|c_1^{(k)} - c_1^{(l)}| < \varepsilon, \quad k - l > \mu_1.$$

Then $\|\mathbf{z}_k - \mathbf{z}_l\|_\theta = |c_1^{(k)} - c_1^{(l)}| < \varepsilon$ and

$$\|\mathbf{z}_k - \mathbf{z}_l\|_\xi \geq \|k\mathbf{y} - l\mathbf{y}\|_\xi - \|\mathbf{z}_k - k\mathbf{y}\|_\xi - \|\mathbf{z}_l - l\mathbf{y}\|_\xi$$

$$\geq (k - l)\|\mathbf{y}\|_\xi - \|\|\mathbf{z}_k - k\mathbf{y}\|\| - \|\|\mathbf{z}_l - l\mathbf{y}\|\| > \mu_1\|\mathbf{y}\|_\xi - 2\tau.$$

Now μ_1 can be chosen so that $\mu_1\|\mathbf{y}\|_\xi - 2\tau > \mu$, and we can set $\mathbf{x} = \mathbf{z}_k - \mathbf{z}_l$. ∎

Let $\gamma_4 = \|[0, 0, \dots, 0, 1]^t\|_\xi$, $\gamma_1 = |\mathbf{e}_1^* \mathbf{e}_1|^{-1}$. Applying Lemma 7, choose a vector $\mathbf{x} = [x_0, \dots, x_{m-1}]^t \in \mathbf{Z}^m$ such that

$$\|\mathbf{x}\|_\theta \leq \gamma_1/2 \tag{19}$$

and

$$\|\mathbf{x}\|_\xi \geq \frac{2K\gamma_4}{|\xi| - 1}. \tag{20}$$

Set $c = \max\{K; |x_i|, i \leq m - 1\}$. Our goal is to show that there are infinitely many remainders in the division of polynomials associated with $LF(Z(U, c))$ by P.

LEMMA 8 . — *Let* $\mathbf{x} = [x_0, x_1, \dots, x_{m-1}]^t \in \mathbf{Z}^m$ *be a vector satisfying (19) and such that* $|x_i| \leq c$. *Then one can find* $J = J(P)$ *such that for* $n \geq J$ *there is a word* $f = f_0 \dots f_n \in Z(U, c)$ *with* $f_i = x_i$ *for* $i \leq m - 1$ *and* $|f_i| \leq K$ *for* $i \geq m$.

Roughly speaking, the lemma is true for the following reason. The estimate on $\|\mathbf{x}\|_\theta$ implies that $|S| = |x_0 u_n + \dots + x_{m-1} u_{n-m+1}|$ is not too large, so the canonical representation of $|S|$ terminates before getting to u_{n-m+1}. Then we can append to $\mathbf{x}$ this representation, reversing all signs if $S > 0$. The resulting sequence will be exactly what is needed in Lemma 8.

Proof of Lemma 8.

Fix η such that $1 < |\xi| < \eta < \theta$. It follows from (2) that for some $D > 0$,

$$|u_j - b_1\theta^j| \leq D\eta^j, \quad j \geq 0. \tag{21}$$

Consider $S = x_0 u_n + \dots + x_{m-1} u_{n-m+1}$. We have

$$|S| \leq b_1|x_0\theta^n + \dots + x_{m-1}\theta^{n-m+1}| + cD|\eta^n + \dots + \eta^{n-m+1}|$$

Next we use (17) and (19) to get

$$|S| \leq b_1\theta^{n-m+1}\gamma_1^{-1}||\mathbf{x}||_\theta + cDm\eta^n \leq \frac{1}{2}b_1\theta^{n-m+1} + cDm\eta^n \qquad (22)$$

Since $\theta > \eta$ one can find $J = J(P) > 0$ such that for $n \geq J$,

$$\frac{1}{2}b_1\theta^{n-m+1} > cDm\eta^n + D\eta^{n-m+1}.$$

Then from (22) and (21)

$$|S| < b_1\theta^{n-m+1} - D\eta^{n-m+1} \leq u_{n-m+1}.$$

It follows that the canonical representation of $|S|$ has the form $g_0 \cdots g_p$, where $p \leq n - m$. Now we form the sequence

$$f = x_0 \cdots x_{m-1} \underbrace{0 \cdots 0}_{n-m-p} \pm g_0 \cdots \pm g_p.$$

The signs for $g_i, i \leq p$, are chosen opposite to the sign of S (of course, if $S = 0$ no work was needed at all). By construction, $f \in Z(U, c)$ and $|f_i| \leq K$, $i \geq m$. ∎

LEMMA 9 . — *Suppose that* $\mathbf{x} = [x_0, \ldots, x_{m-1}]^t \in \mathbf{Z}^m$ *satisfies (20). Let* $f_m, f_{m+1}, \ldots, f_n$ *be any integers such that* $|f_i| \leq K$. *Then the sequence*

$$\mathbf{r}^{(m-1)} = \mathbf{x}, \quad \mathbf{r}^{(k)} = M\mathbf{r}^{(k-1)} + \mathbf{f}_k, \ k \geq m,$$

is such that the norms $||\mathbf{r}^{(k)}||_\xi$ *are strictly increasing in* k.

Proof. We have from (12) for $k \geq m - 1$:

$$||\mathbf{r}^{(k+1)}||_\xi \geq ||M\mathbf{r}^{(k)}||_\xi - ||\mathbf{f}_{k+1}||_\xi \geq |\xi|\,||\mathbf{r}^{(k)}||_\xi - K\gamma_4$$

(recall that $\gamma_4 = ||[0, \ldots, 0, 1]^t||_\xi$). By assumption, $||\mathbf{r}^{(m-1)}||_\xi = ||\mathbf{x}||_\xi \geq 2K\gamma_4(|\xi| - 1)^{-1}$, hence

$$||\mathbf{r}^{(m)}||_\xi \geq |\xi|||\mathbf{r}^{(m-1)}||_\xi - \frac{|\xi| - 1}{2}||\mathbf{r}^{(m-1)}||_\xi = \frac{1 + |\xi|}{2}||\mathbf{r}^{(m-1)}||_\xi.$$

Similarly,

$$||\mathbf{r}^{(k+1)}||_\xi \geq \frac{1 + |\xi|}{2}||\mathbf{r}^{(k)}||_\xi, \ k \geq m.$$

Thus, the norms $||\mathbf{r}^{(k)}||_\xi$ increase geometrically. ∎

We can apply Lemma 9 to the sequence $f_0 \ldots f_n \in Z(U, c)$ constructed in Lemma 8. By Lemma 2 the vectors $\mathbf{r}^{(k)}$ correspond to remainders in the division of polynomials associated with left factors by P. Lemma 9 implies that all these remainders are distinct, and there are $n - m + 2$ of them. Since n can be arbitrarily large, the total number of remainders involved is infinite. So we have found a $c > 0$ such that $Z(U, c)$ is not recognizable by a finite automaton, and so, by Proposition 1, $\nu_{U,C}$ is not computable by a finite 2-tape automaton on $C = \{0, \cdots, c\}$. If normalization was computable by a finite 2-tape automaton on an alphabet containing A'_U, it would be the same on A'_U, and the automaton would be letter-to-letter because P has a dominant root. By Proposition 2 it would be so on any alphabet, a contradiction. This proves Theorem 3.   ∎

CONCLUDING REMARKS.

1.— We have considered only the case when P has a dominant root. It may happen that P has no dominant root, but normalization in the system U is nevertheless a function computable by a 2-tape automaton. Let us take for instance $P(X) = X^2 - 3$ and

$$u_{n+2} = 3u_n, \ u_0 = 1, \ u_1 = 2.$$

The sequence U is thus defined by

$$u_{2p} = 3^p, \ u_{2p+1} = 2 \cdot 3^p.$$

The canonical alphabet is $A_U = \{0, 1\}$. Set $V = \{f \in A_U^* \mid |f| \text{ is even}\}$. Then $L(U) = 1A_U^* \setminus A_U^* 11V$ is recognizable by a finite automaton. It can be shown by a direct computation that normalization in the system U is computable by a letter-to-letter 2-tape automaton on any alphabet [F1].

On the other hand, let $\theta = \sqrt{3}$. It is not a Perron number, $d(1, \theta)$ is not eventually periodic, and normalization in base θ is never computable by a 2-tape automaton.

2.— Theorem 3 can be extended to some cases of reducible P. For instance, suppose that the minimal polynomial of the linear recurrence can be written as $P = P_1 P_2 \cdots P_k$ where every P_i is a minimal polynomial of a Perron number λ_i with $\lambda_1 > \lambda_2 > \cdots > \lambda_k$. Then normalization is not computable by a finite 2-tape automaton, unless $k = 1$ and λ_1 is a Pisot number. The general case of reducible P remains open.

References

[BF]   D. Berend and Ch. Frougny, Computability by finite automata and Pisot bases. *Math. Systems Theory* **27**, 1994, 274–282.

[Be1]   A. Bertrand, Développements en base de Pisot et répartition modulo 1. *C.R.Acad. Sc.*, Paris **285** (1977), 419–421.

[Be2]   A. Bertrand-Mathis, Répartition modulo un des suites exponentielles et systèmes dynamiques symboliques. Thèse d'Etat, Université Bordeaux 1, 1986.

[E]   S. Eilenberg, *Automata, Languages and Machines*, vol. A, Academic Press, 1974.

[ECHLPT] D. Epstein, J. Cannon, D. Holt, S. Levy, M. Paterson and W. Thurston, *Word processing in groups*. Jones and Bartlett Publishers, 1992.

[Fra]   A.S. Fraenkel, Systems of numeration. *Amer. Math. Monthly* **92(2)** (1985), 105–114.

[F1]   Ch. Frougny, Systèmes de numération linéaires et automates finis. Thèse d'Etat, Université Paris 7, LITP Report 89-69, 1989.

[F2]   Ch. Frougny, Representations of numbers and finite automata. *Math. Systems Theory* **25** (1992), 37–60.

[FSa]   Ch. Frougny and J. Sakarovitch, Automatic conversion from Fibonacci to golden mean, and generalizations. Submitted.

[Ho]   M. Hollander, Greedy numeration systems and recognizability. Submitted.

[HU]   J. E. Hopcroft and J. D. Ullman, *Introduction to Automata Theory, Languages, and Computation*, Addison-Wesley, 1979.

[L]   D. Lind, The entropies of topological Markov shifts and a related class of algebraic integers. *Ergod. Th. & Dynam. Sys.* **4** (1984), 283–300.

[P]   W. Parry, On the β-expansions of real numbers. *Acta Math. Acad. Sci. Hungar.* **11** (1960), 401–416.

[PT]   A. Pethö and R. Tichy, On digit expansions with respect to linear recurrences. *J. of Number Theory* **33** (1989), 243–256.

[R]   A. Rényi, Representations for real numbers and their ergodic properties. *Acta Math. Acad. Sci. Hungar.* **8** (1957), 477–493.

[Sc]   K. Schmidt, On periodic expansions of Pisot numbers and Salem numbers. *Bull. London Math. Soc.* **12** (1980), 269–278.

[Sh]   J. Shallit, Numeration Systems, Linear Recurrences, and Regular Sets. *I.C.A.L.P.* 92, Wien, Lecture Notes in Computer Science **623**, 1992, 89–100.

[T]   W. Thurston, *Groups, tilings, and finite state automata.* AMS Colloquium Lecture Notes, Boulder, 1989.

The structure of ergodic transformations conjugate to their inverses

Geoffrey R. Goodson*

Department of Mathematics, Towson State University, Towson, MD 21204, USA

Abstract. Let T be an ergodic automorphism defined on a standard Borel probability space for which T and T^{-1} are isomorphic. We study the structure of T in the case that there are conjugating automorphisms S for which the weak closure of the set $\{S^{2n} : n \in \mathbb{Z}\}$ is compact in the centralizer $C(T)$ of T. This is the situation, for example, if S is of finite order. These results are applied to automorphisms which have the weak closure property, and we see that they have implications concerning the multiplicity of any Lebesgue component in the spectrum of T. In addition, we construct a continuum of examples, each of which is isomorphic to its inverse, has continuous spectrum and has maximal spectral multiplicity equal to two, having a singular component in its spectrum, together with a Lebesgue component of multiplicity two.

0. Introduction. Let T be an ergodic transformation defined on a standard Borel probability space $(X, \mathcal{F}, \mu)$ having simple spectrum. It was shown in [3] that if T is isomorphic to its inverse T^{-1} and the conjugating automorphism is S, i.e. $TS = ST^{-1}$ then $S^2 = I$, the identity automorphism. It follows that if a transformation T has a conjugating automorphism S for which $S^2 \neq I$, then T has non-simple spectrum. It was shown in [2] that in this case, the essential values of the multiplicity function of T must be even (or infinity) in the orthogonal complement of the subspace $\{f \in L^2(X, \mu) : f \circ S^2 = f\}$. In this paper we show that more can be said when the weak closure of the squares of powers of any conjugation are compact in the centralizer of T. For example, suppose that T has the weak closure property, i.e. the weak closure of the powers of T is equal to the centralizer $C(T)$ of T, then if T is isomorphic to its inverse it is known that every conjugation S between T and T^{-1} satisfies $S^4 = I$. Consequently, if there are conjugations of order 4, T can be represented as a $\mathbb{Z}_2$-extension of an ergodic transformation T_0.

Generally, our results enable us to determine the structure of T on $H_0^\perp$,

---

*Partially supported by the TSU Faculty Development Office

the orthocomplement of the subspace H_0 on which T_0 acts. In particular it follows that if T is group extension of T_0 which is conjugate to its inverse via a conjugation S satisfying $S^2 \neq I$, and having a Lebesgue component in $H_0^\perp$, then that Lebesgue component must occur with even (or infinite) multiplicity. This, in part, explains why the Lebesgue component of known examples occurs with even (or infinite) multiplicity.

Section 1 gives some preliminaries necessary for the subsequent discussion. In Section 2 we analyse a special case of the main results of [2], in particular, we examine the situation when there is a conjugation S of finite order, or more generally, when the weak closure of S^2 is compact in $C(T)$. It is here that we see that T can be represented as a compact abelian group extension. In Section 3 we make a study of the structure of compact abelian group extensions which are isomorphic to their inverses. Finally, in the last section we construct a continuum of non-isomorphic weakly mixing transformations, each being a $\mathbb{Z}_2$ extensions of rank one weakly mixing transformations. In addition these extensions are isomorphic to their inverses, have maximal spectral multiplicity equal to 2, and a Lebesgue component of multiplicity 2 in their spectrum. Furthermore, each conjugation S is of order 4. In particular, we see that a special case of the construction gives the examples due to Mathew and Nadkarni [7]. It should be noted that Mathew and Nadkarni [7] were the first to construct ergodic transformations with a finite Lebesgue component in their spectrum. It was shown in [3] that their examples have the property of being isomorphic to their inverses. It should be noted that Ageev [1], was the first to contruct such examples having continuous spectrum, however, his examples are not isomorphic to their inverses. The main novelty in our examples is that they are all isomorphic to their inverses.

Another important class of examples to which our results apply are the order 2 maps having quasi-discrete spectrum. These are studied in Example 2.

1. Preliminaries.

Let $T : (X, \mathcal{F}, \mu) \to (X, \mathcal{F}, \mu)$ be an ergodic automorphism defined on a non-atomic standard Borel probability space. We denote the identity automorphism and the identity group automorphism by I. The group of all automorphisms $\mathcal{U}(X)$ of $(X, \mathcal{F}, \mu)$ becomes a completely metrizable topological group when endowed with the weak convergence of transformations ($T_n \to T$ if for all $A \in \mathcal{F}$, $\mu(T_n^{-1}(A) \triangle T^{-1}(A)) + \mu(T_n(A) \triangle T(A)) \to 0$ as $n \to \infty$). Denote by $C(T)$ the centralizer (or commutant) of T, i.e. those automorphisms of $(X, \mathcal{F}, \mu)$ which commute with T. Since we are assuming the members of $C(T)$ are invertible, we see that $C(T)$ is a group.

Much of the discussion concerns the set

$$\mathcal{B}(T) = \{S \in \mathcal{U}(X) : TS = ST^{-1}\},$$

whose basic properties are discussed in [3]. In particular we note that $\{S^2 : S \in \mathcal{B}(T)\} \subseteq C(T)$.

Let G be a compact abelian group equipped with Haar measure m and denote by $(X \times G, \mathcal{F}_G, \tilde{\mu})$ the product measure space, where $\tilde{\mu} = \mu \times m$. Let $\phi : X \to G$ be a G–cocycle, then the corresponding G–extension ; $T_\phi : X \times G \to X \times G$ is defined by

$$T_\phi(x, g) = (Tx, \phi(x) + g).$$

T_ϕ preserves the measure $\tilde{\mu}$. In this situation $L^2(X \times G, \tilde{\mu})$ can be written as a direct sum of subspaces, each invariant under U_{T_ϕ},

$$L^2(X \times G, \tilde{\mu}) = \oplus_{\chi \in \hat{G}} L_\chi,$$

(where $L_\chi = \{f \otimes \chi : f \in L^2(X, \mu)\}$), such that $U_{T_\phi}|L_\chi$ is unitarily equivalent to

$$V_{\phi,T,\chi} : L^2(X, \mu) \to L^2(X, \mu); \quad V_{\phi,T,\chi}f(x) = \chi(\phi(x))f(Tx).$$

The spectral properties of T are those of the induced unitary operator defined by

$$U_T : L^2(X, \mu) \to L^2(X, \mu); \quad U_T f(x) = f(Tx), \quad f \in L^2(X, \mu).$$

Generally a unitary operator $U : H \to H$ on a separable Hilbert space H is said to have simple spectrum if there exists $h \in H$ such that $Z(h) = H$, where $Z(h)$ is the closed linear span of the vectors $U^n h$, $n \in \mathbb{Z}$. U is determined up to unitary equivalence by a spectral measure class σ and a $\{1, 2, \ldots, \infty\}$ valued multiplicity function M defined on the circle S^1. The set $\mathcal{M}_U$ of essential spectral multiplicities of U is the set of all σ-essential values of M. The maximal spectral multiplicity (or just multiplicity) of U is $\operatorname{msm}(U) = \sup \mathcal{M}_U$. U has simple spectrum if $\operatorname{msm}(U) = 1$, otherwise U has non-simple spectrum. See [8] for a detailed discussion of the spectral properties of dynamical systems.

2. Conjugation by a Transformation S Whose Weak Closure is Compact

It was shown in [2] that for ergodic T and any conjugation $S \in \mathcal{B}(T)$, either (i) S^2 of finite order (i.e. $\exists\, n$ such that $S^{2n} = I$), or (ii) S is weakly mixing, or (iii) S^2 is aperiodic, but non-ergodic. We wish to examine case (i) in more detail, or more generally, the case where the weak closure of the set $\{S^{2n} : n \in \mathbb{Z}\}$ in $C(T)$ is compact. S^2 cannot be ergodic in this case since from [2], the ergodicity of S^2 implies that S is weakly mixing.

Denote by $e(S)$ the eigenvalue group of the unitary operator U_S determined by S (we sometimes use S and T to denote U_S and U_T). For each $\lambda \in e(S^2)$ define

$$H_\lambda = \{f \in L^2(X, \mu) : S^2(f) = \lambda f\}.$$

Then it is easily seen that for each $\lambda \in e(S^2)$, H_λ is both T and S invariant. Also $H_\lambda \perp H_\mu$ for all $\lambda, \mu \in e(S^2)$, $\lambda \neq \mu$. Furthermore $T|H_\lambda \cong T|H_{\overline{\lambda}}$, for all $\lambda \in e(S^2)$ (where $\cong$ means unitarily equivalent). This is a special case of an argument used in [2] to show that such T have an even multiplicity function on the orthocomplement of the functions invariant under S^2. This special case is proved as follows:

If $f \in H_\lambda$, then $S^2(S\overline{f}) = \overline{\lambda}S(\overline{f})$, so $S(\overline{f}) \in H_{\overline{\lambda}}$. Also

$$(T^n f, f) = (ST^n f, Sf) = (T^{-n}Sf, Sf)$$

$$= (Sf, T^n Sf) = \overline{(T^n Sf, Sf)} = (T^n S\overline{f}, S\overline{f}),$$

for all $n \in \mathbf{Z}$, so that f and $S(\overline{f})$ have the same spectral type, and in particular

$$U_T|Z(f) \cong U_T|Z(S\overline{f}).$$

Since this can be done for each $f \in H_\lambda$, the result follows as in [2].

The case where $\lambda = -1$ is of interest, since if $-1 \in e(S^2)$, then $T|H_{-1}$ has an even multiplicity function. To see this, first recall that $H_{-1} = \{f \in L^2(X, \mu) : S^2(f) = -f\}$, and as before, if $f \in H_{-1}$, then $S(\overline{f}) \in H_{-1}$, and f and $S(\overline{f})$ have the same spectral type. However, in addition we have

$$(T^n f, S\overline{f}) = (ST^n f, S^2\overline{f}) = (T^{-n}Sf, S^2\overline{f}) = (Sf, T^n S^2\overline{f})$$

$$= \overline{(T^n S^2 \overline{f}, Sf)} = (T^n S^2 f, S\overline{f}) = (T^n(-f), S\overline{f}) = -(T^n f, S\overline{f}).$$

It follows that $(T^n f, S\overline{f}) = 0$ for all $n \in \mathbf{Z}$. This implies that $Z(f) \perp Z(S(\overline{f}))$ for all $f \in H_{-1}$ and so $T|Z(f) \cong T|(S(\overline{f}))$ for all $f \in H_{-1}$, and this in turn implies that $T|H_{-1}$ has an even multiplicity function (see [2] for more details).

We complete this section by proving some results about the subspaces H_λ.

Proposition 1 *Suppose that $T : Y \to Y$ is an ergodic automorphism for which the weak closure of the set $\{S^{2n} : n \in \mathbf{Z}\}$ is compact in $C(T)$. If $H_\lambda = \{f \in L^2(Y, \mu) : S^2(f) = \lambda f\}$, then*

$$\bigoplus_{\lambda \in e(S^2)} H_\lambda = L^2(Y, \mu).$$

Proof. This result is clear when S^2 is of finite order, and only of interest when $S^2 \neq I$. In the more general situation we apply a special case of a well known classical result as follows (see Theorem 1.8.1 of del Junco and Rudolph [5], for example).

Suppose that $T : Y \to Y$ is an ergodic automorphism and G is a compact subgroup of the centralizer $C(T)$ of T, then T is isomorphic to a compact group extension (which we shall also denote by T), $T : X \times G \to X \times G$ of the form

$$T(x, g) = (T_0 x, \phi(x) + g)$$

for some ergodic map $T_0 : X \to X$ and some measurable cocycle $\phi : X \to G$ (where we write the group operation additively). Under this isomorphism, the action of G may be represented by $\sigma_h : X \times G \to X \times G$, where

$$\sigma_h(x, g) = (x, g + h), \quad \text{for some} \ \ h \in G.$$

Corollary 1 *Let T be an ergodic automorphism for which there exists $S \in \mathcal{B}(T)$ with the property that the weak closure of the set $\{S^{2n} : n \in \mathbb{Z}\}$ is compact in $C(T)$. Then under the above isomorphism, T can be represented as a compact abelian group extension $T : X \times G \to X \times G$ (where G is isomorphic to the weak closure of $\{S^{2n} : n \in \mathbb{Z}\}$), $T(x, g) = (T_0 x, \phi(x) + g)$ where $T_0 : X \to X$ is ergodic, and $\phi : X \to G$ is measurable. Furthermore, S can be represented as an automorphism $S : X \times G \to X \times G$ satisfying*

$$S(x, g) = (kx, \psi(x) + g) \quad and \quad S^2(x, g) = (x, g + h),$$

for some $h \in G$, $k \in \mathcal{B}(T_0)$ and measurable cocycle $\psi : X \to G$.

Proof. Everything, except for the form of S follows from the previous discussion, so we may assume that $S^2(x, g) = (x, g + h)$ for some $h \in G$ such that the set $\{nh : n \in \mathbb{Z}\}$ is dense in G; therefore the rotation $R(g) = g + h$ acts ergodically on G. We show that the first co-ordinate of S is independent of g (we call such an automorphism a G–map; see Section 3 for the definition).

Suppose that $S(x, g) = (k(x, g), \psi(x, g))$ for some measurable maps $k : X \times G \to X$ and $\psi : X \times G \to G$. Then

$$S^2(x, g) = (k \circ S(x, g), \psi \circ S(x, g)).$$

Hence $k \circ S(x, g) = x$ and $\psi \circ S(x, g) = g + h$. These equations imply that for $x \in X$ and $g \in G$,

$$k \circ S^2(x, g) = k(x, g), \quad \text{and} \quad \psi \circ S^2(x, g) = \psi(x, g) + h.$$

But $S^2(x, g) = (x, g + h)$, so we must have

$$k(x, g + h) = k(x, g), \quad \text{and} \quad \psi(x, g + h) = \psi(x, g) + h,$$

for all $x \in X$ and $g \in G$. It follows from the ergodicity of the rotation R that k is independent of g, so we write $k(x) = k(x, g)$. Also, if $\psi(x) = \psi(x, 0)$, then $\psi(x, g) = \psi(x) + g$. $\square$

To complete the proof of Proposition 1, we recall that any $f \in L^2(X \times G, \tilde{\mu}) = \oplus_{\chi \in \hat{G}} L_\chi$ can be represented as a direct sum

$$f(x, g) = \sum_{\chi \in \hat{G}} \chi(g) f_\chi(x).$$

It follows that if $f_\chi(x, g) = \chi(g) f_\chi(x)$, then $f_\chi \circ S^2(x, g) = f_\chi(x, g + h) = \chi(h) f_\chi(x, g)$. In particular, $f_\chi \in H_{\chi(h)}$, so that $L_\chi \subseteq H_{\chi(h)}$. On the other hand, to see that $H_{\chi(h)} \subseteq L_\chi$, let $f \in H_{\chi(h)}$, then $f(x, g + h) = \chi(h) f(x, g)$ for a.e. $(x, g) \in X \times G$. Again using the ergodicity of R and the fact that its eigenfunctions are of the form $c \cdot \gamma(g)$ for some constant c and $\gamma \in \hat{G}$, with corresponding eigenvalues $\gamma(h)$, we deduce that $f(x, g)$ is of the form $f(x, g) = \chi(g) f_\chi(x)$ a.e., where $f_\chi \in L_\chi$, and the result follows. □

In the case where T has the weak closure property we can say even more. In particular that every member of $C(T)$ is the lift of some member of $C(T_0)$. Recall that T has the weak closure property if the weak closure of the powers of T is equal to $C(T)$. It was shown in [3] that for such a T, if $S \in \mathcal{B}(T)$, then $S^4 = I$. It may be that S is of order 2 (for example if T has rank one), but examples of T were constructed in [3] for which every $S \in \mathcal{B}(T)$ is of order 4. Such examples must be rigid, for if not, their centralizer would be trivial, and an order 4 conjugation would be impossible.

Theorem 1 *Suppose that T is an ergodic transformation isomorphic to its inverse and having the weak closure property. If there exists a conjugation of order 4, then every conjugation is of order 4, and T can be represented as a $\mathbf{Z}_2$ extension of an ergodic map T_0. Furthermore, every $S \in \mathcal{B}(T)$ can be represented in the form*

$$S(x, j) = (kx, \psi(x) + j), \quad \text{for some} \quad k \in \mathcal{B}(T_0) \quad \text{satisfying} \quad k^2 = I,$$

and every $\hat{\phi} \in C(T)$ is the lift of some $\phi \in C(T_0)$. As a consequence, $\hat{\phi}(x, j) = (\phi(x), u(x) + j)$ for some measurable $u : X \to \mathbf{Z}_2$.

Proof. Let $\hat{S} \in \mathcal{B}(T)$ be a conjugation of order 4, then Corollary 1 implies that T can be represented as a $\mathbf{Z}_2$-extension of an ergodic transformation T_0, and $\hat{S}^2(x, j) = (x, j + 1)$. However, it was shown in [3] that the set $\{S^2 : S \in \mathcal{B}(T)\}$ is a singleton set. It follows that $S^2(x, j) = (x, j + 1)$ for all $S \in \mathcal{B}(T)$. Now the same method of proof as used in Corollary 1 implies that every $S \in \mathcal{B}(T)$ is of the form $S(x, j) = (kx, \psi(x) + j)$ for some $k \in \mathcal{B}(T_0)$ and measurable $\psi : X \to \mathbf{Z}_2$.

To complete the proof, simply note that if $\hat{\phi} \in C(T)$, then $S^{-1} \hat{\phi} \in \mathcal{B}(T)$ for any $S \in \mathcal{B}(T)$, and so is of the form described in the last paragraph. Since S is also of this form, the result follows. □

3. The Structure of Group Extensions Which Are Conjugate to Their Inverses.

As before we assume that $T : X \rightarrow X$ is an ergodic automorphism, and $T_\phi : X \times G \rightarrow X \times G$ is a compact abelian group extension.

Our first proposition is a general result concerning the structure and spectrum of compact abelian group extensions. This result is related to Theorems 4 and 5 of [4], and is applicable to the situation described in Section 2. Recall that an automorphism $S : X \times G \rightarrow X \times G$ is a G–map, if it factors into the form $S(x, g) = (k(x), \psi(x, g))$ for some measurable transformations $k : X \rightarrow X$ and $\psi : X \times G \rightarrow G$.

If $S \in \mathcal{B}(T_\phi)$ is a G–map with $k^2 = I$, then by Theorem 5 of [3], S can be written in the form $S(x, g) = (k(x), u(x) + v(g))$ for some measurable $u : X \rightarrow G$ and continuous group automorphism $v : G \rightarrow G$ which satisfy

(i) $Tk = kT^{-1}$, and $k^2 = I$,

(ii) $v^2 = I$,

(iii) $\phi(kTx) + v(\phi(x)) = u(x) - u(Tx)$.

Furthermore $S^2(x, g) = (k^2(x), u(kx) + v(u(x)) + g) = (x, h + g)$ for some constant $h \in G$ and the operators $V_{\phi,T,\chi}$ and $V_{\phi,T,\chi \circ v}^{-1}$ are unitarily equivalent via the operator $Wf(x) = \chi(u(x))f(kx)$ (recall that $V_{\phi,T,\chi}f(x) = \chi(\phi(x))f(Tx)$, for $f \in L^2(X, \mu)$, and also $v(h) = h$, so if $\{nh : n \in \mathbb{Z}\}$ is dense in G, then $v = I$).

Proposition 2 *Suppose that $T_\phi : X \times G \rightarrow X \times G$ is an ergodic compact group extension of the ergodic map $T : X \rightarrow X$ and $S \in \mathcal{B}(T_\phi)$ is the G–map described above. Then*

(i) $V_{\phi,T,\chi}$ is unitarily equivalent to $V_{\phi,T,\overline{\chi} \circ v}$, for all $\chi \in \widehat{G}$; $\chi \neq 1$.

(ii) If $\chi(h) \neq 1$, and $\chi = \overline{\chi} \circ v$, then for each $f \in L^2(X, \mu)$, f and $W(f)$ have the same spectral types and generate orthogonal cyclic subspaces of $L^2(X, \mu)$.

(iii) If $\{nh : n \in \mathbb{Z}\}$ is dense in G, and if

$$H_\chi = \{f \in L^2(X, \mu) : f(S^2(x, g)) = \chi(h)f(x, g)\},$$

then $L_\chi = H_\chi$ for all $\chi \in \widehat{G}$.

Proof. (i) The unitary operators $V_{\phi,T,\chi}$ and $V_{\phi T^{-1},T^{-1},\overline{\chi} \circ v}$ ($= V_{\phi,T,\chi \circ v}^{-1}$) are unitarily equivalent via the unitary operator $W(f)(x) = \chi(u(x))f(k(x))$, for each $\chi \in \widehat{G}$, $\chi \neq 1$.

Now let $f \in L^2(X, \mu)$, then for $n \in \mathbb{Z}$ we have

$$(V_{\phi,T,\chi}^n f, f) = (WV_{\phi,T,\chi}^n f, Wf) = (V_{\phi,T,\chi \circ v}^{-n} Wf, Wf)$$

$$= (Wf, V_{\phi,T,\chi \circ v}^n Wf) = \overline{(V_{\phi,T,\chi \circ v}^n Wf, Wf)} = (V_{\phi,T,\overline{\chi} \circ v}^n \overline{Wf}, \overline{Wf}).$$

This says that $V_{\phi,T,\chi}|Z(f) \cong V_{\phi,T,\overline{\chi} \circ v}|Z(\overline{Wf})$ for every $f \in L^2(X, \mu)$. This is enough to prove that $V_{\phi,T,\chi}$ and $V_{\phi,T,\overline{\chi} \circ v}$ are unitarily equivalent for all $\chi \in \widehat{G}$, $\chi \neq 1$.

(ii) Suppose now that $\chi \in \widehat{G}$ satisfies $\chi = \overline{\chi} \circ v$, $\chi(h) \neq 1$. Then since $u(kx) + v(u(x)) = h$, it follows that $\chi(u(kx))\overline{\chi}(u(x)) = \chi(h)$.

As before, for $f \in L^2(X, \mu)$, f and $\overline{Wf}$ have the same spectral type with respect to $V_{\phi,T,\chi}$. However, if $n \in \mathbf{Z}$ then

$$(V_{\phi,T,\chi}^n \overline{Wf}, f) = (W V_{\phi,T,\chi}^n \overline{Wf}, Wf)$$

$$= (V_{\phi,T,\chi \circ v}^{-n} W(\overline{Wf}), Wf) = (W(\overline{Wf}), V_{\phi,T,\chi \circ v}^n Wf)$$

$$= (\chi(u(x))\overline{\chi}(u(kx))\overline{f(k^2 x)}, V_{\phi,T,\overline{\chi}}^n Wf) = (\chi(h)\overline{f}, V_{\phi,T,\overline{\chi}}^n Wf)$$

$$= \chi(h)\overline{(V_{\phi,T,\overline{\chi}}^n Wf, \overline{f})} = \chi(h)(V_{\phi,T,\chi}^n \overline{Wf}, f).$$

Since $\chi(h) \neq 1$, this implies that $(V_{\phi,T,\chi}^n \overline{Wf}, f) = 0$ for all $n \in \mathbf{Z}$ and every $f \in L^2(X, \mu)$, and this in turn implies that with respect to the operator $V_{\phi,T,\chi}$, $Z(f) \perp Z(\overline{Wf})$ for every $f \in L^2(X, \mu)$.

(iii) Since $\{nh : n \in \mathbf{Z}\}$ is dense in G, $\chi(h) \neq 1$ for every non-trivial character $\chi \in \widehat{G}$, and the result follows in a similar way to the proof of Proposition 1. $\qquad\Box$

Remarks. Part (i) of the above shows that even in the case where $h = 0$ (and hence $S^2 = I$), T_ϕ will have even multiplicity in the subspace $\oplus_\chi L_\chi$, where the sum is taken over those χ for which $\chi \neq \overline{\chi} \circ v$. However, in this case examples suggest that there will always exist other $S \in \mathcal{B}(T_\phi)$ for which $S^2 \neq I$.

It follows from (ii) above that T_ϕ will have even multiplicity function in the orthocomplement of the subspace $\{F : F \circ S^2 = f\}$. In addition, (iii) shows that in this case, this orthocomplement will be $L_{\chi_0}^\perp$, the orthogonal complement of the subspace on which T acts.

Corollary 2 *Suppose there exists a G–map $S \in \mathcal{B}(T_\phi)$, $S(x, g) = (k(x), \psi(x, g))$, for which $k^2 = I$ and $S^2 \neq I$. Then if* $\mathrm{msm}(T_\phi) = 2$, *every member of $C(T_\phi)$ and $\mathcal{B}(T_\phi)$ is a G–map.*

Proof. Since $\mathrm{msm}(T_\phi) = 2$, we must have $\mathrm{msm}(V_{\phi,T,\chi}) = 2$ (since it is even). It follows that $\mathrm{msm}(T)$ is either 1 or 2, and in either case its maximal spectral type must be singular with respect to that of $V_{\phi,T,\chi}$. An application of Proposition 3 (below) now gives the result. $\qquad\Box$

We finish this section with some new conditions under which such a map S is a G–map, and we also attempt to explain why for those examples known to have a Lebesgue component in their spectrum, that component occurs with even (or infinite) multiplicity.

Proposition 3 *Suppose that T_ϕ is a group extension with corresponding direct sum decomposition $L^2(X \times G, \tilde{\mu}) = H_0 \oplus H^+$, where $T_\phi|H_0 = T$, and σ_0, σ^+ are the restrictions of the maximal spectral type of T_ϕ to H_0 and H^+ respectively. If $\sigma_0 \perp \sigma^+$, then any $S \in \mathcal{B}(T_\phi)$ is a G–map.*

Proof. Since $ST_\phi = T_\phi^{-1}S$, both H_0 and $H_1 = S(H_0)$ are T_ϕ invariant. Suppose that $H_1 \cap H^+ \neq \{0\}$, then there exists $f \in H_1 \cap H^+$ with $\sigma_f \ll \sigma_0$; $\sigma_f \ll \sigma^+$, which is impossible if $\sigma_0 \perp \sigma^+$. It follows that $S(H_0) = H_0$, and hence that the first coordinate of $S(x, g)$ is independent of g, i.e. we may write

$$S(x, g) = (kx, \psi(x, g))$$

for measurable $k : X \to X$, satisfying $Tk = kT^{-1}$ and $\psi : X \times G \to G$. Since S is an automorphism, and $k(A) \times G = S(A \times G)$ for all measurable subsets A of X, we see that k is measure preserving and hence is also an automorphism. $\square$

Corollary 3 *Let T_ϕ be a compact group extension of a simple map T which also has simple spectrum. Suppose there exists $S \in \mathcal{B}(T_\phi)$ with $S^2 \neq I$. If $T_\phi|H_0$ has a Lebesgue component in its spectrum, then it must occur with even (or infinite) multiplicity.*

Proof. Since T is a simple map, every $S \in \mathcal{B}(T_\phi)$ is a G–map, and the simplicty of spectrum of T implies that every member of $\mathcal{B}(T)$ is an involution (see [3]). Therefore Proposition 2 is applicable, and the result follows. $\square$

4. Examples.

Example 1 *A class of examples which are weakly mixing, isomorphic to their inverses and have a spectrum consisting of a Lebesgue component of multiplicity two and a continuous singular component:*

The construction is based in part on the examples due to Mathew and Nadkarni [7] and also those in [3] and the proofs use some ideas due to Ageev [1]. We start by constructing a collection of rank one transformations using cutting and stacking. One such transformation is defined inductively for each sequence $\omega = \{j_n\}_0^\infty$, where $j_n \in \{0, 1\}$. To simplify the construction we shall assume that $j_0 = 0$. At the nth stage of the construction we have a sequence of partitions $\zeta(n) = \{A_i^{(n)} : 1 \leq i \leq q_n\}$, $n = 0, 1, \ldots$, satisfying:

(i) $\zeta(n) \to \epsilon$ as $n \to \infty$.

(ii) $TA_i^{(n)} = A_{i+1}^{(n)}$, for $i = 1, \ldots q_n - 1$.

(iii) $A_i^{(n)} = A_i^{(n+1)} \cup A_{i+q_n+j_n}^{(n+1)}$, and $q_{n+1} = 2q_n + j_n$, $n \geq 0$.

These transformation, which we shall denote by T_ω, are constructed inductively as follows:

$\zeta(n+1)$ is obtained from $\zeta(n)$ by cutting the tower in half (into two equal subcolumns), putting a single spacer on top of the left hand column (in the case that $j_n = 1$), or not using a spacer (in the case that $j_n = 0$). The induction is started by defining $q_0 = 1$, (i.e. starting with a single interval, say $[0, 1)$ and then normalising when the construction is complete to obtain a probability space X).

The construction gives rise to a Lebesgue probability space in the usual way such that each $T_\omega : X \to X$ is an ergodic (since rank one) automorphism. They are easily seen to be isomorphic to their inverses via the map $k : X \to X$ defined on $\zeta(n)$ by $kA_i^{(n)} = A_{q_n-i+1}^{(n)}$ (i.e. the flip map on the levels of $\zeta(n)$). Since $\zeta(n)$ is an increasing sequence of partitions, k is well defined as a point map.

It can be seen that if $j_n = 0$ for all n, then T_ω is the familiar von Neumann-Kakutani adding machine, having discrete spectrum and eigenvalues the 2^nth roots of unity. Similarly, if $j_n = 1$ for all n then T_ω also has discrete spectrum. We claim that if $\omega = \{j_n\}$ is a sequence which differs from a constant sequence in infinitely many places, then T_ω is weakly mixing. (The usual proof seems to work in this case). Finally, we mention that a recent result of Klemes and Reinhold [6] implies that T_ω has singular spectrum for every sequence ω. We summarize these facts in the following proposition.

Proposition 4 *For each sequence ω, the map T_ω is rank one (hence is ergodic and has simple spectrum), isomorphic to its inverse and has singular spectrum. If the sequence ω differs from a constant infinitely often, then T_ω is weakly mixing.*

Definition of the cocycle $\phi_\omega : X \to \mathbb{Z}_2$

We now construct our examples, which we will denote by T_{ϕ_ω}, or just by T_ϕ if the sequence ω is understood. Thus

$$T_{\phi_\omega}(x, j) = (T_\omega(x), \phi_\omega(x) + j).$$

We need to find a map $u_\omega : X \to \mathbb{Z}_2$ for which the equation

$$\phi_\omega(kTx) + \phi_\omega(x) = u_\omega(x) - u_\omega(Tx)$$

is satisfied, for then the map $S_\omega(x, j) = (k(x), u_\omega(x) + j)$ will conjugate T_{ϕ_ω} to its inverse. Our construction will show that $S_\omega^2(x, j) = (x, 1 + j)$, and we remark that this is enough to give the ergodicity of T_{ϕ_ω} as the following proposition shows. We shall omit the subscript ω from now on.

Proposition 5 *Suppose that T has simple spectrum, $k \in \mathcal{B}(T)$ and $\phi : X \to \mathbb{Z}_2$ is a cocycle for T. If there exists a solution $u : X \to \mathbb{Z}_2$ to the functional equation*

$$\phi(kTx) + \phi(x) = u(x) - u(Tx), \tag{1}$$

then $u(x) + u(kx) = h$ is a constant a.e., and if $h \neq 0$ then T_ϕ is ergodic.

Proof. Replacing x by kTx in equation 1 gives

$$\phi(x) + \phi(kTx) = u(kTx) - u(kx)$$

(T simple spectrum implies $k^2 = I$).
It follows that

$$u(x) - u(Tx) = u(kTx) - u(kx)$$

and T ergodic implies $u(x) + u(kx) = h$ a.e., a constant.

Now suppose that T_ϕ is not ergodic and that $h \neq 0$. Then ϕ is a coboundary, i.e. there exists $v : X \to \mathbb{Z}_2$ for which

$$\phi(x) = v(Tx) - v(x).$$

Thus $\phi(kTx) = v(kx) - v(kTx)$ and again from equation 1 we deduce

$$v(kx) - v(kTx) + v(Tx) - v(x) = u(x) - u(Tx)$$

or

$$u(x) + v(x) - v(kx) = u(Tx) + v(Tx) - v(kTx)$$

and again T ergodic implies that $u(x) + v(x) - v(kx) = c$, a constant. Replacing x by kx and adding gives

$$u(x) + u(kx) = 2c$$

and in $\mathbb{Z}_2$ this implies that $h = 0$, a contradiction. $\qquad\square$

Corollary 4 *Under the conditions of Proposition 5 (so in particular, equation 1 holds with $h \neq 0$), ϕ is a weakly mixing cocycle with respect to T (i.e. T_ϕ has no eigenvalues besides those of T), so if T is weakly mixing, then so is T_ϕ.*

Proof. Just apply Proposition 11 of [3]. $\qquad\square$

Now we define our cocycles. Denote by $\{\epsilon_n\}$ a sequence for which $\epsilon_n \in \mathbb{Z}_2$, $n = 1, 2, \ldots$. Denote $1 - \epsilon_n$ by ϵ'_n. We define the cocycle ϕ and the map u inductively as follows. Start the induction by defining $u(x) = 0$ if $x \in A_1^{(1)} = [0, 1/2)$, and $u(x) = 1$ if $x \in A_2^{(1)} = [1/2, 1)$ (recall $j_0 = 0$), and define $\phi(x) = \epsilon_1$ if $x \in A_1^{(2)} = [0, 1/4)$ and $\phi(x) = \epsilon'_1$ if $x \in A_3^{(2)} = [1/4, 1/2)$.

We split the inductive step in the construction into two cases

Case (i) $j_{n-1} = 0$

In this case $q_n = 2q_{n-1}$ and $\zeta(n)$ is constructed from $\zeta(n - 1)$ by stacking without the use of a spacer. Suppose that ϕ has been defined on all the levels of $\zeta(n)$ except for

$$A_{q_{n-1}}^{(n)} = A_{q_{n-1}}^{(n+1)} \cup A_{q_{n-1}+q_n+j_n}^{(n+1)}$$

and the top level (in this case u is defined on all the levels of $\zeta(n)$), and that the equation

$$\phi(kTx) + \phi(x) = u(x) - u(Tx) \tag{2}$$

holds where it makes sense. Now define $\phi(x) = \epsilon_n$ for $x \in A_{q_{n-1}}^{(n+1)}$, $\phi(x) = \epsilon_n'$ for $x \in A_{q_{n-1}+q_n+j_n}^{(n+1)}$. It is now necessary to show that equation 2 holds again where it makes sense. Notice that

$$u(x) = 1, \quad x \in A_{q_{n-1}}^{(n)}, \quad u(x) = 0, \quad x \in A_{q_{n-1}-1}^{(n)} \cup A_{q_{n-1}+1}^{(n)},$$

and that

$$\phi(x) = \epsilon_1, \quad x \in A_{q_{n-1}-1}^{(n)} \quad \text{and} \quad \phi(x) = \epsilon_1', \quad x \in A_{q_{n-1}+1}^{(n)},$$

(since $j_0 = 0$), so it is a simple matter to check that equation 2 extends to all allowable levels.

Case (ii) $j_{n-1} = 1$

In this case $q_n = 2q_{n-1} + 1$ and the spacer is the set

$$A_{q_{n-1}+1}^{(n)} = A_{q_{n-1}+1}^{(n+1)} \cup A_{q_{n-1}+q_n+j_n+1}^{(n+1)}.$$

This time u is defined on all the levels of $\zeta(n)$ except for the spacer, and ϕ is defined on all the levels except the top level, the spacer and the level below the spacer. Now we define ϕ on the spacer by:

$$\phi(x) = \epsilon_n, \quad x \in A_{q_{n-1}+1}^{(n+1)} \quad \phi(x) = \epsilon_n', \quad x \in A_{q_{n-1}+q_n+j_n+1}^{(n+1)}.$$

We define ϕ on the level below the spacer by

$$\phi(x) = 0, \quad x \in A_{q_{n-1}}^{(n)}.$$

Now the value of u on the spacer is determined by that value which makes equation 2 correct. In fact, let

$u =$ value of $u(x)$ on $A_{q_{n-1}}^{(n)}$, $u' = 1 - u =$ value of $u(x)$ on $A_{q_{n-1}+2}^{(n)}$,

$u_1 =$ value of $u(x)$ on $A_{q_{n-1}+1}^{(n+1)}$, $u_2 =$ value of $u(x)$ on $A_{q_{n-1}+q_n+j_n+1}^{(n+1)}$.

If we now define $u_1 = u - \epsilon_n'$; and $u_2 = u_1'$, then we can see that equation 2 will be satisfied for all appropriate levels.

At the end of the process involving Cases (i) and (ii) we see that ϕ is constant on all the levels of $\zeta(n)$ except for the top level and one level in the middle, where it takes two values. Now we continue the induction by stacking, and we eventually see that ϕ and u are defined almost everywhere in such a way that equation 2 also holds almost everywhere.

Demonstration that $V_{\phi,T,\chi}$ has a Lebesgue component of multiplicity 2 in its spectrum.

As before, $V_{\phi,T,\chi} : L^2(X,\mu) \rightarrow L^2(X,\mu)$ denotes the unitary operator defined by

$$V_{\phi,T,\chi}f(x) = \chi(\phi(x))f(Tx),$$

where $\chi : \mathbb{Z}_2 \rightarrow S^1$ is the character $\chi(j) = \underline{(-1)^j}$. The proof of Proposition 2 shows that given any $h \in L^2(X,\mu)$; h and $\overline{W(h)}$ have identical spectral types and generate orthogonal cyclic subspaces (where $Wf(x) = \chi(u(x))f(k(x))$, and we are assuming that equation 1 holds).

Choose $h = \chi_{A_1^{(n+1)}} = \chi_1^{(n+1)}$, (where χ_A denotes the characteristic function of the set A). Then

$$\overline{Wh}(x) = \chi(u(x))h(k(x)) = -\chi_{q_{n+1}}^{(n+1)},$$

since u is constant on $A_{q_n}^{(n+1)}$ (and takes the value zero on $A_1^{(n+1)}$ and 1 on $A_{q_{n+1}}^{(n+1)}$, since $u(x) + u(kx) = 1$).

It follows from the above remarks that

$$(V_{\phi,T,\chi}^k \chi_1^{(n+1)}, \chi_1^{(n+1)}) = (V_{\phi,T,\chi}^k \chi_{q_{n+1}}^{(n+1)}, \chi_{q_{n+1}}^{(n+1)})$$

and

$$(V_{\phi,T,\chi} \chi_1^{(n+1)}, \chi_{q_{n+1}}^{(n+1)}) = 0 \quad \forall \ k \in \mathbb{Z}, \quad n \in \mathbf{N}.$$

It suffices therefore to show that

$$(V_{\phi,T,\chi}^k \chi_1^{(n+1)}, \chi_1^{(n+1)}) = 0 \quad \forall \ k \neq 0.$$

Clearly $(V_{\phi,T,\chi}^k \chi_1^{(n+1)}, \chi_1^{(n+1)}) = 0$ if $1 \leq k < q_n$, so let us fix $l \geq 0$ for which $q_{n+l} \leq k < q_{n+l+1}$.

Note that $A_1^{(n+1)}$ can be written as a union of the form

$$A_1^{(n+1)} = \cup_{l=0}^{\infty} A_{1+q_{n+l}+j_{n+l}}^{n+l+1},$$

so it suffices to prove the following lemma, which is based on a result of Ageev [1].

Lemma 1 *If $q_n \leq k < q_{n+1}$ and $i \leq q_{n+1}$, then*

$$(V_{\phi,T,\chi}^k \chi_i^{(n+1)}, \chi_j^{(n+1)}) = 0.$$

Proof. We have that $q_{n+1} = 2q_n + j_n$ and $i \leq q_{n+1}$.

Case (i) Suppose that $i > k$, then $0 < i - k \leq 2q_n + j_n - q_n = q_n + j_n$

Now since $i - k > 0$, $V_{\phi,T,\chi}^k \chi_i^{(n+1)} = u(\phi^{(k)}(x))\chi_{i-k}^{(n+1)}$, and since $i - k \leq q_n + j_n$ and u is constant on every level except for the middle, $u(\phi^{(k)}(x))$ takes the value $+1$ on half of the set $A_{i-k}^{(n+1)}$ and the value -1 on the other half (i.e. on the sets $A_{i-k}^{(n+2)}$ and $A_{i-k+q_{n+1}+j_{n+1}}^{(n+2)}$, which are of equal measure). Clearly we may assume that $j = i - k$. It follows that

$$(u(\phi^{(k)}(x))\chi_{i-k}^{(n+1)}, \chi_{i-k}^{(n+1)}) = 0$$

in this case.

Case (ii) $k \geq i$.

Note that we can write

$$\chi_i^{(n+1)} = \sum_{l=1}^{\infty} \chi_{i+q_{n+l}+j_{n+l}}^{(n+l+1)}.$$

But $q_{n+1} - k > 0$, so that $i + q_{n+l} + j_{n+l} > k$, and then

$$V_{\phi,T,\chi}^k \chi_{i+q_{n+l}+j_{n+l}}^{(n+l+1)} = u(\phi^{(k)}(x))\chi_{i-k+q_{n+l}+j_{n+l}}^{(n+l+1)}.$$

Now $i - k + q_{n+l} + j_{n+l} \leq q_{n+l} + j_{n+l}$, so again $u(\phi^{(k)}(x))$ takes the values $+1$ and -1 on sets of equal measure (i.e. the sets $A_{i-k+q_{n+l}+j_{n+l}}^{(n+l+2)}$ and $A_{i-k+q_{n+l}+j_{n+l}+q_{n+l+1}+j_{n+l+1}}^{(n+l+2)}$), and again

$$(V_{\phi,T,\chi}^k \chi_{i+q_{n+l}+j_{n+l}}^{(n+l+1)}, \chi_j^{(n+1)}) = 0,$$

since χ_j^{n+1} is constant on the elements of the partition $\zeta(n + l + 1)$.   □

Remarks 1. The method of Ageev [1] now shows that $V_{\phi,T,\chi}$ has a homogeneous Lebesgue spectrum of multiplicity equal to two. The ergodicity and weak mixing of T_ϕ can also be deduced using this and the fact that T_ω is weak mixing (for appropriate ω).

2. By choosing ω suitably (as in [1]), the maps T_ϕ^n, $n = 1, 2, \ldots$, give rise to automorphisms with a Lebesgue component of multiplicity equal to $2n$ in their spectrum.

3. It now follows from Proposition 3 that every member of $C(T_\phi)$ and $\mathcal{B}(T_\phi)$ is a G–map, i.e. the lift of members of $C(T)$ and $\mathcal{B}(T)$ respectively.

Example 2 *Maps with quasi-discrete spectrum:*

In [3], the general case of an order 2 map with quasi-discrete spectrum was studied. Such maps were shown to be isomorphic to their inverses, and the general form of the conjugation was stated without proof. Here we include some of the details, and show that the results of Sections 2 and 3 apply.

Let $(X, +)$ and $(G, .)$ be compact abelian groups; $T : X \to X$ an ergodic rotation, $Tx = x + \alpha$, and $\phi : X \to G$ a cocycle which is also a continuous group homomorphism, We also assume that ϕ is a weakly mixing cocycle with respect to T (i.e. there are no eigenvalues for T_ϕ besides those of T).

Suppose that $S : X \times G \to X \times G$ is a conjugation between T_ϕ and its inverse, then it follows from the remarks at the start of Section 3 that $S(x, g) = (kx, u(x)v(g))$, where $u : X \to G$ is measurable , $v : G \to G$ is a continuous group automorphism and $k : X \to X$ is measure preserving, satisfying:

(i) $Tk = kT^{-1}$

(ii) $\phi(kTx) \cdot v(\phi(x)) = u(x)/u(Tx)$.

From (i) it follows that $k(x) = \beta - x$, $x \in X$, for some $\beta \in X$, (see [4]). It is easy to see that if $v = I$ and $u(x) = r \cdot w(x)$, the product of a continuous group homomorphism $w : X \to G$, and a constant $r \in G$ satisfying $\phi(\alpha - \beta) = w(\alpha)$, then (i) and (ii) hold, i.e.

$$\text{if } \quad S(x, g) = (\beta - x, r \cdot w(x)g) \quad \text{then} \quad S \in \mathcal{B}(T_\phi). \qquad (3)$$

We show below that all the members of $\mathcal{B}(T_\phi)$ are of this form.

Proposition 6 *The members of $\mathcal{B}(T_\phi)$ are given by equation (3) above.*

Proof. We are assuming that ϕ is a cocycle which is weakly mixing with repect to T. Suppose that (ii) above holds. Then

$$\phi(\beta - \alpha - x)v(\phi(x)) = \frac{u(x)}{u(Tx)}.$$

$$\phi(\beta - \alpha)\frac{v\phi(x)}{\phi(x)} = \frac{u(x)}{u(Tx)}.$$

We see that this is only possible if:

(i) $(\chi \circ v)/\chi = 1$ for all $\chi \in \hat{G}$, so that $v = I$.

(ii) $\chi \circ \phi(\alpha - \beta)$ is an eigenvalue for $Tx = x + \alpha$, and

(iii) $\chi \circ u(x)$ is the corresponding eigenfunction.

(iii) implies that $\chi \circ u(x) = c \cdot \gamma(x)$ for some $c \in S^1$ and some character $\gamma \in \hat{X}$. Also (ii) implies that $\chi \circ \phi(\alpha - \beta) = \gamma(\alpha)$. It follows that $c = \chi(r)$ for some $r \in G$ and that $r^{-1}u(x) = w(x)$ is a group homomorphism which satisfies $w(\alpha) = \phi(\alpha - \beta)$. $\qquad \square$

We examine a special case in more detail: Let $T_\phi : X \times G \to X \times G$ be defined by

$$T_\phi(x, g) = (x + \alpha, \phi(x)g),$$

where $Tx = x + \alpha$ is ergodic, and the character groups $\hat{X}$ and $\hat{G}$ are torsion free (for example, $X = G = [0, 1)$, the unit circle with usual Haar measure). We also assume that the cocycle ϕ is weakly mixing. Then it is not hard to see that the spectrum of T_ϕ consists of a discrete component and a countable Lebesgue component.

From above, if $S \in \mathcal{B}(T_\phi)$, then

$$S^2(x, g) = (x, r^2\omega(\beta)g), \quad \beta \in X, \quad r \in G,$$

and where $\omega : X \to G$ is a group homomorphism which satisfies $\omega(\alpha) = \alpha - \beta$. Notice that S^2 is never ergodic, and if $r^2\omega(\beta)$ is of finite order in G, then S^2 will be of finite order. On the other hand, if $r^2\omega(\beta)$ is of infinite order, then S^2 will be aperiodic with ergodic components of measure zero on which S^2 behaves as a rotation by $r^2\omega(\beta)$. In both cases, the weak closure of the set $\{S^{2n} : n \in \mathbb{Z}\}$ will be compact in $C(T)$. In the case of $X = G = [0, 1)$ for example, all integer orders are possible for S^2, and also S^2 may be aperiodic.

Remark The above shows that order 2 maps with quasi-discrete spectrum can have conjugations S for which S^2 is of infinite order and such that the weak closure of the powers is compact. In a similar way, a wide variety of examples with this property can be constructed using more general compact abelian group extensions, where the group G is infinite.

References

[1] O.N. Ageev. Dynamical systems with a Lebesgue component of even multiplicity in the spectrum. *Mat. Sb.* **136** (1988), 307-319.

[2] G.R. Goodson, M. Lemańczyk. Transformations conjugate to their inverses have even essential values. To appear in Proc. A.M.S., 1995.

[3] G.R. Goodson, A. del Junco, M. Lemańczyk, D.J. Rudolph. Ergodic transformations conjugate to their inverses by involutions. To appear in Ergodic Theory and Dynamical Systems, 1995.

[4] G.R. Goodson. Functional equations associated with the spectral properties of compact group extensions. Ergodic Theory and its Connections with Harmonic Analysis. Proceedings of the 1993 Alexandria Conference, Cambridge University Press, 1994, 309-328.

[5] A. del Junco, D.J. Rudolph. On ergodic actions whose self–joinings are graphs. *Ergodic Theory and Dynamical Systems.* **7** (1987), 531-557.

[6] I. Klemes, K. Reinhold. Rank one transformations with singular spectral type. Preprint, 1994.

[7] J. Mathew, M.G. Nadkarni. A measure preserving transformation whose spectrum has a Lebesgue component of multiplicity two. *Bull. Lond. Math. Soc.* **16** (1984), 402-406.

[8] M. Queffelec. *Substitution Dynamical Systems, Spectral Analysis.* Lecture Notes in Math. **1294** , 1987.

[9] D.J. Rudolph. *Fundamentals of Measurable Dynamics.* Oxford University Press, 1990.

Approximation by periodic transformations and diophantine approximation of the spectrum

A. Iwanik*

Institute of Mathematics, Technical University of Wrocław, Poland

Abstract

It is shown that an ergodic transformation T with discrete spectrum admits cyclic approximation with speed $o(f(n))$ iff its eigenvalues admit simultaneous diophantine approximation with speed $o(f(n))$. Consequently, for any such T we obtain a cyclic approximation with speed $o(1/n)$ (actually, there exist "flat towers" with complement $o(1/n)$). All ergodic rotations of the d-torus allow the speed

$$l(n)/n^{1+1/d}$$

where $l(n) \to \infty$ is arbitrary. On the other hand, in any monothetic group arbitrarily large speeds occur.

For an arbitrary automorphism T we obtain a necessary spectral condition for periodic approximation with a given speed. Our condition improves that given by Katok and Stepin and allows us to estimate the Hausdorff dimension of the spectrum for skew products.

1 Introduction

The method of approximation by periodic transformations introduced by Katok and Stepin has proved useful in studying ergodic properties of skew products. If group extensions of rotations are considered as e.g. in [7], [2], [8], [9], [5], it becomes important to know how well the basic rotation can be approximated. The main aim of the present paper is to clarify the connection between the speed of periodic approximation of an ergodic rotation and the speed of diophantine approximation of its spectrum. Roughly speaking,

---

*Supported by KBN grant PB 666/2/91 and MRT grant 90 R 0433.

the two speeds are the same (Theorems 2 and 3). In the case of a rational spectrum this was proved by Stepin [12], the irrational rotations of the circle were treated in [4]. The present work does not rely on either of these two results. Our method is similar to that in [4], but it avoids the use of continued fractions; we use instead a lemma of del Junco [3] as a base for our construction.

We also majorize the size of the spectrum of any automorphism admitting periodic approximation. This will allow us to estimate the Hausdorff dimension of the continuous part of the spectrum.

In Section 3 we show that any ergodic transformation with discrete spectrum commutes with one admitting any given speed of cyclic approximation. In Section 4 we find a connection between the cyclic and diophantine approximations for ergodic rotations. Actually, Theorem 3 is more general as it gives the speed of diophantine approximation for eigenvalues of any automorphism. In Section 5 we consider an arbitrary automorphism T and prove that if T allows periodic approximation with speed $o(f(n))$ then its spectrum admits simultaneous diophantine approximation also with speed $o(f(n))$. This improves upon an early estimation of Katok and Stepin [7] and allows us in Section 6 to majorize by $1/2$ the Hausdorff dimension of the spectrum of "most" Anzai skew products.

2   Definitions and notation

Let (X, μ) be a nonatomic Lebesgue probability space. If $\xi = \{C_0, \dots, C_{h-1}\}$ is a finite collection of disjoint measurable subsets then we write $h(\xi) = h$, $\mu(\xi) = \mu(C_0 \cup \dots \cup C_{h-1})$. If in addition X is endowed with a metric, we put

$$\operatorname{diam} \xi = \max\{\operatorname{diam} C_0, \dots, \operatorname{diam} C_{h-1}\}.$$

Denote by ϵ the partition of X into singletons. Given a sequence ξ_n we write $\xi_n \to \epsilon$ if for every measurable set A there exist sets A_n such that each A_n is a union of pieces of ξ_n and $\mu(A_n \triangle A) \to 0$. If X is a separable metric space and μ (an extension of) a Borel measure then $\xi_n \to \epsilon$ whenever $\mu(\xi_n) \to 1$ and $\operatorname{diam} \xi_n \to 0$.

Let T be an automorphism of (X, μ), i.e. an invertible measure preserving transformation. We will consider properties of the unitary operator $U_T f(x) = f(Tx)$ acting on $L^2(\mu)$. The eigenvalues, maximal spectral measure, etc., of U_T will be referred to as eigenvalues, etc., of T. It will be convenient to understand the spectral measure as defined on the unit interval rather than the unit circle; the obvious passage is via $x \to \exp(2\pi i x)$. Also, we will identify the torus group $\mathbf{T}$ as the interval $[0, 1)$ endowed with the compact topology identifying 0 and 1, and addition modulo 1.

We recall the basic notion of periodic approximation (see [7] or [2]). Let $0 < f(n) \downarrow 0$. The automorphism T is said to admit an *approximation of the first type by periodic transformations (a.p.t.I) with speed $f(n)$* if there exists a sequence of partitions $\xi_n = \{C_0, \ldots, C_{h_n-1}\}$ and a sequence of automorphisms T_n preserving ξ_n such that

$$\xi_n \to \epsilon$$

and

$$\sum_{j=0}^{h_n-1} \mu(TC_j \triangle T_n C_j) < f(h_n).$$

If in addition T_n permutes ξ_n cyclically, we say that T admits *cyclic a.p.t. with speed $f(n)$*. If, on the other hand, T_n has period q_n, the right hand side of the last condition is replaced by $f(q_n)$, and $U_{T_n} \to U_T$ for the strong operator convergence topology, then T is said to admit *a.p.t.II with speed $f(n)$*. Note that the convergence of operators implies $q_n \to \infty$ if T is not periodic.

Clearly for the same speed $f(n)$ we have

$$\text{cyclic a.p.t.} \Rightarrow \text{a.p.t.I} \quad \text{and} \quad \text{a.p.t.II}$$

As in [7], we denote by $d(T)$ the least upper bound of the numbers $r > 0$ such that T admits cyclic a.p.t. with speed $1/n^r$.

A collection $\xi = \{C_0, \ldots, C_{h-1}\}$ of disjoint measurable sets is called a Rokhlin tower if $TC_{j-1} = C_j$, $j = 1, \ldots, h - 1$. Note that if X is a metric space and T is an isometry then diam ξ = diam C_0. This will be the case for group rotations as we always choose an invariant metric. We say that T admits *flat towers (f.t.) with complement $f(n)$* if there exists a sequence of Rokhlin towers ξ_n such that $\xi_n \to \epsilon$ and

$$1 - \mu(\xi_n) < f(h_{\xi_n}).$$

By distributing the uncovered part of the space among the levels of the tower and defining a new periodic transformation T_n it is easy to see that

$$\text{f.t. with speed } f(n)/4 \Rightarrow \text{cyclic a.p.t. with speed } f(n)$$

Conversely, if T admits cyclic a.p.t. with speed $f(n) = o(1/n)$ then by chopping off the error term from each cell of the partition and then removing the corresponding parts from the other pieces we obtain

$$\text{cyclic a.p.t. with speed } f(n) \Rightarrow \text{f.t. with complement } nf(n).$$

If x is a real number, we denote by $\|x\|$ its distance to the nearest integer. Note that $0 \le \|x\| \le 1/2$ and $\|lx\| = l\|x\|$ whenever l is a positive integer such that $l\|x\| \le 1/2$.

Let A be a set of real numbers and $0 < f(n) \to 0$. We say that A admits *simultaneous diophantine approximation with speed* $f(n)$ if there exists a sequence $q_n \to \infty$ such that for every $x \in A$ one can find integers p_n for which the inequality

$$|x - p_n/q_n| < f(q_n)$$

holds for all sufficiently large n. In other words, $\|q_n x\| < q_n f(q_n)$ for n large enough.

It should be remarked that if A is a subgroup of $\mathbf{R}$ generated by a subset admitting simultaneous diophantine approximation with speed $o(f(n))$, then A also admits that speed.

It easily follows from the Dirichlet pigeonhole principle that any set $\{x_1, \ldots, x_m\}$ of at most m numbers admits the speed $1/n^{1+1/m}$. Indeed, for any $K \geq 1$ we can find $1 \leq k \leq K^m$ such that

$$\|kx_j\| \leq 1/K,$$

for $j = 1, \ldots, m$. Consequently, every countable set of real numbers admits the speed $1/l(n)n$ if $l(n)$ grows slower than any positive power of n.

We need one more definition. Let X be a compact metric space and $T: X \to X$ be a continuous mapping which is minimal, i.e. every T-orbit is dense (this is the case for ergodic compact group rotations). Given $\varepsilon > 0$ we can find a positive integer K such that for any $x \in X$ and any ball V of radius ε we have $T^k x \in V$ for some $0 \leq k < K$. The least such number K will be denoted $K(\varepsilon)$.

3   Well approximable rotations

It was proved by Stepin that a transformation admits any speed of periodic approximation iff it has discrete rational spectrum (see [12]; this will also follow from our Theorems 2 and 4 below). Consequently, if the spectrum is not rational, the speed of approximation must be limited. More specifically, we are going to show that for any ergodic T with discrete spectrum and any given speed $f(n)$ there exists an S which commutes with T and admits cyclic a.p.t. with speed $f(n)$. In other words, in every compact metrizable monothetic group there exists a rotation which admits cyclic a.p.t. with any given speed. First we prove a lemma.

Lemma 1 *Let $0 < f(n) \to 0$ and a positive integer d be fixed. Then there exists a residual subset A of $\mathbf{T}^d$ such that whenever H is a finite cyclic group with generator h, and $\alpha \in A$, the rotation $T_{(h,\alpha)}$ of $H \times \mathbf{T}^d$ by (h, α) admits cyclic a.p.t. with speed $f(n)$.*

Proof. It suffices to find an A for a fixed $H = \{g_0, \ldots, g_{m-1}\}$ with a generator h. Let

$$q_1 < q_2 < \ldots$$

be prime numbers greater than m. For any element

$$\beta = \left(\frac{p_1}{q_{n+1}}, \ldots, \frac{p_d}{q_{n+d}}\right) \in \mathbf{T}^d,$$

where $1 \le p_i < q_{n+i}$ $(i = 1, \ldots, d)$, there exists a partition ξ of $H \times \mathbf{T}^d$ into $q = mq_{n+1} \cdots q_{n+d}$ rectangles of the form

$$\{g_{j_0}\} \times \left[\frac{j_1}{q_{n+1}}, \frac{j_1+1}{q_{n+1}}\right) \times \ldots \times \left[\frac{j_d}{q_{n+d}}, \frac{j_d+1}{q_{n+d}}\right)$$

where $0 \le j_0 < m$, $0 \le j_i < q_{n+i}$ $(i = 1, \ldots, d)$. Clearly ξ is cyclically permuted by the rotation $T_{(h,\beta)}$ and diam $\xi \to 0$ as $n \to \infty$. We obtain that for every α which is sufficiently close to infinitely many such β's the rotation $T_{(h,\alpha)}$ will admit cyclic a.p.t. with any preassigned speed. More explicitly, we arrange the numbers β in a sequence $\beta^1, \beta^2, \ldots$ and choose a sufficiently small neighborhood V^k around each $\beta^k = (\beta_1^k, \ldots, \beta_d^k)$, where $\beta_i^k = p_i^k/q_{n(k)}$, in order to satisfy

$$2\sum_{i=1}^d q_{n(k)+i} |\alpha_i - \beta_i^k| < f(q)$$

whenever $\alpha \in V^k$. It is clear that

$$A = \bigcap_n \bigcup_{k=n}^\infty V^k$$

is a dense G_δ subset of $\mathbf{T}^d$ and for every $\alpha \in A$ the rotation $T_{(h,\alpha)}$ admits cyclic a.p.t. with speed $f(n)$.

Theorem 1 *Let $0 < f(n) \downarrow 0$ and let G be a compact metrizable monothetic group. Then there exists a rotation T_α of G which admits cyclic a.p.t. with speed $f(n)$. If G is connected then the set of such α's is residual in G.*

Proof. G is a projective limit of a sequence of groups G_n of the form $H \times \mathbf{T}^d$ with continuous homomorphisms

$$\varphi_n : G \to G_n.$$

Denote by G^0 the connected component of 0 in G. Since φ_n is an open mapping, $\varphi_n(G^0)$ coincides with the connected component G_n^0, isomorphic with $\mathbf{T}^d$, in G_n. Now fix a topological generator γ of G and write

$$\varphi_n(\gamma) = (h, c)$$

where $h \in H$, $c \in \mathbf{T}^d$. Let $A_n \subset \mathbf{T}^d$ be a residual set as in Lemma 1. The translated set $A_n - c$ is also residual, therefore $B_n = \varphi_n^{-1}(A_n - c)$ is residual in G^0 and so is $B = \bigcap B_n$. For any $\beta \in B$ we have

$$\varphi_n(\gamma + \beta) = (h, a)$$

where $a \in A_n$. Therefore letting $\alpha = \gamma + \beta$ we obtain that the rotation of G_n by $\varphi_n(\alpha)$ admits the required speed of approximation. By passing to a limit (as e.g. in [3], Lemma 2.1), we obtain the required speed for the rotation T_α. Finally note that if G is connected then B is residual in G.

4   Approximation and eigenvalues

Our starting point is the following lemma which was a decisive step in del Junco's proof [3] that ergodic transformations with discrete spectrum are rank-1.

Lemma 2 ([3], Prop. 1.3) *Let H be a finite cyclic group and d be a positive integer. For every ergodic rotation of the group $H \times \mathbf{T}^d$ there exists a sequence of Rokhlin towers ξ_n such that diam $\xi_n \to 0$ and $\mu(\xi_n) \to 1$.*

In fact del Junco obtains $1 - \mu(\xi_n) = O(1/\sqrt{h(\xi_n)})$. We will construct towers which cover the space within a smaller error.

Theorem 2 *Let T be an ergodic transformation with discrete spectrum and let*

$$e^{2\pi i \alpha_0}, e^{2\pi i \alpha_1}, \dots$$

be the eigenvalues of T. If the set $\{\alpha_0, \alpha_1, \dots\}$ admits simultaneous diophantine approximation with speed $o(f(n))$ then T admits f.t. with complement $o(f(n))$.

Proof. By a structure argument as in the proof of Theorem 1 we observe that T is a limit of factors identified as rotations of finite dimensional groups. Consequently, we reduce the proof to the case where T is of the form

$$T(x_0, \dots, x_d) = (x_0 + \beta_0, \dots, x_d + \beta_d),$$

an ergodic rotation of $H \times \mathbf{T}^d$, and there exists a sequence $q_n \to \infty$ such that

$$\|q_n \beta_j\| = o(q_n f(q_n)) \quad (j = 0, \dots, d).$$

We may fix $l_n \to \infty$ with $\|q_n \beta_j\| = o(1/l_n)$ and $1/l_n = o(q_n f(q_n))$. Now if $\varepsilon_n \to 0$ sufficiently slowly, we have

$$l_n \|q_n \beta_j\| < \varepsilon_n$$

and

$$K(\varepsilon_n)/l_n = o(q_n f(q_n)).$$

By Lemma 2 there exist Rokhlin towers ξ_n such that

$$h(\xi_n) > l_n q_n, \quad q_n/h(\xi_n) = o(f(q_n)), \quad 1 - \mu(\xi_n) = o(f(q_n)),$$

and diam $\xi_n < \varepsilon_n$. We construct a new tower η_n of height q_n by selecting for its base certain levels of ξ_n which are ε_n–close to the origin. This is done by choosing packs of $l_n q_n$ consecutive levels with possible gaps of length $< K(\varepsilon_n)$ between the packs. The gaps are needed in order to return to the ε_n–neighborhood of 0. Each pack is then broken into l_n consecutive smaller towers of height q_n each. The base of η_n is defined as the union of the bases of these small towers. The last small tower (i.e. the one at the top of ξ_n) is possibly incomplete; the total error is therefore

$$1 - \mu(\eta_n) \le 1 - \mu(\xi_n) + \frac{K(\varepsilon_n) + q_n}{h(\xi_n)} + \frac{K(\varepsilon_n)}{l_n q_n} = o(f(q_n)).$$

Since the base of η_n is contained in a small neighborhood of zero, we have diam $\eta_n \to 0$ and consequently $\eta_n \to \epsilon$.

By Theorem 1 the maximal value of $d(T)$ for rotations in a given monothetic group is always ∞. Now we obtain lower bounds. Our first corollary is now an immediate consequence of the Dirichlet principle.

Corollary 1 *Let* $0 < l(n) = o(n^\varepsilon)$ *for every* $\varepsilon > 0$. *Then every ergodic trasformation* T *with discrete spectrum admits f.t. with complement* $1/l(n)n$. *In particular,* $d(T) \ge 1$.

The same argument yields

Corollary 2 *Let* $l(n) \to \infty$ *be arbitrary. Every ergodic rotation* T *of* $\mathbf{T}^d$ ($d = 1, 2, \ldots$) *admits f.t. with complement* $l(n)/n^{1+1/d}$. *In particular,* $d(T) \ge 1 + 1/d$.

It follows from a classical result of Khintchine [6] that almost every d–tuple $\alpha_1, \ldots, \alpha_d$ (in the sense of Lebesgue measure) admits simultaneous diophantine approximation with speed

$$o\left(\frac{1}{n^{1+1/d} \log^{1/d} n}\right)$$

but not

$$\frac{1}{n^{1+1/d} \log^\gamma n}$$

for $\gamma > 1/d$. In particular, we obtain

Corollary 3 *Almost every rotation of* $\mathbf{T}^d$ *admits f.t. with complement*

$$o(\frac{1}{n^{1+1/d}\log^{1/d}n})$$

For sharper and more general results on simultaneous diophantine approximations see [6], [11]; for simultaneous approximation of algebraic numbers see e.g. [10].

So far we have shown that good approximation of eigenvalues implies good cyclic approximation of the rotation. To prove the converse we need a lemma.

Lemma 3 *Let* $\theta > 1/2\pi$, $0 < g(n) \to 0$, $q_n \to \infty$, $\alpha \in \mathbf{R}$. *If*

$$\limsup_{n\to\infty} \sup_{l\geq 1} (|e^{2\pi i l q_n \alpha} - 1| - lg(q_n)) = 0$$

then $||q_n\alpha|| < \theta g(q_n)$ *for all sufficiently large* n.

Proof. By assumption there exists a sequence $\delta_n \to 0$ such that

$$|e^{2\pi i l q_n \alpha} - 1| < \delta_n + lg(q_n)$$

uniformly in l. In particular, $||q_n\alpha|| \to 0$. We choose $1/2\pi < \theta' < \theta$ and let $c = c(\theta')$ be sufficiently large in order to ensure $x < \theta'|e^{2\pi i x} - 1|$ whenever $0 \leq x \leq 1/c$. Now define

$$l = l(n) = [\frac{1}{c||q_n\alpha||}].$$

Note that $l \geq 1$ and $l||q_n\alpha|| \leq 1/2$, which implies $l||q_n\alpha|| = ||lq_n\alpha||$. Consequently,

$$l||q_n\alpha|| < \theta'(\delta_n + lg(q_n)).$$

On dividing by l we obtain

$$||q_n\alpha|| < \theta'\delta_n/l + \theta'g(q_n) < 2c\theta'\delta_n||q_n\alpha|| + \theta'g(q_n),$$

so

$$||q_n\alpha|| < \frac{\theta'g(q_n)}{1 - 2c\theta'\delta_n} < \theta g(q_n)$$

for all sufficiently large n.

Theorem 3 *Let* T *be an ergodic transformation with eigenvalues*

$$e^{2\pi\alpha_0}, e^{2\pi\alpha_1}, \ldots.$$

If T *admits a.p.t.II with speed* $f(n)$ *then the set* $\{\alpha_0, \alpha_1, \ldots\}$ *admits simultaneous diophantine approximation with speed* $\theta f(n)$ *for any* $\theta > 1/2\pi$.

Proof. Let χ_j, $|\chi_j| = 1$, be an eigenfunction pertaining to the eigenvalue $e^{2\pi\alpha_j}$. Fix a sequence $0 < \delta_n \to 0$ and a number $m \geq 1$. By assumption there exists a sequence of q_n–periodic tranformations T_n with $q_n \to \infty$ preserving partitions $\xi_n = \{C_0, \ldots, C_{h_n-1}\} \to \epsilon$ such that

$$\sum_{i=0}^{h_n-1} \mu(TC_i \Delta T_n C_i) < f(q_n)$$

and

$$\|\chi_j - \chi_{jn}\|_1 < \delta_n$$

for $j = 1, \ldots, m$, where each χ_{jn} is of the form $\sum \lambda_k 1_{C_k}$, $|\lambda_k| = 1$. A direct estimation gives

$$\|\chi_{jn} T_n^{-1} - \chi_{jn} T^{-1}\|_1 \leq \sum_i \mu(T_n C_i \Delta T C_i) < f(q_n).$$

Clearly the same holds for $\chi_{jn} T_n^{-(k-1)}$ in place of χ_{jn} so we obtain by induction

$$
\begin{aligned}
\|\chi_{jn} T_n^{-k} - \chi_{jn} T^{-k}\|_1 &\leq \|\chi_{jn} T_n^{-k+1} T_n^{-1} - \chi_{jn} T_n^{-k+1} T^{-1}\|_1 \\
&\quad + \|\chi_{jn} T_n^{-k+1} - \chi_{jn} T^{-k+1}\|_1 \\
&< f(q_n) + (k-1)f(q_n) = kf(q_n)
\end{aligned}
$$

for $k = 1, 2, \ldots$ Now we have

$$
\begin{aligned}
|e^{2\pi i k\alpha_j} - 1| &= \|\chi_j T^{-k} - \chi_j\|_1 \\
&\leq \|\chi_j T^{-k} - \chi_{jn} T^{-k}\|_1 + \|\chi_{jn} T^{-k} - \chi_{jn} T_n^{-k}\|_1 \\
&\quad + \|\chi_{jn} T_n^{-k} - \chi_{jn}\|_1 + \|\chi_{jn} - \chi_j\|_1.
\end{aligned}
$$

In particular, for $k = lq_n$ $(l = 1, 2, \ldots)$ we get

$$|e^{2\pi i lq_n \alpha_j} - 1| < 2\delta_n + lq_n f(q_n) .$$

In other words, the assumption of Lemma 3 is satisfied (with no loss of generality we may assume $f(n) = o(1/n)$ so $g(n) = nf(n) \to 0$) and the assertion follows.

By taking $\alpha = (\sqrt{5} - 1)/2$ we obtain immediately

Corollary 4 *There exists an irrational rotation of* **T** *which does not admit a.p.t.II with speed* θ/n^2 *for any* $\theta < 2\pi/\sqrt{5}$.

By Khintchine's theorem and Corollary 2 we have

Corollary 5 $d(T) = 1 + 1/d$ *for almost every rotation of* $\mathbf{T}^d$.

We conclude this section with a remark on the range of $d(T)$ for rotations of tori. We already know that $\min d(T) = 1 + 1/d$ for the ergodic rotations T in $\mathbf{T}^d$. On the other hand, by Theorem 1, $\max d(T) = \infty$. It is easy to see by using continued fractions that the $d(T)$ assumes all values between 2 and ∞ for ergodic rotations on any torus.

5   Approximation and spectral measure

In this section we also consider the continuous part of the spectrum. We will prove that if an automorphism T admits a sufficiently good periodic approximation then its spectral measure must be concentrated on well approximable numbers. A relation of this kind was already indicated by Katok and Stepin [7], Thm. 3.1, but the estimation obtained therein was too weak to allow an application to skew products over arbitrary rotations. In the theorem below we improve upon this early estimate by ridding it of exponent $1/4$. First we need the following lemma which plays an analogous role as Lemma 3 did in the proof of Theorem 3.

Lemma 4 *Let $a(n) \to \infty$ and $f(n) = o(1/n)$. Let σ be a probability measure on $[0, 1)$ satisfying the condition*

$$\liminf_{n\to\infty} \sup_{l \geq 1} \left(|\hat{\sigma}(ln) - 1| - ln f(n) \right) = 0.$$

Then there exists a set B such that $\sigma(B) = 1$ and B admits simultaneous diophantine approximation with speed $a(n)f(n)$.

Proof. Choose a sequence $q_n \to \infty$ such that

$$|\hat{\sigma}(lq_n) - 1| < \delta_n + lq_n f(q_n), \quad l = 1, 2, \ldots$$

for some sequence $0 < \delta_n \to 0$. To simplify the notation we write $q_n f(q_n) = \rho_n$. In addition, we may and will assume $\rho_n = o(\delta_n)$ and $a(q_n)\delta_n \to \infty$. We observe that

$$
\begin{aligned}
|\hat{\sigma}(lq_n) - 1| &\geq \int (1 - \cos 2\pi lq_n x) d\sigma(x) \\
&= \int (1 - \cos 2\pi ||lq_n x||) d\sigma(x) \\
&\geq 8 \int ||lq_n x||^2 d\sigma(x)
\end{aligned}
$$

since $1 - \cos 2\pi y \geq 8y^2$ for $|y| \leq 1/2$. Consequently, for any $L \geq 1$ we obtain

$$\int \sum_{l=1}^{L} ||lq_n x||^2 d\sigma(x) < L\delta_n/8 + L^2 \rho_n/16.$$

Choose $b_n \to \infty$ such that $b_n^3 < a(q_n)^2 \delta_n/4$, $b_n \delta_n \to 0$, and $b_n \rho_n = o(\delta_n)$. Now, on a set $A(L, n) \subset [0, 1)$ of σ measure at least $1 - 1/b_n$ we have

$$\sum_{l=1}^{L} ||lq_n x||^2 < b_n(L\delta_n/8 + L^2\rho_n/16).$$

Let $L = L_n = [\delta_n/b_n\rho_n]$ and denote $A_n = A(L_n, n)$. Since $\sigma(A_n) \to 1$, we may assume, by choosing a further subsequence if necessary, that $\sum(1 - \sigma(A_n)) < \infty$. The sets

$$B_n = \bigcap_{j=n}^{\infty} A_j$$

increase to a set B of σ measure 1 and on each B_n we have

$$\frac{1}{L_n}\sum_{l=1}^{L_n} ||lq_n x||^2 < (b_n + 1)\delta_n/8 \to 0,$$

so for a fixed $x \in B$ and n sufficiently large we obtain

$$\frac{1}{L_n}\sum_{l=1}^{L_n} ||lq_n x||^2 < 1/256.$$

Denote $q_n x = y$. The frequency of the l's satisfying $||ly|| < 1/8$ must be greater than $3/4$ in $\{1, \dots, L_n\}$. Suppose on the other hand that $||ky|| \geq 1/4$ for some $1 \leq k \leq L_n/2$. This leads to a contradiction because we would then have $||(k + l)y|| > 1/8$ for more than a half of the l's in the initial interval $\{1, \dots, [L_n/2]\}$, whence $||ly|| > 1/8$ for more than a quarter of all l's in $\{1, \dots, L_n\}$. This argument shows that

$$||y||, ||2y||, \dots, ||[L_n/2]y||$$

are all less than $1/4$, which implies

$$||ly|| = l||y||$$

for $l = 1, \dots, [L_n/2]$ by easy induction. Therefore, for n large enough we have

$$\left(\sum_{l=1}^{[L_n/2]} l^2\right)||q_n x||^2 \leq \sum_{l=1}^{L_n} ||lq_n x||^2 < b_n(L_n\delta_n/8 + L_n^2\rho_n/16)$$

implying

$$||q_n x||^2 < 25b_n\left(\frac{b_n^2\rho_n^2}{8\delta_n} + \frac{b_n\rho_n^2}{16\delta_n}\right) < 4b_n^3\rho_n^2/\delta_n < a(q_n)^2\rho_n^2,$$

which gives

$$||q_n x|| < a(q_n)q_n f(q_n)$$

and ends the proof of the lemma.

Theorem 4 *If T admits a.p.t.II with speed $o(f(n))$ then the maximal spectral measure of T is concentrated on a set admitting simultaneous diophantine approximation with speed $o(f(n))$.*

Proof. Choose a sequence $a(n) \to \infty$ and a function $g(n)$ such that

$$a(n)g(n) = o(f(n))$$

and T admits approximation with speed $g(n)$. Denote by σ the maximal spectral measure—without loss of generality assume that it is a probability measure on $[0, 1)$. By a theorem of Alekseev [1] there exists a function χ such that $\|\chi\|_2 = 1$, $|\chi| < C < \infty$, and

$$\hat{\sigma}(k) = (\chi T^{-k}, \chi).$$

Choose a sequence $0 < \delta_n \to 0$ and q_n–periodic approximations T_n which preserve partitions $\xi_n = \{C_0, \ldots, C_{h_n-1}\} \to \epsilon$ and satisfy both

$$\sum_{j=0}^{h_n-1} \mu(TC_j \Delta T_n C_j) < g(q_n)$$

and

$$\|\chi - \chi_n\|_2 < \delta_n,$$

where χ_n is of the form $\sum \lambda_j 1_{C_j}$. We may clearly assume $\|\chi_n\|_2 = 1$, $|\lambda_j| < C$, and $h_n f(h_n) = o(\delta_n)$. For any integer k we have

$$
\begin{aligned}
|\hat{\sigma}(k) - 1| &\leq |(\chi T^k, \chi) - (\chi_n T^k, \chi)| + |(\chi_n T^k, \chi) - (\chi_n T^k, \chi_n)| \\
&\quad + |(\chi_n T^k, \chi_n) - (\chi_n T_n^k, \chi_n)| + |\chi_n T_n^k, \chi_n) - (\chi_n, \chi_n)| \\
&< 2\delta_n + |\chi_n T^k - \chi_n T_n^k, \chi_n)| + |(\chi_n T_n^k - \chi_n, \chi_n)|.
\end{aligned}
$$

The last term vanishes if $k = lq_n$, so

$$|\hat{\sigma}(lq_n) - 1| < 2\delta_n + |(\chi_n T^{-lq_n} - \chi_n T_n^{-lq_n}, \chi_n)|.$$

To estimate the second term on the right first note that

$$
\begin{aligned}
|(\chi_n T^{-1} - \chi_n T_n^{-1}, \chi_n)| &= |\sum_i \int_{C_i} \sum_j \bar{\lambda}_i \lambda_j (1_{TC_j} - 1_{T_n C_j}) d\sigma| \\
&\leq C^2 \sum_i \sum_j \mu(C_i \cap (TC_j \Delta T_n C_j)) \\
&< C^2 g(q_n).
\end{aligned}
$$

The same calculation works with $\chi_n T$ instead of χ_n and the sets C_j replaced by $T_n^{k-1} C_j$ so we obtain by induction

$$
\begin{aligned}
|\chi_n T^{-k} - \chi_n T_n^{-k}, \chi_n)| &\leq |(\chi_n T^{-k+1} T^{-1} - \chi_n T_n^{-k+1} T^{-1}, \chi_n)| \\
&\quad + |((\chi_n T_n^{-k+1}) T^{-1} - (\chi_n T_n^{-k+1}) T_n^{-1}, \chi_n)| \\
&< (k-1) C^2 g(q_n) + C^2 g(q_n) = C^2 k g(q_n)
\end{aligned}
$$

for $k = 1, 2, \ldots$ In particular, for $k = lq_n$ we have

$$|\hat{\sigma}(lq_n) - 1| < 2\delta_n + C^2 lq_n g(q_n)$$

uniformly in $l = 1, 2, \ldots$. Now apply Lemma 4 to obtain simultaneous approximation with speed

$$a(n)C^2 g(n) = o(f(n)).$$

Corollary 6 *Let $r > 1$. If T admits a.p.t.II with speed $1/n^r$ then its maximal spectral measure is concentrated on a set of Hausdorff dimension $\gamma \leq 1/r$. In particular, $\gamma \leq 1/d(T)$.*

Proof. It suffices to observe that if B admits simultaneous diophantine approximation with speed $1/n^{r'}$ for any $r' < r$ then its Hausdorff dimension is bounded by $1/r$. Choose q_n as in the definition; we may assume $q_n \geq 2^n$. Fix $0 < r' < r$. It follows that the set B is covered by

$$V_N \cup V_{N+1} \cup \ldots$$

where V_n is a union of intervals $I_{n0}, \ldots, I_{n,q_n-1}$ of length $2/q_n^{r'}$ each. For any $t > 1/r'$ we have

$$\sum_{j=0}^{q_n-1} |I_{nj}|^t < 2/q_n^s$$

where $s = r't - 1 > 0$. Now the intervals I_{nj}, for $n \geq N$, cover B and their lengths satisfy

$$\sum_{n=N}^{\infty} \sum_j |I_{nj}|^t \leq \sum_{n=N}^{\infty} 2/q_n^s \to 0$$

as $N \to \infty$. Consequently, B has Hausdorff dimension $\gamma \leq t$, so $\gamma \leq 1/r$.

It should be noted that the last inequality in Corollary 6 can be strict. To see this it suffices to find an automorphism T possessing no cyclic a.p.t. with speed $1/n$ (whence $d(T) \leq 1$) but whose maximal spectral measure is concentrated on a set admitting simultaneous diophantine approximation with speed $1/n^r$ for any $r > 0$ (whence $\gamma = 0$). This is done in the following example. For basic properties of Gaussian automorphisms see e.g. [2].

Example. There exists an ergodic Gaussian automorphism T which possesses no cyclic a.p.t. with speed θ/n for $\theta < 2$ but the maximal spectral measure of T is concentrated on a set admitting simultaneous diophantine approximation with speed $1/n^r$ for any r.

Define $k_1 = 1$, $k_n = nk_{n-1} + 1$. If x has binary expansion of the form

$$x = 0.a_1 0 a_2 000000 a_3 0 \ldots,$$

where the digits a_n occur at positions k_n, respectively, then

$$||2^{k_n} x|| < 2^{-nk_n}.$$

Consequently,

$$G = \{x \in \mathbf{T} : ||2^{k_n} x|| = O(2^{-nk_n})\}$$

is an uncountable Borel subgroup of $\mathbf{T}$. Choose any nonatomic symmetric probability measure ν on G and define

$$\sigma = \sum_{j=1}^{\infty} 2^{-j} \nu^j$$

where $\nu^j = \nu * \ldots * \nu$ (j times). Now let T be a Gaussian automorphism with covariance function $\hat{\sigma}(n)$. The maximal spectral measure of T is then equivalent to

$$e^\sigma = \delta_0 + \sigma + \sigma^2/2! + \ldots$$

Notice that σ^{k+1} is absolutely continuous with respect to σ^k, which implies that T has infinite spectral multiplicity. Therefore T does not admit cyclic a.p.t. with speed θ/n if $\theta < 2$ (see [2]). On the other hand, it is clear that e^σ is concentrated on G and G admits simultaneous diophantine approximation with speed $1/n^r$ for all r.

6   Application to skew products

Corollary 6 has the following application. Consider an ergodic rotation T of $\mathbf{T}^d$ and a measurable mapping (a "cocycle")

$$\phi : \mathbf{T}^d \to G$$

where G is any compact metrizable monothetic group (e.g. a torus). The skew product

$$T_\phi(x, y) = (Tx, y + \phi(x))$$

is said to be a weakly mixing extension of T if it is ergodic and the only eigenvalues of T_ϕ are those of T. Note that $d(T_\phi) \leq d(T)$ by Theorems 2 and 3. In [5] it was proved that for a residual set of ϕ's (with respect to the topology of convergence in measure where the functions equal a.e. are identified) the skew product T_ϕ is a weakly mixing extension and satisfies $d(T_\phi) \geq d(T)$. By Corollary 2, $d(T) \geq 1 + 1/d$, so we obtain

Corollary 7 *Let T be an ergodic rotation of the d−torus. For a residual set of cocycles φ the skew product $T_φ$ is a weakly mixing rank-1 extension with spectral measure concentrated on a set of Hausdorff dimension $\leq d/(d+1)$.*

References

[1] V. M. Alekseev, *The existence of a bounded function of a maximal spectral type*, Vestnik Moscow Univ. **5** (1958), 13–15

[2] I. P. Cornfeld, S. V. Fomin, Ya. G. Sinai, *Ergodic Theory*, Springer-Verlag, New York 1982

[3] A. del Junco, *Transformations with discrete spectrum are stacking transformations*, Can. J. Math. **28** (1976), 836–839

[4] A. Iwanik, *Cyclic approximations of irrational rotations*, Proc. Amer. Math. Soc. **121** (1994), 691–695

[5] A. Iwanik, J. Serafin, *Most monothetic extensions are rank-1*, Colloq. Math. **66** (1993), 63–76

[6] A. Khintchine, *Zur metrischen Theorie der diophantischen Approximationen*, Math. Z. **24** (1926), 706–714

[7] A. B. Katok, A. M. Stepin, *Approximations in ergodic theory* , Uspekhi Mat. Nauk **22**, no. 5 (1967), 81–106

[8] E. A. Robinson, *Ergodic measure preserving transformations with arbitrary spectral multiplicities*, Invent. Math. **72** (1983), 299–314

[9] E. A. Robinson, *Non-abelian extensions have nonsimple spectrum*, Comp. Math. **65** (1988), 155–170

[10] W. Schmidt, *Approximation to algebraic numbers*, Enseignement Math. **17** (1971), 187–253

[11] V. G. Sprindzhuk, *Metric Theory of Diophantine Approximations*, Winston–Wiley, New York 1979

[12] A. M. Stepin, *Spectrum and approximation of metric automorphisms by periodic transformations*, Funct. Anal. Appl. **1**, no. 2 (1967), 77–80

Invariant σ–algebras for $\mathbb{Z}^d$–actions and their applications

B. Kaminski

Abstract

We give a review of basic properties of invariant σ–algebras for $\mathbb{Z}^d$–actions on a Lebesgue probability space and some applications of them.

1 Introduction

Invariant σ–algebras are useful tools for solving a series of important problems in ergodic theory. In one – dimensional case they have been applied, among other things, for classification problems and for the investigation of dynamical systems with completely positive entropy.

In this paper we present a review of applications of invariant σ–algebras in the multidimensional case. Now the role of time plays the lexicographical order in the group $\mathbb{Z}^d$, $d \geq 2$.

The direct generalization of the concept of invariance to the multidimensional case is not satisfactory for valuable applications. The proper analogue is the so called strong invariance including, beyond the simple extension of the one–dimensional invariance, the continuity condition strictly connected with the rank of the group of multidimensional integers.

The central place in the theory of invariant σ–algebras is taken by perfect σ–algebras. The proof of their existence is based on the methods of relative ergodic theory.

In the review we give applications of perfect σ–algebras to the theory of Kolmogorov $\mathbb{Z}^d$–actions, to the description of the spectrum of $\mathbb{Z}^d$–actions with positive entropy and to find a connection between the monequililorium entropy and the Conze–Katznelson–Weiss entropy.

We also give examples of applications of relatively perfect σ–algebras to describe maximal factors and principal factors. As a by-product we present

403

an axiomatic definition of entropy which is a multidimensional analogue of the axiomatic definition of Rokhlin.

In order to avoid complicated notations we cosider only the two-dimensional case but all considerations may be extended to the general case.

2   Basic concepts and results

Let $(X, \mathcal{B}, \mu)$ be a Lebesgue probability space and let $\mathcal{N}$ be the trivial sub-algebra of $\mathcal{B}$. We denote by $\mathcal{Z}$ the set of all countable measurable partitions of X with finite entropy. It is well known (cf. [Rok 2]) that $\mathcal{Z}$ is a Polish space with respect to the metric ρ defined as follows

$$\rho(P, Q) = H(P \mid Q) + H(Q \mid P), \ P, Q \in \mathcal{Z}.$$

For a given σ–algebra $\mathcal{A} \subset \mathcal{B}$ we denote by $L^2_0(\mathcal{A})$ the subspace of $L^2(X, \mu)$ consisting of $\mathcal{A}$ - measurable functions f with $\int_X f d\mu = 0$.

We equip the group $\mathbf{Z}^2$ with the lexicographical order $\prec$. An ordered pair (A, B) of subsets of $\mathbf{Z}^2$ is said to be a cut if $A \neq \emptyset$, $B \neq \emptyset$, $g \prec h$, $g \in A$, $h \in B$ and $A \cup B = \mathbf{Z}^2$.

A cut (A, B) is called a gap if A doesn't contain the largest element and B the lowest element.

Let Φ be a $\mathbf{Z}^2$–action on $(X, \mathcal{B}, \mu)$. We denote by Φ^g the automorphism of $(X, \mathcal{B}, \mu)$ corresponding to $g \in \mathbf{Z}^2$. The automorphisms

$$T = \Phi^{(1,0)}, \ S = \Phi^{(0,1)}$$

are called the standard automorphisms of Φ.

2.1   Bernoulli and Conze actions

Let $(Y, \mathcal{F}, \lambda)$ be a Lebesgue probability space and let

$$(X, \mathcal{B}, \mu_\lambda) = \sqcap_{g \in \mathbf{Z}^2}(Y_g, \mathcal{F}_g, \lambda_g)$$

where

$$Y_g = Y, \ \mathcal{F}_g = \mathcal{F}, \ \lambda_g = \lambda, \ g \in \mathbf{Z}^2.$$

A $\mathbf{Z}^2$–action Φ on $(X, \mathcal{B}, \mu_\lambda)$ defined by

$$(\Phi^g x)(h) = x(g + h), \ g, h \in \mathbf{Z}^2$$

is called the Bernoulli action determined by λ.

Let now φ be an automorphism of $(Y, \mathcal{F}, \lambda)$ and let T and S_φ be automorphisms of $(X, \mathcal{B}, \mu)$ defined as follows

$$(Tx)(n) = x(n + 1), \ (S_\varphi x)(n), \ x \in X, \ n \in \mathbf{Z}$$

The $\mathbf{Z}^2$-action Φ_φ given by the formula

$$\Phi_\varphi^g = T^i \circ S_\varphi^j, \ g = (i,j) \in \mathbf{Z}^2$$

is called the Conze $\mathbf{Z}^2$-action determined by φ.

2.2 Gaussian actions

Let

$$(X,\mathcal{B}) = \sqcap_{g \in \mathbf{Z}^2}(\mathbf{R}_g, \mathcal{B}_g)$$

where $\mathbf{R}_g = \mathbf{R}$ and $\mathcal{B}_g$ is the σ-algebra of Borel sets of $\mathbf{R}$, $g \in \mathbf{Z}^2$.

Let ρ be a finite symmetric measure on the two-dimensional torus $\mathbf{T}^2$ and let μ_ρ be the Gaussian measure on $\mathcal{B}$ determined by ρ.

The $\mathbf{Z}^2$-action Φ on $(X, \mathcal{B}, \mu_\rho)$ defined by the same formula as in the case of Bernoulli actions is called the Gaussian action with spectral measure ρ.

2.3 Relative entropy and relative Pinsker σ-algebra

Let Φ be a $\mathbf{Z}^2$-action on $(X, \mathcal{B}, \mu)$ and let $\mathcal{F} \subset \mathcal{B}$ be a totally invariant σ-algebra (a factor σ-algebra) i.e.

$$\Phi^g \mathcal{F} = \mathcal{F}, \ g \in \mathbf{Z}^2.$$

We denote by $\Phi_\mathcal{F}$ the factor $\mathbf{Z}^2$-action determined by $\mathcal{F}$.
Let $A \subset \mathbf{Z}^2$ and $P \in \mathcal{Z}$. We put

$$P(A) = \vee_{g \in A} \Phi^g P.$$

In particular we define

$$P_\Phi = P(\mathbf{Z}^2), \ P^- = P_\phi^- = P(\mathbf{Z}_-^2)$$

where $\mathbf{Z}_-^2$ stands for the set of negative elements of $\mathbf{Z}^2$ (with respect to $\prec$).
The number $h(P, \Phi \mid \mathcal{F})$ defined by the formula

$$h(P, \Phi \mid \mathcal{F}) = \lim_{n \to \infty} \frac{1}{\mid A_n \mid} H(P(A_n) \mid \mathcal{F})$$

where (A_n) is an arbitrary Følner sequence of subsets of $\mathbf{Z}^2$ is said to be the mean relative entropy of P with respect to Φ and $\mathcal{F}$.

Applying the idea of the proof of Theorem 1 of [S] one can give a simple proof of the correctness of this definition and the equality

$$h(P, \Phi \mid \mathcal{F}) = H(P \mid P^- \vee \mathcal{F}).$$

We define the relative entropy of Φ with respect to $\mathcal{F}$ as

$$h(\Phi \mid \mathcal{F}) = \sup\{h(P, \Phi \mid \mathcal{F}), \ P \in \mathcal{F}\}$$

and the relative Pinsker σ–algebra of Φ with respect to $\mathcal{F}$ as the smallest σ–algebra $\pi(\Phi \mid \mathcal{H})$ containing all partitions $P \in \mathcal{Z}$ with $h(P, \Phi \mid \mathcal{F}) = 0$ (c.f. [Ka2]).

It is clear that $h(\Phi \mid \mathcal{N})$ and $\pi(\Phi \mid \mathcal{N})$ coincide with the entropy $h(\Phi)$ and the Pinsker σ–algebra $\pi(\Phi)$ of Φ, respectively.

Using the concept of relative entropy B.Kitchens and and K.Schmidt defined in [KS] a collection of measurable isomorphism invariants for $\mathbb{Z}^2$-actions.

2.4 Invariant σ- algebras

Let Φ be a $\mathbb{Z}^2$- action on a Lebesgue space $(X, \mathcal{B}, \mu)$ and let T and S be standard automorphisms of Φ.

A σ–algebra $\mathcal{A} \subset \mathcal{B}$ is said to be invariant if $\Phi^g \mathcal{A} \subset \mathcal{A}$ for every $g \in \mathbb{Z}^2_-$.

This condition can be written, by the use of the standard automorphisms of Φ, as follows

$$S^{-1}\mathcal{A} \subset \mathcal{A}, \ T^{-1}\mathcal{A}_S \subset \mathcal{A}$$

where

$$\mathcal{A}_S = \bigvee_{n=-\infty}^{+\infty} S^n \mathcal{A}.$$

The concept of an invariant σ–algebra of a $\mathbb{Z}^2$–action is too weak in order to obtain valuable applications. We shall use the concept of a strong invariant σ–algebra in the sequel.

A σ–algebra $\mathcal{A}$ is called strong invariant if it is invariant and satisfies the following continuity condition:

$$\bigvee_{g \in A} \Phi^g \mathcal{A} = \bigcap_{g \in B} \Phi^g \mathcal{A}$$

for every cut (A, B) being a gap.

It is easy to see that an invariant σ–algebra $\mathcal{A}$ satisfies the continuity condition iff

$$\bigcap_{n=0}^{\infty} S^{-n}\mathcal{A} = T^{-1}\mathcal{A}_S.$$

The following examples show that there exist $\mathbb{Z}^2$–actions having invariant σ–algebras which are not strongly invariant.

Example 1. ([Ka2]) Let φ be an automorphism of a Lebesgue probability space $(Y, \mathcal{F}, \lambda)$ for which there exists a nontrivial finite measurable partition Q generating a factor with zero entropy.

Consider the Conze $\mathbb{Z}^2$–action Φ_φ determined by φ. Let $P = \pi_0^{-1}Q$ where π_0 is the projection on the zero coordinate. The past σ–algebra

$$\mathcal{A} = P^- = P_S^- \vee (P_S)_{\overline{T}}$$

is invariant but it is not strongly invariant.

Example 2. Let Φ be a Gaussian $\mathbb{Z}^2$–action determined by the spectral measure

$$\rho = \delta \times \lambda.$$

where δ is the discrete measure on T concentrated on $z = 1$ and λ is the Lebesgue measure.

Let T and S be the standard automorphisms of Φ, i.e.

$$(Tx)(m,n) = x(m+1,n), \ (Sx)(m,n) = x(m,n+1), \ (m,n) \in \mathbb{Z}^2.$$

We denote by $\mathcal{C}$ the σ–algebra defined by $\mathcal{C} = \pi_l^{-\infty}(\mathcal{B}^\infty)$ where $\pi_o x = x(0,0), \ x \in X$.

The σ–algebra

$$\mathcal{A} = \mathcal{C} \vee \mathcal{C}^- = \mathcal{C} \vee \mathcal{C}_S^- \vee (\mathcal{C}_S)_{\overline{T}}.$$

is invariant but it is not strongly invariant because, by the definition of ρ,

$$\bigcap_{n=0}^{\infty} S^{-n}\mathcal{A} = \bigcap_{n=0}^{\infty} (S^{-n}\mathcal{C}_S^- \vee (\mathcal{C}_S)_{\overline{T}}) = \bigcap_{\backslash=l}^{\infty} S^{-\backslash}\mathcal{C}_S^- \vee (\mathcal{C}_S)_{\overline{T}} = \mathcal{C} \vee (\mathcal{C}_S)_{\overline{T}} \neq (\mathcal{C}_S)_{\overline{T}} = T^{-\infty}\mathcal{A}_S$$

It is easy to see that for actions with zero entropy the strongly invariant σ–algebras are totally invariant.

Let P be a countable measurable partition of X. The invariant σ–algebra

$$\mathcal{A}_P = P \vee P^- = P \vee P_S^- \vee (P_S)_{\overline{T}}$$

is said to be the past σ-algebra generated by P.

For actions with positive entropy every σ- algebra $\mathcal{A}_P$ where P is a partition generating a Bernoulli factor is strongly invariant. This is an easy consequence of the relative version of the Kolmogorov zero-one law(c.f.[Ka2])

A partition P is said to be regular if the σ–algebra $\mathcal{A}_P$ is strongly invariant. The following two theorems proved in [Ka6] give information about the set of regular partitions.

Theorem 1. If S is ergodic, then for every partition $P \in \mathcal{Z}$ and $\varepsilon > 0$ there exists a regular partition $Q \in \mathcal{Z}$ with $P \leq Q$ and $\rho(P,Q) < \varepsilon$.

The proof of this theorem is based on the relative generator theorem (c.f [Ros]).

Theorem 2. If S is ergodic, and $h(\Phi) > 0$ then for every partition $P \in \mathcal{Z}$ and $\varepsilon > 0$ there exists a nonregular partition $Q \in \mathcal{Z}$ with $P \leq Q$ and $\rho(P, Q) < \varepsilon$.

In order to prove this theorem one applies the relative version of the Sinai theorem (c.f [T1]) and the fact that Bernoulli actions are not coalescent (c.f. [Ka1]).

The Theorem 2 cannot be extended to actions with zero entropy. It is shown in [Ka6] that if Φ is a $\mathbf{Z}^2$-action with $h(T) = 0$ and S ergodic then every partition from $\mathcal{Z}$ is regular.

The most importint class of invariant σ-algebras form the so called perfect σ-algebras.

Let Φ be a $\mathbf{Z}^2$-action on a Lebesgue space $(X, \mathcal{B}, \mu)$ with the standard automorphisms T and S.

A σ-algebra $\mathcal{A} \subset \mathcal{B}$ is said to be perfect if it is strongly invariant and

(i) $\quad \bigvee_{g \in \mathbf{Z}^2} \Phi^g \mathcal{A} = \bigvee_{n=-\infty}^{+\infty} T^n \mathcal{A}_S = \mathcal{B}$,

(ii) $\quad \bigcap_{g \in \mathbf{Z}^2} \Phi^g \mathcal{A} = \bigcap_{n=0}^{\infty} T^{-n} \mathcal{A}_S = \pi(\Phi)$,

(iii) $\quad h(\Phi) = H(\mathcal{A} \mid \mathcal{A}^-) = H(\mathcal{A} \mid S^{-1}\mathcal{A})$.

The following result plays the main role in the theory of invariant σ-algebras.

Theorem 3. ([Ka2]) For every $\mathbf{Z}^2$-action there exists a perfect σ-algebra.

Using the same idea of the proof one shows the relative version of Theorem 3.

Theorem 4. ([Ka4]) For every $\mathbf{Z}^2$-action Φ, and every factor σ-algebra $\mathcal{F}$ of Φ there exists a strongly invariant σ-algebra $\mathcal{A} \supset \mathcal{F}$ such that

(i) $\quad \bigvee_{g \in \mathbf{Z}^2} \Phi^g \mathcal{A} = \mathcal{B}$

(ii) $\quad \bigcap_{g \in \mathbf{Z}^2} \Phi^g \mathcal{A} = \pi(\Phi \mid \mathcal{F})$,

(iii) $\quad h(\Phi \mid \mathcal{F}) = H(\mathcal{A} \mid \mathcal{A}^-)$.

Any σ-algebra $\mathcal{A}$ having properties given in Theorem 4 is said to be relatively perfect with respect to $\mathcal{F}$.

It is easy to check that if a partition $P \in \mathcal{Z}$ is a regular generator for Φ then the σ-algebra $\mathcal{A}_P$ is perfect.

It would be interesting to find an answer to the following.

Problem 1. Let $\mathcal{A}$ be a perfect σ-algebra of a $\mathbf{Z}^2$-action Φ with $h(\Phi) < \infty$. Does there exist a partition $P \in \mathcal{Z}$ with $\mathcal{A} = \mathcal{A}_P$?

A partial positive result is contained in [Ka3].

3   Applications of invariant σ–algebras

3.1 K–actions and the spectral theory of Z^d–actions

The most important applications of invarint σ–algebras are connected with the theory of Kolmogorov **Z**^{2}–actions and the spectral theory of **Z**^{2}–actions.

A **Z**^{2}–action Φ is said to be a Kolmogorov one (K–action) if there exists a strongly invariant σ–algebra $\mathcal{A}$ with

$$\bigvee_{g\in \mathbf{Z}^2} \Phi^g \mathcal{A} = \mathcal{B}, \quad \bigcap_{g\in \mathbf{Z}^2} \Phi^g \mathcal{A} = \mathcal{N}.$$

An easy consequence of Theorem 3 is

Theorem 5. ([Ka2]). Φ is a K–action iff Φ has a completely positive entropy.

Hence one obtains all properties of K–actions which are multidimensional analogues of the well known properties of K–automorphisms.

The application of invariant σ–algebras to the spectral theory of **Z**^{2}–actions is contained in the following.

Theorem 6. ([Ka2]). If a **Z**^{2}–action Φ has a positive entropy then it has a countable Lebesgue spectrum in the subspace $L^2(X,\mu) \ominus L^2(\pi(\Phi))$.

A simple proof of this theorem based on the methods of harmonic analysis allowing to extend it to **Z**$^\infty$–actions is given in [KL].

An extension of Theorem 6 to arbitrary abelian finitely generated group is shown in [Ka7].

An immediate consequence of Theorem 6 is the fact that **Z**^{2}–actions with completely positive entropy have countable Lebesgue spectrum.

This result was a positive answer to the question of Conze ([C]).

It is worth to point out that it is still an open and interesting problem of Thouvenot which is an extension of the Conze question.

Problem 2. Has any action of a countable abelion group with a completely positive entropy a countable Haar spectrum?

It follows easily from Theorem 6 that every **Z**^{2}–action with a singular spectrum or a spectrum with finite multiplicity has zero entropy.

Another corollary, shown in [FK] (see also [Ru]), says that every Gaussian **Z**^{2}–action with a singular spectral measure has zero entropy. The extension of this result to actions of an arbitrary abelian (infinite) finitely generated group and to actions of **Z**$^\infty$ is proved in [Ka7] and [KL] respectively.

3.2   Connection between nonequilibrium entropy and entropy of a Z^2-action

Let $(X, \mathcal{B}, \mu)$ be a Lebesgue probability space and let $M(\mu)$ denote the set of probability measures ν on $\mathcal{B}$ absolutely continuous with respect to μ and such that

$$\int_X \eta(\rho(x))\mu(dx) < \infty$$

where ρ is the Radon–Nikodym derivative of ν with respect to μ and

$$\eta(x) = \begin{cases} x \log x & , \ x > 0, \\ 0 & , \ x = 0. \end{cases}$$

Let Φ be a $\mathbb{Z}^2$-action on $(X, \mathcal{B}, \mu)$ with the standard automorphisms T and S.

Let $\mathcal{A}$ be a strongly invariant σ-algebra and $\nu \in M(\mu)$. The number

$$h(\nu, \mathcal{A}) = -\int_X \eta(E(\rho|\mathcal{A})(x))\mu(dx)$$

is called the nonequilibrium entropy of ν given $\mathcal{A}$.

In order to connect this concept with the Conze–Katznelson–Weiss entropy of Φ we introduce the concept of an increase of the nonequilibrium entropy which is a two–dimensional analogue of the entropy increase defined by Goldstein and Penrose ([G], [GP]).

We put

$$\begin{aligned} \Delta^\Phi(\nu, \mathcal{A}) &= h(\nu, \mathcal{A}^-) - h(\nu, \mathcal{A}) \\ &= h(\nu, S^{-1}\mathcal{A}) - h(\nu, \mathcal{A}), \end{aligned}$$

and

$$\Delta^\phi = \sup_{\nu \in M(\mu)} \Delta^\Phi(\nu, \mathcal{A}).$$

In the view of Theorem 3 we may introduce the following definition. The quantity

$$\Delta^\phi = \inf \Delta^\phi(\mathcal{A})$$

where $\mathcal{A}$ runs over all strongly invariant σ-algebras with $H(\mathcal{A} \mid \mathcal{A}^-) = h(\Phi)$ is called the entropy increase for Φ.

The following two results are proved in [BK].

Theorem 7. If Φ is ergodic then

$$\Delta^\Phi = f(h(\Phi))$$

where $f : [0, +\infty] \longrightarrow [0, +\infty]$ is defined as follows $f(0) = 0$, $f(+\infty) = +\infty$ and $f(x) = \log(n+1)$ for $x \in (\log n, \ \log(n+1)]$, $n \geq 1$.

A sequence $\mathcal{G} = (\mathcal{G}_\backslash)$ of subgroups of $\mathbb{Z}^2$ is called admissible if G_n is of a finite index k_n, $n \geq 1$ and $k_n \longrightarrow +\infty$.
We denote by Φ^n the restiction of Φ to G_n.

Theorem 8. If Φ is ergodic then for every admissible sequence $\mathcal{G} = (\mathcal{G}_\backslash)$ we have

$$\lim_{n \to \infty} \frac{\Delta^n}{k_n} = h(\Phi).$$

3.3   Maximal factors

Thouvenot in [T2] introduced the definition of a relative K-automorphism (with finite entropy) with respect to some factor σ-algebra $\mathcal{F}$.

In [L], [ME] and [O] the factor σ-algebra for which the automorphism is a relative K-automorphism is called maximal.

The Thouvenot's definition for arbitrany $\mathbb{Z}^2$-action may be written as follows.

A $\mathbb{Z}^2$-action Φ is a relative K-action with respect to $\mathcal{F}$ (or $\mathcal{F}$ is a maximal factor σ-algebra of Φ) if for every factor σ-algebra $\mathcal{H} \supset \mathcal{F}$ such that $h(\Phi_\mathcal{H}) = h(\Phi_\mathcal{F})$ it holds $\mathcal{H} = \mathcal{F}$.

For the first glance it seems that this concept has no connection with the traditional meaning of a K-action according to which there should exist special exhaustive σ-algebras for the considered action.

The following theorem proved in [Ka5] shows, however, that in fact the concept of a relative K-action has also its characterization by exhaustive σ-algebras.

Theorem 9. A $\mathbb{Z}^2$-action Φ is a relative K-action with respect to $\mathcal{F}$ iff there exists an exhaustive σ-algebra $\mathcal{A} \supset \mathcal{F}$ with

$$\bigcap_{g \in \mathbb{Z}^2} \Phi^g \mathcal{A} = \mathcal{F}.$$

¿From this theorem one easily obtains the following.

Corollary. If a $\mathbb{Z}^2$-action Φ has a positive entropy, $\mathcal{F}$ and $\mathcal{H}$ are factor σ-algebras of Φ which are independent, $\mathcal{F} \vee \mathcal{H} = \mathcal{B}$ and the factor-action $\Phi_\mathcal{H}$ is a K-action then $\mathcal{F}$ is maximal.

3.4   Principal factors

The following concept is a two-dimensional analogue of the definition of a principal factor for automorphisms introduced by Rokhlin ([Rok1]).

A factor σ-algebra $\mathcal{F}$ of a $\mathbb{Z}^2$-action Φ is principal if every strongly invariant σ-algebra $\mathcal{A} \supset \mathcal{F}$ is totally invariant, i.e. a factor σ-algebra of Φ.

An action Ψ, being a factor of Φ, is called a principal factor if every factor σ–algebra $\mathcal{F}$, such that the actions Ψ and $\Phi_{\mathcal{H}}$ are isomorphic, is principal. The existence of a relatively perfect σ–algebra allows to show

Theorem 10. ([Ka4]). If an action Ψ is a principal factor of Φ then $h(\Psi) = h(\Phi)$. If $h(\Phi) < \infty$ then the converse theorem is also true.

There is given an example in [Ka4] which shows that the above theorem fails to be true if we replace the strong invariance by the invariance in the definition of a principal factor.

Applying Theorem 10 we may give an axiomatic definition of the entropy of a Z^2–action which is a multidimensional analogue of the Rokhlin definition (c.f. [Rok2]).

Let Act Z^2 denote the set of all ergodic Z^2–actions on Lebesgue spaces. We denote by Φ_b the Bernoulli Z^2–action defined by a diserete probabilistic measure concentrated on a two–element set and admitting the same value on each element.

Corollary. ([Ka4]). If $H :$ Act $Z^2 \longrightarrow [0, +\infty]$ is a function such that $H(\Phi_b) = \log 2$ and for any Φ, $\Psi \in Z^2$ we have

(i)   if Ψ is a factor of Φ then $H(\Phi) \leq H(\Psi)$,

(ii) If Ψ is a principal factor of Φ then $H(\Phi) = H(\Psi)$,

(iii) $H(\Phi \times \Psi) = H(\Phi) + H(\Psi)$ then $H(\Phi)$ coincides with entropy $h(\Phi)$.

The proof is based on the multidimensional version of the Sinai theorem ([Ki]) and on Theorem 10.

References

[BK] M.Binkowska, B.Kaminski, *Entropy increase for* Z^d*-actions*, to appear in Israel J. Math.

[C] J.P.Conze, *Entropie d'un groupe abélien de transformations*, Z.Wahr. verw. Geb. 25 (1972), 11–30.

[FK] S.Ferenci, B.Kaminski, *Zero entropy and directional Bernoullicity of a Gaussian* Z^2 *– action*, to appear in Proc. Amer. Math. Soc.

[G] S.Goldstein, *Entropy increase in dynamical systems*, Israel J. Math. 38 (1981), 241–256.

[GP] S.Goldstein, O.Penrose, *A nonequilibrium entropy for dynamical systems*, J. Stat. Phys. 24 (1981), 325–343.

[Ka1] B.Kaminski, *Some properties of coalescent automorphisms of a Lebesgue space*, Commentationes Math. 21 (1979), 95–99.

[Ka2] – , *The theory of invariant partitions for* Z^d*– actions*, Bull. Polish. Acad. Sci. Math. 29 (1981), 349–362.

[Ka3] – , *A representation theorem for perfect partitions of* Z^2 *– actions with finite entropy*, Colloquium Math. 56 (1988), 121–127.

[Ka4] – , *An axiomatic definition of the entropy of a* Z^d*-action on a Lebesgue space*, Studia Math. 96 (1990), 135–144.

[Ka5] – , *Decreasing nets of* σ*– algebras and their applications to ergodic theory*, Tôhoku Math. Journal, 43 (1991), 263–274.

[Ka6] – , *Generators of perfect* σ*– algebras of* Z^d*-actions*, Studia Math. 99 (1991), 1–10.

[Ka7] – , *Spectrum of positive entropy multidimensional dynamical systems with a mixed time*, preprint.

[Ki] J.C.Kieffer, *The isomorphism theorem for generalized Bernoulli Schemes*, Studies in Probability and Ergodic Theory, Advances in Mathematics Supplementary Studies, 2 (1978), 251–267.

[KL] B.Kaminski, P.Liardet, *Spectrum of multidimensional dynamical systems with positive entropy*, Studia Math. 108 (1994), 77–85.

[KS] B.Kitchens, K.Schmidt, *Mixing sets and relative entropies for higher-dimensional Markov shifts*, Ergod. Th. & Dynam. Sys. 13 (1993), 705-735.

[L] D.Lind, *The structure of skew products with ergodic group automorphisms*, Israel J. Math. 28 (1977), 205–248.

[ME] N.F.G. Martin, J.W.England, *Mathematical theory of entropy*, Addison–Wesley Publishing Company, 1981.

[O] D.S. Ornstein, *Factors of Bernoulli shifts*, Israel J. Math. 21 (1975), 145–153.

[P] W.Parry, *Topics in ergodic theory*, Cambridge University Press, Cambridge, 1981.

[Rok1] V.A.Rokhlin, *An axiomatic definition of the entropy of a transformation with invariant measure*, Dokl. Akad. Nauk SSSR, 148 (1963), 779–781 (Russian).

[Rok2] – , *Lectures on the entropy theory of transformations with invariant measure*, Uspehi Mat. Nauk, 22 (1967), 3–56 (Russian)

[Ros] A.Rosenthal, *Uniform generators for ergodic finite entropy free actions of amenable groups*, Probab. Theory Related Fields 77 (1988), 147–166.

[Ru] T.de la Rue, *Entropie d'un système dynamique gaussien; Cas d'une action de Z^d*, C. R. Acad. Sci. Paris, Série I, 317 (1993), 191–194.

[S] A.W.Safonov, *Informational pasts in groups*, Izv. Akad. Nauk SSSR, 47 (1983), 421–426.

[T1] J.P.Thouvenot, *Quelques propriétés des systèmes dynamiques qui se décomposent en un produit de deux systèmes dont l'un est un schema de Bernoulli*, Israel J. Math. 21 (1975), 177–207.

[T2] – , *Une classe de systèmes pour lesquels la conjecture de Pinsker est vraie*, ibid. 21(1975), 208–214.

Faculty of Mathematics and Computer Science
Nicholas Copernicus University
ul. Chopina 12/18
87-100 Torun, POLAND

LARGE DEVIATIONS FOR PATHS
AND CONFIGURATIONS COUNTING

YURI KIFER

Institute of Mathematics
Hebrew University
Jerusalem, Israel

ABSTRACT. Large deviations bounds are obtained for numbers of different paths in deterministic and random graphs via the corresponding problems for subshifts of finite type. The corresponding results for numbers of geodesic paths on manifolds of negative curvature are obtained, as well. Also I discuss large deviations bounds for configuration counting in deterministic and random multidimensional subshifts of finite type.

1. INTRODUCTION.

Consider a directed graph Γ with m vertices. It is determined by a matrix $B = (b_{ij}, \ i.j = 1, ..., m)$ such that $b_{ij} = 1$ if there is an arrow from i to j and $b_{ij} = 0$, otherwise. Assume that there exists n_0 such that B^k is a matrix with strictly positive entries for all $k \geq n_0$ which means that for any pair of vertices there is a path connecting them and the graph is aperiodic, i.e. the common divider of lengts of all cycles is 1. I shall be interested in the large deviations bounds as $n \to \infty$ for the number of paths which contain $n + 1$ vertices and connect two fixed vertices. A model problem is the following. Assign a length $\ell(e)$ to each edge e of Γ. For a path $x = (x_0, ..., x_n)$, $x_{i-1}x_i = e_i \in \Gamma$, i.e. $b_{x_{i-1}x_i} = 1$, let $\ell(x) = \sum_{i=1}^{n} \ell(e_i)$. Let $\Pi_n(a, b)$ denotes the space of paths with $n + 1$-vertices which connect vertices a and b. I am interested, in particular, in the behavior as $n \to \infty$ of

$$\frac{1}{n} \log \left(|\Pi_n(a, b)|^{-1} |\{x \in \Pi_n(a, b) : \frac{1}{n}\ell(x) \in [\alpha, \beta] \subset \mathbb{R}_+\}| \right)$$

1991 *Mathematics Subject Classification.* Primary:60F10, Secondary:58F15, 05C80.
Key words and phrases. Large deviations, subshifts of finite type, geodesics, random graphs..
Partially supported by US-Israel BSF

Typeset by $\mathcal{A}_{\mathcal{M}}\mathcal{S}$-TEX

where $|A|$ denotes the number of elements in a finite set A. The results will be obtained via the general theorem on large deviations from [Ki1] together with the thermodynamic formalism for subshifts of finite type. It turns out that similar arguments yield large deviations bounds for the number of geodesic paths connecting two points on a compact manifold of negative curvature which also will be considered in this paper. The case of closed geodesics from this point of view was considered in [Ki2] and some generalizations of it were studied in [Po].

There is a natural generalization of the above to certain random graphs. This set up includes an ergodic measure preserving transformation θ of a probability space (Ω, P), a $\mathbb{Z}_+$-valued random variable m satisfying

$$(1.1) \qquad 0 < \int \log m \, dP < \infty,$$

and a mesurable family of matrices $B(\omega) = \{b_{ij}(\omega); \, i = 1, ..., m(\omega); j = 1, ..., m(\theta\omega)\}$, $\omega \in \Omega$ such that each $b_{ij}(\omega)$ equals 0 or 1 and there exists a $\mathbb{Z}_+$-valued random variable $n_0 = n_0(\omega) < \infty$ so that for each $n \geq n_0$ and P-a.a.$\omega \in \Omega$ the matrix

$$(1.2) \qquad B(\omega)B(\theta\omega) \cdots B(\theta^n\omega) \text{ is positive.}$$

Next, one can consider randomly evolving graphs (or networks) which are determined by the sets of vertices $V(\omega) = \{1, 2, ..., m(\omega)\}$ so that there is a directed edge between $k \in V(\omega)$ and $l \in V(\theta\omega)$ if and only if $b_{kl}(\omega) = 1$. The space of permissible paths with $n \leq \infty$ edges in such networks can be written in the form

$$(1.3) \qquad \begin{aligned} X_B^\omega(n) &= \{x = (x_i; \, x_i \in \{1, ..., m(\theta^i\omega)\}, \, i = 0, 1, 2, ..., n), \\ & \qquad b_{x_i, x_{i+1}} = 1 \, \forall i = 0, 1, ..., n-1\}. \end{aligned}$$

A typical problem which I shall deal with is the following. Assign to each edge e of the network its length $\ell(e)$ and consider the length of a path $x \in X_B^\omega(n)$ given by $\ell(x) = \sum_{i=0}^{n-1} \ell(x_i x_{i+1})$. Set

$$\Pi_n^\omega(a, b) = \{x = (x_i) \in X_B^\omega(n) : x_0 = a, \, x_n = b\}.$$

I shall be concerned with the asymptotic behavior as $n \to \infty$ of

$$(1.4) \qquad n^{-1} \log \left(|\Pi_n^\omega(a, b)|^{-1} |\{x \in \Pi_n^\omega(a, b) : n^{-1}\ell(x) \in [\alpha, \beta] \subset \mathbb{R}\}| \right).$$

The results will be obtained employing the general large deviations theorem from [Ki1] and the thermodynamic formalism for random subshifts of finite type considered in [BG].

I shall discuss also multidimensional versions of these results corresponding to counting of configurations in deterministic and random multidimensional subshifts of finite type. The upper bound of large deviations is derived here via [Ki1] similarly to the one dimensional case but the lower bound requires additional arguments due to nonuniqueness of equilibrium states (phase transitions) even for good functions. I shall describe here the results which where obtained in [EKW] and [Ki3] via large deviations bounds for Gibbs measures.

2. DETERMINISTIC CASE.

Let Γ be a finite directed graph (network) with vertices $V = \{1, ..., m\}$ described in Introduction. For $n = 1, 2, ..., \infty$, the space of paths of length n can be written in the form

$$X_B(n) = \{x = (x_0, x_1, ..., x_n); x_i \in V \; \forall i = 0, ..., n$$
$$\text{and } b_{x_i x_{i+1}} = 1 \, \forall i = 0, ..., n-1\}.$$

I denote also $X_B(\infty)$ by X_B. Then X_B is a closed subset of $V^{\mathbb{Z}_+ \cup \{0\}}$ taken with the product topology while V is taken with the discrete topology. The set X_B is invariant under the shift σ acting by $(\sigma x)_i = x_{i+1}$ and the pair (X_B, σ) is called a subshift of finite type. Denote by $\mathcal{P}(X_B)$ the space of probability measures on X_B taken with the topology of weak convergence, by $\mathcal{I}_B \subset \mathcal{P}(X_B)$ the subspace of σ- invariant measures, and by $C_{\alpha_0, ..., \alpha_{n-1}}$ the cylinder set $\{x \in X_B : x_i = \alpha_i \, \forall i = 0, ..., n-1\}$. Recall (see [Bow]) that the entropy $h_\mu(\sigma)$ of σ with respect to $\mu \in \mathcal{I}_B$ and the topological pressure $Q_{X_B}(g)$ of a continuous function g on X_B can be obtained by the formulas

$$(2.1) \quad h_\mu(\sigma) = \lim_{n \to \infty} -\frac{1}{n} \sum_{(\alpha_0, ..., \alpha_n) \in X_B(n)} \mu(C_{\alpha_0, ..., \alpha_n}) \log \mu(C_{\alpha_0, ..., \alpha_n})$$

and

$$(2.2) \quad Q_{X_B}(g) = \lim_{n \to \infty} \frac{1}{n} \log \sum_{(\alpha_0, ..., \alpha_n) \in X_B(n)} \exp(S_n g)(x_\alpha)$$

where $S_n g(x) = \sum_{i=0}^{n-1} g(\sigma^i x)$ and x_α is an arbitrary point from $C_{\alpha_0, ..., \alpha_n}$. One has also the variational principle

$$(2.3) \quad Q_{X_B}(g) = \sup_{\mu \in \mathcal{I}_B} \left(\int g \, d\mu + h_\mu(\sigma) \right).$$

The entropy $h_\mu(\sigma)$ is upper semicontinuous in μ in this set up and so the supremum in (2.3) is always attained. If g is Hölder continuous with respect to the metric $d_\beta(x,y) = \beta^N$, $N = \min\{n \geq 0 : x_n \neq y_n\}$, $\beta \in (0,1)$ then this supremum is attained at a unique measure $\mu_g \in \mathcal{I}_B$ which is called an equilibrium state or a Gibbs measure. If $g \equiv 0$ then $\mu_0 = \mu_{\max}$ is called the measure with maximal entropy and $Q_{X_B}(0) = h_{\text{top}}(\sigma)$ is called the topological entropy. Denote by $\Pi_n(a,b)$ the set of all $(x_0, x_1, ..., x_n) \subset X_B(n)$ with $x_0 = a$ and $x_n = b$. Set $\Pi_n = \cup_a \Pi_n(a,a)$ and $\zeta_x^n = \frac{1}{n} \sum_{i=0}^{n-1} \delta_{\sigma^i x}$ where δ_x is the unit mass at x and so $\zeta_x^n \in \mathcal{P}(X_B)$.

2.1. Theorem. *For any $a,b \in V$ and each closed $K \subset \mathcal{P}(X_B)$,*

$$(2.4) \quad \limsup_{n\to\infty} \frac{1}{n} \log |\Pi_n(a,b)|^{-1} |\{\alpha \in \Pi_n(a,b) : \zeta_{x_\alpha}^n \in K\}| \leq - \inf_{\nu \in K} I(\nu)$$

and for each open $U \subset \mathcal{P}(X_B)$,

$$(2.5) \quad \liminf_{n\to\infty} \frac{1}{n} \log |\Pi_n(a,b)|^{-1} |\{\alpha \in \Pi_n(a,b) : \zeta_{x_\alpha}^n \in U\}| \geq - \inf_{\nu \in U} I(\nu)$$

where $x_\alpha, \alpha = (\alpha_0, ..., \alpha_n)$ is any point from the cylinder C_α (and the result does not depend on the choice of x_α) and $I(\nu) = h_{\text{top}}(\sigma) - h_\nu(\sigma)$ if $\nu \in \mathcal{I}_B$ and $I(\nu) = \infty$, otherwise. The result remains true if $\Pi_n(a,b)$ is replaced by Π_n.

Proof. Pick up an arbitrary point $x_\alpha^{(n)}$ from each cylinder set C_α with $\alpha = (\alpha_0, ..., \alpha_n) \in \Pi_n(a,b)$ and denote by $G_n(a,b)$ the set of such points. For any $Y \subset X_B$ set $\eta_n(Y) = |G_n(a,b)|^{-1} |Y \cap G_n(a,b)|$ which determines a sequence of measures $\eta_n \in \mathcal{P}(X_B)$. Since I assume that all high powers of B have positive entries it follows easily from (2.2) (cf. [Bow], Chapter 1) that for any continuous function g on X_B,

$$
\begin{aligned}
& \lim_{n\to\infty} \frac{1}{n} \log \int \exp(n \int g d\zeta_x^n) d\eta_n(x) \\
(2.6) \quad & = \lim_{n\to\infty} \frac{1}{n} \log |\Pi_n(a,b)|^{-1} \sum_{\alpha \in \Pi_n(a,b)} e^{S_n g(x_\alpha^{(n)})} \\
& = \lim_{n\to\infty} \frac{1}{n} \log |X_B(n)|^{-1} \sum_{\alpha \in X_B(n)} e^{S_n g(x_\alpha^{(n)})} \\
& = Q_{X_B}(g) - h_{\text{top}}(\sigma) \\
& = \sup_{\nu \in \mathcal{I}_B} \left(\int g d\nu - (h_{\text{top}} - h_\nu(\sigma)) \right).
\end{aligned}
$$

By Theorem 2.1 from [Ki1] this together with the uniqueness of equilibrium states for Hölder continuous functions yield (2.4) and (2.5). Replacing in the above proof $\Pi_n(a,b)$ by Π_n I arrive at the same limit in (2.6), and so the same conclusion holds true, completing the proof of Theorem 2.1. $\square$

By the general contraction principle argument (see [DZ], Section 4.2) I obtain

2.2. Corollary. *For each continuous function g on X_B and any numbers $r_1 < r_2$,*

$$
(2.7) \quad \limsup_{n\to\infty} \frac{1}{n} \log |\Pi_n(a,b)|^{-1}|\{\alpha \in \Pi_n(a,b) : \frac{1}{n}(S_n g)(x_\alpha) \in [r_1, r_2]\}|
$$
$$
\leq - \inf_{r\in[r_1,r_2]} J(r)
$$

and

$$
(2.8) \quad \liminf_{n\to\infty} \frac{1}{n} \log |\Pi_n(a,b)|^{-1}|\{\alpha \in \Pi_n(a,b) : \frac{1}{n}(S_n g)(x_\alpha) \in (r_1, r_2)\}|
$$
$$
\geq - \inf_{r\in(r_1,r_2)} J(r)
$$

where $J(r) = \inf\{I(\nu) : \int g d\nu = r\}$ if a $\nu \in \mathcal{P}(X_B)$ satisfying the condition in braces exists and $J(r) = \infty$, otherwise.

If $g(x)$ depends only on x_0 and x_1 then g is continuous, and so it satisfies the condition of Corollary 2.2. In particular, if $g(x) = \ell(x_0 x_1)$ is the length of the edge $x_0 x_1$ of the graph Γ I arrive at the deterministic version of the case mentioned in Introduction. Then $(S_n \ell)(x) = \sum_{i=0}^{n-1} \ell(x_i, x_{i+1})$ is the length of a path in $X_B(n)$ and Corollary 2.2 yields large deviations bounds for the number of paths with n vertices in Γ leading from a to b with the length per vertex sandwiched between r_1 and r_2. Observe that always $I(\nu) \geq 0$ and $I(\nu) = 0$ if and only if $\nu = \mu_{\max}$, and so $J(r) = 0$ if and only if $r = \int g d\mu_{\max}$. Thus as $n \to \infty$ except for an exponentially small proportion of $\alpha \in \Pi_n(a,b)$ the measure $\zeta_{x_\alpha}^n$ belongs to a neighborhood of $\mu_{\max}$ and $(S_n g)(x_\alpha)$ belongs to a neighborhood of $\int g d\mu_{\max}$.

Next, I shall consider corresponding counting problems for geodesic paths on a compact Riemannian manifold M with negative curvature. Similar results for closed geodesics were obtained in [Ki2] and [Po]. For $a, b \in M$ denote by $\Pi_r(a,b)$ the set of geodesic paths γ with length $\ell(\gamma)$ not exceeding r connecting a and b and parametrized by length. Let $\tilde{M}$ be the universal cover of M with the induced metric and $p : \tilde{M} \to M$ be the natural projection. Fix $\tilde{a}_0 \in p^{-1}a$ then $\Pi_r(a,b)$ is in one-to-one correspondence with the set of geodesics $\tilde{\Pi}_r(a,b)$ on $\tilde{M}$ with lengths not exceeding r and connecting $\tilde{a}_0$ with points from $p^{-1}b$. For each $\Gamma \subset \Pi_r(a,b)$ set

$$
(2.9) \quad \nu_{a,b}^{(r)}(\Gamma) = |\Pi_r(a,b)|^{-1}|\Gamma \cap \Pi_r(a,b)|,
$$

which defines the uniform (counting) probability measure on $\Pi_r(a,b)$ and where, again, $|A|$ is the number of elements in A. Denote by $T^{(1)}M$ the unit tangent bundle of M then for any $\gamma \in \Pi_r(a,b)$ we have $(\gamma(t), \dot\gamma(t)) \in T^{(1)}_{\gamma(t)}M$ where $\dot\gamma(t)$ is the unit tangent vector to γ at $\gamma(t)$. I thank M. Pollicott who pointed to me out that the following result is a consequence of [Ru2].

2.3. Proposition. *For any continuous function g on $T^{(1)}M$,*

(2.10)

$$\lim_{r\to\infty} \frac{1}{r} \log \int \exp\left(\int_0^{\ell(\gamma)} g(\gamma(t), \dot\gamma(t))dt \right) d\nu^{(r)}_{a,b}(\gamma)$$

$$= \lim_{r\to\infty} \frac{1}{r} \log \sum_{\gamma \in \Pi_r(a,b)} \exp \int_0^{\ell(\gamma)} g(\gamma(t), \dot\gamma(t))dt - \lim_{r\to\infty} \frac{1}{r} \log |\Pi_r(a,b)|$$

$$= Q(g) - Q(0)$$

where $Q(g)$ is the topological pressure of the geodesic flow f^t on $T^{(1)}M$ corresponding to the function g (see [BR]) and, accordingly, $Q(0) = h_{\text{top}}$ is the topological entropy of f^1.

Proof. The result follows from the general theorem in [Ru2] by taking the Radon measure μ on $\tilde{M}$ appearing there such that for any open set $U \subset \tilde{M}$,

(2.11) $$\mu(U) = \sum_{n=1}^{\infty} \frac{1}{\ell(\gamma)} \sum_{\gamma \in \tilde\Pi_n(a,b)} \sum_i \ell(\gamma_i(U))$$

where $\{\gamma_i(U), i = 1,2,...\}$ are connected components of the intersection $\gamma \cap U$. $\square$

Now Theorem 2.1 from [Ki1] together with Proposition 2.3 and the uniqueness of equilibrium states for Hölder continuous functions in the variational principle for the geodesic flow f^t on $T^{(1)}M$ (see [BR]) yield the following large deviations result.

2.4. Theorem. *Let ζ_γ denotes the uniform measure on $(\gamma, \dot\gamma)$, $\gamma \in \Pi_r(a, b)$, i.e.*

(2.12) $$\zeta_\gamma = \frac{1}{\ell(\gamma)} \int_0^{\ell(\gamma)} \delta_{\gamma(t), \dot\gamma(t)} dt,$$

which belongs to the space $\mathcal{P}(T^{(1)}M)$ of probability measures on $T^{(1)}M$. Then for any closed $K \subset \mathcal{P}(T^{(1)}M)$,

(2.13) $$\limsup_{r\to\infty} \frac{1}{r} \log \nu^{(r)}_{a,b} \{\gamma \in \Pi_r(a,b) : \zeta_\gamma \in K\} \le - \inf_{\eta \in K} I(\eta)$$

and for any open $U \subset \mathcal{P}(T^{(1)}M)$,

$$(2.14) \qquad \liminf_{r \to \infty} \frac{1}{r} \log \nu_{a,b}^{(r)} \{ \gamma \in \Pi_r(a,b) : \zeta_\gamma \in U \} \geq - \inf_{\eta \in U} I(\eta)$$

where $I(\eta) = h_{\text{top}} - h_\eta$ *if* η *is a probability invariant measure of the geodesic flow* f^t *and* $I(\eta) = \infty$, *otherwise, and* h_η *is the entropy of* f^1 *with respect to* η.

3. Paths in random networks.

Next, I shall consider a random network $\mathcal{N}$ which is determined by a probability space (Ω, P) with a P-preserving map $\theta : \Omega \to \Omega$, a random variable $m(\omega) \in \mathbb{Z}_+$, $\omega \in \Omega$ satisfying (1.1) and a random matrix $B(\omega) = (b_{ij}(\omega); i = 1, ..., m(\omega); j = 1, ..., m(\theta\omega))$ with 0–1 entries. Set $V(\omega) = \{1, 2, ..., m(\omega)\}$ and connect by an arrow $i \in V(\omega)$ to $j \in V(\theta\omega)$ if and only if $b_{ij}(\omega) = 1$. This determines the space of permissible paths in $\mathcal{N}$. I assume that for any ω there exists $n_0 = n_0(\omega)$ such that for each $k \geq n_0$ the matrix $B^k(\omega) = B(\omega)(\theta\omega) \cdots B(\theta^{k-1}\omega)$ has strictly positive entries. For $n = 1, 2, ..., \infty$ the space of paths of length n can be written in the form $X_B^\omega(n) = \{x = (x_0, ..., x_n) : x_i \in V(\theta^i\omega) \forall i = 0, 1, ..., n \text{ and } b_{x_i x_{i+1}}(\theta^i\omega) = 1 \forall i = 0, 1, ..., n-1\}$ and I denote $X_B^\omega(\infty)$ by just X_B^ω. Similarly to [BG] I shall consider the compact metric space $X = \prod_{i=0}^\infty \bar{\mathbb{Z}}_+$, where $\bar{\mathbb{Z}}_+ = \mathbb{Z}_+ \cup \{\infty\}$ is the one-point compactification of $\mathbb{Z}_+$, with the metric

$$\rho(x,y) = \sum_{i=0}^\infty 2^{-i} |x_i^{-1} - y_i^{-1}|; \quad x = (x_0, x_1, ...), y = (y_0, y_1, ...); \quad x_i, y_i \in \mathbb{Z}_+$$

and $\frac{1}{\infty} = 0$. Then all $X_B^\omega, \omega \in \Omega$ become compact subsets of X. Set $X_B = \{(x,\omega) \in X \times \Omega : x \in X_B^\omega\}$ and denote by σ the shift on X, i.e. $(\sigma x)_i = x_{i+1}$, and by $\tau = (\sigma, \theta)$ the skew product transformation. Then X_B is τ-invariant. Let μ be a τ-invariant probability measure on X_B whose marginal on Ω is P, i.e. $d\mu(x,\omega) = d\mu^\omega(x)dP(\omega)$ where $\mu^\omega, \omega \in \Omega$ are probability measures on X_B^ω satisfying $\sigma\mu^\omega = \mu^{\theta\omega}$. The space of such μ will be denoted by $\mathcal{I}_B$. Set $C_{\alpha_0,...,\alpha_{n-1}}^\omega = \{x \in X_B^\omega : x_i = \alpha_i, \forall i = 0, 1, ..., n-1\}$ which will be called a cylinder set. Then the relativized entropy of τ with respect to $\mu \in \mathcal{I}_B$ can be obtained by the P-a.s. limit (cf. [Bo],[BG],[KK]),

$$(3.1) \quad h_\mu^{(r)}(\tau) = \lim_{n \to \infty} -\frac{1}{n} \sum_{(\alpha_0,...,\alpha_n) \in X_B^\omega(n)} \mu^\omega(C_{\alpha_0,...,\alpha_n}^\omega) \log \mu^\omega(C_{\alpha_0,...,\alpha_n}^\omega).$$

Denote by $L_{X_B}^1(\Omega, \mathcal{C})$ the space of measurable families $g^\omega, \omega \in \Omega$ of continuous functions g^ω on X_B^ω such that $\int \|g^\omega\| dP(\omega) < \infty$ where $\|g^\omega\| =$

$\sup_{x \in X_B^\omega} \|g^\omega(x)\|$. I denote by $L^1(\Omega, \mathcal{C})$ the corresponding space of measurable families of functions on the whole X. It turns out (see [BG]) that any family g^ω from $L^1_{X_B}(\Omega, \mathcal{C})$ can be extended to a family $\tilde{g}^\omega$ from $L^1(\Omega, \mathcal{C})$ so that $\|g^\omega\| = \|\tilde{g}^\omega\|$. The relativized topological pressure of $g \in L^1_{X_B}(\Omega, \mathcal{C})$ can be obtained by the P-a.s. limit (cf. [Bo],[BG],[KK]),

$$(3.2) \qquad Q_{X_B}^{(r)}(g) = \lim_{n \to \infty} \frac{1}{n} \log \sum_{(\alpha_0, \ldots, \alpha_n) \in X_B^\omega(n)} \exp(S_n g)(x_\alpha, \omega)$$

where $(S_n g)(x_\alpha, \omega) = \sum_{i=0}^{n-1} g^{\theta^i \omega}(\sigma^i x)$ and x_α is an arbitrary point from $C_{\alpha_0, \ldots, \alpha_n}^\omega$. One has also the relativized variational principle (see [Bo]),

$$(3.3) \qquad Q_{X_B}^{(r)}(g) = \sup_{\mu \in \mathcal{I}_B} \left(\int g d\mu + h_\mu^{(r)}(\tau) \right).$$

Again, the relativized entropy here is upper semicontinuous in μ, and so the supremum in (3.3) is attained on some measure μ_g which is unique if the family $g^\omega, \omega \in \Omega$ is equi Hölder continuous (see [BG], [KK]). As before, $\mu_{\max} = \mu_0$ is called the measure with maximal entropy and $h_{\text{top}}^{(r)}(\tau) = Q_{X_B}(0)$ is called the topological entropy. Denote by $\Pi_n^\omega(a, b)$ the set of all $(x_0, x_1, \ldots, x_n) \subset X_B^\omega(n)$ with $x_0 = a \in V(\omega)$ and $x_n = b \in V(\theta^n \omega)$. Set $\Pi_n^\omega = \cup_a \Pi_n^\omega(a, a)$ if $a \in V(\omega) \cap V(\theta^n \omega)$ and $\zeta_{x,\omega}^n = \frac{1}{n} \sum_{i=0}^{n-1} \delta_{(\sigma^i x, \theta^i \omega)}$ where $\delta_{(x,\omega)}$ is the unit mass at x and so $\zeta_{x,\omega}^n \in \mathcal{P}(X \times \Omega)$ and $\zeta_{x,\omega}^n \in \mathcal{P}(X_B)$ if $x \in X_B^\omega$.

3.1 Theorem. *(cf.[Ki3]) Suppose that Ω is a compact metric space and consider $\mathcal{P}(X \times \Omega)$ with the topology of weak convergence. Then with probability one for any $a \in V(\omega)$, $b_n \in V(\theta^n \omega)$, and each closed $K \subset \mathcal{P}(X \times \Omega)$,*
$$(3.4)$$
$$\limsup_{n \to \infty} \frac{1}{n} \log \left(|\Pi_n^\omega(a, b_n)|^{-1} |\{\alpha \in \Pi_n^\omega(a, b_n) : \zeta_{x_\alpha, \omega}^n \in K\}| \right) \leq - \inf_{\nu \in K} I(\nu)$$

and for each open $U \subset \mathcal{P}(X \times \Omega)$,
$$(3.5)$$
$$\liminf_{n \to \infty} \frac{1}{n} \log \left(|\Pi_n^\omega(a, b_n)|^{-1} |\{\alpha \in \Pi_n^\omega(a, b_n) : \zeta_{x_\alpha, \omega}^n \in U\}| \right) \geq - \inf_{\nu \in U} I(\nu)$$

where $x_\alpha, \alpha = (\alpha_0, \ldots, \alpha_n)$ is any point from the cylinder C_α^ω (and the result does not depend on the choice of x_α) and $I(\nu) = h_{\text{top}}^{(r)}(\tau) - h_\nu^{(r)}(\tau)$ if $\nu \in \mathcal{I}_B$ and $I(\nu) = \infty$, otherwise. The result remains true if $\Pi_n^\omega(a, b)$ is replaced by Π_n^ω.

Proof. Pick up an arbitrary point $x_{\alpha,\omega}$ from each (ω, n)-cylinder set C_α^ω with $\alpha = (\alpha_0, \ldots, \alpha_n) \in \Pi_n^\omega(a, b)$ and denote by $G_n^\omega(a, b)$ the set of such points.

For any $Y \subset X$ set $\eta_n^\omega(Y) = |G_n^\omega(a, b)|^{-1}|Y \cap G_n^\omega(a, b)|$, $a \in V(\omega)$, $b \in V(\theta^n\omega)$ which determines a sequence of measures $\eta_n^\omega \in \mathcal{P}(X_B^\omega) \subset \mathcal{P}(X)$. Next, I shall show that with probability one for any $g \in L_{X_B}^1(\Omega, \mathcal{C})$ and $a \in V(\omega)$, $b_n \in V(\theta^n\omega)$,

$$
\lim_{n \to \infty} \frac{1}{n} \log \int \exp\left(n \int g d\zeta_{x,\omega}^n\right) d\eta_n^\omega(x)
$$

$$
(3.6) \qquad = \lim_{n \to \infty} \frac{1}{n} \log |\Pi_n^\omega(a, b_n)|^{-1} \sum_{\alpha \in \Pi_n^\omega(a, b_n)} e^{S_n g(x_{\alpha, \omega}, \omega)}
$$

$$
= \lim_{n \to \infty} \frac{1}{n} \log |X_B^\omega(n)|^{-1} \sum_{\alpha \in X_B^\omega(n)} e^{S_n g(x_{\alpha, \omega}, \omega)} = Q_{X_B}^{(r)}(g) - h_{\text{top}}^{(r)}(\tau)
$$

$$
= \sup_{\nu \in \mathcal{I}_B} \left(\int g d\nu - (h_{\text{top}}^{(r)} - h_\nu^{(r)}(\tau))\right).
$$

The first equality in (3.6) follows by the definition of the measures η_n^ω, the third follows by (3.2), and the fourth follows by (3.3). Thus it remains to establish the second equality in (3.6). Denote by $L_{X_B}^{1,\kappa}(\Omega, \mathcal{C})$, $\kappa > 0$ the space of measurable families g^ω, $\omega \in \Omega$ of continuous functions g^ω on X_B such that $\int (\|g^\omega\| + \|g^\omega\|_\kappa) dP(\omega) < \infty$ where

$$
(3.7) \qquad \|g^\omega\|_\kappa = \sup\left\{ \frac{|g^\omega(x) - g^\omega(y)|}{(\rho(x, y))^\kappa} : x, y \in X_B^\omega, x \neq y \right\}.
$$

For each $N \in \mathbb{Z}_+$ set $\Psi_N = \{\omega : n_0(\omega) \leq N\}$. Assuming $n > 2N$ define the map

$$
\phi = \phi_{a,b}^{\omega, n, N} : \Pi_n^\omega(a, b) \to X_B^{\theta^N \omega}(n - 2N)
$$

such that if $\alpha = (\alpha_0, ..., \alpha_n) \in \Pi_n^\omega(a, b)$ then $\phi\alpha = (\beta_0, ..., \beta_{n-2N})$ with $\beta_i = \alpha_{i+N} \; \forall i = 0, ..., n - 2N$. If $\omega, \theta^{n-N}\omega \in \Psi_N$ then for any $\beta \in X_B^{\theta^N \omega}(n - 2N)$,

$$
(3.8) \qquad 1 \leq |\phi^{-1}\beta| \leq \prod_{i=0}^N (m(\theta^i\omega)m(\theta^{n-i}\omega)).
$$

I can write

$$
(3.9) \qquad \sum_{\alpha \in \Pi_n^\omega(a, b)} e^{S_n g(x_{\alpha, \omega}, \omega)} = \sum_{\beta \in X_B^{\theta^N \omega}(n - 2N)} \sum_{\alpha \in \phi^{-1}\beta} e^{S_n g(x_{\alpha, \omega}, \omega)}.
$$

By the definition of the metric ρ if $g \in L_{X_B}^{1,\kappa}(\Omega, \mathcal{C})$ and $\alpha \in \phi^{-1}\beta$ then,

$$
(3.10)
$$

$$
\left| S_n g(x_{\alpha, \omega}, \omega) - S_{n-2N} g(x_{\beta, \theta^N \omega}, \theta^N \omega) \right|
$$

$$
\leq \sum_{i=0}^N (\|g^{\theta^i\omega}\| + \|g^{\theta^{n-i}\omega}\|) + \sum_{i=N}^{n-N} 2^{-(n-N-i-1)\kappa} \|g^{\theta^i\omega}\|_\kappa.
$$

By the ergodic theorem with probability one $\lim_{n\to\infty} n^{-1}(\|g^{\theta^n \omega}\| + \|g^{\theta^n \omega}\|_\kappa)$ $= 0$ provided $g \in L^{1,\kappa}_{X_B}(\Omega, \mathcal{C})$, and so the right hand side of (3.10) divided by n tends to zero P-a.s as $n \to \infty$. Let $n_i = n_i^{(N)}(\omega)$ be the consecutive integers for which $\theta^{n_i - N}\omega \in \Psi_N$. Assuming also that $\omega \in \Psi_N$ I derive from (3.8)-(3.10) that

(3.11)
$$\lim_{n_i \to \infty} \frac{1}{n_i}\left(\log \sum_{\alpha \in \Pi^\omega_{n_i}(a, b_{n_i})} e^{S_{n_i} g(x_{\alpha,\omega}, \omega)} \right.$$
$$\left. - \log \sum_{\beta \in X^{\theta^N \omega}_B(n_i - 2N)} e^{S_{n_i - 2N} g(x_{\beta, \theta^N \omega}, \theta^N \omega)} \right) = 0.$$

Since (3.2) holds true for P-a.a.ω it remains true with probability one for $\theta^N \omega$ in place of ω. Observe, that P-a.a. ω belong to some Ψ_N and by the ergodic theorem with probability one,

(3.12)
$$\lim_{i \to \infty} \frac{n_{i+1}(\omega)}{n_i(\omega)} = 1.$$

Let $n_i(\omega) \le n < n_{i+1}(\omega)$. Denote by π_k the projection which maps any $\beta \in X^\omega_B(m)$, $m \ge k$ to $\alpha \in X^\omega_B(k)$ such that $\alpha_i = \beta_i \, \forall i = 0, 1, ..., k$. Set $\Pi^\omega_k(a) = \{\alpha \in X^\omega_B(k) : \alpha_0 = a\}$. By the definition of $N(\omega)$ and $n_i(\omega)$ it is clear that for any $b_k \in V(\theta^k \omega)$, $k \ge n_i$ and $\alpha \in \Pi^\omega_{n_i - N}(a)$,

(3.13)
$$1 \le \left| \pi^{-1}_{n_i - N} \alpha \cap \Pi^\omega_k(a, b_k) \right| \le \prod_{i = n_i - N}^{k} m(\theta^i \omega).$$

Also for any $k \ge n_i$ I can write

(3.14)
$$\sum_{\beta \in \Pi^\omega_k(a, b_k)} e^{S_k g(x_{\beta,\omega}, \omega)} = \sum_{\alpha \in \Pi^\omega_{n_i - N}(a)} \sum_{\beta \in \pi^{-1}_{n_i - N} \alpha \cap \Pi^\omega_k(a, b_k)} e^{S_k g(x_{\beta,\omega}, \omega)}.$$

Applying (3.13) and (3.14) both with $k = n_i$ and with $k = n$ I derive from (3.12) and the ergodic theorem similarly to (3.10) that (3.11) remains true if n_i is replaced by n. Applying this also with $g \equiv 0$ I complete the proof of (3.6) for $g \in L^{1,\kappa}_{X_B}(\Omega, \mathcal{C})$. For any $g \in L^1_{X_B}(\Omega, \mathcal{C})$ take a sequence g_k converging as $k \to \infty$ to g in $L^1_{X_B}(\Omega, \mathcal{C})$. Since by the ergodic theorem with probability one

$$\frac{1}{n}|S_n g(x_{\alpha,\omega}, \omega) - S_n g_k(x_{\alpha,\omega}, \omega)| \le \frac{1}{n} \sum_{i=0}^{n-1} \|g^{\theta^i \omega} - g_k^{\theta^i \omega}\|$$

$$\xrightarrow[n\to\infty]{} \int \|g^\omega - g_k^\omega\| dP(\omega)$$

I derive (3.6) for any $g \in L^1_{X_B}(\Omega, C)$. In particular, (3.6) holds true for any continuous function g on $X \times \Omega$. I can replace also the measure $\eta^\omega_n \in \mathcal{P}(X)$ in (3.6) by the measures $\tilde{\eta}^\omega_n \in \mathcal{P}(X \times \Omega)$ so that for any measurable $\Gamma = \{(x, \omega) : x \in \Gamma^\omega\} \subset X \times \Omega$ one has $\tilde{\eta}^\omega_n(\Gamma) = \eta^\omega_n(\Gamma^\omega)$. In this set up (3.4) follows from Theorem 2.1 in [Ki1]. The inequality (3.5) also follows from Theorem 2.1 in [Ki1] taking into account the uniqueness of equilibrium states (i.e. the measures maximizing the right hand side of (3.6)) for any $g \in L^{1,\kappa}_{X_B}(\Omega, C)$ (see [BG],[KK]). $\square$

By the general contraction principle (see [DZ]) I derive from Theorem 3.1

3.2. Corollary. *For any continuous function g on $X \times \Omega$ and any numbers $r_1 < r_2$,*
(3.15)
$$\limsup_{n \to \infty} \frac{1}{n} \log |\Pi^\omega_n(a, b_n)|^{-1} |\{\alpha \in \Pi^\omega_n(a, b_n) : \frac{1}{n}(S_n g)(x_\alpha, \omega) \in [r_1, r_2]\}|$$
$$\leq - \inf_{r \in [r_1, r_2]} J(r)$$

and
(3.16)
$$\liminf_{n \to \infty} \frac{1}{n} \log |\Pi^\omega_n(a, b)|^{-1} |\{\alpha \in \Pi^\omega_n(a, b) : \frac{1}{n}(S_n g)(x_\alpha, \omega) \in (r_1, r_2)\}|$$
$$\geq - \inf_{r \in (r_1, r_2)} J(r),$$

for all $a \in V(\omega), b_n \in V(\theta^n \omega)$ and $x_\alpha \in C^\omega_\alpha$, where $J(r) = \inf\{I(\nu) : \int g d\nu = r\}$ if a $\nu \in \mathcal{P}(X \times \Omega)$ satisfying the condition in braces exists and $J(r) = \infty$, otherwise.

In particular, if we assign to each edge e of the graph its length $\ell(e)$ and set $g^\omega(x) = \ell(x_0 x_1)$ then taking into account that such g is continuous on $X \times \Omega$ I conclude that Corollary 3.2 yields large deviations for the average length of paths with n vertices as discussed in Introduction.

3.3. Remark. The uniqueness of equilibrium states for functions $g \in L^{1,\kappa}_{X_B}(\Omega, C)$ is proved in [BG] assuming that $B^{n_0}(\omega)$ is a positive matrix for some nonrandom n_0 but it was shown in [KK] that an arbitrary random $n_0 = n_0(\omega)$ will suffice, as well. In fact, the only essential use of this condition is to show that for any $\varepsilon > 0$ and P-a.a.ω there exists $N(\varepsilon, \omega)$ such that for each $x \in X^\omega$ and every $n \geq N(\epsilon, \omega)$ the set $\sigma^{-n}x$ is ε dense in $X^{\theta^{-n}\omega}$. But if $\tilde{k}$ is sufficiently large then $P\{n_0(\omega) \leq \tilde{k}\} > 0$, and so for P-a.a.ω there exists $k_0(\omega) = \min\{k \geq \tilde{k} : n_0(\theta^{-k}\omega) \leq \tilde{k}\}$. Then for any $n = j + k_0(\omega), j \in \mathbb{Z}_+$ and each $y \in X^{\theta^{-n}\omega}$ there exists $\tilde{y} \in X^{\theta^{-n}\omega}$ such that $\tilde{y}_i = y_i$ for $i = 0, 1, ..., j$ and $\tilde{y}_i = x_{i-n}$ for $i = n, n+1, ...$, i.e. $\tilde{y} \in \sigma^{-n}x$, which yields the required assertion. Still, the direct extension of [BG] to this situation encounters technical difficulties and we employ in [KK] a different machinery. Observe that this improvement of [BG] yields also the random Perron-Frobenius theorem for random nonnegative matrices $D(\omega)$ from [BG] under the condition that for P-a.a.ω there exists a random $n_0 = n_0(\omega)$ (and not fixed as in [BG]) such that $D(\theta^{n_0}\omega) \cdots D(\theta\omega)D(\omega)$ is a positive matrix. Note also that one can avoid using the uniqueness of equilibrium states for the lower large deviations bound which becomes crucial in the case of multi-dimensional (deterministic and random) subshifts of finite type (see [EKW] and [Ki3]).

3.4. Remark. The thermodynamic formalism for random subshifts of finite type yields via Theorem 2.1 from [Ki1] the large deviations bounds for all Gibbs measures constructed for functions from $L_{X_B^\omega}^{1,\kappa}(\Omega, C)$ and, in fact, employing the machinery from [EKW] and [Ki3] one obtains large deviations bounds for a larger class of Gibbs measures. Among simple examples of Gibbs measures are random Markov measures which can be constructed in the following way. Let p_i^ω; $i = 1, ..., m(\omega)$ and p_{ij}^ω; $i = 1, ..., m(\omega), j = 1, ..., m(\theta\omega)$ be measurable in ω families of probability vectors and probability matrices, i.e. $\sum_{i=1}^{m(\omega)} p_i^\omega = 1$ and $\sum_{j=1}^{m(\theta\omega)} p_{ij}^\omega = 1 \, \forall i$, such that $\sum_{i=1}^{m(\omega)} p_i^\omega p_{ij}^\omega = p_j^{\theta\omega}$. By the Kolmogorov extension theorem I can define a measurable in ω family of measures $\nu_p^\omega \in \mathcal{P}(X_B^\omega)$ such that $\nu_p^\omega(C_{\alpha_0,...,\alpha_{n-1}}^\omega) = p_{\alpha_0}^\omega p_{\alpha_0\alpha_1}^\omega \cdots p_{\alpha_{n-2}\alpha_{n-1}}^{\theta^{n-2}\omega}$ for any cylinder $C_{\alpha_0,...,\alpha_{n-1}}^\omega$. These measures satisfy $\sigma\nu_p^\omega = \nu_p^\omega$, and so the measure $\nu_p \in \mathcal{P}(X \times \Omega)$ given by $d\nu_p(x,\omega) = d\nu_p^\omega(x)dP(\omega)$ is $\tau = (\sigma, \theta)$-invariant. It is easy to see that ν_p is a Gibbs measure corresponding to the function $g^\omega(x) = \log p_{x_0x_1}^\omega$. Then Theorem 2.1 from [Ki1] yields the following large deviations behavior as $n \to \infty$ (which follows from [Ki3], as well),

$$\nu_p^\omega \left\{ x \in X_B^\omega : \zeta_{x,\omega}^n \in G \right\} \asymp \exp\left(-n \inf_{\eta \in G} \tilde{I}(\eta) \right)$$

where $\tilde{I}(\eta) = -\int \log p_{x_0x_1}^\omega d\eta^\omega(x)dP(\omega) - h_\eta^{(r)}(\tau)$ if $\eta \in \mathcal{P}(X \times \Omega)$ is τ-invariant and $\tilde{I}(\eta) = \infty$, otherwise. The sign $\asymp$ means the logarithmic equivalence in the sense of the upper bound for $\limsup_{n \to \infty}$ and the lower bound for $\liminf_{n \to \infty}$ as in (3.4) and (3.5). This result can be naturally called large deviations for Markov chains in random environment.

4. Counting of configurations.

I shall start with the deterministic set up which includes a finite set Q taken with the discrete topology and called the alphabet (which may represent, for instance, the spin values etc.), the set $Q^{\mathbb{Z}^d}$ considered with the product topology (making it compact) of all maps (configurations) $x : \mathbb{Z}^d \to Q$, the shifts σ^m, $m \in \mathbb{Z}^d$ of $Q^{\mathbb{Z}^d}$ acting by the formula $(\sigma^m x)_n = x_{n+m}$ where $x_k \in Q$ is the value of $x \in Q^{\mathbb{Z}^d}$ on $k \in \mathbb{Z}^d$, and a closed in the product topology subset X of $Q^{\mathbb{Z}^d}$ called the space of (permissible) configurations which is supposed to be shift invariant, i.e. $\sigma^m X = X$ for every $m \in \mathbb{Z}^d$. The pair (X, σ) is called a subshift and if $X = Q^{\mathbb{Z}^d}$ it is called the full shift. I assume that (X, σ) is a subshift of finite type which means that there exist a finite set $F \subset \mathbb{Z}^d$ (which I call a window) and a set $\Psi \subset Q^F$ such that

$$(4.1) \qquad X = X_{(F,\Psi)} = \{x \in Q^{\mathbb{Z}^d} : (\sigma^m x)_F \in \Psi \text{ for every } m \in \mathbb{Z}^d\}$$

where $(x)_R = x_R$ denotes the restriction of $x \in Q^{\mathbb{Z}^d}$ to $R \subset \mathbb{Z}^d$. The set $\Psi \subset Q^F$ is the collection of permissible (allowed) words or configurations on F (visible through F).

For $a = (a_1, \ldots, a_d) \in \mathbb{Z}^d$, $a_i > 0$, $1 \le i \le d$ set $\Lambda(a) = \{i \in \mathbb{Z}^d : 0 \le i_k < a_k, 1 \le k \le d\}$ and I shall write $a \to \infty$ if $a_1, \ldots, a_d \to \infty$. Let also $\mathbb{Z}^d(a)$ be the subgroup of $\mathbb{Z}^d$ generated by $(a_1, 0, ..., 0), ..., (0, ..., 0, a_d)$. The collection $\Pi_a = \{x \in X : \mathbb{Z}^d(a)x = x\}$, which is clearly finite, is called the set of a-periodic points. I shall define next the notions of the weak and strong specifications. The weak specification means in our circumstances that there exists $N > 0$ such for any subsets $R_i \subset \mathbb{Z}^d$ which are N apart and for any permissible configurations ξ_i on R_i one can find $x \in X$ such that $x_{R_i} = \xi_i$. One sais that the strong specification holds true if there exists $N > 0$ such for any subsets $R_i \subset \Lambda(a)$ such that all $R_i + \mathbb{Z}^d(a)$ are N apart and for any permissible configurations ξ_i on R_i one can find $x \in \Pi_a$ such that $x_{R_i} = \xi_i$. For any $x \in X$ and a finite $\Lambda \subset \mathbb{Z}^d$ introduce the probability measures

$$(4.2) \qquad \zeta_x^\Lambda = |\Lambda|^{-1} \sum_{m \in \Lambda} \delta_{\sigma^m x}, \qquad x \in X,$$

where δ_x is the unit mass at x. The following result from [EKW] gives the bounds for large deviations from the set of measures with maximal entropy for occupational measures sitting on periodic orbits. Define $\nu_a \in \mathcal{P}(X)$ by

$$(4.3) \qquad \nu_a(\Gamma) = |\Pi_a|^{-1}|\Gamma \cap \Pi_a|, \qquad \Gamma \subset X,$$

which is the uniform distribution on Π_a, and $\zeta^a : \Pi_a \to \mathcal{P}(X)$ by

$$(4.4) \qquad \zeta_x^a = \zeta_x^{\Lambda(a)}.$$

4.1. Theorem. *Suppose that (X, σ) is a subshift of finite type satisfying the strong specification. Then for any closed $K \subset \mathcal{P}(X)$,*

$$(4.5) \qquad \limsup_{a \to \infty} |\Lambda(a)|^{-1} \log \nu_a \{x : \zeta_x^a \in K\} \leq - \inf_{\eta \in K} J(\eta)$$

and for any open $G \subset \mathcal{P}(X)$,

$$(4.6) \qquad \liminf_{a \to \infty} |\Lambda(a)|^{-1} \log \nu_a \{x : \zeta_x^a \in G\} \geq - \inf_{\eta \in G} J(\eta)$$

where

$$J(\eta) = \begin{cases} h_{\text{top}} - h_\eta & \text{if } \eta \text{ is shift invariant} \\ \infty, & \text{otherwise} \end{cases}$$

and $h_{\text{top}} = \sup\{h_\eta : \eta \text{ is shift invariant}\}$ is the topological entropy of the subshift (X, σ).

The upper bound (4.5) can be derived from [Ki1] similarly to one dimensional subshifts of finite type taking into account that for any continuous function g on X,

$$\lim_{a \to \infty} |\Lambda(a)|^{-1} \log \int \exp(|\Lambda(a)| \int g d\zeta_x^a) d\nu_a(x)$$

$$= \lim_{a \to \infty} |\Lambda(a)|^{-1} \log \sum_{x \in \Pi_a} \exp\left(\sum_{m \in \Lambda(a)} g(\sigma^m x) \right) - \lim_{a \to \infty} |\Lambda(a)|^{-1} \log |\Pi_a|$$

$$= Q(g) - h_{\text{top}}$$

where $Q(g)$ is the topological pressure of g (see [Ru1]). The lower bound (4.6) is obtained in [EKW] by proving first a general large deviations theorem for all Gibbs measures (which requires only the weak specification) and then comparing $\nu_a\{x : \zeta_x^a \in G\}$ with the corresponding quantity for a measure with maximal entropy which is a Gibbs measure corresponding to the potential identically equal to 0. This lower bound for Gibbs measures cannot be obtained via Theorem 2.1 from [Ki1] for $d > 1$ in view of nonuniqueness of equilibrium states (phase transitions) and it requires different arguments (see [EKW]).

Next, I shall consider a random set up which is partially motivated by the spin-glasses models. Let $(\Omega, \mathcal{F}, P)$ be a probability space with a complete σ-algebra $\mathcal{F}$ and an ergodic P-preserving $\mathbf{Z}^d$-action which we denote by $\theta^m, m \in \mathbf{Z}^d$. The model includes also a finite random alphabet $Q_\omega, \omega \in \Omega$ which without loss of generality will be set as $Q_\omega = \{1, 2, \ldots, k(\omega)\}$ where $k(\omega)$ depends measurably on ω. Fix a nonrandom finite subset $F \subset \mathbf{Z}^d$ (which I call "the window") and a random set $\Psi_\omega \subset \prod_{m \in F} Q_{\theta^m \omega} \subset \mathbf{Z}_+^F$

of permissible words or configurations on F (visible through F) such that $\{\omega : \Psi_\omega = \xi\} \in \mathcal{F}$ for any fixed configuration $\xi \in \mathbb{Z}_+^F$ on F. Let $\sigma^m, m \in \mathbb{Z}^d$ be the shift on $\mathcal{X} = \mathbb{Z}_+^{\mathbb{Z}^d}$ acting by the formula $(\sigma^m x)_l = x_{l+m}$ where $x = (x_l, l \in \mathbb{Z}^d, x_l \in \mathbb{Z}_+ = \{1, 2, \dots\})$. Define

$$X^\omega = X_{(F, \Psi)}^\omega = \{x \in \mathbb{Z}_+^{\mathbb{Z}^d} : (\sigma^m x)_F \in \Psi_{\theta^m \omega}$$

(4.7)   for all $m \in \mathbb{Z}^d\}$

where $x_F = (x_m, m \in F)$ and assume that Ψ_ω's are chosen so that with probability one $X^\omega \neq \emptyset$. Clearly,

(4.8) $$\sigma^m X^\omega = X^{\theta^m \omega}$$

and the collection $(\Omega, P, \theta; X, \sigma)$, where $X \subset \mathcal{X} \times \Omega$ satisfies $\{x : (x, \omega) \in X\} = X^\omega$, will be called a random multidimensional subshift of finite type. If $\Psi_\omega = \prod_{m \in F} Q_{\theta^m \omega}$ the above is called a full shift. Using the one point compactification of $\mathbb{Z}_+$ and the product topology on $\mathcal{X}$ we make $\mathcal{X}$ compact and then $X^\omega, \omega \in \Omega$ become compact subsets of $\mathcal{X}$. I shall need also the notion of the specification (in the dynamical systems sense) which means in our circumstances that there exists a random variable $N = N(\omega) > 0$ such that for any subsets $R_i \subset \mathbb{Z}^d$ satisfying: $\min_{n \in R_i} \|m - n\| \geq N(\theta^m \omega)$ for each $m \in R_j$, all $i \neq j$ and every j except for j equal to some index j_0, and for any $\xi_i \in X_{R_i}^\omega$ one can find $x \in X^\omega$ such that $x_{R_i} = \xi_i$ for all i. A special index j_0 in this definition makes it a bit nonsymmetrical but it is natural in view of the form of standard thermodynamical formalism constructions where one has to use the specification in order to produce configurations with given restrictions on a finite set and on an infinite set which is the complement of another bigger finite set. Since I do not know a natural counterpart of a-periodic points in my random set up I have here only one notion of specification.

As with any new notion one has to exhibit first a nontrivial example satisfying the conditions. Observe that unless many nonrandom finite configurations are permissible with probability one the specification condition can be satisfied very rarely with a nonrandom N, especially if (Ω, P) is a product space so that random choices in different sites $m \in \mathbb{Z}^d$ are independent. Using a well known result from the theory of percolation I shall produce next a two dimensional random subshift of finite type satisfying the specification. Let $d = 2$, $\Omega = S^{\mathbb{Z}^2}$, $S = \{1, 2, 3\}$ be a product space, $P = \{p_1, p_2, p_3\}^{\mathbb{Z}^2}$, $p_1 + p_2 + p_3 = 1$ be a product measure, and θ be the shift on Ω, i.e. $(\theta^m \omega)_l = \omega_{l+m}$ for each $\omega = (\omega_l, l \in \mathbb{Z}^2) \in \Omega$. I choose also the alphabet $Q_\omega = \{1, 2\}$ for all ω, the window $F = \{l = (l_1, l_2) \in$

$\mathbb{Z}^2 : 0 \le l_i \le 1, i = 1,2\}$, and the permissible configurations Ψ_ω such that if $\omega_0 = 1$ (respectively, $\omega_0 = 2$) then Ψ_ω permits all configurations which have no two 1's (respectively, no two 2's) in adjacent sites $(l = (l_1, l_2)$ and $m = (m_1, m_2)$ are adjacent if $|l_1 - m_1| + |l_2 - m_2| = 1$) and if $\omega_0 = 3$ then Ψ_ω permits all configurations. Now I define an associated model of the site percolation so that $m \in \mathbb{Z}^2$ will be open for $\omega \in \Omega$ if $\omega_m = 1$ or $\omega_m = 2$ and closed if $\omega_m = 3$. I shall say that for $\omega \in \Omega$ there exists an open path $\gamma = \{m^{(i)}, i = 0, 1, ...\}$ from m to infinity if γ is infinite, each $m^{(i)} \in \mathbb{Z}^2$ is open for $\omega, m^{(0)} = m$, $m^{(i)} \ne m^{(j)}$ if $i \ne j$, and each pair $m^{(i)}, m^{(i+1)}$ are neighbours in the sense that either one or both of their coordinates differ by 1. It is well known (see [Gr], Section 1.4) that if $p_1 + p_2$ is small enough then with probability one there exists no open path to infinity for any starting point $m \in \mathbb{Z}^2$. Therefore with probability one there exists $N = N(\omega)$ such that each $m \in \mathbb{Z}^2$ is surrounded by a loop $\eta_m = (l^{(i)}, i = 1, ..., j_m(\omega))$ of closed for ω sites $l^{(i)}$ containing in the square centered at m with the side equal $N(\theta^m \omega)$ and such that $l^{(i)}$ and $l^{(i+1)}, i = 1, ..., j_m(\omega); l^{(j_m(\omega))} = l^{(1)}$ are adjacent in the sense that one of their coordinates coincides and the other differs by 1. It is easy to see that this yields the specification property with such $N(\omega)$.

For any $\omega \in \Omega, x \in X^\omega$, and a finite $\Lambda \subset \mathbb{Z}^d$ define the empirical (or occupational) measures by

(4.9)	$$\zeta_{x,\omega}^\Lambda = |\Lambda|^{-1} \sum_{m \in \Lambda} \delta_{(\sigma^m x, \theta^m \omega)} \in \mathcal{P}(X).$$

The following result from [Ki3] gives large deviations bounds in counting of configurations which is new in the deterministic set up, as well.

4.2. Theorem. *Let $\Omega, P, \theta, X, \sigma$ be a multidimensional subshift of finite type satisfying the specification where Ω is a compact metric space and suppose that*

(4.10)	$$1 < \int \log k dP < \infty.$$

Then with probability one for each closed $K \subset \mathcal{P}(\mathcal{X} \times \Omega)$,
(4.11)
$$\limsup_{a \to \infty} |\Lambda(a)|^{-1} \log \left(|X_{\Lambda(a)}^\omega|^{-1} |\{\xi \in X_{\Lambda(a)}^\omega : \zeta_{x_\xi, \omega}^{\Lambda(a)} \in K\}| \right)$$
$$\le - \inf_{\nu \in K} J(\nu)$$
and for any open $G \subset \mathcal{P}(\mathcal{X} \times \Omega)$,
(4.12)
$$\liminf_{a \to \infty} |\Lambda(a)|^{-1} \log \left(|X_{\Lambda(a)}^\omega|^{-1} |\{\xi \in X_{\Lambda(a)}^\omega : \zeta_{x_\xi, \omega}^{\Lambda(a)} \in G\}| \right)$$
$$\ge - \inf_{\nu \in G} J(\nu)$$

where x_ξ is any point from $\Xi^\omega_{\Lambda(a)}(\xi) = \{x \in X^\omega : x_{\Lambda(a)} = \xi\}$ (and the result does not depend on the choice of x_ξ) and $J(\nu) = h^{(r)}_{top} - h^{(r)}_\nu$ if $\nu \in \mathcal{P}_P(X)$ is (σ,θ)-invariant and $J(\nu) = \infty$, otherwise, where $h^{(r)}_{top}$ and $h^{(r)}_\nu$ are the relativized topological and metric entropies of the skew product transformation $\tau = (\sigma,\theta)$.

Again, the upper bound (4.11) follows via [Ki1] taking into account that for any continuous function g on $\mathcal{X} \times \Omega$ with probability one,

$$\lim_{a\to\infty} |\Lambda(a)|^{-1} \log \left(|X^\omega_{\Lambda(a)}|^{-1} \exp \left(\sum_{m\in\Lambda(a)} g_{\theta^m\omega}(\sigma^m x) \right) \right) = Q^{(r)}(g) - h^{(r)}_{top}$$

where $Q^{(r)}(g)$ is the relativized topological pressure of the skew product transformation $\tau = (\sigma,\theta)$ corresponding to the function g (see [Ki3]). As in the deterministic case with $d > 1$ phase transitions do not permit the use of Theorem 2.1 from [Ki1] for the lower bound (4.12) which requires different arguments (see [Ki3]).

Next, by the general contraction principle (see [DZ]) Theorem 4.2 yields

4.3. Corollary. *For every continuous function g on $\mathcal{X} \times \Omega$ and any $r_1, r_2 \in [-\infty, \infty], r_1 < r_2$ with probability one*

(4.13)
$$\limsup_{a\to\infty} |\Lambda(a)|^{-1} \log \left(|X^\omega_{\Lambda(a)}|^{-1} |\{\xi \in X^\omega_{\Lambda(a)} : \right.$$
$$|\Lambda(a)|^{-1} S_{\Lambda(a)} g(x_\xi, \omega) \in [r_1, r_2]\}| \bigg)$$
$$\leq - \inf_{c\in[r_1,r_2]} \tilde{J}(c)$$

and

(4.14)
$$\liminf_{a\to\infty} |\Lambda(a)|^{-1} \log \left(|X^\omega_{\Lambda(a)}|^{-1} |\{\xi \in X^\omega_{\Lambda(a)} : \right.$$
$$|\Lambda(a)|^{-1} S_{\Lambda(a)} g(x_\xi, \omega) \in (r_1, r_2)\}| \bigg)$$
$$\geq - \inf_{c\in(r_1,r_2)} \tilde{J}(c)$$

where, again, $(S_\Lambda g)(x, \omega) = \sum_{m\in\Lambda} g^{\theta^m\omega}(\sigma^m x)$, x_ξ is an arbitrary point from $\Xi^\omega_{\Lambda(a)}(\xi)$, and $\tilde{J}(c) = \inf\{J(\nu) : \int g d\nu = c\}$ if a ν satisfying the condition in braces exists and $\tilde{J}(c) = \infty$, otherwise.

Taking here an appropriate function $g^\omega(x)$ Corollary 4.3 yields large deviations for the number of configurations from $X^\omega_{\Lambda(a)}$ having the per site average energy sandwiched between certain constants (see [Ki3]).

REFERENCES

[Bo] T. Bogenschutz, *Entropy, pressure, and a variational principle for random dynamical systems*, Random&Comp.Dyn. **1** (1992), 99-116.

[Bow] R.Bowen, *Equilibrium States and the Ergodic Theory of Anosov Diffeomorphisms*, Lecture Notes in Math., 470, Springer-Verlag, Berlin, 1975.

[BG] T. Bogenschutz and V. M. Gundlach, *Ruelle's transfer operator for random subshifts of finite type*, Preprint (1993).

[BR] R.Bowen and D.Ruelle, *The ergodic theory of Axiom A flows*, Invent. Math. **29** (1975), 181-202.

[DZ] A.Dembo and O.Zeitouni, *Large Deviations Techniques and Applications*, Jones and Bartlett, Boston, 1993.

[EKW] A.Eizenberg, Y.Kifer, and B.Weiss, *Large deviations for Z^d-actions*, Comm. Math. Phys. **164** (1994), 433-454.

[Gr] G. Grimmett, *Percolation*, Springer-Verlag, New York, 1989.

[Ki1] Y. Kifer, *Large deviations in dynamical systems and stochastic processes*, Trans. Amer. Math. Soc. **321** (1990), 505-524.

[Ki2] Y.Kifer, *Large deviations, averaging and periodic orbits of dynamical systems*, Comm. Math. Phys. **162** (1994), 33-46.

[Ki3] Y.Kifer, *Multidimensional random subshifts of finite type and their large deviations*, Preprint (1994).

[KK] K. Khanin and Y. Kifer, *Thermodynamic formalism for random transformations and statistical mechanics*, Preprint (1994).

[Po] M. Pollicott, *Large deviations, Gibbs measures and closed orbits for hyperbolic flows*, Preprint (1993).

[Ru1] D. Ruelle, *Thermodynamic Formalism*, Addison Wesley, New York, 1978.

[Ru2] D. Ruelle, *The pressure of the geodesic flow on a negatively curved manifold*, Bol. Soc. Bras. Mat. **12** (1981), 95-100.

INSTITUTE OF MATHEMATICS, HEBREW UNIVERSITY OF JERUSALEM, GIVAT RAM, JERUSALEM 91904, ISRAEL
E-mail address: kifer@math.huji.ac.il

A ZETA FUNCTION FOR $\mathbb{Z}^d$-ACTIONS

D. A. LIND

ABSTRACT. We define a zeta function for $\mathbb{Z}^d$-actions α that generalizes the Artin-Mazur zeta function for a single transformation. This zeta function is a conjugacy invariant, can be computed explicitly in some cases, and has a product formula over finite orbits. The analytic behavior of the zeta function for $d \geq 2$ is quite different from the case $d = 1$. Even for higher-dimensional actions of finite type the zeta function is typically transcendental and has natural boundary a circle of finite radius. We compute the radius of convergence of the zeta function for a class of algebraic $\mathbb{Z}^d$-actions. We conclude by conjecturing a general description of the analytic behavior of these zeta functions and discussing some further problems.

1. INTRODUCTION

Let $\phi\colon X \to X$ be a homeomorphism of a compact space and $p_n(\phi)$ denote the number of points in X fixed by ϕ^n. We assume that $p_n(\phi)$ is finite for all $n \geq 1$. Artin and Mazur [1] introduced a zeta function $\zeta_\phi(s)$ for ϕ defined by

$$(1.1) \qquad \zeta_\phi(s) = \exp\left(\sum_{n=1}^\infty \frac{p_n(\phi)}{n} s^n\right).$$

Bowen and Lanford [2] showed that if ϕ is a shift of finite type, then $\zeta_\phi(s)$ is a rational function. This was extended to the case when ϕ is an Axiom A diffeomorphism by Manning [11] and to finitely presented systems by Fried [4]. If ϕ is expansive then an elementary argument shows that $\zeta_\phi(s)$ has radius of convergence at least $\exp(-h(\phi))$, where $h(\phi)$ is the topological entropy of ϕ. Finally, the zeta function has the product formula

$$(1.2) \qquad \zeta_\phi(s) = \prod_\gamma \frac{1}{1 - s^{|\gamma|}},$$

where the product is over all finite orbits γ of ϕ and $|\gamma|$ denotes the number of points in γ.

The purpose of this paper is to define an analogous zeta function for $\mathbb{Z}^d$-actions generated by d commuting homeomorphisms. First some notation. Let α be an action of $\mathbb{Z}^d$ on X. For $\mathbf{n} \in \mathbb{Z}^d$ let $\alpha^{\mathbf{n}}$ denote the element of this action corresponding to $\mathbf{n}$. Denote the set of finite-index subgroups of $\mathbb{Z}^d$ by

Supported in part by NSF Grant DMS 9303240

433

$\mathcal{L}_d$. For $L \in \mathcal{L}_d$ let $p_L(\alpha)$ be the number of points in X fixed by α^n for all $n \in L$. The index $|\mathbf{Z}^d/L|$ of L in $\mathbf{Z}^d$ is denoted by $[L]$. Note that when $d = 1$ we have $\mathcal{L}_1 = \{n\mathbf{Z} : n \geq 1\}$. We generalize (1.1) to $d \geq 1$ by replacing the sum over $n \geq 1$ by the sum over $L \in \mathcal{L}_d$. This leads to our definition of the zeta function of α as

$$(1.3) \qquad \zeta_\alpha(s) = \exp\left(\sum_{L \in \mathcal{L}_d} \frac{p_L(\alpha)}{[L]} s^{[L]} \right).$$

In this paper we establish some basic properties of the zeta function of a $\mathbf{Z}^d$-action, and compute this function explicitly for a number of examples. Even the trivial $\mathbf{Z}^2$-action on a point has the "interesting" zeta function

$$\prod_{n=1}^{\infty} \frac{1}{1 - s^n} = \sum_{n=1}^{\infty} p(n)\, s^n,$$

where $p(n)$ is the number of partitions of n. This formula goes back to Euler, and was the starting point for proving the asymptotic formula

$$p(n) \sim \exp\left(\pi \sqrt{2n/3} \right)$$

developed by Hardy and Ramanujan [5]. We prove an analogue of the product formula (1.2) in §5, and provide evidence for the conjecture that when $d \geq 2$ the zeta function is typically meromorphic for $|s| < \exp(-h(\alpha))$ and has the circle $|s| = \exp(-h(\alpha))$ as its natural boundary.

Another approach to generalizing the zeta function to $\mathbf{Z}^d$-actions was taken by Mathiszik in his dissertation [12], which was never published. There he counts multiples of the period of a point, not the lattices contained in the stabilizer of the periodic point, leading to a different function. The definition (1.3) was stated independently by Ward in the preprint [16].

2. DEFINITION AND BASIC PROPERTIES

Let α be a $\mathbf{Z}^d$-action on X, and $\mathcal{L}_d$ denote the collection of finite-index subgroups (or *lattices*) in $\mathbf{Z}^d$. For $L \in \mathcal{L}_d$ put $[L] = |\mathbf{Z}^d/L|$, and let

$$p_L(\alpha) = |\{x \in X : \alpha^n x = x \text{ for all } n \in L\}|,$$

which we will assume to be finite for all $L \in \mathcal{L}_d$.

Definition 2.1. The *zeta function* of the $\mathbf{Z}^d$-action α is defined as

$$\zeta_\alpha(s) = \exp\left(\sum_{L \in \mathcal{L}_d} \frac{p_L(\alpha)}{[L]} s^{[L]} \right).$$

When $d = 1$ note that $\mathcal{L}_1 = \{n\mathbf{Z} : n \geq 1\}$ and that $p_{n\mathbf{Z}}(\alpha) = p_n(\phi)$, where $\phi = \alpha^1$ is the generator of α. Hence the zeta function of a $\mathbf{Z}$-action generated by ϕ is just the Artin-Mazur zeta function of ϕ.

The zeta function is obviously a conjugacy invariant for $\mathbf{Z}^d$-actions.

For $A \in GL(d, \mathbf{Z})$ define the $\mathbf{Z}^d$-action α^A by $(\alpha^A)^{\mathbf{n}} = \alpha^{A\mathbf{n}}$. The following shows that the zeta function is independent of a choice of basis for $\mathbf{Z}^d$.

Lemma 2.2. Let α be a $\mathbf{Z}^d$-action and $A \in GL(d, \mathbf{Z})$. Then $\zeta_\alpha(s) = \zeta_{\alpha^A}(s)$.

Proof. For $L \in \mathcal{L}_d$ let $AL = \{A\mathbf{n} : \mathbf{n} \in L\}$. Then $L \longleftrightarrow AL$ is a bijection from $\mathcal{L}_d$ to itself, and $[AL] = [L]$. Clearly

$$p_L(\alpha^A) = p_{AL}(\alpha),$$

and the result now follows. $\square$

If $Y \subset X$ is α-invariant, let $\alpha|Y$ denote the restriction of α to Y.

Lemma 2.3. Let $Y, Z \subset X$ be compact α-invariant sets. Then

$$\zeta_{\alpha|Y \cup Z}(s) = \frac{\zeta_{\alpha|Y}(s)\zeta_{\alpha|Z}(s)}{\zeta_{\alpha|Y \cap Z}(s)}.$$

In particular, if Y and Z are disjoint, then

$$\zeta_{\alpha|Y \cup Z}(s) = \zeta_{\alpha|Y}(s)\zeta_{\alpha|Z}(s).$$

Proof. For $L \in \mathcal{L}_d$ it is elementary that

$$p_L(\alpha|Y \cup Z) = p_L(\alpha|Y) + p_L(\alpha|Z) - p_L(\alpha|Y \cap Z),$$

and the result follows from the definition of the zeta function. $\square$

3. Examples

In this section we compute the zeta function for some specific $\mathbf{Z}^2$-actions. For this we require a parameterization of $\mathcal{L}_2$. A convenient one for our purposes is

$$(3.1) \qquad \mathcal{L}_2 = \left\{ \begin{bmatrix} a & b \\ 0 & c \end{bmatrix} \mathbf{Z}^2 : a \geq 1, \ c \geq 1, \ 0 \leq b \leq a - 1 \right\}.$$

The Hermite normal form for integral matrices, discussed in the next section, shows that this gives a complete listing of the lattices in $\mathbf{Z}^2$, with each lattice listed exactly once.

We also use the familiar power series

$$(3.2) \qquad -\log(1 - t) = \sum_{n=1}^{\infty} \frac{t^n}{n}.$$

Example 3.1. Let α be the trivial $\mathbf{Z}^2$-action on a single point. Then $p_L(\alpha) = 1$ for all $L \in \mathcal{L}_2$. Hence using the parameterization (3.1) of $\mathcal{L}_2$ we see that

$$\zeta_\alpha(s) = \exp\left(\sum_{a=1}^{\infty} \sum_{c=1}^{\infty} \sum_{b=0}^{a-1} \frac{1}{ac} s^{ac} \right) = \exp\left(\sum_{a=1}^{\infty} \sum_{c=1}^{\infty} \frac{(s^a)^c}{c} \right)$$

$$= \exp\left(\sum_{a=1}^{\infty} -\log(1 - s^a) \right)$$

$$= \prod_{a=1}^{\infty} \frac{1}{1 - s^a}.$$

We denote this zeta function by $\pi_2(s)$. As noted in §1, $\pi_2(s)$ is the classical generating function for the partition function, i.e.,

$$\pi_2(s) = \sum_{n=1}^{\infty} p(n)\, s^n.$$

It is known that this function is analytic in $|s| < 1$ and has the unit circle $|s| = 1$ as its natural boundary [5].

Example 3.2. Let α be the full $\mathbf{Z}^2$ k-shift. This means that $X = \{0, 1, \ldots, k-1\}^{\mathbf{Z}^2}$ and α is the natural shift action of $\mathbf{Z}^2$ on X. Suppose $L \in \mathcal{L}_2$. Then a point in X that is fixed by all $\alpha^{\mathbf{n}}$ for $\mathbf{n} \in L$ is determined by its coordinates in a fundamental domain for L, and all choices there are possible. Hence

$$p_L(\alpha) = k^{[L]}.$$

Since $p_L(\alpha)$ depends only on the index of L, a calculation similar to Example 3.1 shows that here

$$\zeta_\alpha(s) = \prod_{a=1}^{\infty} \frac{1}{1 - (ks)^a} = \pi_2(ks).$$

Note that $h(\alpha) = \log k$, so that $\zeta_\alpha(s)$ is analytic in $|s| < \exp(-h(\alpha)) = 1/k$ and has the circle $|s| = \exp(-h(\alpha))$ as its natural boundary.

Example 3.3. Let $X = \{0, 1, \ldots, k-1\}^{\mathbf{Z}}$, and σ be the shift $\mathbf{Z}$-action on X. Define the $\mathbf{Z}^2$-action α on X by $\alpha^{(m,n)} = \sigma^n$, so that α is generated by the identity map and σ. This $\mathbf{Z}^2$-action is clearly conjugate to the $\mathbf{Z}^2$ shift action β on

$$Y = \left\{ y \in \{0, 1, \ldots, k-1\}^{\mathbf{Z}^2} : y_{i,j} = y_{i+1,j} \text{ for all } i, j \in \mathbf{Z} \right\}.$$

We will compute $\zeta_\beta(s) = \zeta_\alpha(s)$.

Let

$$L = \begin{bmatrix} a & b \\ 0 & c \end{bmatrix} \mathbf{Z}^2,$$

where $a \geq 1$, $c \geq 1$, and $0 \leq b \leq a - 1$. A point $y \in Y$ that is fixed by $\beta^{\mathbf{n}}$ for all $\mathbf{n} \in L$ is determined by its coordinates $y_{0,0}, y_{0,1}, \ldots, y_{0,c-1}$, and the choices for these are arbitrary. Hence $p_L(\beta) = k^c$. Thus

$$\zeta_\alpha(s) = \zeta_\beta(s) = \exp\left(\sum_{a=1}^{\infty} \sum_{c=1}^{\infty} \sum_{b=0}^{a-1} \frac{k^c}{ac} s^{ac}\right)$$

$$= \exp\left(\sum_{a=1}^{\infty} \sum_{c=1}^{\infty} \frac{(ks^a)^c}{c}\right)$$

$$= \exp\left(\sum_{a=1}^{\infty} -\log(1 - ks^a)\right)$$

$$= \prod_{a=1}^{\infty} \frac{1}{1 - ks^a} = \frac{1}{(1 - ks)(1 - ks^2)(1 - ks^3)\ldots}.$$

Observe that $h(\alpha) = 0$, but the factor $1 - ks$ in the denominator shows that $\zeta_\alpha(s)$ has a simple pole at $1/k$. This shows that the radius of convergence of ζ_α may be strictly smaller that $\exp(-h(\alpha))$. In fact, the product development above for $\zeta_\alpha(s)$ shows that it has poles at

$$\left\{\frac{1}{\sqrt[n]{k}} e^{2\pi i j/n} : 0 \leq j \leq n - 1, \ n \geq 1\right\}.$$

Hence $\zeta_\alpha(s)$ has infinitely many poles inside the unit disk, and these poles cluster to the unit circle. It follows that $\zeta_\alpha(s)$ is meromorphic in $|s| < \exp(-h(\alpha)) = 1$ and has natural boundary $|s| = \exp(-h(\alpha)) = 1$.

Example 3.4. Let $\mathbb{F}_2 = \{0, 1\}$ be the field with two elements, and

$$X = \left\{x \in \mathbb{F}_2^{\mathbb{Z}^2} : x_{i,j} + x_{i+1,j} + x_{i,j+1} = 0 \text{ for all } i, j \in \mathbb{Z}\right\}.$$

Then X is a compact group with coordinate-wise operations, and it is invariant under the natural $\mathbb{Z}^2$ shift action α. This action was originally investigated by Ledrappier [8], who showed that it was mixing, but not mixing of higher orders.

Let

$$L = \begin{bmatrix} a & b \\ 0 & c \end{bmatrix} \mathbb{Z}^2,$$

where $a \geq 1$, $c \geq 1$, and $0 \leq b \leq a - 1$. We compute $p_L(\alpha)$ using some linear algebra over $\mathbb{F}_2$. A point $x \in X$ that is L-invariant must have horizontal period a, so is determined by an element y from the vector space $\mathbb{F}_2^a$. Let I_a denote the $a \times a$ identity matrix and P_a the $a \times a$ permutation matrix corresponding to the cyclic shift of elementary basis vectors of $\mathbb{F}_2^a$. The condition of L-periodicity of x translates to the condition

$$(I_a + P_a)^c \, y = P_a^{-b} \, y.$$

Hence

(3.3) $p_L(\alpha) = \left| \ker\left((I_a + P_a)^c - P_a^{-b}\right) \right|.$

For example, if $a = 3$, $b = 1$, and $c = 1$, then

$$P_a = \begin{bmatrix} 0 & 0 & 1 \\ 1 & 0 & 0 \\ 0 & 1 & 0 \end{bmatrix}$$

and

$$(I_a + P_a)^c - P_a^{-b} = \begin{bmatrix} 1 & 1 & 1 \\ 1 & 1 & 1 \\ 1 & 1 & 1 \end{bmatrix}.$$

The kernel of the latter has dimension two over $\mathbf{F}_2$, so that for this L we have $p_L(\alpha) = 4$.

Formula (3.3) allows computation of $\zeta_\alpha(s)$ to as many terms as we like. A calculation using *Mathematica* gives the first terms to be

$$\zeta_\alpha(s) = 1+s + 2s^2 + 4s^3 + 6s^4 + 9s^5 + 16s^6 + 24s^7 + 35s^8 + 54s^9$$
$$+ 78s^{10} + 110s^{11} + 162s^{12} + 226s^{13} + 317s^{14} + 446s^{15} + 612s^{16} + .$$

Such calculations do not, however, reveal the analytic properties of $\zeta_\alpha(s)$. For this we next employ other ideas to show that $\zeta_\alpha(s)$ has radius of convergence one and the unit circle as natural boundary.

We first estimate the size of $p_L(\alpha)$ in terms of $[L]$.

Lemma 3.5. Let α be the $\mathbf{Z}^2$ shift action on

$$X = \{x \in \mathbf{F}_2^{\mathbf{Z}^2} : x_{i,j} + x_{i+1,j} + x_{i,j+1} = 0 \quad \text{for all } i, j \in \mathbf{Z}\}.$$

Then there is a constant C such that for all $L \in \mathcal{L}_2$ we have that

(3.4) $p_L(\alpha) \le 2^{C\sqrt{[L]}}.$

Proof. Let $L \in \mathcal{L}_2$. We first show that there is a nonzero vector $\mathbf{v} \in L$ such that $\|\mathbf{v}\| \le \theta\sqrt{[L]}$, where $\|\cdot\|$ denotes the Euclidean norm on $\mathbf{R}^2$ and $\theta = \sqrt{4/\pi}$. Let B_r be the ball of radius r in $\mathbf{R}^2$ around 0, and A be an integral matrix such that $A\mathbf{Z}^2 = L$. If $B_r \cap L = \{0\}$, then $A^{-1}(B_r)$ is a convex, symmetric region in $\mathbf{R}^2$ that does not contain any nonzero vectors in $\mathbf{Z}^2$. By Minkowski's theorem [6, Thm. 37] we have that

$$\text{area}\left(A^{-1}(B_r)\right) < 4.$$

Since $\det A = [L]$, it follows that

$$\frac{\pi r^2}{[L]} < 4, \quad \text{or} \quad r < \theta\sqrt{[L]}.$$

Hence the ball of radius $\theta\sqrt{[L]}$ must contain a nonzero vector $\mathbf{v}$ in L.

Next, consider the strip S of width 2 about the line segment $[0, \mathbf{v}]$. If $x \in X$ is L-periodic, then its coordinates in S determine by periodicity those in the strip of width 2 about the line $\mathbf{R}\mathbf{v}$, which in turn determine the rest of the coordinates by use of the defining relations for points in X and L-periodicity. Hence the number of L-periodic points in X is bounded above by the number of possible configurations in S. Since the the number of lattice points in S is bounded by a constant times $\sqrt{[L]}$, the estimate (3.4) now follows. $\square$

Continuing with Example 3.4, the parameterization (3.1) of $\mathcal{L}_2$ shows that the number of $L \in \mathcal{L}_2$ with $[L] = n$ equals $\sigma(n)$, the sum of the divisors of n. Trivially $\sigma(n) \leq 1 + 2 + \cdots + n < n^2$, so that

$$\sum_{\{L \in \mathcal{L}_2 : [L] = n\}} p_L(\alpha) \leq n^2 \cdot 2^{C\sqrt{n}}.$$

It follows from the Hadamard formula that the series for $\zeta_\alpha(s)$ has radius of convergence one.

We next show that $\zeta_\alpha(s)$ has the unit circle as natural boundary. For this, we first claim that the estimate (3.4) is sharp in the sense that no power of $[L]$ strictly less than $1/2$ will work for all L. Let

$$L_n = (2^n - 1)\mathbf{Z} \oplus (2^n - 1)\mathbf{Z},$$

so that $[L_n] = (2^n - 1)^2$. The expansion of $(1 + t)^{2^n - 1} \bmod 2$ is $\sum_{j=0}^{2^n - 1} t^j$. It then follows from (3.3) that $p_{L_n}(\alpha)$ equals the number of elements in $\mathbf{F}_2^{2^n - 1}$ whose coordinates sum to 0, which is half the cardinality of $\mathbf{F}_2^{2^n - 1}$. Hence

$$(3.5) \qquad p_{L_n}(\alpha) = \frac{1}{2} 2^{2^n - 1} = \frac{1}{2} 2^{\sqrt{[L_n]}},$$

verifying our claim.

Suppose that $\zeta_\alpha(s)$ were rational, say

$$\zeta_\alpha(a) = \frac{\prod_{i=1}^{k}(1 - \lambda_i s)}{\prod_{j=1}^{l}(1 - \mu_j s)}.$$

It follows from calculus that we would then have

$$(3.6) \qquad \sum_{\{L \in \mathcal{L}_2 : [L] = n\}} p_L(\alpha) = \sum_{l=1}^{l} \mu_j^n - \sum_{i=1}^{k} \lambda_i^n.$$

Now $\zeta_\alpha(s)$ is analytic in $|s| < 1$, so that $|\mu_j| \leq 1$ for all j. Furthermore, $\zeta_\alpha(s)$ does not vanish in $|s| < 1$, being the exponential of a convergent power series there, so that $|\lambda_i| \leq 1$ for all i. Thus (3.6) would imply that $p_L(\alpha) \leq k + l$ for all $L \in \mathcal{L}_2$, contradicting (3.5). This proves that $\zeta_\alpha(s)$ is not rational.

The product formula, which is proved in §5, shows that the Taylor series of $\zeta_\alpha(s)$ has integer coefficients (see Corollary 5.5). A theorem of Carlson [3] (cf. Polya [13]) asserts that an analytic function whose Taylor series has integer coefficients and radius of convergence one is either rational or has the unit circle as natural boundary. For $\zeta_\alpha(s)$ we have just ruled out the first alternative. Hence $\zeta_\alpha(s)$ is analytic in $|s| < 1$ and has the circle $|s| = 1$ as its natural boundary.

4. CALCULATION FOR TRIVIAL $\mathbf{Z}^d$-ACTIONS

In this section we explicitly compute the zeta function $\pi_d(s)$ of the trivial $\mathbf{Z}^d$-action on a point. For $d = 1$, equation (3.2) shows that

$$\pi_1(s) = \frac{1}{1 - s}.$$

In Example 3.1 we showed that

$$\pi_2(s) = \prod_{n=1}^{\infty} \frac{1}{1 - s^n}.$$

The functions $\pi_d(s)$ play a central role in the product formula in §5.

Let α be the trivial $\mathbf{Z}^d$-action on a point. Then $p_L(\alpha) = 1$ for all $L \in \mathcal{L}_d$. Hence

$$\pi_d(s) = \zeta_\alpha(s) = \exp\left(\sum_{L \in \mathcal{L}_d} \frac{1}{[L]} s^{[L]} \right) = \exp\left(\sum_{n=1}^{\infty} \frac{e_d(n)}{n} s^n \right),$$

where

$$(4.1) \qquad e_d(n) = \left| \{ L \in \mathcal{L}_d : [L] = n \} \right|.$$

To compute $e_d(n)$ we use the *Hermite normal form* of an integer matrix (see [10, Thm. 22.1]).

Theorem 4.1 (Hermite). Every lattice in $\mathcal{L}_d$ has a unique representation as the image of $\mathbf{Z}^d$ under a matrix having the form

$$(4.2) \qquad \begin{bmatrix} a_1 & b_{12} & b_{13} & \cdots & b_{1d} \\ 0 & a_2 & b_{23} & \cdots & b_{2d} \\ 0 & 0 & a_3 & \cdots & b_{3d} \\ \vdots & \vdots & \vdots & \ddots & \vdots \\ 0 & 0 & 0 & \cdots & a_d \end{bmatrix},$$

where $a_i \geq 1$ for $1 \leq i \leq d$ and $0 \leq b_{ij} \leq a_i - 1$ for $i + 1 \leq j \leq d$. $\square$

This result provides a convenient parameterization of $\mathcal{L}_d$, generalizing that of $\mathcal{L}_2$ we described in equation (3.1). It also shows how to inductively compute the $e_d(n)$ as follows.

	$n=1$	2	3	4	5	6	7	8	9	10
$d=1$	1	1	1	1	1	1	1	1	1	1
2	1	3	4	7	6	12	8	15	13	18
3	1	7	13	35	31	91	57	155	130	217
4	1	15	40	155	156	600	400	1395	1210	2340
5	1	31	121	651	781	3751	2801	11811	11011	24211
6	1	63	364	2667	3906	22932	19608	97155	99463	246078

TABLE 1. Values of $e_d(n)$

Proposition 4.2. Let $e_d(n)$ be the number of lattices in $\mathbf{Z}^d$ having index n. Then

$$(4.3) \qquad e_d(n) = \sum_{k|n} e_{d-1}\left(\frac{n}{k}\right) k^{d-1}.$$

Proof. Suppose that $L \in \mathcal{L}_d$ is the image of $\mathbf{Z}^d$ under the matrix in (4.2). Then $[L] = a_1 a_2 \ldots a_{d-1} a_d$, so that a_1 divides $[L]$. Let $k = a_1$. Each of the $b_{12}, b_{13}, \ldots, b_{1d}$ can assume the values $0, 1, \ldots, k-1$, giving k^{d-1} choices for the top row. There are $e_{d-1}(n/k)$ choices for the remaining part of the matrix. Summing over all divisors k of n gives (4.3). $\square$

For example, we have trivially that

$$e_0(n) = \begin{cases} 1 & \text{if } n=1, \\ 0 & \text{if } n \geq 2. \end{cases}$$

Then (4.3) with $d=1$ gives $e_1(n)=1$ for all $n \geq 1$, and next with $d=2$ that $e_2(n) = \sigma(n)$, the sum of the divisors of n. Table 1 lists some values of $e_d(n)$.

It follows from (4.3) that the sequence $\{e_d(n) : n \geq 1\}$ is the Dirichlet product of $\{e_{d-1}(n) : n \geq 1\}$ with the sequence $\{n^{d-1} : n \geq 1\}$. Inductively this shows that $\{e_d(n) : n \geq 1\}$ is the Dirichlet product of the sequences $\{n^0\}$, $\{n^1\}, \ldots, \{n^{d-1}\}$. Now the Dirichlet generating function of $\{n^k\}$ is

$$\sum_{n=1}^{\infty} \frac{n^k}{n^s} = \sum_{n=1}^{\infty} \frac{1}{n^{s-k}} = \zeta(s-k),$$

where $\zeta(s)$ denotes the Riemann zeta function (see [6, §17.5]). Since the Dirichlet generating function of the Dirichlet product of sequences is the product of their Dirichlet generating functions, we see that

$$\sum_{n=1}^{\infty} \frac{e_d(n)}{n^s} = \zeta(s)\zeta(s-1)\ldots\zeta(s-d+1).$$

This provides an explicit representation of the $e_d(n)$.

We next estimate the growth rate of $e_d(n)$ as $n \to \infty$.

Proposition 4.3. Let $e_d(n)$ be defined as in (4.1). Then

(4.4) $$e_d(n) \leq n^{d+1}.$$

Proof. Since $e_1(n) = 1$ for all $n \geq 1$, the estimate (4.4) is certainly valid for $d = 1$. Assume inductively that it holds when $d \geq 2$ is replaced by $d - 1$. Then by (4.3),

$$e_d(n) = \sum_{k|n} e_{d-1}\left(\frac{n}{k}\right) k^{d-1} \leq \sum_{k|n} \left(\frac{n}{k}\right)^d k^{d-1} \leq n^d \sum_{k=1}^n \frac{1}{k} \leq n^{d+1},$$

completing the inductive step and the proof. $\square$

To compute $\pi_d(s)$, we make use of the following observation.

Lemma 4.4. Let β be a $\mathbf{Z}^{d-1}$-action and α be the $\mathbf{Z}^d$-action defined by

$$\alpha^{(n_1, n_2, \ldots, n_d)} = \beta^{(n_2, n_3, \ldots, n_d)}.$$

Then

$$\zeta_\alpha(s) = \prod_{k=1}^\infty \zeta_\beta\left(s^k\right)^{k^{d-2}}.$$

Proof. Consider a $(d-1) \times (d-1)$ matrix

$$B = \begin{bmatrix} a_2 & b_{23} & \ldots & b_{2d} \\ 0 & a_3 & \ldots & b_{3d} \\ \vdots & \vdots & \ddots & \vdots \\ 0 & 0 & \ldots & a_d \end{bmatrix}$$

in Hermite normal form. We can form all $d \times d$ matrices A in Hermite normal form containing B as the lower right principal minor by appending to B a top row of the form $[k \ b_{12} \ b_{13} \ \ldots \ b_{1d}]$, where $k \geq 1$ and $0 \leq b_{1j} \leq k - 1$, and making the remaining entries in the left column 0, forming

$$A = \begin{bmatrix} k & b_{12} & b_{13} & \ldots & b_{1d} \\ 0 & a_2 & b_{23} & \ldots & b_{2d} \\ 0 & 0 & a_3 & \ldots & b_{3d} \\ \vdots & \vdots & \vdots & \ddots & \vdots \\ 0 & 0 & 0 & \ldots & a_d \end{bmatrix}.$$

Note that since $\alpha^{(1,0,\ldots,0)}$ is the identity, we have that

$$p_{A\mathbf{Z}^d}(\alpha) = p_{B\mathbf{Z}^{d-1}}(\beta)$$

for all choices of k and the b_{1j}. Hence

$$
\zeta_\alpha(s) = \exp\left(\sum_{M \in \mathcal{L}_{d-1}} \sum_{k=1}^{\infty} \sum_{b_{12}=0}^{k-1} \cdots \sum_{b_{1d}=0}^{k-1} \frac{p_M(\beta)}{k[M]} s^{k[M]} \right)
$$

$$
= \exp\left(\sum_{M \in \mathcal{L}_{d-1}} \sum_{k=1}^{\infty} \frac{k^{d-1} p_M(\beta)}{k[M]} s^{k[M]} \right)
$$

$$
= \prod_{k=1}^{\infty} \exp\left(\sum_{M \in \mathcal{L}_{d-1}} \frac{p_M(\beta)}{[M]} (s^k)^{[M]} \right)^{k^{d-2}}
$$

$$
= \prod_{k=1}^{\infty} \zeta_\beta(s^k)^{k^{d-2}} \qquad \square
$$

This lemma implies, for example, that if α is the $\mathbf{Z}^2$-action generated by the identity on X and $\phi \colon X \to X$, then

$$
\zeta_\alpha(s) = \prod_{k=1}^{\infty} \zeta_\phi(s^k).
$$

We used this in Example 3.1 to compute that

$$
\pi_2(s) = \prod_{k=1}^{\infty} \pi_1(s^k) = \prod_{k=1}^{\infty} \frac{1}{1 - s^k},
$$

and also in Example 3.3. A general version of this argument leads to the following formula for $\pi_d(s)$.

Theorem 4.5. Let $\pi_d(s)$ denote the zeta function of the trivial $\mathbf{Z}^d$-action on a point. Then the Taylor series of $\pi_d(s)$ has radius of convergence one. Furthermore, $\pi_d(s)$ has the product expansion

$$
(4.5) \qquad \pi_d(s) = \prod_{n=1}^{\infty} \frac{1}{(1 - s^n)^{e_{d-1}(n)}},
$$

where $e_d(n)$ is defined in (4.1) and the product converges for $|s| < 1$.

Proof. By Proposition 4.3 we know that $1 \le e_d(n) \le n^{d+1}$. Hence the series

$$
\psi(s) = \sum_{n=1}^{\infty} \frac{e_d(n)}{n} s^n
$$

has radius of convergence one. Furthermore, $\lim_{s \to 1^-} \psi(s) = \infty$. Hence the Taylor series of $\pi_d(s) = e^{\psi(s)}$ has radius of convergence one.

We have already seen that (4.5) is valid for $d = 2$. Suppose inductively that it is valid when $d \ge 3$ is replaced by $d - 1$. Lemma 4.4 applies to the trivial $\mathbf{Z}^d$-action α and the trivial $\mathbf{Z}^{d-1}$-action β. Hence using the lemma and Proposition 4.2, we obtain

$$
\pi_d(s) = \prod_{k=1}^{\infty} \pi_{d-1}^{\cdot}(s^k)^{k^{d-2}} = \prod_{k=1}^{\infty} \prod_{n=1}^{\infty} (1 - s^{kn})^{-e_{d-2}(n)\, k^{d-2}}
$$

$$
= \prod_{m=1}^{\infty} (1 - s^m)^{-\sum_{k|m} e_{d-2}(m/k)\, k^{d-2}} = \prod_{m=1}^{\infty} (1 - s^m)^{-e_{d-1}(m)},
$$

where the estimate (4.4) shows that our manipulations are valid when $|s| < 1$. This verifies (4.5) for d, completing the proof. $\square$

Proposition 4.6. For $d \geq 2$ the function $\pi_d(s)$ has the unit circle as natural boundary.

Proof. Theorem 4.5 shows that the Taylor series for $\pi_d(s)$ has integer coefficients and radius of convergence one. Also, an easy consequence of Proposition 4.2 is that $e_d(n) \geq n$ for $d \geq 2$, so that $e_d(n) \to \infty$ as $n \to \infty$. The same argument as at the end of Example 3.4 shows that $\pi_d(s)$ is not a rational function for $d \geq 2$, and the same use of Carlson's theorem as there then proves that $\pi_d(s)$ has the unit circle as natural boundary. $\square$

5. THE PRODUCT FORMULA

Our goal is this section is to generalize the product formula (1.2) for single transformations to $\mathbf{Z}^d$-actions α. We begin by considering the case when α has exactly one finite orbit.

Proposition 5.1. Let γ be a finite set and α be a transitive $\mathbf{Z}^d$-action on γ. Then
$$
\zeta_\alpha(s) = \pi_d(s^{|\gamma|}).
$$

Proof. Let $H = \{\mathbf{n} \in \mathbf{Z}^d : \alpha^{\mathbf{n}} x = x \text{ for all } x \in \gamma\}$. Since γ is finite, $H \in \mathcal{L}_d$. Now α acts transitively on γ, so that H is also the stabilizer subgroup of each element of γ, so that $|\gamma| = [H]$. Hence for $L \in \mathcal{L}_d$ we have that

$$
p_L(\alpha) = \begin{cases} |\gamma| & \text{if } L \subseteq H, \\ 0 & \text{if } L \nsubseteq H. \end{cases}
$$

Since $[H]/[L] = 1/|H/L|$, we see that

$$
\zeta_\alpha(s) = \exp\left(\sum_{\{L \in \mathcal{L}_d : L \subseteq H\}} \frac{[H]}{[L]} s^{[L]} \right)
$$

$$
= \exp\left(\sum_{\{L \in \mathcal{L}_d : L \subseteq H\}} \frac{1}{|H/L|} (s^{[H]})^{|H/L|} \right)
$$

$$
= \pi_d(s^{[H]}) = \pi_d(s^{|\gamma|}). \quad \square
$$

In order to determine the validity of the product formula, we need to first find the radius of convergence of the Taylor series for $\zeta_\alpha(s)$. To do this, we introduce the following quantity.

Definition 5.2. Let α be a $\mathbf{Z}^d$-action. Define the *growth rate of periodic points of α* as

$$g(\alpha) = \limsup_{[L] \to \infty} \frac{1}{[L]} \log p_L(\alpha) = \lim_{n \to \infty} \sup_{[L] \geq n} \frac{1}{[L]} \log p_L(\alpha).$$

Theorem 5.3. Let α be a $\mathbf{Z}^d$-action, and assume as usual that $p_L(\alpha)$ is finite for every $L \in \mathcal{L}_d$. Then $\zeta_\alpha(s)$ has radius of convergence $\exp(-g(\alpha))$. In particular, if there is a number $\theta > 1$ for which $p_L(\alpha) \leq \theta^{[L]}$ for all $L \in \mathcal{L}_d$, then $\zeta_\alpha(s)$ is analytic for $|s| < \theta^{-1}$.

Proof. By Proposition 4.3, the number of $L \in \mathcal{L}_d$ with index n is polynomial in n. It then follows easily from the Hadamard formula that

$$\psi(s) = \sum_{L \in \mathcal{L}_d} \frac{p_L(\alpha)}{[L]} s^{[L]}$$

has radius of convergence $\rho = \exp(-g(\alpha))$. Also, since the Taylor coefficients of $\psi(s)$ are nonnegative, we see that $\lim_{s \to \rho^-} \psi(s) = \infty$. It now follows that $\zeta_\alpha(s) = e^{\psi(s)}$ has radius of convergence $\rho = \exp(-g(\alpha))$. $\square$

We are ready for the main result of this section.

Theorem 5.4 (Product Formula). Let α be a $\mathbf{Z}^d$-action for which $g(\alpha) < \infty$. Then

(5.1) $$\zeta_\alpha(s) = \prod_\gamma \pi_d(s^{|\gamma|}),$$

where the product is taken over all finite orbits γ of α and this product converges for all $|s| < \exp(-g(\alpha))$.

Proof. Enumerate the finite orbits of α as $\gamma_1, \gamma_2, \ldots$. Lemma 2.3 and Proposition 5.1 show that (5.1) is valid for the restriction of α to $\gamma_1 \cup \gamma_2 \cup \ldots \gamma_n$ for all $n \geq 1$. Since everything in sight converges absolutely for $|s| < \exp(-g(\alpha))$, a standard argument shows that letting $n \to \infty$ proves (5.1) in general. $\square$

Corollary 5.5. Let α be a $\mathbf{Z}^d$-action for which $\zeta_\alpha(s)$ has a positive radius of convergence. Then the Taylor series for $\zeta_\alpha(s)$ has integer coefficients.

Proof. Theorem 4.5 shows that the Taylor series of $\pi_d(s)$ has integer coefficients. The corollary now follows from the product formula. $\square$

6. ALGEBRAIC EXAMPLES

In this section we investigate the zeta function for some $\mathbf{Z}^d$-actions of algebraic origin. A general framework for studying such actions was first given by Kitchens and Schmidt [7]; see [15] for a systematic exposition.

Let $R_d = \mathbf{Z}[u_1^{\pm 1}, \ldots, u_d^{\pm 1}]$ be the ring of Laurent polynomials in d commuting variables. For $\mathbf{u} = (u_1, \ldots, u_d)$ and $\mathbf{n} = (n_1, \ldots, n_d) \in \mathbf{Z}^d$ put $\mathbf{u}^{\mathbf{n}} = u_1^{n_1} \ldots u_d^{n_d}$. A polynomial $f \in R_d$ then has the form

$$f(\mathbf{u}) = \sum_{\mathbf{n} \in \mathbf{Z}^d} c_f(\mathbf{n}) \mathbf{u}^{\mathbf{n}},$$

where $c_f(\mathbf{n}) \in \mathbf{Z}$ and all but finitely many of the $c_f(\mathbf{n})$ are zero. Let $\mathbf{T} = \mathbf{R}/\mathbf{Z}$ be the additive torus and $\mathbf{S} = \{e^{2\pi i t} : t \in \mathbf{T}\}$ be its multiplicative counterpart. For each $f \in R_d$ put

$$X_f = \left\{ x \in \mathbf{T}^{\mathbf{Z}^d} : \sum_{\mathbf{n} \in \mathbf{Z}^d} c_f(\mathbf{n}) x_{\mathbf{k}+\mathbf{n}} = 0 \text{ for all } \mathbf{k} \in \mathbf{Z}^d \right\}.$$

Then X_f is a compact group under coordinate-wise operations. The $\mathbf{Z}^d$ shift action α_f on X_f is defined by $(\alpha_f^{\mathbf{n}} x)_{\mathbf{k}} = x_{\mathbf{k}+\mathbf{n}}$. Each $\alpha_f^{\mathbf{n}}$ is clearly a continuous automorphism of X_f. In the framework of [7] the system (X_f, α_f) corresponds to the R_d-module $R_d/\langle f \rangle$.

Let us say that $f \in R_d$ is *expansive* if it has no zeros on $\mathbf{S}^d$. For example, $p(u, v) = 3 + u + v$ is expansive since $|3 + u + v| \geq 3 - |u| - |v| \geq 1$ provided $|u| = |v| = 1$. Similarly, $q(u, v) = 3 - u - v$ is also expansive. There is a natural notion of expansiveness for $\mathbf{Z}^d$-actions. A special case of work of Schmidt [14] is that α_f is expansive if and only if f is expansive.

We next turn to entropy (see Appendix A of [9] for definitions). For $0 \neq f \in R_d$ define the *Mahler measure* of f to be

$$M(f) = \exp\left(\int_{\mathbf{S}^d} \log |f| \right) = \exp\left(\int_0^\infty \cdots \int_0^\infty \log |f(e^{2\pi i t_1}, \ldots, e^{2\pi i t_d})| \, dt_1 \ldots dt_d \right).$$

By [9, Thm. 3.1], the topological entropy of α_f is given by $h(\alpha_f) = \log M(f)$. For instance, it is easy to verify that for $p(u, v) = 3 + u + v$ and $q(u, v) = 3 - u - v$ we have that $M(p) = M(q) = 3$, so that $h(\alpha_p) = h(\alpha_q) = \log 3$.

For $L \in \mathcal{L}_d$ put $\|L\| = \min\{\|\mathbf{n}\| : \mathbf{n} \in L \setminus \{0\}\}$. Then $\|L\| \to \infty$ means that "L goes to infinity in all directions." By [9, Thm. 7.1], if f is expansive then

(6.1) $$h(\alpha_f) = \lim_{\|L\| \to \infty} \frac{1}{[L]} \log p_L(\alpha_f).$$

However, there are other ways for $[L] \to \infty$ without $\|L\| \to \infty$, and the growth rate for periodic points can be strictly larger for these. For instance, in Example 3.3 we have that

$$\lim_{n \to \infty} \frac{1}{n} \log p_{\mathbf{Z} \oplus n\mathbf{Z}}(\alpha) = \log k,$$

while it is easy to verify that

$$\lim_{\|L\|\to\infty} \frac{1}{[L]} \log p_L(\alpha) = h(\alpha) = 0.$$

For the $\mathbf{Z}^d$-actions α_f we will explicitly determine the growth rate of periodic points in terms of the Mahler measure of f on compact subgroups of $\mathbf{S}^d$.

Let $\mathcal{C}_d$ denote the collection of all compact subgroups of $\mathbf{S}^d$, and $\mathcal{K}_d \subset \mathcal{C}_d$ be the subset of infinite compact subgroups of $\mathbf{S}^d$. For $K \in \mathcal{C}_d$ define the *Mahler measure of f over K* to be

$$M_K(f) = \int_K \log|f|,$$

where the integral is with respect to Haar measure on K. By definition, $M(f) = M_{\mathbf{S}^d}(f)$.

Theorem 6.1. Let $f \in R_d$ be an expansive polynomial and α_f be the associated $\mathbf{Z}^d$-action. Then the growth rate of periodic points for α_f is given by

$$(6.2) \qquad g(\alpha_f) = \sup_{K \in \mathcal{K}_d} \log M_K(f).$$

Furthermore, there is a $K_0 \in \mathcal{K}_d$ for which $g(\alpha_f) = \log M_{K_0}(f)$.

Before proving this result let us give two examples.

Example 6.2. (a) Let $d = 2$, $R_2 = \mathbf{Z}[u^{\pm 1}, v^{\pm 1}]$, and $p(u,v) = 3 + u + v$. As noted above, $p(u,v)$ is expansive. In order to compute $g(\alpha_p)$, we first observe that if $f(\mathbf{u}) = \sum_{\mathbf{n} \in \mathbf{Z}^d} c_f(\mathbf{n})\mathbf{u}^{\mathbf{n}} \in R_d$ has $|c_f(\mathbf{0})| = |f(\mathbf{0})| \geq \sum_{\mathbf{n} \neq \mathbf{0}} |c_f(\mathbf{n})|$, then Jensen's formula shows that $M(f) = |c_f(\mathbf{0})|$. Hence $M(p) = 3$, so that $h(\alpha_p) = \log 3$. Observe that

$$M_{\mathbf{S} \times \{1\}}(p) = M(4 + u) = 4 = M_{\{1\} \times \mathbf{S}}.$$

We show that the subgroups $\mathbf{S} \times \{1\}$ and $\{1\} \times \mathbf{S}$ give maximal Mahler measure for p over all subgroups in $\mathcal{K}_2$, so that by Theorem 6.1 we have that $g(\alpha_p) = \log 4$ and so $\zeta_{\alpha_p}(s)$ has radius of convergence $1/4$.

To see this, first note that any $K \in \mathcal{K}_2$ is either all of $\mathbf{S}^2$ or is 1-dimensional. In the latter case the connected component of the identity in K is the image of a rational line under the the exponential map $\mathbb{R}^2 \to \mathbf{S}^2$. Hence there is a vector $(a,b) \in \mathbf{Z}^2 \setminus \{0\}$ and a finite subgroup Ω of roots of unity in $\mathbf{S}^2$ such that

$$K = \bigcup_{t \in \mathbb{R}} \bigcup_{(\omega, \eta) \in \Omega} (\omega e^{2\pi i a t}, \eta e^{2\pi i b t}).$$

Hence

$$M_K(p) = \left(\prod_{(\omega, \eta) \in \Omega} M(3 + \omega u^a + \eta u^b) \right)^{1/|\Omega|}$$

Our observation above using Jensen's formula shows that

(6.3) $$M(3 + \omega u^a + \eta u^b) = \begin{cases} 3 & \text{if } a \neq 0 \text{ and } b \neq 0, \\ |3 + \omega| & \text{if } a = 0 \text{ and } b \neq 0, \\ |3 + \eta| & \text{if } a \neq 0 \text{ and } b = 0. \end{cases}$$

Thus $M_K(p)$ is the geometric mean of numbers all of which are less that or equal to 4, so that $M_K(p) \leq 4$ for all $K \in \mathcal{K}_2$. This proves that $g(\alpha_p) = \log 4$.

Hence $\zeta_{\alpha_p}(s)$ has radius of convergence $1/4$. It is conjectured that it has the circle of radius $1/3$ as natural boundary. Also, (6.3) shows that $\mathbb{S} \times \{1\}$ and $\{1\} \times \mathbb{S}$ are the only subgroups K in $\mathcal{K}_2$ for which $M_K(p) = 4$.

(b) Again let $d = 2$, but now use $q(u, v) = 3 - u - v$. As before, $q(u, v)$ is expansive and $M(q) = 3$. We claim that $M_K(q) \leq 3$ for all $K \in \mathcal{K}_2$. By (6.3), we need only consider subgroups of the form $\mathbb{S} \times \Omega_n$ or $\Omega_n \times \mathbb{S}$, where Ω_n is the group of nth roots of unity in $\mathbb{S}$. But for these,

$$M_{\mathbb{S} \times \Omega_n}(q) = M_{\Omega_n \times \mathbb{S}}(q) = \left(\prod_{k=0}^{n-1} M(3 - e^{2\pi i k/n} - u) \right)^{1/n}$$

$$= \prod_{k=0}^{n-1} |3 - e^{2\pi i k/n}|^{1/n} = (3^n - 1)^{1/n} \leq 3.$$

This proves that $g(\alpha_q) = \log 3$, so that $\zeta_{\alpha_q}(s)$ has radius of convergence $1/3$. It is conjectured that it has the circle of radius $1/3$ as natural boundary.

Proof of Theorem 6.1. Denote the annihilator of $L \in \mathcal{L}_d$ by $K_L = L^\perp \in \mathcal{C}_d$. By duality, $|K_L| = [L]$. Proposition 7.4 of [9] shows that

$$p_L(\alpha_f) = \prod_{\omega \in K_L} |f(\omega)|.$$

Hence

$$\frac{1}{[L]} \log p_L(\alpha_f) = \frac{1}{|K_L|} \sum_{\omega \in K_L} \log |f(\omega)| = \log M_{K_L}(f).$$

It is easy to see that $\mathcal{C}_d$ is compact with respect to the Hausdorff metric on compact subsets of $\mathbb{S}^d$. Since f is expansive, it follows that $\log |f|$ is continuous on $\mathbb{S}^d$, so that the function $K \mapsto \log M_K(f)$ is continuous on $\mathcal{C}_d$.

Suppose that $K \in \mathcal{K}_d$. Then there are lattices $L_n \in \mathcal{L}_d$ for which $K_{L_n} \to K$ in the Hausdorff metric. Since K is infinite, $[L_n] \to \infty$, so that

$$\log M_K(f) = \lim_{n \to \infty} \log M_{K_{L_n}}(f) = \lim_{n \to \infty} \frac{1}{[L_n]} \log p_{L_n}(\alpha_f) \leq g(\alpha_f).$$

This proves that $\sup_{K \in \mathcal{K}_d} \log M_K(f) \leq g(\alpha_f)$.

To prove the opposite inequality, choose $L_n \in \mathcal{K}_d$ with $[L_n] \to \infty$ and

$$\frac{1}{[L_n]} \log p_{L_n}(\alpha_f) \to g(\alpha_f).$$

Since $\mathcal{C}_d$ is compact, the subgroups $K_{L_n} \subseteq \mathbb{S}^d$ have a convergent subsequence $\{K_{L_{n_j}}\}$ converging to some $K_0 \in \mathcal{C}_d$, where clearly $K_0 \in \mathcal{K}_d$ since $[L_n] \to \infty$. Then

$$\log M_{K_0}(f) = \lim_{j\to\infty} \log M_{K_{L_{n_j}}}(f) = \lim_{n\to\infty} \frac{1}{[L_n]} \log p_{L_n}(\alpha_f) = g(\alpha_f).$$

This shows that $\sup_{K\in\mathcal{K}_d} \log M_K(f) = g(\alpha_f)$, and also that the supremum is attained at K_0. $\quad\Box$

7. QUESTIONS AND PROBLEMS

In the examples of $\mathbf{Z}^d$-actions α considered thus far, the zeta function $\zeta_\alpha(s)$ may have poles s with $|s| < \exp(-h(\alpha))$ but $\zeta_\alpha(s)$ has natural boundary $|s| = \exp(-h(\alpha))$. The quantity $h(\alpha)$ enters as the growth rate

$$(7.1) \qquad\qquad \rho = \limsup_{\|L\|\to\infty} \frac{1}{[L]} \log p_L(\alpha)$$

as L goes to infinity in all directions. This leads to the main conjecture regarding the analytic behavior of $\zeta_\alpha(s)$.

Conjecture 7.1. Let α be a $\mathbf{Z}^d$-action and ρ be defined by (7.1). Then $\zeta_\alpha(s)$ is meromorphic in $|s| < e^{-\rho}$ and has the circle of radius $e^{-\rho}$ as natural boundary.

When $d = 1$ the zeta function of a finitely determined system ϕ can be specified by a finite amount of data, namely the zeros and poles of the rational function $\zeta_\phi(s)$.

Problem 7.2. For "finitely determined" $\mathbf{Z}^d$-actions α such as shifts of finite type, is there a reasonable finite description of $\zeta_\alpha(s)$?

When $\alpha = \alpha_f$ is one of the algebraic examples described in §6, the numbers $M_K(f)$ for $K \in \mathcal{K}_d$ appear to play an essential role in describing $\zeta_\alpha(s)$. How much do these number tell about the polynomial $f \in R_d$?

Problem 7.3. Suppose that $f, g \in R_d$, and that $M_K(f) = M_K(g)$ for all $K \in \mathcal{K}_d$. What is the relationship between f and g? Must f and g differ by multiplicative factors that are monomials or generalized cyclotomic polynomials?

Problem 7.4. Compute explicitly the zeta function of the algebraic examples α_f discussed in §6.

Our zeta function $\mathbf{Z}^d$-actions can be generalized, as for a single transformation, to a "thermodynamic" setting by introducing a weight function $\theta\colon X \to (0, \infty)$, and defining

$$\zeta_{\alpha,\theta}(s) = \exp\left(\sum_{L\in\mathcal{L}_d} \left\{\sum_{x\in\text{fix}_L(\alpha)} \prod_{\mathbf{k}\in\mathbb{Z}^d/L} \theta(\alpha^{\mathbf{k}}x)\right\} \frac{s^{[L]}}{[L]}\right),$$

where $\mathrm{fix}_L(\alpha)$ is the set of points fixed by $\alpha^{\mathbf{n}}$ for all $\mathbf{n} \in L$. We obtain the usual zeta function ζ_α by using the weight function $\theta \equiv 1$. The following question was suggested to us by David Ruelle.

Problem 7.5. Compute explicitly the thermodynamic zeta function for the 2-dimensional Ising model, where α is the $\mathbf{Z}^2$ shift action on the space of configurations.

REFERENCES

1. M. Artin and B. Mazur, On periodic points, Annals Math. **81** (1965), 82–99.
2. R. Bowen and O. Lanford, Zeta functions of restrictions of the shift transformation, Proc. AMS Symp. Pure Math. **14** (1970), 43–49.
3. F. Carlson, Über Potenzreihen mit ganzzahligen Koeffizienten, Math. Zeit. **9** (1921), 1-13.
4. D. Fried, Finitely presented dynamical systems, Ergodic Th. & Dynam. Syst. **7** (1987), 489–507.
5. G. H. Hardy and S. Ramanujan, Asymptotic formulae for the distribution of integers of various types, Proc. Royal Soc. A **95**, 144–155.
6. G. H. Hardy and E. M. Wright, *An Introduction to the Theory of Numbers*, Fifth Ed., Oxford, 1988
7. B. Kitchens and K. Schmidt, Automorphisms of compact groups, Ergodic Th. & Dynam. Syst. **9** (1989), 691–735.
8. F. Ledrappier, Un champ markovien peut être d'entropie nulle et mélangeant, C. R. Acad. Sc. Paris Ser. A **287** (1978), 561–562.
9. D. Lind, K. Schmidt, and T. Ward, Mahler measure and entropy for commuting automorphisms of compact groups, Inventiones Math. **101** (1990), 593–629.
10. C. MacDuffie, *The Theory of Matrices*, Chelsea, New York, 1956.
11. A. Manning, Axiom A diffeomorphisms have rational zeta functions, Bull. London Math. Soc. **3** (1971), 215–220.
12. Bernd Mathiszik, Zetafunktionen und periodische Orbits in Gittermodellen, Dissertation, Martin-Luther-Universität Halle-Wittenberg, 1987.
13. G. Polya, Sur les séries entières à coefficients entiers, Proc. London Math. Soc. (2) **21** (1922), 22-38.
14. K. Schmidt, Automorphisms of compact groups and affine varieties, Proc. London Math. Soc. **61** (1990), 480–496.
15. K. Schmidt, *Dynamical Systems of Algebraic Origin*, preprint, University of Warwick, May, 1994.
16. T. Ward, Zeta functions for higher-dimensional actions, preprint, University of East Anglia, March, 1994.

DEPARTMENT OF MATHEMATICS, UNIVERSITY OF WASHINGTON, SEATTLE, WA 98195
E-mail address: lind@math.washington.edu

THE DYNAMICAL THEORY OF TILINGS AND QUASICRYSTALLOGRAPHY

E. ARTHUR ROBINSON, JR.

ABSTRACT. A tiling x of R^n is *almost periodic* if a copy of any patch in x occurs within a fixed distance from an arbitrary location in x. Periodic tilings are almost periodic, but aperiodic almost periodic tilings also exist; for example, the well known Penrose tilings have this property. This paper develops a generalized symmetry theory for almost periodic tilings which reduces in the periodic case to the classical theory of symmetry types. This approach to classification is based on a *dynamical theory of tilings*, which can be viewed as a continuous and multidimensional generalization of symbolic dynamics.

1. INTRODUCTION

The purpose of this this paper is to describe a natural generalization of the standard theory of symmetry types for periodic tilings to a larger class of tilings called *almost periodic tilings*. In particular, a tiling x of R^n is called *almost periodic* [1] if a copy of any patch which occurs in x re-occurs within a bounded distance from an arbitrary location in x. Periodic tilings are clearly almost periodic since any patch occurs periodically, but there are also many aperiodic examples of almost periodic tilings—the most famous being the *Penrose tilings*, discovered in around 1974 by R. Penrose [18].

Ordinary symmetry theory is based on the notion of a symmetry group—the group of all rigid motions leaving an object invariant. The symmetry groups of periodic tilings are characterized by the fact that they contains a lattice of translations as a subgroup. In contrast, for aperiodic tilings the symmetry group contains no translations, and it is typically empty. Thus a generalization of symmetry theory to almost periodic tilings must be based on different considerations. In this paper we describe a generalization of symmetry theory that uses ideas from dynamical systems theory, applied 'tiling dynamical systems'. The simplest example of a tiling dynamical system consist of a translation invariant set of tilings, equipped with a compact metric

---

Partially supported by NSF DMS–9007831.
[1]Such tilings are also sometimes called *repetitive* or said to satisfy the *local isomorphism* property. The term *almost periodic* is used in the same sense as in topological dynamics.

topology (analogous to the product topology), with R^n acting on it by translation. In the periodic case, this shift is a transitive action. More generally, almost periodic tilings generate minimal shifts. It turns out that minimality makes it possible to develop a 'symmetry theory' for almost periodic tilings[2] (called the theory of *quasisymmetry types*) which is closely analogous to the classical theory of symmetry types for periodic tilings (see [28]). In this paper, we establish a formalism for this theory and describe its relation to symmetry theory. We also discus some of the differences between the periodic and almost periodic cases. In particular, it turns out that there is no crystallographic restriction theorem (see [28]) for almost periodic tilings.

Part of the interest in a symmetry theory for almost periodic tilings comes from their connection with the theory of quasicrystals, a new form of solid matter discovered in 1985 by D. Schectman *et al* [27]. Although there is little agreement on the precise definition of a quasicrystal, roughly speaking a quasicrystal is a solid which unlike a crystal, is not made of a periodic array of atoms, but nevertheless has enough spatial order to produce sharp Bragg peaks in its diffraction pattern (see [13]). In particular, quasicrystals can have rotational 'symmetries' which are forbidden for ordinary crystals by the crystallographic restriction. For example, the Schectman quasicrystal has a diffraction pattern with five-fold rotational symmetry (reminiscent of a Penrose tiling). The possibility of an 'almost periodic crystal' based on a 3-dimensional Penrose-like tiling was first suggested by A. Mackay [14] several years before Schectman's discovery. Following Schectman's discovery, almost periodic tiling models for quasicrystals quickly became popular (see for example [13], [12]), although such models have always been somewhat controversial. We avoid this controversy here, noting only that a good symmetry theory for almost periodic tilings represents an important first step in any reasonable symmetry theory for 'not quite periodic' structures.

Tiling dynamical systems are interesting as a subject in themselves. We view them as symbolic dynamical systems which are *multidimensional* and have *continuous* 'time' (i.e., R^n acts instead of Z^n). In particular, they represent a promising new source of examples for dynamical systems theory. There are already several nontrivial applications of tilings to dynamical problems (see [25], [26], [16], [11]) and the concept of a tiling dynamical system provides a uniform foundation for all of these. In the first part of this paper we set up the basic framework for a topological theory of tiling dynamical systems (we briefly describe some connections to ergodic theory in later sections). However, it is not the intention of *this* paper to develop the general theory of tiling dynamical systems. Rather, our goal here is to exploit dynamical theory as a tool for classifying almost periodic tilings.

This paper is structured as follows: In Section 2 we set up a framework for tiling dynamical systems, and also define quasisymmetry groups, a kind of

---

[2]This idea was inspired by G. Mackey's theory of virtual groups, [15].

group which plays a role similar to the symmetry group in classical symmetry theory. In Section 3 we study the algebraic properties of quasisymmetry groups, proving that they satisfy an analogue of Bieberbach's second theorem. We define almost periodic tilings in Section 4, and we discuss their connection with minimal dynamical systems. In Section 5 we show how to associate a quasisymmetry group and a 'point group' to an almost periodic tiling, and show that in the periodic case, the quasicrystallographic point group is the same as the classical point group. However, as we show in Section 6, quasisymmetry groups alone do not provide a very strong symmetry theory. To compensate for this we define quasisymmetry types in Section 7 in terms of certain dynamical properties of the corresponding shift and in Section 8 we prove the central result: that quasisymmetry types reduce to symmetry types in the periodic case. In section Section 9 we discuss a geometric interpretation of quasisymmetry types, showing quasisymmetry theory to be similar to symmetry theory on a geometric level. The remaining sections discuss how quasisymmetry theory relates to various other properties of tilings and tiling dynamical systems: inflation, the spectrum and invariant measures. We conclude by discussing some additional examples.

2. TILING SPACES AND TILING DYNAMICS

A *tile in* R^n is a homeomorphic image of a closed ball in R^n. A *tiling* x in R^n is a collection of tiles with disjoint interiors. The *support* of a tiling x, denoted $supp(x)$, is the union of its tiles. We will mostly be interested in tilings x of R^n (i.e., $supp(x) = R^n$). Most of the examples that we consider also satisfy the following two additional hypotheses: (i) *protofiniteness*: Let p be a set of translationally incongruent tiles, called *prototiles*. A p-tiling is a tiling by translations of the tiles in p; tilings which are p-tilings for some p will be called *protofinite*. The set of all p-tilings of R^n will be denoted by X_p. We will always assume p is such that X_p is nonempty (see Section 13). (ii) *finite type*: A *patch* q in a tiling x is a finite set of tiles with simply connected support. Given prototiles p, let $f = \{q_1, \ldots q_k\}$ be a finite set of p-tiling patches with $card(q_i) \geq 2$. Let $X_{p,f}$ be the set of tiles $x \in X_p$ such that every pair of adjacent tiles τ_1, τ_2 in x belong to a patch q that is a translation of some $q_i \in f$. We call f a *finite type condition* and refer to any $x \in X_{p,f}$ as a *tiling of finite type*, (the term *Markov tiling* is used in Rudolph [26] for a special case).

There is a natural metrizable topology on sets of tilings of R^n in which two tilings are close if they nearly agree on a large cube around 0 (see [19], [25], [26], [21]). To describe a metric for this topology, let H denote the Hausdorff metric[3], i.e., for $\omega_1, \omega_2 \subseteq R^n$ compact

$$H(\omega_1, \omega_2) = max\Big\{inf\{\epsilon_1 : \omega_1 \subseteq N_{\epsilon_1}(\omega_2)\}, inf\{\epsilon_2 : \omega_2 \subseteq N_{\epsilon_2}(\omega_1)\}\Big\},$$

---

[3]My thanks to H. Furstenberg for suggesting the use of the Hausdorff metric in this context.

where $N_\epsilon(\omega) = \cup_{v \in \omega} B_\epsilon(v)$. Then for two tilings x and y of R^n let

(1) $$h(x, y) = \inf\{\epsilon : H(\partial_\epsilon(x), \partial_\epsilon(y)) \leq \epsilon\},$$

where $\partial_\epsilon(x) = \partial(C_{1/\epsilon}) \cup \cup_{\tau \in x} \left(\partial(\tau) \cap C_{1/\epsilon}\right)$ with $C_t = \{(v_1, \ldots, v_n) : |v_i| \leq t\}$. One can easily verify that h is a metric.

Proposition 2.1. For any set p of prototiles, the set X_p is compact in the metric h.

This result is stated in Rudolph [26]. A proof of this is given in [21].

A compact translation invariant set X of tilings will be called a *shift space* or 'shift'. It is easy to see that given p and f, both X_p and $X_{p,f}$ are shifts. For a tiling x of R^n, let $O(x) = \{T^t x : t \in R^n\}$ denote the *orbit* of x with respect to *translation*, and let $\overline{O(x)}$ denote the *orbit closure*. We will always assume that $\overline{O(x)}$ *is compact*. In this case we say the tiling x is *protocompact*. It follows that $\overline{O(x)}$ is a shift.

The group $M(n)$ of *rigid motions* of R^n is a semidirect product of the group of translations and the orthogonal group $O(n)$; for any $S \in M(n)$ there exists unique $t \in R^n$ and $U \in O(n)$ with $S = T^t U$, where T^t denotes translation by $t \in R^n$. Similarly, the affine motions of R^n, denoted $A(n)$, consist of (unique) products of translations and invertible linear transformations. The natural action of $M(n)$ on R^n induces an action of $M(n)$ on the tilings of R^n. It is easy to see that this action is continuous.

Definition 2.2. A *quasisymmetry group* is a closed subgroup G of $M(n)$ with $R^n \subseteq G$.

By a *dynamical system* we mean a continuous left action of a locally compact group on a compact metric space. For a quasisymmetry group G, and a G-invariant shift space of tilings X, the natural action L of G on X defines a dynamical system. We call such a dynamical system a *tiling dynamical system*. Here are two important special cases: (i) The *smallest* quasisymmetry group leaving X invariant is R^n. The action of R^n on a shift X by translation will be called the *shift* action (usually denoted T). (ii) For $x \in X$, let $G_{X,x} = \{R \in M(n) : Rx \in X\}$. Then

$$G_X = \bigcap_{x \in X} G_{X,x},$$

is the largest quasisymmetry group leaving X invariant. We call G_X the *the quasisymmetry group of* X and refer to the action of G_X on X (usually denoted Q) as the *quasi-shift* action.

3. The algebraic properties of quasisymmetry groups

In the previous section, a quasisymmetry group was defined to be a closed subgroup of $M(n)$ containing R^n. In this section we show that the algebraic properties of quasisymmetry groups are similar to the properties of the symmetry groups of periodic tilings (i.e., *space groups* or *crystallographic groups*, see [28]). The results in this section are purely algebraic.

Recall that the *symmetry group* of a tiling x of R^n is given by $G'_x = \{R \in M(n) : Rx = x\}$. A tiling x is periodic if the *translation group* $Z_x = G'_x \cap R^n$ is isomorphic to Z^n. In this case, Z_x is normal and maximal abelian in G'_x. The following result is known as *Bieberbach's second theorem*.

Theorem 3.1. (Bieberbach [3], [4]) Let G'_x and G'_y be the symmetry groups for the periodic tilings x and y and suppose $\varphi : G'_x \to G'_y$ is a group isomorphism. Then there exists $S \in A(n)$ such that $\varphi(R) = SRS^{-1}$.

Our main result in this section is the following:

Theorem 3.2. Let G and G' be quasisymmetry groups and suppose $\varphi : G \to G'$ is a topological group isomorphism. Then there exists $S \in A(n)$ such that $\varphi(R) = SRS^{-1}$.

Lemma 3.3. Any quasisymmetry group G is a semidirect product of R^n and a closed subgroup H of $O(n)$. In particular, G is generated by $H \cup R^n$, and R^n is normal in G.

Proof: If $S \in G$ then $S = TU$ uniquely for $T \in R^n$ and $U \in O(n)$. Since $R^n \subseteq G$, $T \in G$ and it follows that $U \in G$, so that G is a semidirect product. Now $ST^tS^{-1} = T^{Ut}$, so that R^n is normal. $\square$

We will identify $H = G \cap O(n)$ and G/R^n, and refer to H as the *point group* of G. Now we state our analogue of Bieberbach's second theorem for quasisymmetry groups.

Lemma 3.4. For any $U \in O(n)$ and $t \in R^n$ there exists $v \in R^n$ such that $T^v(T^tU)T^{-v} = T^{t'}U$, with $Ut' = t'$.

Proof: We have $T^vT^tUT^{-v} = T^{v-Uv+t}U$, so that $t' = v - Uv + t$. Thus, to show $Ut' = t'$, it suffices to find v such that $-(U - I)^2v = (U - I)t$. Let W be the eigenspace for U corresponding to the eigenvalue 1, and let $W^\perp$ be the ortho-complement of W, so that the decomposition $R^n = W \oplus W^\perp$ is orthogonal and U–invariant. This implies $(U - I)W^\perp \subseteq W^\perp$, and since $(U - I)|_{W^\perp}$ is nonsingular, it follows that $(U - I)W^\perp = W^\perp$.

To find v, we first write $t = t_0 + t^\perp$, where $t_0 \in W$ and $t^\perp \in W^\perp$, and let $v = -((U - I)|_{W^\perp})^{-1}t^\perp$. Since $v \in W^\perp$, and since $(U - I)t_0 = 0$, it follows that

$$-(U - I)^2 v = -(U - I)((U - I)|_{W^\perp})v$$
$$= (U - I)t^\perp$$
$$= (U - I)t.$$

$\square$

Lemma 3.5. If G is a quasisymmetry group, then $R^n \subseteq G$ is the unique maximal normal abelian subgroup of G.

Proof of Lemma 3.5: Let $K \subseteq G$ be a normal abelian subgroup and suppose $T^t U \in K$. We will show $U = I$, so that $K \subseteq R^n$. Using Lemmas 3.4 and 3.3 we conjugate $T^t U$ by a translation and assume without loss of generality that $Ut = t$. Since K is normal, Lemma 3.3 implies $T^v T^t U T^{-v} = T^{v-Uv+t}U \in K$, and K abelian implies

$$(2) \qquad\qquad I = [T^t U, T^{v-Uv+t}U] = T^{-(U-I)^2 v + Ut}.$$

Since $Ut = t$, (2) implies $(U - I)^2 v = 0$. Letting $W = ker(U - I)$, write $v = v_0 + v^\perp$ where $v_0 \in W$ and $v^\perp \in W^\perp$. Then $0 = (U - I)^2 v = (U - I)^2(v_0 + v^\perp) = (U - I)^2 v^\perp$, and since $U - I$ is nonsingular on $W^\perp$, we have $v^\perp = 0$. Hence $Uv = v$ and $U = I$. $\square$

Lemma 3.6. Let H be a subgroup of $M(n)$ and suppose $w : H \to R^n$. Then $P = \{T^{w(U)}U : U \in H\}$ is a subgroup of $M(n)$ if and only if w satisfies the *cocycle equation*

$$(3) \qquad\qquad w(UV) = w(U) + Uw(V),$$

for all $U, V \in H$.

This follows from a direct computation.

Lemma 3.7. Suppose P is a compact subgroup of $M(n)$. Then there exists $v \in R^n$ such that for all $T^t U \in P$ with $U \in O(n)$, $T^v(T^t U)T^{-v} = U$. In particular, if $T^t U \in P$ then $t = v - Uv$.

Proof: First we show that if $T^{t_1}U, T^{t_2}U \in P$ then $t_1 = t_2$. Since $(T^{t_2}U)^{-1} = T^{-U^{-1}t_2}U^{-1}$, it follows $T^{t_1-t_2} = (T^{t_1}U)(T^{t_2}U)^{-1}$, which implies $T^{t_1-t_2} \in P$. It follows that $t_1 = t_2$, since otherwise $T^{t_1-t_2}$ would generate an infinite discrete subgroup of P.

Define a homomorphism $\pi : M(n) \to O(n)$ by $\pi(T^t U) = U$, and define $\pi' : M(n) \to R^n$ by $\pi'(T^t U) = t$. Let $H = \pi(P)$. By the first step, $\pi|_P : P \to H$ is an isomorphism so that H is a compact. Thus there exists $w : H \to R^n$ continuous such that $T^{w(U)}U = \pi^{-1}(U)$, i.e. $w(U) = (\pi' \circ \pi^{-1})(U)$. Since P is a group, w satisfies (3).

Define

$$(4) \qquad v = -\int_H w(V)\,d\mu(V),$$

where μ denotes normalized Haar measure. It follows that

$$(5) \qquad \int_H w(UV)\,d\mu(V) = \int_H w(V)\,d\mu(V) = -v,$$

and

$$(6) \qquad \int_H (w(U) + Uw(V))\,d\mu(V) = w(U) - Uv,$$

so that by (4), (5) and (6) we obtain $-v = w(U) - Uv$. This implies $w(U) = Uv - v$ (i.e., w is a *coboundary*) and

$$T^v(T^{w(U)}U)T^{-v} = T^{w(U)+Uv-v}U = U.$$

$\square$

Lemma 3.8. Suppose $\varphi : G \to G'$ is a topological group isomorphism of quasisymmetry groups such that there exists $\gamma : G \to R^n$ with

$$(7) \qquad \varphi(T^sU) = T^{\gamma(T^sU)}U,$$

and such that

$$(8) \qquad \varphi(T^s) = T^s.$$

Then there exists $v \in R^n$ such that $\gamma(T^sU) = s + v - Uv$. In particular,

$$(9) \qquad \varphi(T^sU) = T^v(T^sU)T^{-v}.$$

Proof: Let $H = G \cap O(n)$, and define $P = \{T^{\gamma(T^sU)-s}U : U \in H, s \in R^n\}$. Note that $\gamma = \pi'\circ\varphi$, and $\pi'(T^{s_1}T^{s_2}U) = s_1+s_2$. Since $\varphi(U) = \varphi(T^{-s})\varphi(T^sU)$, (8) implies $\gamma(U) = -s + \gamma(T^sU)$, and hence $P = \{T^{\gamma(U)}U : U \in H\}$. Since γ is continuous, P is compact, and

$$\begin{aligned}
\gamma(UV) &= \pi'(\varphi(U)\varphi(V)) \\
&= \pi'(T^{\gamma(U)}UT^{\gamma(V)}) \\
&= \pi'(T^{\gamma(U)+U\gamma(V)}UV) \\
&= \gamma(U) + U\gamma(V),
\end{aligned}$$

so that γ satisfies (3). It follows that P is a compact subgroup of $M(n)$. Thus $\gamma(T^sU) - s = v - Uv$ by Lemma 3.7, and (9) follows using (7). $\square$

Proof of Theorem 3.2: Lemma 3.5 implies $\varphi(R^n) = R^n$ and thus $\varphi|_{R^n}$ is a topological group isomorphism. It follows that there exists $G \in GL(n)$ such that

$$(10) \qquad \varphi(T^t) = GT^tG^{-1} = T^{Gt}.$$

For $T \in R^n$ and $U \in O(n)$, we write $\varphi(TU) = \varphi_1(TU)\varphi_2(TU)$, where $\varphi_1(TU) \in R^n$ and $\varphi_2(TU) \in O(n)$. Then $\varphi_2(TU) = (\pi \circ \varphi)(TU)$ and $\varphi_1(TU) = \varphi(TU)\varphi_2(TU)^{-1}$. Since $\pi : G \to O(n)$ is a homomorphism, (10) implies that

$$
\begin{aligned}
\varphi_2(T^{t_1}U)\varphi_2(T^{t_2}U)^{-1} &= (\pi \circ \varphi)(T^{t_1}U) \cdot ((\pi \circ \varphi)(T^{t_2}U))^{-1} \\
&= (\pi \circ \varphi)((T^{t_1}U)(T^{t_2}U)^{-1}) \\
&= \pi(\varphi(T^{t_1-t_2})) \\
&= \pi(T^{G(t_1-t_2)}) = I,
\end{aligned}
$$

and thus $\varphi_2(TU) = \varphi_2(U)$ for all $T \in R^n$. By (10), $\varphi(T^{Uv}) = T^{GUv}$. Hence if we apply φ to the relation $(T^tU)T^v(T^tU)^{-1} = T^{Uv}$, we obtain

$$
\begin{aligned}
T^{GUv} &= \varphi((T^tU)T^v(T^tU)^{-1}) \\
&= T^{Gt}\varphi(U)T^{Gv}\varphi(U)^{-1}T^{-Gt} \\
&= T^{Gt}\varphi_1(U)\varphi_2(U)T^{Gv}\varphi(U)_2^{-1}\varphi_1(U)^{-1}T^{-Gt} \\
&= T^{Gt}T^{(\varphi_2(U)G)v}T^{-Gt} \\
(11) \qquad\qquad &= T^{(\varphi_2(U)G)v}
\end{aligned}
$$

Replacing v with $G^{-1}v$ in (11) and applying π', it follows that $\varphi_2(U) = GUG^{-1}$, which implies $GUG^{-1} \in O(n)$. Moreover $\varphi(T^t) = \varphi_1(T^t)\varphi_2(T^t) = \varphi_1(t)$, which implies

$$
(12) \qquad\qquad \varphi_1(T^t) = T^{Gt}.
$$

Now let us define $\Phi : G \to M(n)$ by

$$
(13) \qquad\qquad \Phi(T^tU) = GT^tUG^{-1} = T^{Gt}GUG^{-1}.
$$

Putting $G'' = \Phi(G)$, we observe that G'' is a quasisymmetry group since Φ is an isomorphism with $\Phi(T^t) = T^t$. Let $\varphi' : G'' \to G'$ be defined by $\varphi' = \varphi \circ \Phi^{-1}$. Then $\varphi'(T^sU) = \varphi(T^{G^{-1}s}G^{-1}UG) = \varphi_1(T^{G^{-1}s}G^{-1}UG)U$, and since $im(\varphi_1) \subseteq R^n$, it follows that $\varphi'(T^sU) = T^{\gamma(T^sU)}U$, where $\gamma(T^sU) = \varphi_1(T^{G^{-1}s}G^{-1}UG)$. Thus γ is continuous and φ' satisfies (7). Now by (12), $\gamma(T^s) = \pi'(\varphi_1(T^{G^{-1}s})) = s$, so that (7) implies(8). By Lemma 3.8 there exists $v \in R^n$ such that $\gamma(T^sU) = s + v - Uv$. This implies $\varphi(T^sU) = T^vT^sUT^{-v}$, and since $\varphi = \varphi' \circ \Phi$, (13) implies

$$
\varphi(T^tU) = (\varphi' \circ \Phi)(T^tU) = (T^vG)(T^tU)(T^vG)^{-1}.
$$

$\square$

4. ALMOST PERIODIC TILINGS

Having studied the algebraic properties of symmetry groups, we now proceed to our main goal: to study the symmetry properties of individual almost periodic tilings x. The basic set-up will be as follows. We let $G_x = G_{\overline{O(x)}}$ and then study the dynamical properties of the quasi-shift action of G_x on $\overline{O(x)}$.

Recall that a dynamical system (i.e. a group action) is called *transitive* if has a single orbit. A dynamical system is called *minimal* if every orbit is dense. It is called *topologically transitive* if some point has a dense orbit. The shift on $\overline{O(x)}$ is always topologically transitive. It is easy to see that a tiling x is a periodic tiling if and only if the shift on $\overline{O(x)}$ is transitive, and in this case, $O(x) = \overline{O(x)}$ (i.e., $O(x)$ is closed), and $O(x)$ is isometric to an n–dimensional torus. Note also that any periodic tiling is of finite type. This discussion illustrates the fundamental relation between periodicity (i.e., periodic points) and transitive actions. The following well known result of W. Gottschalk shows that almost periodicity plays a similar role in the theory of minimal actions.

For $r > 0$, a set $Z \subset R^n$ is called r–dense if for any $t \in R^n$, $B_r(t) \cap Z \neq \phi$. A set Z is called *relatively dense* if it is r–dense for some $r > 0$. Suppose T is an action of R^n on a compact metric space X. A point $x \in X$ is called an *almost periodic* point for T if for any $U \subseteq X$ open, the set $\{t \in R^n : T^t x \in U\}$ is relatively dense.

Theorem 4.1. (Gottschalk [7]) For an action T of R^n on a compact metric space X the following are equivalent: (i) $x \in X$ is almost periodic, (ii) the restriction of T to $\overline{O(x)}$ is minimal, (iii) every $y \in \overline{O(x)}$ is almost periodic, and (iv) $\overline{O(y)} = \overline{O(x)}$ for $y \in \overline{O(x)}$.

We call a tiling x of R^n *almost periodic* if it is an almost periodic point for the shift on $\overline{O(x)}$. A tiling x will be called *rigidly almost periodic* if for any patch q which occurs in x, there exists $r > 0$ such that for any $v \in R^n$, the tiling x contains a translation of q inside $B_r(v)$. Given a tiling x, a large patch in x centered at $0 \in R^n$ defines a neighborhood of x; it consists of the tilings which have nearly the same patch around 0 that x does. If x is rigidly almost periodic, this patch repeats relatively densely throughout x. It follows that rigid almost periodicity implies almost periodicity. For tilings of finite type, one can show that the converse is also true (see [21]).

Two tilings x and y are said to be of the same *species*[4] if $\overline{O(x)} = \overline{O(y)}$. By (iv) of Lemma 4.1, if x is almost periodic, this is equivalent to $y \in \overline{O(x)}$. Notice that if x and y are protofinite and made from different prototiles then they automatically belong to different species. However, tilings by the same prototiles can also belong to different species.

One can show that two finite type almost periodic tilings x and y belong to the same species if and only if every patch q occurring in x also occurs in

---

[4]This terminology comes from [19].

y (such x and y are sometimes said to be *locally isomorphic*, see [21]). The following well known example will illustrate some of these ideas.

(a)   (b)

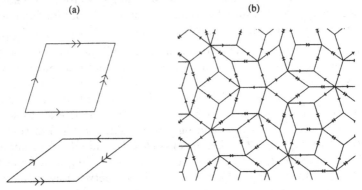

Figure 1: (a) Two marked Penrose prototiles. (b) Part of a Penrose tiling of the plane.

Example 1: (Rhombic Penrose tilings) Consider the two rhombic tiles (see Figure 1a) marked with arrows. Let p be the set of 20 marked prototiles obtained by rotating the tiles in Figure 1a by multiples of $2\pi/10$. These prototiles are laid edge-to-edge, subject to the following *matching rule*: the arrows on adjacent edges must match. This enforces a finite type condition. One can show that it is possible to tile the entire plane in this way [18] (see Figure 1b). The resulting tiles are called *rhombic Penrose tilings*.

Once the tilings are constructed we ignore the markings (so that p really consists of 10 prototiles). The finite type condition f can be taken to be the set of all 'pictures' of the vertex configurations that occur. It is well known that every rhombic Penrose tiling is aperiodic and almost periodic, and that any two rhombic Penrose tilings are of the same species (see [8]). This implies the following result.

Corollary 4.2. The Penrose shift $X_{p,f}$ is a minimal aperiodic shift of finite type.

This result might seem a little surprising from the point of view of 1-dimensional symbolic dynamics (where any minimal subshift of finite type consists of a single periodic orbit), but turns out to be a fairly common multidimensional phenomenon. The earliest examples of this phenomenon consist of discrete 2-dimensional subshifts of finite type in the form of aperiodic 'Wang' tilings of the plane (see [8]), due to M. Berger [2]. Berger was studying the 'Tiling Problem': *Given a set p of prototiles, determine whether $X_p \neq \phi$.* He showed that in general, the tiling problem is undecidable, [2]. This fact is closely related to the existence of aperiodic minimal shifts of finite type (see [11]). Using a construction similar to Berger's, S. Mozes recently showed that, up to an almost 1:1 extension, most 2-dimensional discrete *substitution* dynamical systems are actually subshifts of finite type. Since a typical 2-dimensional substitution system is minimal and aperiodic, this result provides an abundant source of examples of minimal aperiodic 2-dimensional discrete subshifts of finite type. Mozes also showed that *weak mixing* is possible for such examples [16].

Comment There is a clear qualitative similarity between discrete substitution dynamical systems and tilings which satisfy an inflation (see Section 10). An interesting open question is whether there is tiling version of the theorem of Mozes. Specifically, is every shift of tilings satisfying an inflation rule 'nearly' (i.e., up to almost 1:1 extension) a shift of finite type?

5. THE QUASISYMMETRY GROUP AND POINT GROUP OF AN ALMOST PERIODIC TILING

Recall that G_x denotes the quasisymmetry group of the shift $\overline{O(x)}$. Let Q_x denote the quasi-shift action of G_x on $\overline{O(x)}$, and let T_x denote the shift action of R^n on $\overline{O(x)}$. The next result shows that for almost periodic tilings the group G_x depends only on the species of x. It should be noted that the proposition is not generally true for tilings which are not almost periodic since it is essentially a corollary of Theorem 4.1.

Proposition 5.1. If x is an almost periodic tiling of R^n then $G_y = G_x$ for any $y \in \overline{O(x)}$.

First we note that if $y \in \overline{O(x)}$ and $S \in A(n)$ then $Sy \in \overline{O(Sx)}$. In this case, the map $S : \overline{O(x)} \to \overline{O(Sx)}$ is a homeomorphism, and $\overline{O(Sx)} = S(\overline{O(x)})$. The proposition follows directly from the next lemma.

Lemma 5.2. Let x be an almost periodic tiling of R^n and suppose that $Rx \in \overline{O(x)}$ for some $R \in M(n)$. If $y \in \overline{O(x)}$ then $Ry \in \overline{O(x)}$.

Proof: By Lemma 4.1 the shift T_x on $\overline{O(x)}$ is minimal, and since $Rx \in \overline{O(x)}$, it follows that $\overline{O(Rx)} = \overline{O(x)}$. Since $y \in \overline{O(x)}$, this implies $Ry \in \overline{O(Rx)}$. □

We define the *point group*[5] of an almost periodic tiling x, by $H_x = G_x \cap O(n)$. Note that this is the point group (in the sense of Section 3) of the quasisymmetry group G_x. By Proposition 5.1, $H_y = H_x$ for all $y \in \overline{O(x)}$.

For an (unmarked) rhombic Penrose tiling x the point group is generated by the reflection F through the horizontal axis and rotation R_θ by $\theta = 2\pi/10$, i.e., $H_x = D_{10}$, the dihedral group of order 20. The corresponding quasisymmerty group is generated by D_{10} and R^2. Note that by the *crystallographic restriction* D_{10} cannot be the point group of any periodic tiling of R^2 This was the observation that started the theory of quasicrystals.

---

[5]This definition is implicit in Niizeki [17].

Lemma 5.3. For a periodic tiling x, the classical point group H'_x and the quasicrystallographic point group H_x are isomorphic.

Proof: Define $\psi : H'_x \to H_x$ by $\psi(RZ_x) = RR^n$. Then Z_x is a subgroup of R^n and of G'_x. It follows that ψ is a well defined surjection, and it suffices to show ψ is 1:1. If $RZ_x \in ker(\psi)$, then $\psi(RZ_x) = RR^n$, and it follows that $RR^n = R^n$, so that $R \in R^n$. But since $RZ_x \in H'_x$, it follows that $R \in G'_x$, so that $R \in G'_x \cap R^n = Z_x$. Thus $ker(\psi)$ is trivial. $\square$

Corollary 5.4. If x is a periodic tiling of R^n, then G_x is generated by $G'_x \cup R^n$. In particular, for any $R \in G_x$ there exists $Q \in G'_x$ and $T \in R^n$ such that $R = TQ$.

6. THE INSUFFICIENCY OF A PURELY ALGEBRAIC THEORY

The main invariant in the symmetry theory of periodic tilings is the isomorphism type of the symmetry group. One might expect quasisymmetry groups to play the same role for almost periodic tilings. However, this turns out not to be the case. We show in this section that it is not even possible to recover the symmetry type of a periodic tiling from its quasisymmetry group.

Proposition 6.1. There exist two periodic tilings x and y of R^2 with different (non-isomorphic) symmetry groups G'_x and G'_y such that the quasisymmetry groups G_x and G_y are isomorphic.

Proof: Consider the two periodic tilings x and y of R^2 in Figure 2, which have $G'_x = pm$ and $G'_y = pg$ (see [8] or [28]), so that in particular, the symmetry groups of x and y are not isomorphic.

(a)   (b)

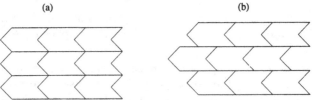

Figure 2: Parts of two periodic tilings with symmetry groups (a) *pm*, and (b) *pg*.

On the other hand, it is easy to see that the quasisymmetry groups G_x and G_y are both isomorphic to the semidirect product of R^2 and $Z/2$. $\square$

The proof of Proposition 6.1 shows that the distinction between pure reflections and glide reflections is lost in quasisymmetry groups. This is because quasisymmetry groups are always semidirect products. In the next section we show how to recover the lost information in terms of dynamics.

QUASICRYSTALLOGRAPHY

7. QUASISYMMETRY TYPES FOR ALMOST PERIODIC TILINGS

Two actions L_1 and L_2 of a locally compact group G on compact metric spaces X_1 and X_2 are said to be *topologically conjugate*[6] if there exists a homeomorphism $\eta : X_1 \to X_2$ such that for all $S \in G$ and $x \in X_1$, $\eta(L_1^S x) = L_2^S \eta(x)$. Topological conjugacy is the standard notion of isomorphism in topological dynamics, but it is not well suited to symmetry theory since it is not invariant under rotation or rescaling. Thus we make the following modification. Let L be a G action on X and let $\varphi : G_1 \to G$ be a topological group homomorphism. Define a G_1 action $L \circ \varphi$ by $(L \circ \varphi)^S x = L^{\varphi(S)} x$. If L_1 and L_2 are continuous actions of isomorphic locally compact groups G_1 and G_2, we say L_1 and L_2 are *rescale topologically conjugate* if there exists a (topological group) isomorphism $\varphi : G_1 \to G_2$ such that L_1 and $L_2 \circ \varphi$ are topologically conjugate G_1 actions.

Definition 7.1. Let x and y be almost periodic tilings of R^n. We say x and y *have the same quasisymmetry type* if (i) G_x is isomorphic to G_y and (ii) the corresponding quasi-shift actions Q_x and Q_y on $\overline{O(x)}$ and $\overline{O(y)}$ are rescale topologically conjugate (so that in particular, $\overline{O(x)}$ and $\overline{O(y)}$ are homeomorphic).

Note that a necessary condition for x and y to have the same quasisymmetry type is for G_x and G_y to be isomorphic. However, by Proposition 6.1 this is not sufficient. Clearly a sufficient condition for having the same quasisymmetry type is to have the same species. The next example shows that this is not necessary.

Example 2: (*Kites-and-darts Penrose tilings*) Consider the two marked prototiles shown in Figure 3a, together with their rotations by multiples of $2\pi/10$.

---

[6]Analogous theories of quasisymmetry types can be based *almost topological conjugacy* (see [1]), and *metric isomorphism* (when there is an invariant measure; see Section 12). It should be noted that for both of these alternatives, Theorem 8.1 (below) still holds.

(a)   (b)

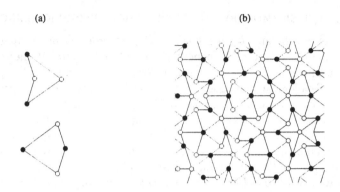

Figure 3: (a) Penrose kites-and-darts prototiles. (b) Part of a kites and darts Penrose tiling.

These prototiles are laid edge-to-edge, subject to the matching condition that the black and white dots on adjacent tiles match, (see Figure 3b). The resulting tilings are called *kites-and-darts Penrose tilings* ([18], see also [8]).

Like the rhombic Penrose tilings, every kites–and–darts Penrose tiling is aperiodic and almost periodic. Moreover, both kinds of Penrose tilings have the same quasisymmetry group. However, because they are made of different prototiles, they belong to different species.

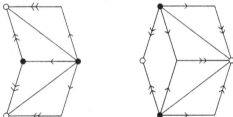

Figure 4: Rhombic tiles to kites-and-darts and back.

There is a standard procedure for converting between the two kinds of Penrose tilings (see [8], [5]). Let $X_{p,f}$ denote the finite type shift of all rhombic Penrose tilings, and let $X_{p',f'}$ denote the finite type shift of all kites–and–darts Penrose tilings. Given $x \in X_{p,f}$, a line is drawn along the major axis of each 'fat' rhombic prototile, as shown in Figure 4. Then the vertices of the resulting kites and darts are colored black and white. Finally, all the lines which do not connect black and white dots (along with any remaining arrows) are erased. We denote this conversion operation by $x' = \eta(x)$.

Proposition 7.2. The two kinds of Penrose tilings (rhombic and kites-and-darts [7]) have the same quasisymmetry type (even though they belong to

---

[7]The same result holds for the third well known kind of Penrose tiling (based on stars, half-stars, rhombuses, and pentagons; see [18], [8]).

different species).

This follows from the fact that φ is reversible, local in its effect, and conjugates the quasi-shift actions Q_x and $Q_{x'}$.

8. QUASISYMMETRY TYPES GENERALIZE SYMMETRY TYPES

Theorem 8.1. If x and y are periodic tilings of R^n then they have the same symmetry type if and only if they have the same quasisymmetry type. In particular, for periodic tilings, symmetry types and quasisymmetry types coincide.

Proof: Suppose there exists a homeomorphism $\eta : \overline{O(x)} \to \overline{O(y)}$ and a topological group isomorphism $\varphi : G_x \to G_y$ such that

$$(14) \qquad \eta \circ Q_x^R = Q_y^{\varphi(R)} \circ \eta \text{ for all } R \in G_x.$$

Since x is periodic, G'_x is a subgroup of G_x and so $\varphi(G'_x)$ is a subgroup of G_y. Suppose $R \in G'_x$. By (14), $Q_y^{\varphi(R)}\eta(x) = \eta(Q_x^R x) = \eta(Rx) = \eta(x)$, so that $\varphi(R) \in G'_{\eta(x)}$. We apply φ^{-1} to (14) to obtain $\eta \circ Q_x^{\varphi^{-1}(R')} = Q_x^{R'} \circ \eta$ for all $R' \in G'_y$. Reversing the steps, it follows that for any $R \in G'_{\eta(x)}$ that $\varphi^{-1}(R) \in G'_x$. Thus G'_x and $G_{\eta(x)}$ are isomorphic. Now, since y is periodic, it follows from Section 4 that $O(y)$ is closed, so that $\overline{O(y)} = O(y) = \{Ty : T \in R^n\}$. Thus $\eta(x) = Ty$ for some $T \in R^n$, and this implies that $T^{-1}G'_yT = G'_{\eta(x)}$. Hence, G'_x and G'_y are isomorphic.

Conversely, suppose x and y are periodic tilings of R^n with G'_x and G'_y isomorphic. By Theorem 3.1 there exists an affine transformation $S \in A(n)$ such that the map $\varphi : G'_x \to G'_y$ defined $\varphi(R) = SRS^{-1}$ is a topological group isomorphism. Write $S = TG$ for $T \in R^n$ and $G \in GL(n)$. Then define $\eta : \overline{O(x)} \to \overline{O(y)}$ by $\eta(T_x^t x) = T_y^{Gt} y$. It follows from Lemma 3.3 that to show that η is a well defined homeomorphism, it suffices to show that $Gt \in Z_y$ whenever $t \in Z_x$. Letting $t \in Z_x$, so that $T_x^t x = x$, and using the fact that $\varphi(T_x^t) \in G'_y$, we obtain $T_y^{Gt} y = ST_y^t S^{-1} y = \varphi(T_y^t)y = y$.

The equation $\varphi(R) = SRS^{-1}$ defines a continuous 1:1 homomorphism $\varphi : G_x \to M(n)$. Using Corollary 5.4, we write $R = TV$, where $T \in R^n$ and $V \in G'_x$. Then $\varphi(R) = \varphi(TV) = \varphi(T)\varphi(V)$, and since $\varphi(T) \in R^n \subseteq G_y$, and $\varphi(V) \in G'_y \subseteq G_y$, we have $\varphi(R) \in G_y$. Using the same argument for φ^{-1}, it follows that $\varphi : G_x \to G_y$ is a topological group isomorphism. Finally, to prove (14), we let $x' = T^t x$, and compute

$$
\begin{aligned}
(Q_y^\varphi)^R(x') &= \varphi(R)\eta(T^t x) \\
&= SRS^{-1}T^{Gt}x \\
&= SRS^{-1}ST^t S^{-1}x \\
&= SRT^t S^{-1}x \\
&= \eta(RT^t x) \\
&= \eta(Q_x^R(x'))
\end{aligned}
$$

$\square$

9. THE GEOMETRY OF QUASISYMMETRY TYPES

Let $R \in M(n)$, and define $N_R = \{S \in A(n) : SRS^{-1} \in M(n)\}$. Clearly N_R is a closed subset of $A(n)$ with $M(n) \subseteq N_R$. Given $\lambda > 0$, let $M_\lambda v = \lambda v$, let $S(n) = \{M_\lambda : \lambda > 0\}$, and note that $S(n) \subseteq N_R$ for any R. Let $C(n)$ denote the closed subgroup of $A(n)$ generated by $S(n) \cup M(n)$.

Lemma 9.1. Let x be an almost periodic tiling of R^n, let $R \in G_x$, and let $S \in N_R$. Then $SRS^{-1} \in G_{Sx}$.

Proof: By definition, $SRS^{-1} \in M(n)$, and $(SRS^{-1})(Sx) = S(Rx)$. Since $R \in G_x$, it follows that $Rx \in \overline{O(x)}$. Thus $S(Rx) \in \overline{O(Sx)}$, and $SRS^{-1} \in G_{Sx}$. $\square$

For a quasisymmetry group G let $N_G = \cap_{R \in G} N_R$, a closed subgroup of $A(n)$ containing $C(n)$. By Lemma 9.1, $S \in N_{G_x}$ implies $SG_x S^{-1} \subseteq G_{Sx}$.

Theorem 9.2. Two almost periodic tilings x and y of R^n have the same quasisymmetry type if and only if (i) there exists $S \in N_{G_x}$ with $G_y = SG_x S^{-1}$, and (ii) the quasi-shift systems Q_{Sx} and Q_y are topologically conjugate.

Proof: By Definition 7.1 there exists an isomorphism $\varphi : G_x \to G_y$ so that the G_x actions Q_y^φ and Q_x are topologically conjugate via some $\eta : \overline{O(x)} \to \overline{O(y)}$. By Theorem 3.2, there exists $S \in A(n)$ such that $\varphi(R) = SRS^{-1}$, so that $G_y = SG_x S^{-1}$ and $S \in N_{G_x}$. Also, $\eta(Rx) = SRS^{-1}\eta(x)$, and thus $\overline{O(Sx)} = S(\overline{O(x)})$. Define a homeomorphism $\omega : \overline{O(Sx)} \to \overline{O(y)}$ by $\omega = \eta \circ S^{-1}$ and suppose that $R \in G_y$, so that $R = SR'S^{-1}$ for some $R' \in G_x$. Let $z \in \overline{O(Sx)}$ and let $x' \in \overline{O(x)}$ satisfy $z = Sx'$. Then

$$
\begin{aligned}
(\omega \circ Q_{Sx}^R)(z) &= \omega(Rz) \\
&= (\eta \circ S^{-1})((SR'S^{-1})(Sx')) \\
&= \eta(R'x') \\
&= SR'S^{-1}\eta(x') \\
&= R(\eta \circ S^{-1})(Sx') \\
&= (R \circ \omega)(z) = (Q_y^R \circ \omega)(z).
\end{aligned}
$$

Conversely, if $\omega : \overline{O(Sx)} \to \overline{O(y)}$ is a homeomorphism satisfying $\omega(R(Sx'))$ $= R(\omega(Sx'))$ for all $x' \in O(x)$ and some $S \in N_{G_x}$ with $SG_xS^{-1} = G_y$, then $\eta = \omega \circ S : \overline{O(x)} \to \overline{O(y)}$ conjugates Q_y^φ and Q_x. $\square$

Corollary 9.3. If x is an almost periodic tiling of R^n and $S \in N_{G_x} \cap N_{G_{Sx}}$ then x and Sx have the same quasisymmetry type.

Proof: Since $N_{G_{Sx}}$ is a group, $S \in N_{G_{Sx}}$ implies $S^{-1} \in N_{G_{Sx}}$. It follows that $SN_{G_{Sx}}S^{-1} \subseteq N_{G_x}$, or equivalently, $N_{G_{Sx}} \subseteq S^{-1}N_{G_x}S$. Since $S \in N_{G_x}$, it follows from Lemma 9.1 that $S^{-1}N_{G_x}S \subseteq N_{G_{Sx}}$. $\square$

Corollary 9.4. For all $S \in C(n)$, x and Sx have the same quasisymmetry type.

Proof: $C(n) \subseteq N_{M(n)}$. $\square$

In particular, for tilings of the plane, the quasisymmetry type is invariant under rotation, reflection, translation, and scaling. Certain quasisymmetry types are also invariant under other affine transformations, but in general, affine transformations can create or destroy quasisymmetries.

10. INFLATION

One of the most interesting and celebrated features of Penrose tilings is the fact that they satisfy a kind of 'self similarity' property called an inflation. While many other almost periodic tilings satisfy inflations, there are also many examples that do not. Thus inflations (or their absence) provide important invariants for almost periodic tilings.

A tiling x is said to satisfy an *inflation*[8] with *inflation constant* $\lambda > 0$ if there exists a homomorphism $\psi : \overline{O(x)} \to \overline{O(x)}$, such that for all $ST^t \in G_x$ (with $S \in H_x$)

$$(15) \qquad \psi \circ ST_x^t = ST_x^{\lambda t} \circ \psi.$$

For protofinite tilings, inflations are usually defined in terms of a construction resembling the procedure (described above) for converting between the two kinds of Penrose tilings: First, the prototiles are marked with some 'new edges'. Then some 'old edges' are removed, resulting in a new tiling on a smaller scale. Finally, the small scale tiling is 'scaled up' by λ to obtain a new tiling on the original scale. The Penrose inflation, with $\lambda = (1/2)(\sqrt{5}+1)$, is shown in Figure 5, (see [8] for many other examples).

---

[8]This usage seems natural. Unfortunately, the term *deflation* is also used by some authors to mean the same thing.

Figure 5: Inflation for the rhombic Penrose prototiles.

Proposition 10.1. If x and y are tilings having the same quasisymmetry type, then x satisfies an inflation with constant λ if and only if y satisfies an inflation with the same constant. In particular, the existence of an inflation with a given constant is an invariant of quasisymmetry type.

Proof: Let ψ be the inflation with constant λ satisfied by x. Define an automorphism A_λ of G_x by $A_\lambda(T_x^t U) = T_x^{\lambda t} U$. By the definition of inflation, $\psi \circ Q_x^R = Q_x^{A_\lambda(R)} \circ \psi$ for all $R \in G_x$. Since y has the same quasisymmetry type as x, Theorem 9.2 implies that there exists $S \in N_{G_y}$ such that (i) $G_x = G_{y'} = SG_y S^{-1}$ and, (ii) there exists a homeomorphism $\eta : \overline{O(x)} \to \overline{O(y)}$ such that $\eta \circ Q_x^R = Q_{y'}^R \circ \eta$ for all $R \in H_x$. Let $\varphi(R) = SRS^{-1}$, and note that $S \circ Q_y^R = Q_{y'}^{\varphi(R)} \circ S$. Note also that φ and A_λ commute, since if $R = T^t U$ and $S = T^s G$, then

$$
\begin{aligned}
\varphi(A_\lambda R) &= \varphi(T^{\lambda t} U) \\
&= S \circ T^{\lambda t} U \circ S^{-1} \\
&= T^{G(\lambda t)} S U S^{-1} \\
&= A_\lambda(T^{Gt} S U S^{-1}) \\
&= A_\lambda(\varphi(R)),
\end{aligned}
$$

where $SUS^{-1} \in O(n)$ since $S \in N_{G_y}$. Let $S' = \eta \circ S$ and let $\psi' = (S')^{-1} \psi S'$. Then

$$
\begin{aligned}
(\psi')^{-1} \circ Q_y^R \circ gq' &= (S')^{-1} \psi^{-1} \circ \eta \circ Q_{y'}^{\varphi(R)} \circ \eta^{-1} \circ \psi S' \\
&= (S')^{-1} \psi^{-1} Q_x^{\varphi(R)} \psi S' \\
&= (S')^{-1} Q_x^{A_\lambda(\varphi(R))} S' \\
&= (S')^{-1} Q_x^{\varphi(A_\lambda(R))} S' \\
&= Q_y^{A_\lambda(R)}.
\end{aligned}
$$

Hence G' is an inflation for y with constant λ. $\square$

11. THE SPECTRUM

An important technique in crystallography is to study the diffraction patterns of crystals (and other kinds of solids). For example, the icosahedral point symmetry of the Schectman quasicrystal [27] was initially detected by observing five-fold rotational symmetry (not possible for crystals) in the diffraction pattern. If we assume that (at least approximately) a quasicrystal consists of atoms located at the vertices of an almost periodic tiling x, then one can show [23] that the diffraction pattern is essentially the point spectrum of the corresponding shift action T_x. In this section we show how the symmetries of the point spectrum of the shift T_x relate to the quasisymmetry type of x.

For an almost periodic tiling x of R^n, let $\Sigma_x \subseteq R^n$ denote the (topological) point spectrum of the shift T_x; that is, Σ_x consists of all w such that there exists a continuous complex *eigenfunction* $f \in C(\overline{O(x)})$, $f \neq 0$, with

$$f(T_x^t y) = e^{2\pi i <t,w>} f(y),$$

for all $y \in \overline{O(x)}$. The shift T_x is said to have *discrete spectrum* (or be *weakly mixing*) if the eigenfunctions have a dense span in $C(\overline{O(x)})$ (or $\Sigma_x = \{0\}$). The *symmetry group* of the spectrum Σ_x is given by $H_{\Sigma_x} = \{U \in O(n) : U\Sigma_x = \Sigma_x\}$.

Proposition 11.1. If x is an almost periodic tiling of R^n and $S = TG \in A(n)$ for $T \in R^n$ and $G \in Gl(n)$, then $\Sigma_{Sx} = (G^*)^{-1}\Sigma_x$. Moreover, T_{Sx} has discrete spectrum (or is weakly mixing) if and only if T_x has the same property.

Proof: Let $\chi_w : \overline{O(x)} \to C$ be an eigenfunction for $w \in \Sigma_x$. Then $S^{-1} : \overline{O(Sx)} \to \overline{O(x)}$ is a homeomorphism, and $\chi_w \circ S^{-1} : \overline{O(Sx)} \to C$ is continuous. Thus for $z \in \overline{O(Sx)}$ and $t \in R^n$,

$$
\begin{aligned}
(\chi_w \circ S^{-1})(T_x^t z) &= \chi_w(S^{-1}T_x^t z) \\
&= \chi_w(T_x^{G^{-1}t} S^{-1} z) \\
&= e^{2\pi i <G^{-1}t,w>} \chi_w(S^{-1}z) \\
&= e^{2\pi i <t,(G^*)^{-1}w>}(\chi_w \circ S^{-1})(z),
\end{aligned}
$$

and $(G^*)^{-1}w \in S_{Sx}$. $\square$

Corollary 11.2. If x is an almost periodic tiling of R^n and $U \in H_x$ then $U\Sigma_x = \Sigma_x$. In particular, $H_x \subseteq H_{\Sigma_x}$, i.e., eigenvalues provide obstructions to quasisymmetries.

Proof: $U\Sigma_x = \Sigma_{Ux}$ since $U \in O(n)$, and $U \in H_x$ since $\Sigma_{Ux} = \Sigma_x$. $\square$

One can show [24] that for Penrose tilings x, the shift T_x has discrete spectrum with $\Sigma_x = Z[\zeta]$, where $\zeta = e^{2\pi i/5}$ (here we view $Z[\zeta]$ as a subset of R^2).

The next result is similar to Corollary 11.2. It shows how the spectrum relates to inflations.

Proposition 11.3. If x satisfies an inflation with constant λ, then the spectrum Σ_x of x satisfies $\lambda \cdot \Sigma_x = \Sigma_x$.

12. Symmetries of Invariant Measures

It turns out that in addition to being minimal, the shifts T_x for almost periodic tilings x are frequently also uniquely ergodic. This means that there exists a unique T_x-invariant Borel probability measure on $\overline{O(x)}$ (see [20]). We denote this measure by μ_x. It follows from the ergodic theorem for uniquely ergodic dynamical systems that each patch in such a tiling x occurs with a uniform positive frequency. We will refer to an almost periodic finite type tiling with a uniquely ergodic shift as *strictly almost periodic*. One can easily show show, for example, that Penrose tilings are strictly almost periodic (see [8]).

It has been suggested by C. Radin that one should study the symmetries of the unique invariant measure to understand the symmetries of the corresponding strictly almost periodic tiling. The results in this section show how such symmetries relate to the theory of quasisymmetry types.

Proposition 12.1. Let x be a strictly almost periodic tiling. If $S \in H_x$ then μ_x is S–invariant, (i.e., $\mu_x(S^{-1}E) = \mu_x(E)$ for all Borel sets $E \subseteq \overline{O(x)}$).

This follows from the fact that $\mu_x \circ S^{-1}$ is an invariant Borel measure, and so by unique ergodicity, equal to μ_x. Now let $\overline{O_M(x)} = \overline{\{Rx : R \in M(n)\}}$ denote the orbit closure of x with respect to the action of $M(n)$[9]. We will consider the Borel probability measures on $\overline{O_M(x)}$ which are invariant for the shift action. Note that $\overline{O_M(x)}$ is compact in the metric h, and that $\overline{O_M(x)} = \cup_{R \in M(n)} \overline{O(Rx)}$. Thus for $R \in M(n)$ we can regard $\mu_x \circ R^{-1}$ as a measure on $\overline{O_M(x)}$. In general, however, $H_x \neq O(n)$, so the shift on $\overline{O_M(x)}$ need not be minimal or uniquely ergodic.

Proposition 12.2. Suppose x is strictly almost periodic and let

$$G_{\mu_x} = \{R \in M(n) : \mu_x \text{ is } R\text{–invariant }\},$$

(i.e., G_{μ_x} is the 'symmetry group' for μ_x). Then $G_{\mu_x} = G_x$.

Proof: First note that G_{μ_x} is a quasisymmetry group. Letting $H_{\mu_x} = G_{\mu_x} \cap O(n)$, it suffices to prove $H_x = H_{\mu_x}$. By Corollary 9.4 $H_x = H_{Sx}$ for $S \in H_x$, and by Proposition 12.2, $H_{Sx} \subseteq H_{\mu_x}$, so that $H_x \subseteq H_{\mu_x}$. Now suppose $S \in H_{\mu_x}$. If μ is any ergodic T^t–invariant measure (see [20]), $supp(\mu) \subseteq \overline{O(Rx)}$ for some $R \in O(n)$. Thus by unique ergodicity, $\mu = \mu_{Rx}$, and it follows that $S \in H_{Rx}$, which implies $S \in H_x$. $\square$

---

[9]This space was studied by Radin and Wolff, [21]

13. Other Examples

Suppose $\bar{p}$ is a finite set of incongruent tiles. We say x is a *generalized $\bar{p}$-tiling* if every tile in x is *congruent* (not necessarily by translation) to a tile in $\bar{p}$. One can show that Proposition 2.1 extends to generalized $\bar{p}$-tilings (see [21]), and thus generalized $\bar{p}$-tilings are protocompact. An interesting example recently studied by Radin [22] is called *pinwheel tilings*. For pinwheel tilings, $\bar{p}$ consists of a single $(1, 2, \sqrt{5})$-right triangle, rotations and reflections of which are used to tile the plane. Without going into details, we note that pinwheel tilings are defined by an *inflation* rule, but Radin has shown that after marking the tiles in a certain way, the pinwheel tiling shift is of finite type [22]. The most interesting feature of this example is its point group. In every pinwheel tiling the right–triangular 'prototile' occurs with infinitely many different rotational orientations. This implies that pinwheel tilings are not protofinite, and moreover that[10]

(16) $H_x = O(n).$

Radin has observed that (16) implies T_x is weakly mixing (this follows, for example, from our Corollary 11.2).

In the literature on quasicrystals and almost periodic tilings, there is a standard method for constructing examples of almost periodic tilings. There are two essentially equivalent formalisms for this method (see [6]) called the *cut and project method* and *the dual method*. The idea of this construction goes back to the work of de Bruijn [5], who showed that the Penrose tilings are this type. Tilings obtained by this construction are sometimes called *quasiperiodic tilings* (see [10]). Roughly speaking, the vertex set $b(x)$ of a quasiperiodic tiling x of R^n (at least in the 'typical' case) is obtained by projecting part of the periodic lattice Z^r in R^r, $r > n$, to an irrationally sloped n-dimensional plane L in R^r. Using this method, one can obtain examples of almost periodic tilings with various point groups [29]. For example, by modifying examples in [17] (to break some of the symmetry) one can obtain an almost periodic tiling x of R^2 with an arbitrary proper closed subgroup of $O(2)$ as its point group H_x. This illustrates a crucial fact about almost periodic tilings: there is essentially no crystallographic restriction for almost periodic tilings. This is a major difference between crystallography and quasicrystallography.

Most of the studies of the symmetry properties of quasiperiodic tilings have concentrated (at least implicitly) on studying only their point symmetries, often by studying the symmetries of the spectrum (see [10], [9]). In effect, this means looking only at the information in their quasisymmetry groups. One can also show that the shift T_x for any quasiperiodic tiling x has nontrivial eigenvalues. In particular, the eigenfunctions consist of the periodic

---

[10]It should be noted that for periodic tilings, the point group is always finite. One can also show that if p consists of polygons (in R^2) then any almost periodic $x \in X_p$ must have a finite point group.

functions of R^r restricted to L. It follows that quasiperiodic tilings x generate shifts T_x which are never weakly mixing. Since certain almost periodic tilings (e.g., pinwheel tilings) do have weakly mixing shifts, it follows that although quasiperiodicity implies almost periodicity, the converse is false.

REFERENCES

1. Adler, R. and B. Marcus, Topological entropy and equivalence of dynamical systems, *Memoirs Amer. Math. Soc.* **219**, (1979).
2. Berger R., The undecidability of the domino problem, *Memoirs Amer. Math. Soc.* **66**, (1966).
3. Bieberbach, L., Über die Bewegungsgruppen der Euklidischen Raüme I, *Math. Ann.* **70**, (1911), 297–336.
4. ———, Über die Bewegungsgruppen der Euklidischen Raüme II, *Math. Ann.* **72**, (1912), 400–412.
5. de Bruijn, N. G., Algebraic theory of Penrose's non-periodic tilings of the plane I & II, *Kon. Nederl. Akad. Wetensch.* **A84**, (1981), 39-66.
6. Gähler, F. and J. Rhyner, Equivalence of the generalized grid and projection methods for the construction of quasiperiodic tilings, *J. Phys. A: Math. Gen.* **19**, (1986), 267–277.
7. Gottschalk, W., Orbit-closure decompositions and almost periodic properties, *Bull. Amer. Math. Soc.* **50**, (1944), 915-919.
8. Grünbaum, B., and G. C. Shephard, *Tilings and Patterns*, W. H. Freeman and Company, (1987).
9. Janssen, T., Crystallography of quasicrystals, *Acta Cryst.* **A42**, (1986), 261–271.
10. Katz, A., A Short Introduction to Quasicrystallography, *From Physics to Number Theory*, Springer Verlag, (1992), 478–537.
11. Kitchens, B. and K. Schmidt, Periodic Points, Decidability, and Markov Subgroups, in Dynamical Systems, Proceedings University of Maryland 1986, *Lecture Notes in Math.* **1342**, Springer Verlag, (1988).
12. Kramer, P. and R. Neri, On periodic and non-periodic space fillings of E^m obtained by projection, *Acta Cryst.* **A40**, (1984), 580–587.
13. Levine, D. and P. J. Steinhardt, Quasicrystals: A new class of ordered solid, *Phys. Rev. Lett.*, **54**, (1984), 1520.
14. Mackay, A. L., Crystallography and the Penrose pattern, *Physica* **114A**, (1982), 609–613.
15. Mackey, G. W., Ergodic theory and virtual groups, *Math. Ann.* **55**, (1966), 187–207.
16. Mozes, S., Tilings, substitution systems and dynamical systems generated by them, *J. d'Analyse Math.* **53**, (1989), 139-186.
17. Niizeki, K., A classification of two-dimensional quasi-periodic tilings obtained with the grid method, *J. Phys. A* **21**, (1988), 3333-3345.
18. Penrose, R., Pentaplexy, *Eureka* **39**, (1978), 16–22; reprinted *Mathematical Intelligencer* **2**, (1979), 32–37.
19. Pleasants, P. A. B., Quasicrystallography: some interesting new patterns, *Elementary and Analytic Theory of Numbers*, Banach Center Publications, **17**, Warsaw, (1985), 439-461.
20. Queffélec, M., Substitution Dynamical Systems–Spectral Analysis, *Lecture Notes in Math.* **1294**, Springer Verlag, (1987).
21. Radin, C., and M. Wolff, Space tilings and local isomorphism, *Geometria Dedicata* **42**, (1992), 355-360.
22. Radin, C., The pinwheel tilings of the plane, *Annals of Math.* **139**, (1994), 661-702.

23. Robinson, E. A., The point spectra of quasilattices, in preparation.
24. Robinson, E. A., The dynamics of Penrose tilings, to appear, *Transactions of the American Mathematical Society.*
25. Rudolph, D. J., Rectangular tilings of R^n-actions and free R^n-actions, in Dynamical Systems, Proceedings University of Maryland 1986, *Lecture Notes in Math.* **1342**, Springer Verlag, (1988)
26. Rudolph, D. J., Markov Tilings of R^n and representations of R^n actions, *Contemporary Mathematics* **94**, 271-290.
27. Schectman, D., I. Blech, D. Gratias and J. Cahn, Metallic phase with long range orientational order and no translational symmetry, *Phys. Rev. Lett.* **53**, (1984), 1951-1953.
28. Senechal, M., *Crystalline Symmetries: An Informal Mathematical Introduction*, Adam Hilger, (1990).
29. Socolar, J. , P. Steinhardt, and D. Levine, Quasicrystals with arbitrary orientational symmetry, *Physical Review B* **32**, (1985), 5547-5550.

DEPARTMENT OF MATHEMATICS, THE GEORGE WASHINGTON UNIVERSITY, WASHINGTON, DC 20052

APPROXIMATION OF GROUPS AND GROUP ACTIONS, THE CAYLEY TOPOLOGY

A. M. STEPIN

For Bill Parry on his 60th birthday

We shall have to do with certain topology in the set of finitely generated groups which arises in the problems related to approximation properties of (transformation) groups. This topology gives us new insight concerning the amenable groups and their actions.

Our approach will be illustrated by proving a version of the one-tower Halmos-Kakutani-Rokhlin property for majority of the known amenable groups. Incidentally we shall get a clear (as I hope) understanding of

1) the existence phenomenon of non-elementary amenable groups and
2) the nature of such groups known up to now.

There are 3 sources and 3 constituents of the arguments leading to our construction of the one-tower Halmos-Kakutani-Rokhlin (HKR) property for some non-elementary amenable groups. These (both sources and constituents) are:

1) the notion of local isomorphism (and corresponding topology) for finitely generated groups,
2) Grigorchuk's construction of the finitely generated groups having intermediate growth,
3) Caroline Series' proof of the HKR property for solvable groups.

I. GROUP ACTIONS WITH THE FREE APPROXIMATION PROPERTY AND LOCAL ISOMORPHISM OF GROUPS

A partition ξ of a measure space (X, μ) is (called) nonsingular if the saturation mapping corresponding to ξ transforms the class of measure zero sets into itself. An action T of a countable group G in (X, μ) is said to be approximable if its trajectory partition is the intersection (i.e. the greatest lower bound) of some decreasing sequence of measurable nonsingular partitions. For actions in a probability measure space (X, μ) this approximation

Partially supported by RFFR (grant 93-01-00239).

Typeset by $\mathcal{A}_{\mathcal{M}}\mathcal{S}$-TEX

property is equivalent to the following one (see [1]): given any finite collection $g_1, \ldots, g_s \in G$ and any $\varepsilon > 0$ there exist such transformations $k_1, \ldots, k_s$ in the full group[1] $[T]$ that they generate finite group and

(1) $\mu\{x \mid T_{g_i} x = k_i x, i = 1, \ldots, s\} > 1 - \varepsilon.$

We say that a free action of G is freely approximable if the group generated by the transformations $k_1, \ldots, k_s$ above acts freely.

The existence of approximation for an action T in (X, μ) implies that T is nonsingular ($= \mu$ is T-quasiinvariant). It is known that the groups possessing approximable free action with finite invariant measure are amenable (i.e. they have an invariant Banach mean). The requirement that the (invariant) measure be finite is essential since there exists a conservative approximable free action of the non-amenable group $\mathbb{Z}_2 * \mathbb{Z}_3$ (for example, the natural action of $PSL(2, \mathbb{Z})$ in $\mathbb{R}^1$). Further, the construction of G-action induced by an action of a subgroup $H \subset G$ allows one to get an approximable free action for any countable group G. So the approximation property of a G-action does not impose any algebraic restriction upon G. This is not the case for the freely approximable actions [2].

The free approximation property of group actions is in a close relation with local approximation of the finitely generated groups by finite groups. The notion of local approximation generalizes the well known (in group theory) property called residual finiteness ($=$ intersection of the normal subgroups of finite index in G is trivial). Roughly speaking the local approximation reflects the fact that for finitely generated groups (with the same set of generators) one can talk about their local isomorphism in a neighbourhood of identity with respect to some length metric. For a group G endowed with this metric let G_n denote the ball of radius n centered at the identity element e.

Definition. Two groups with the same finite set of generators are called n-isomorphic if their (marked by generators) Cayley graphs restricted to n-balls centered at e coincide; corresponding identification of n-balls is called n-isomorphism.

This notion naturally gives rise to the topology of local isomorphism (the Cayley topology) in the set of groups having fixed finite collection of generators, with subsets of n-isomorphic groups as the base of neighbourhoods. The local topology found the interesting applications in group theory [3,4]. Two applications to amenability and HKR property will be given below.

We say that a finitely generated (f.g.) group G is locally approximable by finite groups if for any positive integer N the group G is N-isomorphic to

---

[1]The full group $[T]$ for a measurable action T of G is the group of measurable transformations that piecewise coincide with T_g, $g \in G$.

some finite group. This property does not depend on the choice of generators in G.

Proposition 1. *a) If a f.g. group G is residually finite then G possesses the local approximation property by finite groups;*
b) a finitely presented group with the property of local approximation by finite groups is residually finite.

Proof. a) Let $a_1, \ldots, a_s$ be generators of a residually finite group G and G_N be the set of elements in G that can be represented as a product of not more than N factors $a_i^{\pm 1}$, $i = 1, \ldots, s$. For any $g \in G$, $g \neq e$, there exist a finite group K_g and a homomorphism $\varphi_g : G \to K_g$ such that $\varphi_g(g) \neq e$. Consider the homomorphism $\varphi : G \to K_N := \prod_{g \in G_N} K_g$ mapping element $h \in G$ to element $\{\varphi_g(h)\} \in K_N$. Denote by K the subgroup in K_n generated by $\varphi(a_i)$, $i = 1, \ldots, s$. The restriction of φ to $G_{[N/2]}$ is $[N/2]$-isomorphism of G and K.

b) Let $G = \langle a_1, \ldots, a_s; R_i, i = 1, \ldots, n \rangle$ be a finitely presented group and N be the maximal length of the relators R_i, $i = 1, \ldots, n$. If the group G is N-isomorphic to group K and φ is the corresponding mapping of G_N into K then there exists a homomorphism $\psi : G \to K$ that coincides with φ on G_N. To see this it suffices to remark that any relation satisfied in G is also satisfied in K and hence there are no obstacles in extending φ to some homomorphism of G into K.

Suppose now that G possesses the property of local approximation by finite groups. Fix $g \in G$, $g \neq e$, and let L be the length of g w.r.t. generators $a_1, \ldots, a_s$. For any N there exist a finite group K_N and N-isomorphism φ_N of the groups G and K_N. If N is greater than L and the lengths of the relators R_i, $i = 1, \ldots, n$, then (according to the previous remark) φ_N can be extended to a homomorphism $\varphi_g : G \to K_N$ such that $\varphi_g(g) \neq e$.

Requirement that G be finitely presented in part b) of the proposition above is essential. To see this consider the group H of finitely supported permutations of $\mathbb{Z}$ with the natural action of $\mathbb{Z}$ on H; the semidirect product $G = \mathbb{Z} \ltimes H$ has the property of local approximation by finite groups. This group is not however residually finite since the subgroup of even permutations is simple.

As an example of the group without the property of local approximation by finite groups one can take $G = \langle a, b; ab^2a^{-1} = b^3 \rangle$. Indeed, if G were locally approximable by finite groups then G would be residually finite according to proposition 1 but this is not the case.

Theorem (cf. [2]). *A finitely generated group G has an action with free approximation property if and only if G is locally approximable by finite groups.*

Outline of proof. 1) Let T be an approximable action of G in (X, μ), $\mu(X) = 1$, and $g_1, \ldots, g_s \in G$. First we show that for any positive integer N there exist a collection of measurable transformations $k_1, \ldots, k_s \in [T]$ and a point $x \in X$ such that $k_1, \ldots, k_s$ generate a finite group and for any element $w(a_1, \ldots, a_s, a_1^{-1}, \ldots, a_s^{-1})$ of the free group $F(a_1, \ldots, a_s)$ with generators $a_1, \ldots, a_s$ having the length less than N the equality

$$w(T_{g_1}, \ldots, T_{g_s}, T_{g_1}^{-1}, \ldots, T_{g_s}^{-1})x = w(k_1, \ldots, k_s, k_1^{-1}, \ldots, k_s^{-1})x$$

holds. To see this we fix $\varepsilon > 0$ and choose transformations $k_1, \ldots, k_s \in [T]$ that generate a finite group and approximate transformations $T_{g_1}, \ldots, T_{g_s}$ as in (1). Consider subsets

$$B_1 = X \setminus \{x : T_{g_i}^{\pm 1}x = k_i^{\pm 1}x, \, i = 1, \ldots, s\},$$

$$B_{m+1} = X \setminus \{x : T_{g_i}^{\pm 1}x \notin B_m, \, i = 1, \ldots, s\}, \quad m = 1, \ldots, N-1.$$

We have

$$\mu(B_{m+1}) \le \int_{B_m} J \, d\mu,$$

where $J = \sum_{i=1}^{s}(J_i^+ + J_i^-)$ and $J_i^{\pm}$ is the Radon-Nikodym derivative of measure $(T_{g_i}^{\pm 1})\mu$ w.r.t. μ. By the absolute continuity of the set function $\int_A J \, d\mu$ number ε can be chosen at the very beginning so small that

$$\varepsilon + \int_{B_1} J \, d\mu + \cdots + \int_{B_N} J \, d\mu < 1.$$

Any point from $X \setminus \bigcup_{m=1}^{N} B_m$ is required.

2) Suppose now that T is freely approximable. Fix some generators $h_1, \ldots, h_n$ of G and a positive integer N. As in 1) choose transformations $k_1, \ldots, k_n \in [T]$ generating finite group K_N and a point $x \in X$ such that

$$(2) \quad w(T_{h_1}, \ldots, T_{h_n}, T_{h_1}^{-1}, \ldots, T_{h_n}^{-1})x = w(k_1, \ldots, k_n, k_1^{-1}, \ldots, k_n^{-1})x$$

for any element $w(a_1, \ldots, a_n, a_1^{-1}, \ldots, a_n^{-1}) \in F(a_1, \ldots, a_n)$ with the length less than $2N$. We define the mapping φ of the neighbourhood G_{2N} into K_N by sending $w(h_1, \ldots, h_n, h_1^{-1}, \ldots, h_n^{-1})$ to $w(k_1, \ldots, k_n, k_1^{-1}, \ldots, k_n^{-1}) \in K_N$.

Since K_N acts freely it follows from (2) that the mapping φ is well-defined (i.e. $\varphi(g)$ does not depend on the representation of g in the form $w(h_1, \ldots, h_n, h_1^{-1}, \ldots, h_n^{-1})$).

The mapping φ is injective (it follows again from (2) and the fact that the action T is free) and preserves multiplication (that is $\varphi(g_1 g_2) = \varphi(g_1)\varphi(g_2)$ for any pair $g_1, g_2 \in G_N$ such that $g_1 g_2 \in G_N$).

3) To prove that the local approximation of G by finite groups is sufficient for the existence of some G-action with the free approximation property we consider the action T of group G by left translations in (G, μ) where μ is the probability measure equivalent to invariant measure on G. It is clear that the action T is free and approximable. The corresponding full group $[T]$ consists of all permutations acting on G.

Fix $\varepsilon > 0$ and some generators $h_1, \ldots, h_n$ for G. Choose a positive integer N so that $\mu(G_N) > 1 - \varepsilon$. Let K be a finite group which is $(N+1)$-isomorphic to group G and φ be a corresponding mapping of G_{N+1} to K. For an injective mapping $\psi : K \to G$ extending φ^{-1} there are the permutations $k_i : G \to G$, $i = 1, \ldots, n$, such that

a) $k_i |_{\psi(K)} = \psi L_i \psi^{-1}$, where $L_i : K \to K$ is the left translation by $\varphi(h_i)$,

b) the group generated by k_i, $i = 1, \ldots, n$, acts freely in G.

The transformations k_i, $i = 1, \ldots, n$ belong to $[T]$ and generate some finite group (isomorphic to K). The condition $\mu\{x : T_{h_i} x = k_i x, i = 1, \ldots, n\} > 1 - \varepsilon$ is satisfied by the choice of N. It remains to observe that the case of arbitrary set $g_1, \ldots, g_m$ in checking the free approximation property of T reduces to one already considered.

Corollary. *If a finitely presented group G has an action with the free approximation property then G is residually finite.*

Thus the theorem shows that there exist countable groups having no actions with the free approximation property.

II. THE GRIGORCHUK'S GROUPS

In [3] Grigorchuk constructed groups of intermediate growth answering the question raised by Milnor. This section is concerned with the construction mentioned and the proof (based on local topology) of the existence of non-elementary amenable groups.

Fix the alphabet $\{a, e\}$ and consider the pairs of sequences $\omega = (\{b_k\}_1^\infty, \{c_k\}_1^\infty)$ for which b_k and c_k takes values a, e and are not equal to e simultaneously.

The symbol a is identified with the mapping $a : (0, 1] \to (0, 1]$, $x \mapsto x + \frac{1}{2}$ mod 1. Consider the interval exchange transformation $b_\omega : (0, 1] \to (0, 1]$ defined by the formula

$$b_\omega |_{(2^{-k}, 2^{1-k}]} = \begin{cases} \text{transposition of halves } \left(\frac{1}{2^k}, \frac{3}{2^{k+1}}\right], \left(\frac{3}{2^{k+1}}, \frac{1}{2^{k-1}}\right] & \text{if } b_k = a, \\ \text{id} & \text{if } b_k = e; \end{cases}$$

c_ω is defined similarly by the sequence $\{c_k\}_1^\infty$.

Let G_ω be the transformation group generated by a, b_ω and c_ω. The subgroup $H_\omega^{(1)} \subset G_\omega$ consisting of transformations fixing intervals $(0, \frac{1}{2}]$ and $(\frac{1}{2}, 1]$ is generated by the elements b_ω, c_ω, $ab_\omega a$, $ac_\omega a$.

Using notation $\sigma\omega$ for the pair $(\{b_{k+1}\}_1^\infty, \{c_{k+1}\}_1^\infty)$ define homomorphisms (of left and right rewriting) $\varphi_\omega^l, \varphi_\omega^r : H_\omega^{(1)} \to G_{\sigma\omega}$ by the following formulas:

$$\varphi_\omega^l(b_\omega) = b_{\sigma\omega}, \quad \varphi_\omega^l(c_\omega) = c_{\sigma\omega}, \quad \varphi_\omega^l(ab_\omega a) = b_1, \quad \varphi_\omega^l(ac_\omega a) = c_1,$$

$$\varphi_\omega^r(b_\omega) = b_1, \quad \varphi_\omega^r(c_\omega) = c_1, \quad \varphi_\omega^r(ab_\omega a) = b_{\sigma\omega}, \quad \varphi_\omega^r(ac_\omega a) = c_{\sigma\omega}.$$

The group G_ω is periodic[2] if and only if pair (b_k, c_k) takes each value (a, a), (e, a) and (a, e) infinitely many times. The set Ω_0 of such ω's is σ-invariant. To verify periodicity of G_ω, $\omega \in \Omega_0$, it is convenient to represent its elements by words in generators $a, b_\omega, c_\omega, d_\omega = b_\omega c_\omega$ and their inverses. Let $L(g)$ denote the length of an element $g \in G_\omega$ w.r.t. these generators. If $g \in H_\omega^{(1)}$ then $L(\varphi_\omega^l(g)) \leq \frac{1}{2}L(g)$ and $L(\varphi_\omega^r(g)) \leq \frac{1}{2}L(g)$. Suppose that for $g \in G_\omega$ the generator b_ω enters into nonreducible representation of g. If $b_1 = e$ then $L(\varphi_\omega^l(g^2)) < L(g)$ and $L(\varphi_\omega^r(g^2)) < L(g)$. The generator $b_{\sigma\omega}$ enters nonreducible representation of the element $\varphi_\omega^l(g^2)$ if $L(\varphi_\omega^l(g^2)) = L(g)$; the same holds true for $\varphi_\omega^r(g^2)$. Again, $b_2 = e$ implies that the lengths of elements

$$\varphi_{\sigma\omega}^l((\varphi_\omega^l(g^2))^2), \quad \varphi_{\sigma\omega}^r((\varphi_\omega^l(g^2))^2), \quad \varphi_{\sigma\omega}^l((\varphi_\omega^r(g^2))^2), \quad \varphi_{\sigma\omega}^r((\varphi_\omega^r(g^2))^2)$$

are less than $L(g)$. If no, iterate this process till we get $b_n = e$. This will be ultimately the case since $\omega \in \Omega_0$. The described procedure of length decreasing leads to periodicity of g (occurence of c_ω or d_ω in nonreducible representation of g is considered similarly).

Consider now the set M_3 of all groups with three generators, a, b and c, say, and endow M_3 with Cayley (i.e. local) topology. This topology is, obviously, metrizable. Let Γ be the subset in M_3 that consists of groups isomorphic to G_ω, $\omega \in \Omega_0$, under identification $b_\omega \to b$, $c_\omega \to c$. Its closure $\widetilde{\Gamma}$ is a compact metric space homeomorphic to the Cantor set. It is shown in [3] that there exists a countable everywhere dense subset $AS \subset \widetilde{\Gamma}$ which consists of infinite almost solvable groups (in fact, these groups correspond to the sequences ω that eventually stabilize).

Let $\{F_k\}_1^\infty$ be a Følner sequence for $H \in AS$ and n_k denote the radius of F_k (w.r.t. length metric). Take a neighbourhood $U_k \subset \widetilde{\Gamma}$ of H so that any $G \in U_k$ is n_k-isomorphic to H. The set

$$\bigcap_{l=1}^\infty \bigcup_{k>l} \bigcup_{H \in AS} U_k$$

consists of amenable groups.

---

[2]Periodicity of a group means that all its elements are of finite order.

The simplest amenable groups are the elementary ones. Recall that a group is elementary if it belongs to the minimal class of groups 1) containing finite groups and abelian groups and 2) closed w.r.t. operations of group extension and inductive limit, taking subgroup and quotient group. It is known that any finitely generated periodic elementary group is finite. Since the set of periodic groups in $\tilde{\Gamma}$ is residual we see that there exist in $\tilde{\Gamma}$ many nonelementary amenable groups (in fact, they form a residual subset in $\tilde{\Gamma}$). So we get the examples of nonelementary amenable groups without growth estimates.

Remark. The groups G_ω, $\omega \in \Omega_0$, can not be finitely presented. To see this suppose the contrary and let n be the maximal length of relators for G_ω. Choose an almost solvable group H so that H and G_ω are $(n+1)$-isomorphic. This local isomorphism extends to some homomorphism of G_ω onto H, hence H is periodic. This is impossible since H is a finitely generated elementary infinite group.

III. THE ONE-TOWER HKR PROPERTY FOR G_ω'S

We apply here the Cayley topology to the proof that for actions of some G_ω's certain analog of the Halmos-Kakutani-Rokhlin lemma is valid. Countable group G is said to possess HKR property if it has an increasing to G Følner sequence $\{F_n\}$ such that for every free μ-preserving action T of G in the Lebesque space (X, μ), $\mu(X) = 1$, and any $\varepsilon > 0$ there are arbitrarily large $n \in \mathbb{N}$ and a measurable subset $B \subset X$ such that

$$1 - \varepsilon < \mu\Big(\bigcup_{g \in F_n} T_g B \Big) = |F_n| \cdot \mu(B).$$

Proposition 2. *There is a residual subset in $\tilde{\Gamma}$ that consists of groups possessing the HKR property.*

Scheme of argument. The HKR property for almost solvable groups is proved in [5]. Analysis of the proof shows that the construction of Følner sets F_n required in the definition of the HKR property can be made local in the following sense. Let G be a finitely generated almost solvable group; for any $\varepsilon > 0$ and $k \in \mathbb{N}$ there are (G_k, ε)-invariant finite subset $F = F(\varepsilon, k) \subset G$ containing G_k and a number $s = s(\varepsilon, k)$ such that $F \subset G_s$ and for every free action T of a local group[3] G_s in the Lebesque space (X, μ), $\mu(X) = 1$,

---

[3] Action of (a local group) G_s in (X, μ) is a collection $\{T_g, \ g \in G_s\}$ of μ-preserving transformations $T_g : X \to X$ such that $T_{g_1} T_{g_2} = T_{g_1 g_2}$ if $g_1, g_2, g_1 g_2 \in G_s$. An action $\{T_g, \ g \in G_s\}$ is free if for μ-almost all $x \in X$ mapping $g \mapsto gx$, $G_s \to \{T_g x \mid g \in G_s\}$ is bijective.

there exists a measurable subset $B \subset X$ so that

$$1 - \varepsilon < \mu\Big(\bigcup_{g \in F} T_g B \Big) = |F| \cdot \mu(B).$$

For $H \in AS \subset \widetilde{\Gamma}$ let $U_{\varepsilon,k}(H)$ denote the set of groups $G \in \widetilde{\Gamma}$ which are $s(\varepsilon, k)$-isomorphic to H. The set

$$\Gamma' = \bigcap_{l=1}^{\infty} \bigcup_{k>l} \bigcup_{H \in AS} U_{\frac{1}{k},k}(H)$$

is everywhere dense G_δ-subset in $\widetilde{\Gamma}$. Any group $G \in \Gamma'$ possesses the HKR property. Indeed, there exist $k_1 \in \mathbf{N}$ and $H^{(1)} \in AS$ such that $G \in U_{\frac{1}{k_1},k_1}(H^{(1)})$. Choose $F(\frac{1}{k_1}, k_1) \subset H^{(1)}$ and set

$$F_1 = \varphi_{H^{(1)}G}\, F\Big(\frac{1}{k_1}, k_1\Big),$$

where $\varphi_{H^{(1)}G}$ is $s\big(\frac{1}{k_1}, k_1\big)$- isomorphism of $H^{(1)}$ and G. Let numbers k_i, groups $H^{(i)} \in AS$ and subsets $F_i \subset G$, $i = 1, \ldots, n$, be already constructed so that

$$G \in U_{\frac{1}{k_i},k_i}(H^{(i)}),$$

$$k_i > \Big| G_{s(\frac{1}{k_{i-1}}, k_{i-1})} \Big|, \qquad i = 2, \ldots, n,$$

$$F_i = \varphi_{H^{(i)}G}\, F\Big(\frac{1}{k_i}, k_i\Big),$$

where $F\big(\frac{1}{k_i}, k_i\big) \subset H^{(i)}$ and $\varphi_{H^{(i)}G}$ is $s\big(\frac{1}{k_i}, k_i\big)$-isomorphism of $H^{(i)}$ and G. Then there exists $k_{n+1} > \big| G_{s(\frac{1}{k_n}, k_n)} \big|$ such that $G \in U_{\frac{1}{k_{n+1}},k_{n+1}}(H^{(n+1)})$ for some $H^{(n+1)}$ in AS. Now set $F_{n+1} = \varphi_{H^{(n+1)}G}\, F\big(\frac{1}{k_{n+1}}, k_{n+1}\big)$, where $F\big(\frac{1}{k_{n+1}}, k_{n+1}\big) \subset H^{(n+1)}$. Sequence $\{F_n\}$ is the required Følner sequence for G.

It is interesting to find out whether the Ornstein and Weiss construction [6] may give the one-tower HKR property for groups G_ω.

A propos, the shortest way to the isomorphism theorem for Bernoulli actions of G_ω's is given by the following

Proposition 3. *For any* $\omega \in \Omega_0$ G_ω *contains countable locally finite subgroup.*

Inductive argument. Let $x_0, \ldots, x_n \in G_\omega$ be elements of order 2 such that a) the transformation x_0 is identical outside dyadic interval Δ_0 and transposes halves of some dyadic subinterval $\Delta_0' \subset \Delta_0$, b) the transformation

x_k, $k = 1, \ldots, n$, is identical outside of a half Δ_k of the interval Δ'_{k-1} and transposes halves of some dyadic interval $\Delta'_k \subset \Delta_k$. We are going to construct an element $x_{n+1} \in G_\omega$ of order 2 which is identical outside of a half Δ_{n+1} of Δ'_n and transposes halves of some dyadic interval $\Delta'_{n+1} \subset \Delta_{n+1}$. Let r_n denote the dyadic rank of interval Δ'_n and $H_\omega^{(r_n)}$ be subgroup (of finite index) in G_ω fixing the elements of dyadic partition ξ_{r_n} of rank r_n. Consider the homomorphism $\psi_\omega^{(r_n)}$ of restriction to elements of ξ_{r_n},

$$\psi_\omega^{(r_n)} : H_\omega^{(r_n)} \to (G_{\sigma^n \omega})^{2^{r_n}}.$$

Since for a subgroup S of finite index in a finite product intersection of S with any factor is of finite index in corresponding factor, we conclude by induction that there exists subgroup $S_{\sigma^n \omega} \subset G_{\sigma^n \omega}$ such that

$$(S_{\sigma^n \omega})^{2^{r_n}} \subset \psi_\omega^{(r_n)}(H_\omega^{(r_n)});$$

the base and the step of induction use the existence of a subgroup $S_{\sigma \omega} \subset G_{\sigma \omega}$ of finite index satisfying the inclusion $S_{\sigma \omega} \times S_{\sigma \omega} \subset \psi_\omega^{(1)}(H_\omega^{(1)})$ (cf. [3]).

Let $y_{n+1} \in S_{\sigma^n \omega}^{2^{r_n}}$ be a transformation of order 2 (recall that $G_{\sigma^n \omega}$ is 2-group), which is identical outside a half (we denote by Δ_{n+1} this half) of the interval Δ'_n and let $x_{n+1} = [\psi_\omega^{(r_n)}]^{-1}(y_{n+1})$. Since the transformation x_{n+1} is not identical on $[0, 1]$ there exists dyadic interval $\Delta'_{n+1} \subset \Delta_{n+1}$ such that its halves are transposed by x_{n+1}. If we set $x_0 = a$, $\Delta_0 = \Delta'_0 = [0, 1]$ we may conclude by induction that there exists a sequence of transformations x_k, dyadic intervals Δ_k and Δ'_k such that for each $n \in \mathbb{N}$ the conditions a) and b) above are satisfied.

Consider now the groups $K_{s,t}$ generated by the elements $x_s, \ldots, x_t$. Since the transformations $x_{s+1}, \ldots, x_t$ are supported by Δ_{s+1}, the element x_s is of order 2 and $x_s \Delta_{s+1}$ does not intersect Δ_{s+1} mod 0, the group $K_{s,t}$ is a semidirect product of $\mathbb{Z}_2$ and the group generated by elements $x_{s+1}, \ldots, x_t, x_s x_{s+1} x_s, \ldots, x_s x_t x_s$. The latter is isomorphic to $K_{s+1,t} \times K_{s+1,t}$ because the elements $x_s x_{s+1} x_s, \ldots, x_s x_t x_s$ commute with $x_{s+1}, \ldots, x_t$. It follows that $|K_{s,t}| = 2|K_{s+1,t} \times K_{s+1,t}| = 2|K_{s+1,t}|^2$. Since $K_{t,t} \simeq \mathbb{Z}_2$ all of the groups $K_{s,t}$ are finite. Hence subgroup $\bigcup_t K_{0,t}$ is locally finite.

REFERENCES

1. W. Krieger, *On nonsingular transformations of measure space*, Z. Wahrscheinlichkeitstheor. und verw. Geb. **11** (1969), no. 2, 83–87.
2. A. M. Stepin, *Approximability of groups and their actions*, Uspekhi Matem. Nauk **38** (1983), no. 6, 123–124. (in Russian)
3. R. I. Grigorchuk, *Growth degrees of finitely generated groups and theory of invariant means*, Izv. Akad. Nauk SSSR Ser. Mat. **48** (1984), no. 5, 939–985. (in Russian)

4. M. Gromov, *Groups of polynomial growth and expanding maps*, Publ. Math. IHES **53** (1981), 53–73.
5. C. Series, *The Rohlin tower theorem and hyperfiniteness for action of continuous groups*, Isr. J. Math. **30** (1978), no. 1-2, 99–122.
6. D. Ornstein & B. Weiss, *Entropy and isomorphism theorems for action of amenable groups*, J. d'Analyse Math. **48** (1987), 1–141.

Printed in the United States
By Bookmasters